NONLINEAR SYSTEMS IN
AVIATION AEROSPACE AERONAUTICS ASTRONAUTICS

# STABILITY DOMAINS

**Nonlinear Systems In Aviation Aerospace Aeronautics Astronautics**

A series edited by:
S. Sivasundaram
*Embry-Riddle Aeronautical University, Daytona Beach, USA*

---

**Volume 1**
Stability Domains
*L.Gruyitch, J-P. Richard, P. Borne and J-C.Gentina*

**Volume 2**
Advances in Dynamics and Control
*S. Sivasundaram*

**Volume 3**
Optimal Control of Turbulence
*S.S. Sritharan*

NONLINEAR SYSTEMS IN
AVIATION AEROSPACE AERONAUTICS ASTRONAUTICS

# STABILITY DOMAINS



L. GRUYITCH, J-P. RICHARD,
P. BORNE AND J-C. GENTINA

CHAPMAN & HALL/CRC
A CRC Press Company
Boca Raton London New York Washington, D.C.

**Library of Congress Cataloging-in-Publication Data**

Catalog record is available from the Library of Congress



Direct all inquiries to CRC Press LLC, 2000 N.W. Corporate Blvd., Boca Raton, Florida 33431.



No claim to original U.S. Government works
International Standard Book Number 0-415-30848-8
Printed in the United States of America 1 2 3 4 5 6 7 8 9 0
Printed on acid-free paper

# Contents

# Preface

> "At the same time it becomes clear why some stability investigations should not be taken too seriously.
>
> One needs to know the size of the region of asymptotic stability."
>
> J.P. LaSalle and S. Lefschetz[1]

The kind of persistency of dynamical behaviour is important not only for various sciences like mathematics, mechanics, fluid mechanics, thermodynamics, electricity, electronics, chemistry, control, econometrics, biology, medicine, and not only for engineering and science, but also for individuals, social life and state development. An appropriate persistency is the essential sense of the corresponding stability property. Relativeness of the notion of persistence resulted in a number of stability concepts among which was that of practical stability probably initiated by Chetaev in the thirties of the twentieth century.

The above cited comment by LaSalle and Lefschetz, who were probably those who attracted the attention and interest of English speaking scientists to the Lyapunov stability theory, explains well the need to study stability domains in general and stability regions in particular.

This book is aimed at what we consider basic from the theory of stability domains for various direct effective applications and/or for further research. In doing so we express freely and openly our views on stability theory in the framework of the Lyapunov stability concept and practical stability concept.

Authors

---

[1] *Stability by Lyapunov's Direct Method*, Academic Press, New York, 1961, p. 121.

# Notations

Capital script Roman letters denote sets and spaces, capital block Roman letters designate matrices, lower case script Roman letters are used for scalars, lower case block Roman letters represent vectors, most Greek letters denote scalars.

Latin

- $\mathcal{A}$ — a nonempty subset of $\Re^n : \mathcal{A} \neq \emptyset, \mathcal{A} \subset \Re^n$,
- $\overline{\mathcal{A}}$ — the closure of the set $\mathcal{A} \neq \emptyset$,
- $\mathring{\mathcal{A}}$ — the interior of the set $\mathcal{A}$,
- $A$ — a matrix: $A \in \Re^{n\times n}$,
- $\mathsf{a}$ — elementwise positive constant vector, $\mathsf{a} = (a_1\ a_2\ \dots\ a_n)^T$, $a_i \in \Re^+$, $i = 1, 2, \dots, n$,
- $\mathcal{B}_\mathsf{a}$ — (where $\mathsf{a}$ is a vector) the open set centred at $X = 0$, $\mathcal{B}_\mathsf{a} = \{X : \mid X \mid < \mathsf{a}\}$,
- $\mathcal{B}_\mathsf{a}^+$ — (where $a$ is a vector) the open set of positive vectors in $\mathcal{B}_\mathsf{a}$, $\mathcal{B}_\mathsf{a}^+ = \{X : 0 < X < \mathsf{a}\}$,
- $\mathcal{B}_\alpha$ — (where $\alpha$ is a scalar) the open hyperball centred at $X = 0$ with radius $\alpha$: $\mathcal{B}_\alpha = \{X : ||X|| < \alpha\}$,
- $B$ — a matrix $B \in \Re^{n\times m}$,
- $b$ — a bounded vector function $b(X) \in \Re^k$ or $b(X, d) \in \Re^k$, $k \leq n$,
- $\mathsf{b}$ — elementwise positive constant vector, $\mathsf{b} = (b_1\ b_2\ \dots\ b_n)^T$, $b_i \in \Re^+$, $i = 1, 2, \dots, n$,
- $C$ — the family of all the continuous functions, $C(\mathcal{S})$ is the family of all the continuous functions on a set $\mathcal{S}$,
- $C^{(k)}(\mathcal{S})$ — the family of all functions $f$ defined, continuous and $k$-times continuously differentiable in $X \in \mathcal{S}$,
- $C_0(\mathcal{I}_0 \times \mathcal{X}_0)$ — the family of $\underline{X}(t; X_0)$ continuous in $X_0 \in \mathcal{X}_0$ for any $t \in \mathcal{I}_0$,
- $C_t(\mathcal{I}_0 \times \mathcal{X}_0)$ — the family of $\underline{X}(t; X_0)$ continuous in $t \in \mathcal{I}_0$ for any $X_0 \in \mathcal{X}_0$,
- $\mathsf{C}$ — a matrix, $\mathsf{C} \in \Re^{m\times n}$,
- $\mathsf{c}$ — elementwise constant vector, $\mathsf{c} = (c_1\ c_2\ \dots\ c_n)^T$, $c_i \in \Re^+$, $i = 1, 2, \dots, n$
- $\mathcal{D}$ — the asymptotic stability domain of $X = 0$,
- $\mathcal{D}(\mathcal{A})$ — the asymptotic stability domain of a set $\mathcal{A}$,
- $\mathcal{D}_a$ — the attraction domain of $X = 0$,

| | |
|---|---|
| $\mathcal{D}_a(\mathcal{A})$ | the attraction domain of a set $\mathcal{A}$, |
| $\mathcal{D}_p$ | the domain of practical stability with the settling time $\tau_s$ with respect to $\{\tau, \mathcal{X}_A, \mathcal{X}_F, \mathcal{I}\}$: $\mathcal{D}_p = \mathcal{D}_p(\tau, \tau_s, \mathcal{X}_A, \mathcal{X}_F, \mathcal{I})$, |
| $\mathcal{D}_p(\mathcal{A})$ | the domain of practical stability with the settling time $\tau_s$ of a set $\mathcal{A}$ with respect to $\{\tau, \mathcal{X}_A, \mathcal{X}_F, \mathcal{I}\}$: $\mathcal{D}_p(\mathcal{A}) = \mathcal{D}_p(\tau, \tau_s, \mathcal{X}_A, \mathcal{X}_F, \mathcal{I}; \mathcal{A})$, |
| $\mathcal{D}_{pc}$ | the domain of practical contraction with the settling time $\tau_s$ with respect to $\{\tau, \mathcal{X}_F, \mathcal{I}\}$: $\mathcal{D}_{pc} = \mathcal{D}_{pc}(\tau, \tau_s, \mathcal{X}_F, \mathcal{I})$, |
| $\mathcal{D}_{pc}(\mathcal{A})$ | the domain of practical contraction with the settling time $\tau_s$ of a set $\mathcal{A}$ with respect to $\{\tau, \mathcal{X}_F, \mathcal{I}\}$: $\mathcal{D}_{pc}(\mathcal{A}) = \mathcal{D}_{pc}(\tau, \tau_s, \mathcal{X}_F, \mathcal{I}; \mathcal{A})$, |
| $\mathcal{D}_{ps}$ | the domain of practical stability with respect to $\{\tau, \mathcal{X}_A, \mathcal{I}\}$: $\mathcal{D}_{ps} = \mathcal{D}_{ps}(\tau, \mathcal{X}_A, \mathcal{I})$, |
| $\mathcal{D}_{ps}(\mathcal{A})$ | the domain of practical stability of a set $\mathcal{A}$ with respect to $\{\tau, \mathcal{X}_A, \mathcal{I}\}$: $\mathcal{D}_{ps}(\mathcal{A}) = \mathcal{D}_{ps}(\tau, \mathcal{X}_A, \mathcal{I}; \mathcal{A})$, |
| $\mathcal{D}_s$ | the stability domain of $X = 0$, |
| $\mathcal{D}_s(\mathcal{A})$ | the stability domain of a set $\mathcal{A}$, |
| $\mathcal{D}_s(\varepsilon)$ | the largest neighbourhood of $X = 0$ such that $X_0 \in \mathcal{D}_s(\varepsilon)$ implies $\|\underline{X}(t; X_0)\| < \varepsilon$ for all $t \in \Re_+$, |
| $\mathcal{D}_s(\varepsilon; \mathcal{A})$ | the largest neighbourhood of a set $\mathcal{A}$ such that $X_0 \in \mathcal{D}_s(\varepsilon; \mathcal{A})$ guarantees $\rho[\underline{X}(t; X_0),\ \mathcal{A}] < \varepsilon$ for all $t \in \Re_+$, |
| $\mathcal{D}_1(\alpha)$ | rhomboid closed set $\{X \in \Re^k : X = X_e + Y,\ u^T\|Y\| \leq \alpha\}$, |
| $\mathcal{D}_2(\alpha)$ | ellipsoid closed set $\{X \in \Re^k : X = X_e + Y,\ Y^T U^2 Y \leq \alpha\}$, |
| $\mathcal{D}_\infty(\alpha)$ | parallelepipedic closed set $\{X \in \Re^k : X = X_e + Y,\ \|Y\| \leq \alpha u\}$, |
| $d$ | an external disturbance vector (function) with the elements whose values are measured with respect to their total zero values, |
| $\mathcal{E}_{(\cdot)}(\cdot\cdot)$ | an estimate of a stability property domain $\mathcal{D}_{(\cdot)}(\cdot\cdot)$, |
| $\mathcal{F}$ | a scalar function, $\mathcal{F} : \Re \to \Re$, |
| $\underline{\mathcal{F}}_0$ | a family of motions $\underline{X}$, $\underline{\mathcal{F}}_0 = \{\underline{X} : \underline{X}(0; X_0) = X_0\} = \underline{\mathcal{F}}(X_0)$, |
| $f$ | a scalar function, or a vector function, either $f : \mathcal{T} \times \mathcal{X} \to \Re^n$, or $f : \mathcal{X} \times \mathcal{S}_i \to \Re^n$, or $f : \mathcal{X} \to \Re^n$, or $f : \Re^m \to \Re^m$, $f = (f_1\ f_2\ \ldots\ f_k)^T$, $k \in \{m, n\}$, |
| $f_X$ | the Jacobian matrix function of $f$ with respect to $X$, $f_X : \Re^n \to \Re^{n \times n}$, $f_X(X) = \left[\dfrac{\partial f_i(X)}{\partial x_j}\right]$, |
| $g$ | a vector function $g : \Re^n \times \mathcal{S} \to \Re^n$, |
| $H$ | a matrix function, |
| $h$ | a vector function $h : \Re^k \to \Re^{k'}$ or $\Re^k \times \mathcal{S}_d \to \Re^{k'}$ or $\Re^n \times \Re^k \to \Re^{k'}$ or $\Re^k \times \Re^n \times \mathcal{S}_d \to \Re^{k'}$, $k$ and $k' \leq n$, |
| $\mathcal{I}$ | the family of permitted inputs $\mathtt{i}$, |
| $\mathcal{I}_0$ | the maximal time interval on which $\underline{X}(t; X_0)$ is defined: $\mathcal{I}_0 = (l^-, l^+)$, |

| | |
|---|---|
| $i$ | an integer, |
| $I$ | the identity matrix of the appropriate order, |
| $\mathtt{i}$ | an input vector function: $\mathtt{i} : \Re \to \Re^l$, |
| $j$ | an integer, |
| $\mathcal{K}$ | the cone defined by $\{a \in \Re^n_+ : -Ma > q\}$, $M \in \mathcal{M}$, $q \leq 0$, |
| $k$ | an integer, |
| $K$ | a diagonal matrix: $K \in \Re^{m \times m}$, or a possibly non-diagonal matrix $K \in \Re^{n \times n}$, |
| $\mathcal{L}$ | a Lurie matrix sector: $\mathcal{L} = (K_1, K_2) \subset \Re^{m \times m}$, |
| $\mathcal{L}^+(\underline{X})$ | the positive limit set of $\underline{X}$, |
| $\mathcal{L}^-(\underline{X})$ | the negative limit set of $\underline{X}$, |
| $L$ | the Lipschitz constant of $f$ on $\overline{\mathcal{S}}_L$, |
| $\mathcal{M}$ | a matrix sector : $\mathcal{M} = (M_1, M_2) \subseteq \overline{\Re}^{m \times m}$, |
| $\mathcal{M}_0$ | a set of zero measure, $M_0 \subset \Re$, |
| $m$ | an integer, |
| $\mathbf{M}$ | the class of $M$-matrices, $\mathbf{M} = \mathbf{P} \cap \mathbf{Z}$, |
| $-\mathbf{M}$ | the class of $-M$-matrices, $-A \in \mathbf{M}$, |
| $\mathtt{M}$ | a matrix, entries of each can be unbounded: $\mathtt{M} \in \overline{\Re}^{m \times m}$, |
| $\mathcal{N}(\mathcal{A})$ | an open neighbourhood of a set $\mathcal{A}$, $\mathcal{N}(\mathcal{A}) \subseteq \Re^n$, |
| $\mathcal{N}(X)$ | an open neighbourhood of $X$, $\mathcal{N}(X) \subseteq \Re^n$, |
| $\mathcal{N}(\delta, X)$ | $\delta$-neighbourhood of $X$, $\mathcal{N}(\delta, X) = \{y : y \in \Re^n,\ \lVert X - y \rVert < \delta\}$, |
| $\mathcal{N}(\alpha, \mathcal{A})$ | the $\alpha$-neighbourhood of a set $\mathcal{A}$: $\mathcal{N}(\alpha, \mathcal{A}) = \{X : \rho(X, \mathcal{A}) < \alpha\}$, $\mathcal{A} \subset \mathcal{N}(\alpha, \mathcal{A})$ for $\alpha > 0$, |
| $\mathcal{N}_0(\mathcal{L}; \mathcal{S})$ | the family of all the Lurie functions $f$ in a matrix sector $\mathcal{L}$ over a set $\mathcal{S}$, |
| $\mathcal{N}_1(\mathcal{L}; \mathcal{M}; \mathcal{S})$ | the family of all the Lurie functions $f$ in a matrix sector $\mathcal{L}$ over a set $\mathcal{S}$, which are differentiable with Jacobian $f_w$ in a matrix set $\mathcal{M}$ over $\mathcal{S}$, |
| $n$ | an integer, |
| $\mathcal{O}$ | the singleton: $\mathcal{O} = \{X : X = 0\} = \{0\} \subset \Re^n$, |
| $p$ | a vector norm of size $k$, i.e. a vector function whose components are scalar norms, $p : \Re^n \to \Re^n_+$, |
| $P$ | a matrix, $P \in \Re^{n \times m}$, |
| $\mathbf{P}$ | the class of all the square matrices with positive principal minors, |
| $Q$ | a scalar function, $Q : \Re \times \Re^n \to \Re$, |
| $q$ | a scalar variable whose value is measured with respect to its total zero value, |
| $\mathtt{q}$ | a vector variable, $\mathtt{q} \in \Re^k$, $k \leq n$. In Chapter 7, $\mathtt{q} \geq 0$, |
| $q_n$ | an unperturbed (reference, desired, nominal) motion, |
| $\Re$ | the set of all real numbers, $\Re = \{\alpha : -\infty < \alpha < +\infty\} = ]-\infty, +\infty[$, |
| $\Re^n$ | $n$-dimensional metric real vector space, if not otherwise stated $\Re^n$ is the $n$-dimensional Euclidean space, $\Re^1 = \Re$, |

| | |
|---|---|
| $\Re_-$ | $=\{t : t \in \Re, -\infty < t \leq 0\} = ]-\infty, 0]$, |
| $\Re^-$ | the interior of $\Re_-$: $\Re^- = ]-\infty, 0[ = \{t : t \in \Re_-, t < 0\} = \mathring{\Re}_-$, |
| $\Re_0$ | a time interval: $\Re_0 = [t_0, +\infty[$, iff $t_0 = 0$ then $\Re_0 = \Re_+$, |
| $\Re_+$ | $= \{t : t \in \Re, 0 \leq t < +\infty\} = [0, +\infty[$, $\overline{\Re}_+ = [0, +\infty]$, |
| $\Re^+$ | the interior of $\Re_+$: $\Re^+ = ]0, +\infty[ = \{t : t \in \Re_+, t > 0\} = \mathring{\Re}_+$, |
| $\Re_\tau$ | a time interval: $\Re_\tau = [0, \tau[ = \{t : 0 \leq t < \tau\}$, |
| $\Re_{\tau s}$ | a time interval: $\Re_{\tau s} = [0, \tau_s[ = \{t : 0 \leq t < \tau_s\}$, |
| $r$ | a real number, |
| $\mathbf{r}$ | a vector function, $\mathbf{r} : \Re \times \Re^n \to \Re^n$, |
| $\mathcal{S}$ | a nonempty subset of $\Re^n$: $\mathcal{S} \neq \emptyset$, $\mathcal{S} \subset \Re^n$, |
| $\mathring{\mathcal{S}}$ | the interior of $\mathcal{S}$, |
| $\overline{\mathcal{S}}$ | the closure of $\mathcal{S}$, |
| $\partial\mathcal{S}$ | the boundary of $\mathcal{S}$, |
| $\mathcal{S}_d$ | some set of the evolution of a disturbance $d$, allowing for the solution of the perturbed system to exist, |
| $\mathcal{S}_e$ | the set of all the equilibrium points of a given system, |
| $\mathbf{s}$ | elementwise positive constant vector, $\mathbf{s} = (s_1\ s_2\ \ldots\ s_n)^T$, $s_i \in \Re^+$, $i = 1, 2, \ldots, n$, |
| $sat(\cdot)$ | saturation function with a linear part, $sat(\cdot) : \Re \to [-1, 1]$, $sat\ x = sgn\ x$ for $\|x\| \geq 1$, $sat\ x = x$ for $\|x\| \leq 1$, |
| $sgn(\cdot)$ | ideal relay function, $sgn(\cdot) : \Re \to [-1, 1]$, $sgn\ x = \|x\|^{-1}x$ if $x \neq 0$, $sgn\ 0 \in [-1, 1]$, |
| $sign(\cdot)$ | sign function, $sign(\cdot) : \Re \to \{-1, 0, 1\}$, $sign\ x = sgn\ x$ for $x \neq 0$, $sign\ 0 = 0$, |
| $\mathcal{T}$ | a time interval $\mathcal{T} \subseteq \Re$, |
| $\mathcal{T}_{t^*}^{\overset{+}{(-)}}$ | $= \{t_k : t_k \in \Re,\ t_k > t^*\ (t_k < t^*),\ k = 0, 1, \ldots,\ t_k \to t^*$ as $k \to +\infty\}$, |
| $\mathcal{T}_{\tau_s}$ | a time interval: $\mathcal{T}_{\tau_s} = [\tau_s, \tau[ = \{t : \tau_s \leq t < \tau\}$, |
| $t$ | an independent scalar variable denoting time, |
| $t_0$ | an initial moment, $t_0 \in \Re$, |
| $\mathcal{U}$ | a uniquely bounded set, |
| $u$ | a generating function of a uniquely bounded set, |
| $u(M)$ | an importance vector of a $-M$-matrix $M$, $u(M) \geq 0$, |
| $\mathbf{u}$ | either a control vector (function), $\mathbf{u} = (u_1\ u_2\ \ldots\ u_r)^T$, with the elements $u_i$ whose values are measured with respect to their total zero values, or an elementwise positive vector of size $k$ or $n$, |
| $\mathcal{V}^{-1}(\zeta)$ | $= \{X : X \in \Re^n, v(X) = \zeta\}$, |
| $\mathcal{V}_\zeta(\mathcal{A})$ | the largest open connected neighbourhood of $\mathcal{A}$ obeying $v(X) < \zeta$ for $X \in \mathcal{V}_\zeta$ and $v(X) = \zeta$ for $X \in \partial\mathcal{V}_\zeta$ if $\|\zeta\| < +\infty$, |
| $\mathcal{V}_\zeta$ | $= \mathcal{V}_\zeta(\mathcal{O})$, |

| | |
|---|---|
| $v$ | a tentative scalar Lyapunov function, in general $v : \Re \times \Re^n \to \Re$, and $v : \Re^n \to \Re$ in particular, |
| $w$ | a real number, or a vector $w \in \Re^m$, |
| $W$ | a matrix, $W \in \Re^{n \times n}$, |
| $\mathtt{W}$ | an input vector (function) with the values of all its elements measured with respect to their total zero values, $\mathtt{W} : \Re \to \Re^l$, it can represent an external disturbance vector (function) $d$: $\mathtt{W} = d$, or a control vector (function) $\mathtt{u} : \mathtt{W} = \mathtt{u}$, or can be composed of both: $\mathtt{W} = (d^T \ \mathtt{u}^T)^T$, |
| $\mathcal{X}$ | the domain of definition of $f$ with respect to $X$, $\mathcal{X}$ is an open nonempty subset of $\Re^n$: $\mathcal{X} = \mathring{\mathcal{X}} \neq \emptyset$, $\mathcal{X} \subseteq \Re^n$, |
| $\mathcal{X}_A$ | the set of admitted vectors $X$ over a time interval $\mathcal{T}_\tau$ with nonempty interior: $\mathring{\mathcal{X}}_A \neq \emptyset$, $\mathcal{X}_A \subset \Re^n$, |
| $\mathcal{X}_F$ | the set of permitted vectors $X$ over a time interval $\mathcal{T}_{\tau_s}$ with nonempty interior: $\mathring{\mathcal{X}}_F \neq \emptyset$, $\mathcal{X}_F \subset \Re^n$, |
| $\mathcal{X}_0$ | the domain of definition of $\underline{X}$ with respect to $X_0$, $\mathcal{X}_0 \subseteq \Re^n$; also a set of initial states $X_0$ acceptable with respect to both $\mathcal{X}_A$ and $\mathcal{X}_F$, |
| $\mathcal{X}_{0A}$ | a set of initial states $X_0$ acceptable with respect to $\mathcal{X}_A$, |
| $\mathcal{X}_{0F}$ | a set of initial states $X_0$ acceptable with respect to $\mathcal{X}_F$, |
| $x$ | a state variable, the value of which at time $t$ is $x(t)$, $x(t) \in \Re$, |
| $X$ | a vector variable $X = (x_1 \ x_2 \ \dots \ x_n)^T$, its vector value at time $t$ is $X(t)$, $X(t) \in \Re^n$, if it is a dynamical system vector variable then $X(t)$ is the system state (vector) at time $t$, in a particular case $X(t_1) = X(t_2)$ for all $t_1, t_2 \in \Re$ and then $X(t) \equiv X$ is an $n$-dimensional vector: $X \in \Re^n$, $X = x$ iff $n = 1$, |
| $X_0$ | an initial vector value of $X$, $X_0 = X(t_0)$, |
| $\underline{X}$ | is a *motion*, or a *solution*, of a system, $\underline{X} : \Re_0 \times \mathcal{T} \times \mathcal{X}_0 \times \mathcal{S}_i \to \Re^n$, and it denotes a time evolution of $X$ due to an action of an input $\mathtt{i}$ on the system, which at time $t_0$ passes through $X_0$, and whose vector value at time $t$ represents $X(t)$, $\underline{X}(t; t_0, X_0; \mathtt{i}) \equiv X(t)$; $\underline{X}(t_0; t_0, X_0; \mathtt{i}) \equiv X_0$, $\underline{X}$ is also called *an integral curve* of the system, |
| $\underline{X}(t; X_0; \mathtt{i})$ | the simplified notation for $\underline{X}(t; t_0, X_0; \mathtt{i})$ only in the case $t_0 = 0$, $\underline{X}(0; X_0; \mathtt{i}) \equiv X_0$, when $t$ is arbitrary rather than fixed then $\underline{X}(t; X_0; \mathtt{i})$ will be also called motion (solution, integral curve), |
| $\underline{X}(t; X_0)$ | the simplified notation for $\underline{X}(t; X_0; \mathtt{i})$ only if $\mathtt{i}(t) \equiv 0$, $\underline{X}(0; X_0) \equiv X_0$, |
| $\underline{X}(\mathcal{T}; X_0)$ | the set of $\underline{X}(t; X_0)$ for all $t \in \mathcal{T}$, $\underline{X}(\mathcal{T}; X_0) = \{X : \exists t \in \mathcal{T}, X = \underline{X}(t; X_0)\}$, iff $\mathcal{T}$ is the maximal interval $\mathcal{I}_0$ on which $\underline{X}(t; X_0)$ is defined (has a vector value in $\mathcal{X}$) then $\underline{X}(\mathcal{T}; X_0) = \underline{X}(\mathcal{I}_0)$ is a *trajectory* through $X_0$, otherwise $\underline{X}(\mathcal{T}; X_0)$ is an *arc* (*on* $\mathcal{T}$) of the trajectory $\underline{X}(\mathcal{I}_0) : \underline{X}(\mathcal{T}; X_0) \subseteq \underline{X}(\mathcal{I}_0)$ if $\mathcal{T} \subseteq \mathcal{I}_0$, |

$\mathbf{y}$ — the real output vector (function) with the elements whose values are measured with respect to their total zero values, or a vector $\mathbf{y} \in \Re^m$,

$\mathbf{y}_d$ — a desired output vector (function): a desired (prespecified, demanded, reference) output vector (function) $\mathbf{y}$,

$\mathbf{Z}$ — the class of square matrices with non positive off-diagonal elements,

$\mathcal{Z}$ — at most finite set,

$\mathbf{z}$ — a vector, $\mathbf{z} \in \Re^n$.

Greek

$\alpha$ — a scalar,

$\beta$ — a scalar,

$\delta_{ij}$ — Kronecker delta: $\delta_{ij} = 1$ for $i = j$, $\delta_{ij} = 0$ for $i \neq j$,

$\gamma$ — a scalar,

$\emptyset$ — the empty set,

$\lambda(A)$ — the minimal eigenvalue of the matrix $A = A^T$,

$\lambda(M)$ — the importance (eigen) value of a $[-M]$-matrix $M$, $\lambda(M) < 0$,

$\Lambda(A)$ — the maximal eigenvalue of the matrix $A = A^T$,

$\rho$ — a distance function, $\rho(X, \mathcal{A})$ is the distance from $X$ to $\mathcal{A}$: $\rho(X, \mathcal{A}) = inf(||X - y|| : y \in \mathcal{A})$,

$\tau$ — a final time, $\tau \in \Re^+$ or $\tau = +\infty$,

$\tau_a$ — an attraction time: $\tau_a \in \Re^+$,

$\tau_s$ — a settling time: $\tau_s \in ]0, \tau]$,

$\theta$ — a transformed time variable: $d\theta = dt(1 + ||f(X)||)$,

$\psi$ — a scalar function associated with the derivative of $v$, $\psi : \Re^n \to \Re$,

$\omega$ — a scalar function, $\omega : \Re \times \Re \to \Re$,

$$\omega(x, \dot{x}) = \begin{cases} -1, & (x \in \Re^-,\ \dot{x} \in \Re) \text{ or } (x = 0,\ \dot{x} \in \Re^-), \\ 0, & (x = 0,\ \dot{x} = 0), \\ +1, & (x \in \Re^+,\ \dot{x} \in \Re) \text{ or } (x = 0,\ \dot{x} \in \Re^+), \end{cases}$$

$\Omega$ — a matrix function, $\Omega : \Re^n \times \Re^n \to \Re^{n \times n}$, $\Omega(X, \dot{X}) = diag\{\omega(x_1, \dot{x}_1) \quad \omega(x_2, \dot{x}_2) \ \ldots \ \omega(x_n, \dot{x}_n)\}$.

Symbols

$\in$ — belongs to, is member of,

$\notin$ — does not belong to, is not member of,

$\Rightarrow$ — implies,

$\Leftrightarrow$ — if and only if, necessary and sufficient

$\to$ — converges to, tends to,

$\exists$ — exists,

$\exists!$ — exists exactly one,

| | |
|---|---|
| $\mathcal{A}-\mathcal{B}$ | set difference: $\mathcal{A}-\mathcal{B}=\{\alpha:\alpha\in\mathcal{A},\alpha\notin\mathcal{B}\}$ |
| $\mathcal{A}\cup\mathcal{B}$ | set union: $\mathcal{A}\cup\mathcal{B}=\{\alpha:\alpha\in\mathcal{A}$ or $\alpha\in\mathcal{B}$ or both$\}$ |
| $\mathcal{A}\cap\mathcal{B}$ | set intersection: $\mathcal{A}\cap\mathcal{B}=\{\alpha:\alpha\in\mathcal{A}$ and $\alpha\in\mathcal{B}\}$ |
| $\mathcal{A}\subset\mathcal{B}$ | there is $\alpha\in\mathcal{A}$, $\alpha\notin\mathcal{B}$, |
| $\mathcal{A}\subseteq\mathcal{B}$ | permits both $\mathcal{A}\subset\mathcal{B}$ and $\mathcal{A}=\mathcal{B}$, |
| $\forall$ | for every, |
| $\forall_a(\cdot)$ | for every $(\cdot)$ almost everywhere, almost for every $(\cdot)$, |
| $sat\ x$ | $=\begin{cases} x & , \lvert x\rvert\le 1, \\ sign\ x & , \lvert x\rvert\in\Re_+, \end{cases}$ |
| $sgn\ x$ | $\begin{cases} =-1 & , x<0, \\ \in[-1,1] & , x=0, \\ =1 & , x>0 \end{cases}$ |
| $sgn\ X$ | $=(sgn\ x_1\ sgn\ x_2\ \dots\ sgn\ x_n)^T$, |
| $sign\ x$ | $=\begin{cases} sgn\ x & , x\neq 0 \\ 0 & , x=0, \end{cases}$ |
| $sign\ X$ | $=(sign\ x_1\ sign\ x_2\ \dots\ sign\ x_n)^T$, |
| $]\alpha,\beta[$ | an open interval, $]\alpha,\beta[=\{t:t\in\Re,\ \alpha<t<\beta\}$, |
| $[\alpha,\beta]$ | a closed interval, $[\alpha,\beta]=\{t:t\in\Re,\ \alpha\le t\le\beta\}$, |
| $(\alpha,\beta)$ | a general interval that is not completely specified, it can be $]\alpha,\beta]$ or $[\alpha,\beta[$ or $[\alpha,\beta]$, or $]\alpha,\beta[$, |
| $\nabla\nabla\nabla$ | end of a theorem, |
| $\vee$ | (inclusive) logical "OR", $a\vee b$ means either "$a$" or "$b$" or both "$a$" and "$b$", |
| $\mathbf{1}$ | $=(1\ 1\ \dots\ 1)^T\in\Re^n$, |
| $\lvert X\rvert$ | elementwise absolute value of the vector $X=(x_1\ x_2\ \dots\ x_n)^T$, $\lvert X\rvert=(\lvert x_1\rvert\ \lvert x_2\rvert\ \dots\ \lvert x_n\rvert)^T$, |
| $\lvert X\rvert\le\lvert S\rvert$ | $\Leftrightarrow \lvert x_i\rvert\le\lvert s_i\rvert,\ \forall i=1,2,\dots,n$ for $S=(s_1\ s_2\ \dots\ s_n)^T$. |
| $\lvert X\rvert<\lvert S\rvert$ | $\Leftrightarrow \lvert x_i\rvert<\lvert s_i\rvert,\ \forall i=1,2,\dots,n$, |
| $\lVert\cdot\rVert$ | a norm on $\Re^n$, $\lVert\cdot\rVert:\Re^n\to\Re_+$; it is Euclidean norm: $\lVert X\rVert=(X^TX)^{1/2}$ if it is not stated otherwise, |
| $X^k$ | $=(x_1^k\ x_2^k\ \dots\ x_n^k)^T$. |

# General introduction

The notion of **stability** is widely used every day by experts as well as by people not familiar with stability theory. It is utilised to express a kind of **persistency** of a **closeness** of **objects** over a **time interval** and under usually specified **circumstances**. The objects, their closeness, the time interval, the circumstances and the persistency can be defined in various ways, which has led to different stability concepts such as the Lyapunov stability concept [127], the Lagrange stability concept (see [152], [157]), the hyperstability concept by Popov [168], the L-stability concept (see [39]) and the practical stability concept (see [11], [29], [43], [44], [52], [60]–[63], [91], [93], [99], [105], [110], [111], [121], [133], [144]–[146], [150], [156], [179], [194], [205], [207]–[209]).

Roughly speaking, every stability concept should clarify the following questions:

(A) **Whose stability (stability of what) is considered**? The object can be a single state or output of a system. It can be a trajectory, or a motion, or a set, or a whole system. The object whose stability is studied has to be prespecified. In this book it will be a state, or a motion or a set or a system.

(B) **How is the distance between objects (vectors, functions or sets) defined**? The distance is naturally induced by a norm accepted on an appropriate vector space. Euclidean norm will be accepted herein.

(C) **How is the closeness between two objects (vectors, functions or sets) defined**? The closeness is defined via a distance function. Sometimes the distance function itself does not determine the closeness completely. Then the distance function and prescribed bounds on its values determine the closeness.

(D) **Over which time interval is the requested closeness to be realised**? From the mathematical point of view the time interval can be unbounded, like in the Lyapunov stability concept, or bounded as permitted in the practical stability concept. From the engineering (technical) point of view it should be bounded, simply because every man-made product is aimed to be used over a bounded time interval only.

(E) **Under which conditions the closeness should be preserved**? The closeness can be requested only under the influence of initial deviations (initial conditions) as required in the Lyapunov stability concept. It can be demanded only under actions of external excitations as in the L-stability concept, or under both as in the practical stability framework.

(F) **Under which initial conditions the closeness has to be guaranteed?** The closeness can be required only under sufficiently small initial conditions, which characterises the classical Lyapunov stability concept. This is more attractive mathematically than technically. From the engineering, as well as mathematical, point of view it is most important and useful to determine, or at least estimate well, **the largest set of the initial conditions under which the closeness is realisable. Such a set obeying specified requirements will be called the domain of the corresponding stability property.** The purpose of this book is to present the basic theory of stability domains.

# Chapter 1

# Introductory comments on stability concepts

## 1.1 Comments on Lyapunov's stability concept

### 1.1.1 Lyapunov's definition of stability and the definitions of stability in the Lyapunov sense

Lyapunov [127] used $q = [q_1\ q_2\ \dots\ q_k\ \dot{q}_1\ \dot{q}_2\ \dots\ \dot{q}_k]^T$ as the state (vector) of a material system with $k$ degrees of freedom. It satisfies a vector differential equation of the first order and dimension equal to $2k$ (or to $n$ if we replace $2k$ by $n : n = 2k$), which is of the form (1.1):

$$\frac{\mathrm{d}q}{\mathrm{d}t} = r(t, q), \quad r : \Re \times \Re^n \to \Re^n, \quad \Re = ]-\infty, +\infty[. \tag{1.1}$$

By following Lyapunov [127] a particular solution $q_u$ of (1.1) being of special interest to us is denoted by $q_u(t)$ at time $t \in \Re$, $q_u = [q_{u1}\ \ q_{u2}\ \ \dots\ \ q_{uk}\ \ \dot{q}_{u1}\ \ \dot{q}_{u2}\ \ \dots\ \ \dot{q}_{uk}]^T$. It is called "the unperturbed motion of (1.1)". All the other solutions are then called "perturbed motions of (1.1)". The initial (vector) value of such a solution is denoted by $q(t_0)$, and that of the unperturbed motion is $q_u(t_0)$. The difference between them is expressed in terms of $X_0 = [\varepsilon_1\ \ \varepsilon_2\ \ \dots\ \ \varepsilon_k\ \ \varepsilon'_1\ \ \varepsilon'_2\ \ \dots\ \ \varepsilon'_k]^T$, where all $\varepsilon_i$ and $\varepsilon'_i$ are real numbers. Hence, in the vector form:

$$q(t_0) = q_u(t_0) + X_0, \tag{1.2}$$

and in the scalar form:

$$q_j(t_0) = q_{ju}(t_0) + \varepsilon_j, \quad \dot{q}_j(t_0) = \dot{q}_{ju}(t_0) + \varepsilon'_j, \quad \forall j = 1, 2, \dots, k. \tag{1.3}$$

The perturbed motions essentially depend on $X_0$.

Lyapunov [127] used scalar real-valued continuous functions $Q_1, Q_2, \dots, Q_m$ dependent on $q$ and denoted their values taken along the unperturbed motion $q_u$

at an instant $t \in \Re$ by $\mathcal{F}_1(t), \mathcal{F}_2(t), \ldots, \mathcal{F}_m(t)$. This means that $\mathcal{F}_i(t) = Q_i[q_u(t)]$, $\forall i = 1, 2, \ldots, n$.

LYAPUNOV'S DEFINITION OF STABILITY OF MOTION [127]: "Let $\mathcal{L}_1, \mathcal{L}_2, \ldots, \mathcal{L}_m$ be arbitrarily given positive numbers. If for all $\mathcal{L}_s$, irrespective of how small they are, it is possible to select positive numbers $E_1, E_2, \ldots, E_k, E'_1, E'_2, \ldots, E'_k$ so that for all real $\varepsilon_j, \varepsilon'_j$, which satisfy the conditions:

$$|\varepsilon_j| \leq E_j, \quad |\varepsilon'_j| \leq E'_j \quad (j = 1, 2, \ldots, k),$$

and for every $t$, greater than $t_0$, the inequalities:

$$|Q_1 - \mathcal{F}_1| < \mathcal{L}_1, \quad |Q_2 - \mathcal{F}_2| < \mathcal{L}_2, \ldots, |Q_m - \mathcal{F}_m| < \mathcal{L}_m,$$

are satisfied, then the unperturbed motion is stable with respect to quantities $Q_1, Q_2, \ldots, Q_m$; otherwise – it is unstable with respect to the same quantities."

**Comment 1.1** *Lyapunov's definition is related to cases for which the initial instant does not influence the stability property [127]. In such cases we may accept $t_0 = 0$, which holds for time-invariant (also called stationary) systems.*

**Comment 1.2** *Lyapunov's definition demands that inequalities*

$$|Q_i - \mathcal{F}_i| < \mathcal{L}_i, \quad \forall i = 1, 2, \ldots, m,$$

*are satisfied for all $t \in ]t_0, +\infty]$ rather than for all $t \in [t_0, +\infty[$. However, it has been commonly accepted to require that for all $t \in \Re_0$, $\Re_0 = [t_0, +\infty[$ [10], [94]-[121], [218].*

**Comment 1.3** *Lyapunov's definition introduces stability with respect to some continuous functions $Q_i$ dependent on $q$. In a special case $Q_i = q_i$ and $Q_{k+i} = \dot{q}_i$, $i = 1, 2, \ldots, k$. It is interesting to note that Lyapunov himself studied only this special case. A justification for that is clarified by the next theorem, which is a generalised version of theorem 1 by Grujić in [88] (page 7).*

**Theorem 1.1** *Let the unperturbed motion $q_u$ be a differentiable vector function. Let $Q = (\,Q_1 \quad Q_2 \quad \ldots \quad Q_n\,)^T$, $n = 2k$:*

$$a) \qquad X(t) = q(t) - q_u(t), \tag{1.4}$$

$$b) \qquad f(t, X) = r[t, q_u(t) + X] - r[t, q_u(t)], \tag{1.5}$$

*and, in view of (1),*

$$c) \qquad \frac{\mathrm{d}X}{\mathrm{d}t} = f(t, X). \tag{1.6}$$

*Then, stability of $X = 0$ of the system (1.4)-(1.6) with respect to $Q(t, X) \equiv X$ is necessary and sufficient for stability of the unperturbed motion $q_u(.)$ of the system (1.1) with respect to any vector function $Q$, $Q : \Re \times \Re^n \to \Re^n$, which is uniformly continuous in $(t, q) \in \Re \times \Re^n$.* ∇∇∇

**Proof.** *Necessity.* Let $Q(t,q) \equiv q$ be chosen for system (1.1) and let $Q(t,X) \equiv X$ be selected for the system (1.6). Such choices are permissible. The function $Q$ is uniformly continuous. Let the unperturbed motion $q_u$ of the system (1.1) be stable with respect to any function $Q$ that is uniformly continuous in $(t,q) \in \Re \times \Re^n$. Hence:

$$\forall \mathcal{L}_i \in \Re^+, \quad \Re^+ = ]0,+\infty[, \quad \forall i = 1,2,\ldots,n, \quad \exists (E_j, E_j') \in \Re^+ \times \Re^+,$$

such that

$$|\varepsilon_j| = |q_j(t_0) - q_{ju}(t_0)| < E_j,$$

and

$$|\varepsilon_j'| = |\dot{q}_j(t_0) - \dot{q}_{ju}(t_0)| < E_j', \quad \forall j = 1,2,\ldots,k,$$

imply both:

$$|Q_i[t,q(t)] - \mathcal{F}_i(t)| = |q_i(t) - q_{iu}(t)| < \mathcal{L}_i, \quad \forall i = 1,2,\ldots,k,$$

and

$$|Q_{k+i}[t,q(t)] - \mathcal{F}_{k+i}(t)| = |\dot{q}_i(t) - \dot{q}_{iu}(t)| < \mathcal{L}_{k+i}, \quad \forall i = 1,2,\ldots,k,$$

for all $t \in ]t_0,+\infty[$ and every $t_0 \in \Re$. These facts and the definition of $X = (\,x_1 \;\; x_2 \;\; \ldots \;\; x_n\,)^T$ given by (1.4) yield what follows: $\forall \mathcal{L}_i \in \Re^+$, $\forall i = 1,2,\ldots,n$, $\exists (E_j, E_j') \in \Re^+ \times \Re^+$ such that $|x_j(t_0)| < E_j$ and $|x_{k+j}(t_0)| < E_j'$, $\forall j = 1,2,\ldots,k$, imply

$$|x_j(t)| < \mathcal{L}_j, \quad \forall t \in ]t_0,+\infty[, \quad \forall t_0 \in \Re, \quad \forall j = 1,2,\ldots,n,$$

due to (1.4), which proves stability of $X = 0$ of the system (1.4)–(1.6) with respect to $Q(t,X) \equiv X$ in view of (1.5).

*Sufficiency.* Let $Q$ be arbitrary uniformly continuous function in $(t,q) \in \Re \times \Re^n$, which implies the following:

$$\forall t \in \Re, \quad \forall q^* \in \Re^n, \quad \forall \mathcal{L}_i \in \Re^+, \quad \forall i = 1,2,\ldots,n, \quad \exists l_i \in \Re^+,$$

such that

$$|q_j - q_j^*| < l_j \quad \text{and} \quad |\dot{q}_{k+j} - \dot{q}_{k+j}^*| < l_{k+j}, \quad \forall j = 1,2,\ldots,k,$$

guarantee

$$|Q_i(t,q) - Q_i(t,q^*)| < \mathcal{L}_i, \quad \forall i = 1,2,\ldots,n. \tag{1.7}$$

Let $X = 0$ of the system (1.4)–(1.6) be stable with respect to $Q(t,X) \equiv X$. This means what follows:

$$\begin{aligned} &\forall t_0 \in \Re, \quad \forall l_i \in \Re^+, \quad \forall i = 1,2,\ldots,n, \\ &\exists (E_j, E_j') \in \Re^+ \times \Re^+, \quad \forall j = 1,2,\ldots,k, \end{aligned} \tag{1.8}$$

such that

$$|x_j(t_0)| < E_j \quad \text{and} \quad |x_{k+j}(t_0)| < E_j', \quad \forall j = 1,2,\ldots,k,$$

imply:

$$|x_j(t)| < l_j \quad \text{and} \quad |x_{k+j}(t)| < l_{k+j}, \quad \forall t \in ]t_0, +\infty[, \quad \forall j = 1, 2, \ldots, k. \qquad (1.9)$$

Now, (1.5), (1.6), (1.7) and (1.9) prove that for the unperturbed motion $q_u$ of the system (1.1) the following holds:

$$\begin{aligned} &\forall t_0 \in \Re, \quad \forall l_i \in ]0, +\infty[, \quad \forall i = 1, 2, \ldots, n, \\ &\exists (E_j, E_j') \in \Re^+ \times \Re^+, \quad \forall j = 1, 2, \ldots, k, \end{aligned} \qquad (1.10)$$

such that

$$|q_j(t_0) - q_{ju}(t_0)| < E_j \quad \text{and} \quad |\dot{q}_j(t_0) - \dot{q}_{ju}(t_0)| < E_j', \quad \forall j = 1, 2, \ldots, k,$$

guarantee

$$\begin{aligned} &|q_j(t) - q_{ju}(t)| < l_j \quad \text{and} \quad |\dot{q}_j(t) - \dot{q}_{ju}(t)| < l_{k+j}, \\ &\forall t \in ]t_0, +\infty[, \quad \forall j = 1, 2, \ldots, k. \end{aligned} \qquad (1.11)$$

Since all $l_i \in \Re^+$ can be arbitrarily chosen in (1.11) then let them be accepted to obey (1.7). Now, (1.4) with $q^* = q_u$ in it, and (1.11) yield $\forall t_0 \in \Re$, $\forall \mathcal{L}_i \in \Re^+$, $\forall i = 1, 2, \ldots, n$, $\exists (E_j, E_j') \in \Re^+ \times \Re^+$, $\forall j = 1, 2, \ldots, k$, such that

$$|\varepsilon_j| = |q_j(t_0) - q_{ju}(t_0)| < E_j$$

and

$$|\dot{\varepsilon}_j| = |\dot{q}_j(t_0) - \dot{q}_{ju}(t_0)| < E_j', \quad \forall j = 1, 2, \ldots, k,$$

ensure

$$\begin{aligned} &|Q_i[t, q(t)] - Q_i[t, q_u(t)]| = |Q_i[t, q(t)] - \mathcal{F}_i(t)| < \mathcal{L}_i, \\ &\forall t \in ]t_0, +\infty[, \quad \forall i = 1, 2, \ldots, n, \end{aligned}$$

which proves stability of the unperturbed motion $q_u(.)$ of the system (1.1) with respect to the vector function $\boldsymbol{Q}$.

Theorem 1.1 reduces the stability analysis of the unperturbed motion $q_u$ of the system (1.1) to the stability analysis of the zero state $X = 0$ of the system (1.4)–(1.6).

Lyapunov's original definition, Comments 1–3 and Theorem 1.1 explain why the following definition is commonly used as the definition of stability in the Lyapunov sense in case the initial instant $t_0$ is fixed, e.g. $t_0 = 0$. Let $\underline{X}(t; t_0, X_0)$ be a motion of the system (1.6) at time $t$, which passes through $X_0$ at time $t = t_0$, $\underline{X}(t_0; t_0, X_0) \equiv X_0$.

**Definition 1.1** *a) A point $X^* \in \Re^n$ is stable at $t_0 \in \Re$ for the system (1.6) if and only if for every $\varepsilon \in \Re^+$ there is $\delta \in \Re^+$, $\delta = \delta(\varepsilon; t_0; X^*)$, such that $||X_0 - X^*|| < \delta$ implies $||\underline{X}(t; t_0, X_0) - X^*|| < \varepsilon$ for every $t \in \Re_0$.*

*b) The stability is global (in the whole, in the large) if and only if the maximal $\delta(\varepsilon; t_0; X^*)$ denoted by $\delta_M(\varepsilon; t_0; X^*)$ obeying a) tends to $+\infty$ as $\varepsilon \to +\infty$.*

c) *If and only if* $X^* = 0$ *then* $\delta(\varepsilon; t_0; 0)$ *and* $\delta_M(\varepsilon; t_0, 0)$ *are denoted by* $\delta(\varepsilon; t_0)$ *and* $\delta_M(\varepsilon; t_0)$ *respectively.*

**Comment 1.4** *a) The Lyapunov stability concept was originally concerned with stability of a motion and of the origin* $X = 0$*, which was later broadened to stability of a set [10], [218].*

*b) The distance was originally defined by Lyapunov [127] via the absolute values* $|Q_i(t) - \mathcal{F}_i(t)|$*. It is now commonly defined via the Euclidean norm in the literature, e.g. [10], [94], [116], [121], [218].*

*c) The closeness in the Lyapunov sense means that the distance is less than* $\varepsilon$ *for any given* $\varepsilon \in \Re^+$*, [94], [218].*

*d) The closeness in the Lyapunov sense is requested over the infinite time interval* $\Re_0 = [t_0, +\infty[ \subset \Re$*, [10], [94], [116], [121], [218]. This differs from Lyapunov's original requirement for the closeness over the time interval* $]t_0, +\infty]$*.*

*e) The Lyapunov closeness is demanded provided external influences to the system (1.1) are nominal over the time interval* $\Re_0$*. This means the Lyapunov closeness is required for the system (1.6) in its free regime (i.e. under no external disturbance action over the time interval* $\Re_0$*).*

*f) The Lyapunov closeness is demanded for all the initial states* $X_0 = X(t_0)$ *whose distance from* $X = 0$ *is less than some* $\delta = \delta(t_0, \varepsilon) \in \Re^+$*.*

*g) Definition 1.1 broadens the concept of continuity of a function at a point to motions. For a deep analysis of it see Rouche and Mawhin [178, pp. 1–3].*

**Comment 1.5** *Definition 1.1 can be satisfied by arbitrarily small* $\delta \in \Re^+$ *even if the corresponding* $\varepsilon \in \Re^+$ *has been chosen very large. It requires only the existence of* $\delta \in \Re^+$ *obeying its condition. This is a conceptual drawback of the Lyapunov stability concept from the point of view of engineering, technical applications and needs.*

**Example 1.1 [70]** Let $n = 2$, $X = [x_1 \quad x_2]^T$, $\alpha \in \Re^+$ and the system (1.6) take the following specific form:

$$\frac{\mathrm{d}X}{\mathrm{d}t} = (-\alpha + |x_1| + |x_2|)X.$$

The system state portrait is given in Fig. 1.1. The state $X = 0$ is stable. However, if $\alpha$ is small (e.g. $\alpha = 10^{-3}$) or very small (e.g. $\alpha = 10^{-10}$), then the dynamic behaviour of the system is unsatisfactory in the engineering sense even for small initial states ($|x_{10}| + |x_{20}| > 10^{-3}$) or ($|x_{10}| + |x_{20}| > 10^{-10}$), respectively.

**Comment 1.6** *Example 1.1 shows that existence of positive* $\delta$ *obeying Definition 1.1 is not adequate engineering information about the qualitative dynamic properties of the system. Useful information is that about the largest neighbourhood* $\mathcal{D}_s(\varepsilon; t_0)$ *of* $X = 0$ *such that* $\|\underline{X}(t; t_0, X_0)\| < \varepsilon$ *is satisfied for all* $t \in \Re_0$ *iff* $X_0 \in \mathcal{D}_s(\varepsilon; t_0)$ *for any* $\varepsilon \in \Re^+$*. Moreover, we need information about the largest neighbourhood* $\mathcal{D}_s(t_0)$ *of* $X = 0$ *containing all* $\mathcal{D}_s(\varepsilon; t_0)$*, that is that* $\mathcal{D}_s(t_0) = \cup[\mathcal{D}_s(\varepsilon; t_0) : \varepsilon \in \Re^+]$*.*

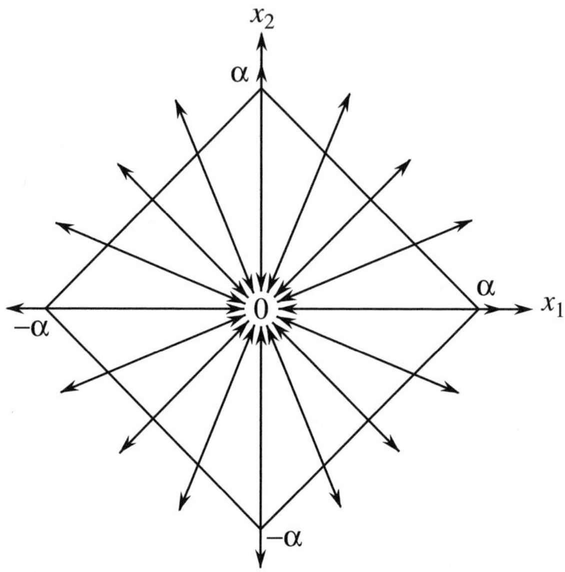


Figure 1.1: The state portrait of the system of Example 1.1.

**Note 1.1** For the sake of simplicity of introducing new concepts this book will deal only with time-invariant continuous-time systems. In the preceding analysis time-varying systems were treated due to the reference to Lyapunov's original considerations and definition.

Motions of time-invariant systems, as well as their properties, do not depend on the initial moment $t_0$. Hence, it is accepted equal to zero:

$$t_0 = 0.$$

This enables the use of the simpler, more concise notation for a system motion, $\underline{X}(.; X_0)$:

$$\underline{X}(t; t_0, X_0) \equiv \underline{X}(t; 0, X_0) \equiv \underline{X}(t, X_0) \equiv X(t).$$

In case the system (1.6) is time-invariant then it will be described by (1.12):

$$\frac{\mathrm{d}X}{\mathrm{d}t} = f(X), \quad f : \Re^n \to \Re^n. \tag{1.12}$$

Its motion, which passes through $X_0$ at $t = 0$, is denoted by $\underline{X}(.; X_0)$. This reduces Definition 1.1 to Definition 1.2. The former holds for time-varying continuous-time systems, while the latter is valid only for time-invariant continuous-time systems.

**Definition 1.2** *a) The state $X = 0$ of the system (1.12) is stable if and only if for every $\varepsilon \in \Re^+$ there is $\delta \in \Re^+$, $\delta = \delta(\varepsilon)$, such that $||X_0|| < \delta$ implies $||\underline{X}(t; X_0)|| < \varepsilon$ for all $t \in \Re_+$.*

*b) The state $X = 0$ of the system (1.12) is globally stable (stable in the whole, stable in the large) if and only if a) holds so that the maximal $\delta(\varepsilon)$ denoted by $\delta_M(\varepsilon)$ tends to $+\infty$ as $\varepsilon$ tends to $+\infty$.*

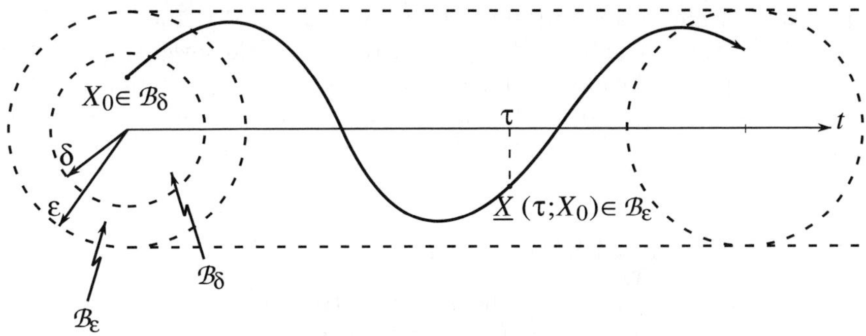


$\forall \varepsilon \in \Re^+, \exists \delta \in \Re^+ \rightarrow X_0 \in \mathcal{B}_\delta \Rightarrow \underline{X}(t; X_0) \in \mathcal{B}_\varepsilon, \forall t \in \Re_+.$

Figure 1.2: Stable $X = 0$.

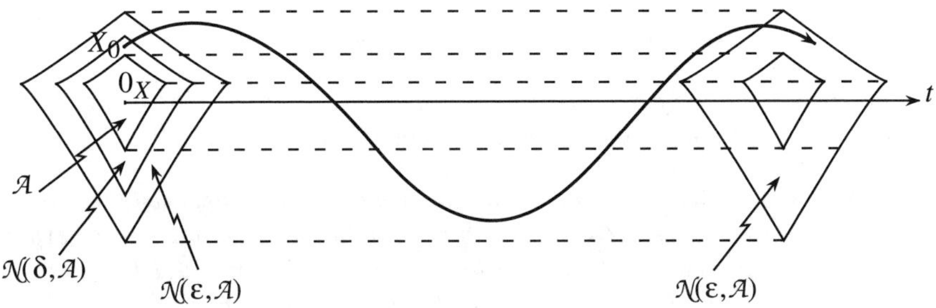


$\forall \varepsilon \in \Re^+, \exists \delta \in \Re^+ \rightarrow X_0 \in \mathcal{N}(\delta, \mathcal{A}) \Rightarrow \underline{X}(t; X_0) \in \mathcal{N}(\varepsilon, \mathcal{A}), \forall t \in \Re_+.$

Figure 1.3: Stable set $\mathcal{A}$.

$\mathcal{D}_s(\varepsilon; t_0)$ reduces to $\mathcal{D}_s(\varepsilon)$, $\mathcal{D}_s(\varepsilon; t_0) \equiv \mathcal{D}_s(\varepsilon)$, for time-invariant systems.

Comments 1.5 and 1.6 are valid also for stability of a set $\mathcal{A}$ defined as follows. The distance function $\rho(.) : \Re^n \times 2^{\Re^n} \rightarrow \Re_+$ is induced by the Euclidean norm, $\rho(X, \mathcal{A}) = inf\,[||X - y|| : y \in \mathcal{A}]$.

**Definition 1.3** *a) A set $\mathcal{A}$, $\mathcal{A} \subset \Re^n$, is stable with respect to motions of the system (1.12), for short: stable, if and only if for every $\varepsilon \in \Re^+$ there is $\delta \in \Re^+$, $\delta = \delta(\varepsilon; \mathcal{A})$, such that $\rho(X_0, \mathcal{A}) < \delta(\varepsilon, \mathcal{A})$ implies $\rho[\underline{X}(t; X_0), \mathcal{A}] < \varepsilon$ for all $t \in \Re_+$.*

*b) The set $\mathcal{A}$ is globally stable (stable in the whole, stable in the large) if and only if a) holds and the maximal $\delta(\varepsilon; \mathcal{A})$ denoted by $\delta_M(\varepsilon; \mathcal{A})$ tends to $+\infty$ as $\varepsilon$ tends to $+\infty$.*

**Comment 1.7** *We wish to know more than just about stability of a set $\mathcal{A}$. We want to get information about what is, for any $\varepsilon \in \Re^+$, the largest neighbourhood $\mathcal{D}_s(\varepsilon; \mathcal{A})$ of the set $\mathcal{A}$ such that $\rho[\underline{X}(t; X_0), \mathcal{A}] < \varepsilon$ for all $t \in \Re_+$ holds iff $X_0 \in \mathcal{D}_s(\varepsilon; \mathcal{A})$, and in addition, to know $\mathcal{D}_s(\mathcal{A}) = \cup[\mathcal{D}_s(\varepsilon; \mathcal{A}) : \varepsilon \in \Re^+]$.*

The preceding definitions are illustrated by Fig. 1.2 and Fig. 1.3. From the engineering point of view $\varepsilon$ can be used to measure accuracy; the lower $\varepsilon$, the higher the accuracy. Stability can be interpreted as a system's ability to maintain as high finite accuracy of its motions as desired (i.e. arbitrarily small $\varepsilon \in \Re^+$) provided initial state accuracy is sufficiently high (i.e. $\delta$ is sufficiently small, $\delta \in \Re^+$), respectively.

### 1.1.2 Definitions of attraction

Although Lyapunov did not define the notion of attraction, he was the first to state general qualitative conditions for convergence of system motions to the zero state (Lyapunov's remark 2 in [127], page 61), which led to definitions of attraction and asymptotic stability of $X = 0$ of the system (1.12). The former is commonly defined as follows (e.g. [10], [94], [218]):

**Definition 1.4** *a) The state $X = 0$ of the system (1.12) is attractive if and only if there is $\Delta > 0$ and for every $\zeta > 0$ there is $\tau \in \Re_+$, $\tau = \tau(X_0; \zeta)$, such that $||X_0|| < \Delta$ implies $lim\,[||\underline{X}(t; X_0)|| : t \to +\infty] = 0$.*

*b) The state $X = 0$ of the system (1.7) is globally attractive (attractive in the whole, in the large) if and only if a) holds for $\Delta = +\infty$.*

**Comment 1.8** *The definition of attraction requires only the existence of $\Delta > 0$ obeying its condition irrespective of whether $\Delta$ is large or small, even very small. In the case of the system (1.12) in the form given in Example 1.1 the state $X = 0$ is attractive. However, this property becomes useless from the engineering point of view as soon as $\alpha$ is small (e.g. $\alpha = 10^{-3}$) or very small (e.g. $\alpha = 10^{-10}$).*

**Comment 1.9** *For engineering purposes it is important to derive, or at least to estimate well, the largest neighbourhood $\mathcal{D}_a$ of $X = 0$ such that $lim\,[||\underline{X}(t; X_0)|| : t \to +\infty] = 0$ holds iff $X_0 \in \mathcal{D}_a$.*

**Definition 1.5** *a) A set $\mathcal{A}$ is attractive with respect to motions of the system (1.12), for short, attractive, if and only if there is $\Delta > 0$ and for every $\zeta > 0$ there is $\tau \in \Re_+$, $\tau = \tau(X_0, \zeta; \mathcal{A})$, such that $\rho(X_0, \mathcal{A}) < \Delta$ implies $\rho[\underline{X}(t; X_0), \mathcal{A}] < \zeta$ for all $t \in ]\tau, +\infty[$, and $lim\,[\underline{X}(t; X_0) : t \to +\infty] = X \in \mathcal{A}$.*

*b) The set $\mathcal{A}$ is globally attractive (attractive in the whole, in the large) if and only if a) holds for $\Delta = +\infty$.*

The preceding definitions are illustrated by Fig. 1.4 and Fig. 1.5.

**Comment 1.10** *We wish to know more than only about the attraction of a set $\mathcal{A}$. We want to get information also about the largest neighbourhood $\mathcal{D}_a(\mathcal{A})$ of the set $\mathcal{A}$ such that for every $\zeta > 0$ there is $\tau \in \Re_+$ obeying that $\rho[\underline{X}(t; X_0), \mathcal{A}] < \zeta$ for all $t \in ]\tau, +\infty[$ iff $X_0 \in \mathcal{D}_a(\mathcal{A})$.*

**Note 1.2** Notice that the condition "for every $\zeta > 0$ there is $\tau \in \Re_+$ such that $\rho(X_0, \mathcal{A}) < \Delta$ implies $\rho[\underline{X}(t; X_0), \mathcal{A}] < \zeta$ for all $t \in ]\tau, +\infty[$" is equivalent to "$lim\{\rho[\underline{X}(t; X_0), \mathcal{A}] : t \to +\infty\} = 0$".

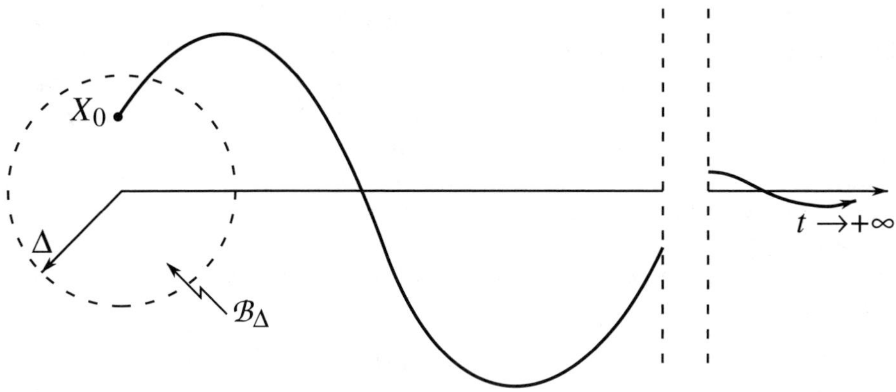


$$\exists \Delta \in \Re^+ \to X_0 \in \mathcal{B}_\Delta \Rightarrow lim\{\rho[\underline{X}(t, X_0), \mathcal{O}] : t \to +\infty\} = 0, \mathcal{O} = \{O_X\}\ \mathcal{O} = \{O_X\}$$

Figure 1.4: Attractive $X = 0$.

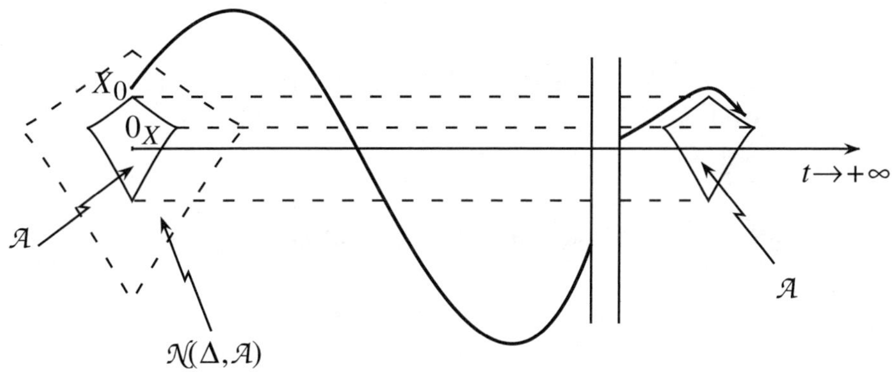


$$\exists \Delta \in \Re^+ \to X_0 \in \mathcal{N}(\Delta, \mathcal{A}) \Rightarrow lim\{\rho[\underline{X}(t; X_0), \mathcal{A}] : t \to +\infty\} = 0$$

Figure 1.5: Attractive set $\mathcal{A}$.

### 1.1.3 Definitions of asymptotic stability

Stability and attraction are in general mutually independent system dynamic properties. It was well illustrated by Vinograd (Hahn 1967 [94], pp. 191–194). Both are often desired, which led to the concept of asymptotic stability.

**Definition 1.6** *a) The state $X = 0$ of the system (1.12) is asymptotically stable if and only if it is both stable and attractive.*

*b) The state $X = 0$ of the system (1.12) is globally asymptotically stable (asymptotically stable in the whole, in the large) if and only if it is both stable and globally attractive.*

*c) The state $X = 0$ of the system (1.12) is completely globally asymptotically stable if and only if it is both globally stable and globally attractive.*

**Comment 1.11** *The notion of asymptotic stability has the same drawback as stability and attraction. The state $X = 0$ of the system considered in Example 1.1 is asymptotically stable. Unfortunately, its asymptotic stability can be useless for engineering purposes in the case where $\alpha$ is small (e.g. $\alpha = 10^{-3}$) or very small (e.g. $\alpha = 10^{-10}$). Engineering applications of the stability analysis need more information about qualitative dynamic properties of the system. They often need data on the largest neighbourhood $\mathcal{D}$ of $X = 0$ which is contained in both $\mathcal{D}_s$ and $\mathcal{D}_a$, that is that on $\mathcal{D} = \mathcal{D}_s \cap \mathcal{D}_a$. Asymptotic stability of $X = 0$ is meaningful for engineering purposes provided only that $\mathcal{D}$ is sufficiently large from the engineering point of view.*

**Definition 1.7** *a) A set $\mathcal{A}$ is asymptotically stable with respect to motions of the system (1.12), in short, asymptotically stable if and only if it is both stable and attractive.*

*b) The set $\mathcal{A}$ is globally asymptotically stable (asymptotically stable in the whole, in the large) if and only if it is both stable and globally attractive.*

*c) The set $\mathcal{A}$ is completely globally asymptotically stable if and only if it is both globally stable and globally attractive.*

**Comment 1.12** *Besides asymptotic stability of a set $\mathcal{A}$ we often wish to know the set $\mathcal{D}(\mathcal{A}) = \mathcal{D}_s(\mathcal{A}) \cap \mathcal{D}_a(\mathcal{A})$.*

### 1.1.4 Definitions of exponential stability

Asymptotic stability does not provide any information on the rate of system motion convergence to $X = 0$, or to the set $\mathcal{A}$. Engineering requests for a higher quality of the system dynamic behaviour demand sufficient rate of the system motion convergence. For this purpose the concept of exponential stability introduced by Krasovskii [116] is important.

**Definition 1.8** *a) The state $X = 0$ of the system (1.12) is exponentially stable if and only if there are $\Delta > 0$ and positive numbers $\beta$ and $\gamma$, $\beta \geq 1$, such that $||X_0|| < \Delta$ implies $||\underline{X}(t; X_0)|| \leq \beta ||X_0|| \, exp\,(-\gamma t)$ for all $t \in \Re_+$.*

*b) The state $X = 0$ of the system (1.12) is globally exponentially stable (exponentially stable in the whole, in the large) if and only if a) holds for $\Delta = +\infty$.*

**Comment 1.13** *Existence of $\Delta, \beta$ and $\gamma$ is not sufficient information for engineering applications. In addition to that, we need the knowledge of the smallest possible value of $\beta$, the largest possible value of $\gamma$ and of the largest neighbourhood $\mathcal{D}_e$ of $X = 0$ such that $||\underline{X}(t; X_0)|| \leq \beta ||X_0|| \, exp\,(-\gamma t)$ for all $t \in \Re_+$ holds iff $X_0 \in \mathcal{D}_e$. There is certainly a trade-off among $\beta$, $\gamma$ and $\mathcal{D}_e$ in general, which is illustrated as follows. For the system of Example 1.1 the set $\mathcal{D}_e$ depends on $\alpha$ and $\gamma$ via $\alpha_1$, $\mathcal{D}_e(\alpha_1) = \{X : |x_1| + |x_2| \leq \alpha_1\}$, where $\alpha_1 = \alpha - \gamma$ should be positive. For given $\alpha$ and $\gamma$, $\alpha > \gamma > 0$, the set $\mathcal{D}_e(\alpha_1)$ is the largest neighbourhood of $X = 0$ such that $X_0 \in \mathcal{D}_e(\alpha_1)$ implies $||\underline{X}(t; X_0)|| \leq \beta ||X_0|| \, exp\,(-\gamma t)$ for all $t \in \Re_+$, where $\beta = \sqrt{2}$.*

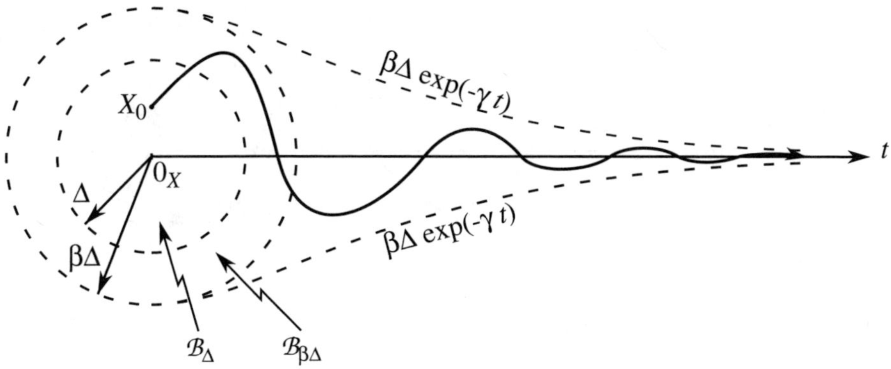


$\exists\Delta > 0,\ \exists(\beta,\gamma) \in \Re^+ \times \Re^+,\ \beta \geq 1\ \rightarrow X_0 \in \mathcal{B}_\Delta \Rightarrow$
$\rho[\underline{X}(t;X_0),\mathcal{O}] \leq \beta\rho(X_0,\mathcal{O})\,exp\,(-\gamma t),\ \ \forall t \in \Re_+.$

Figure 1.6: Exponentially stable $X = 0$.

**Definition 1.9** *a) A set $\mathcal{A}$ is exponentially stable with respect to motions of the system (1.12), for short: exponentially stable, if and only if there are $\Delta > 0$ and positive numbers $\beta$ and $\gamma$, $\beta \geq 1$, such that $\rho(X_0,\mathcal{A}) < \Delta$ implies $\rho[\underline{X}(t;X_0),\mathcal{A}] \leq \beta\rho(X_0,\mathcal{A})\,exp\,(-\gamma t)$ for all $t \in \Re_+$.*

*b) The set $\mathcal{A}$ is globally exponentially stable (exponentially stable in the whole, in the large) if and only if a) holds for $\Delta = +\infty$.*

**Comment 1.14** *In addition to information on exponential stability of a set $\mathcal{A}$ we need information about the largest neighbourhood $\mathcal{D}_e(\mathcal{A})$ of the set $\mathcal{A}$ such that $\rho[\underline{X}(t;X_0),\mathcal{A}] \leq \beta\rho(X_0,\mathcal{A})\,exp\,(-\gamma t)$ for all $t \in \Re_+$ holds iff $X_0 \in \mathcal{D}_e(\mathcal{A})$.*

**Note 1.3** In the sequel the expression "globally" can be everywhere replaced by "in the whole", or by "in the large" as pointed out in the preceding definitions, which are illustrated by Fig. 1.6 and Fig. 1.7.

### 1.1.5 Definitions of absolute stability on $\mathcal{N}_i(\mathcal{L})$

The notion of absolute stability is due to Lurie and Postnykov [125]. It is related to nonlinear systems called after them Lurie–Postnykov systems, for short, Lurie systems, which are described by (1.13) and (1.14):

$$\text{a)} \qquad \frac{\mathrm{d}X}{\mathrm{d}t} = AX + Bf(w), \tag{1.13}$$

$$\text{b)} \qquad w = CX + Df(w), \tag{1.14}$$

where $X \in \Re^n$, $w = [\,w_1 \ \ w_2 \ \ \ldots \ \ w_m\,]^T \in \Re^m$, $f : \Re^m \rightarrow \Re^m$ and the matrices $A$, $B$, $C$ and $D$ have the appropriate dimensions. The vector nonlinearity $f = [\,f_1 \ \ f_2 \ \ \ldots \ \ f_m\,]^T$ belongs to a family $\mathcal{N}_{(.)}(\mathcal{L};\mathcal{S})$ of Lurie nonlinearities on a

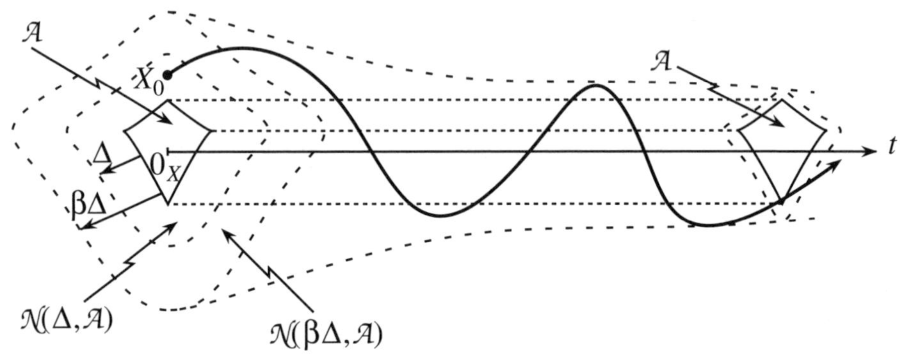


$\exists \Delta > 0, \ \exists(\beta, \gamma) \in \Re^{+} \times \Re^{+}, \ \beta \geq 1 \ \rightarrow X_0 \in \mathcal{N}(\Delta, \mathcal{A}) \Rightarrow$
$\rho[\underline{X}(t; X_0), \mathcal{A}] \leq \beta \rho(X_0, \mathcal{A}) \, exp\,(-\gamma t), \ \ \forall t \in \Re_{+}.$

Figure 1.7: Exponentially stable set $\mathcal{A}$.

(matrix) sector $\mathcal{L}$, $\mathcal{L} = (K_1, K_2)$, $K_i \in \Re^{m \times m}$, $i = 1, 2$. The notation $(K_1, K_2)$ has a general meaning and can be replaced by any of the following:

$$[K_1, K_2], \quad ]K_1, K_2], \quad [K_1, K_2[, \quad \text{or} \quad ]K_1, K_2[.$$

The matrices $K_1$ and $K_2$ are diagonal, $0 \leq K_1 \leq K_2$. The diagonal entries $k_{2j}$ of $K_2$ can be positive numbers or equal to $+\infty$, $K_i = diag\ \{k_{i1}\ k_{i2}\ \ldots\ k_{im}\}$, $i = 1, 2$.

The vector nonlinearity $f$ is in the class (family) $\mathcal{N}_0(\mathcal{L}; \mathcal{S})$ iff the system (1.13), (1.14) has a unique solution through $X = 0$ and:

$$\text{a)} \quad f(CX) \in C(\mathcal{S}),$$

$$\text{b)} \quad f(0) = 0, \tag{1.15}$$

$$\text{c)} \quad \left.\frac{f_j(W)}{w_j}\right|_{W=W(X)} \begin{cases} \in (k_{1j}, k_{2j}), \ \forall X \in \mathcal{S}, \ w_j \neq 0, \\ = 0, \ \forall X \in \mathcal{S}, \ w_j = 0, \ j = 1, 2, \ldots, m. \end{cases}$$

With $\mathcal{N}_d(\mathcal{L}; \mathcal{S})$ we denote all odd $f \in \mathcal{N}_0(\mathcal{L}; \mathcal{S})$, and $\mathcal{N}_1(\mathcal{L}, \mathcal{M}; \mathcal{S})$ designates the family of all $f \in \mathcal{N}_0(\mathcal{L}; \mathcal{S})$ obeying also (1.16) and (1.17):

$$\text{a)} \quad f[W(X)] \in C^{(1)}(\mathcal{S}), \tag{1.16}$$

$$\text{b)} \quad f_W(W)\,\big|_{W=W(X)} \in \mathcal{M}, \quad \forall X \in \mathcal{S}, \tag{1.17}$$

where

$$f_W(W) = \left(\frac{\partial f_i}{\partial w_j}\right) : \Re^m \rightarrow \Re^{m \times m}.$$

$f_W$ is the Jacobian (matrix function) of the vector nonlinearity $f$, and $\mathcal{M} = (M_1, M_2) \subseteq \bar{\Re}^{m \times m}$.

In case $\mathcal{S} = \Re^n$ we use the simplified notations $\mathcal{N}_{(.)}(\mathcal{L})$ and $\mathcal{N}_1(\mathcal{L}, \mathcal{M})$ instead of $\mathcal{N}_{(.)}(\mathcal{L}; \mathcal{S})$ and $\mathcal{N}_1(\mathcal{L}, \mathcal{M}, \mathcal{S})$, respectively.

$\mathcal{N}_{id}(.)$ will denote a set of all odd functions $f$ belonging to $\mathcal{N}_i(.)$.

**Definition 1.10** *The state* $X = 0$ *of the system (1.14) is absolutely stable on* $\mathcal{N}_{(.)}(\mathcal{L})$ *[on* $\mathcal{N}_1(\mathcal{L};\mathcal{M})$*], if and only if it is globally asymptotically stable for every* $f \in \mathcal{N}_{(.)}(\mathcal{L})$ *[for every* $f \in \mathcal{N}_1(\mathcal{L},\mathcal{M})$*], respectively.*

**Comment 1.15** *The requirement for the global asymptotic stability is often unnecessarily stringent from the engineering point of view. There is not a technical system whose initial state can be arbitrary far from the origin. Besides, for many nonlinearities the condition (1.15c) appears too restrictive in the case* $\mathcal{S} = \Re^n$*. For these reasons it is important to consider* $\mathcal{S} \subset \Re^n$*.*

**Definition 1.11** *A set* $\mathcal{A}$ *is absolutely stable on* $\mathcal{N}_{(.)}(\mathcal{L})$ *[or on* $\mathcal{N}_1(\mathcal{L},\mathcal{M})$*], if and only if it is globally asymptotically stable for every* $f \in \mathcal{N}_{(.)}(\mathcal{L})$ *[*$f \in \mathcal{N}_1(\mathcal{L},\mathcal{M})$*], respectively.*

### 1.1.6 Definitions of attraction with finite attraction time

The attraction property clarifies behaviour of motions as $t \to +\infty$. For engineering needs it is more important to ensure attraction as time $t$ converges to a finite time $\tau_a \in \Re^+$.

**Definition 1.12** *a) The state* $X = 0$ *of the system (1.12) is attractive with the finite attraction time* $\tau_a(X_0)$ *if and only if there is* $\Delta > 0$ *and for every* $X_0 \in \Re^n$ *with* $\|X_0\| < \Delta$ *there is* $\sigma_a \in \Re_+$, $\sigma_a = \sigma_a(X_0)$*, such that* $\underline{X}(t;X_0) = 0$ *for all* $t \in [\sigma_a, +\infty[$*, and* $\tau_a(X_0)$ *is the minimal* $\sigma_a(X_0)$ *for the chosen* $X_0$*.*

*b) The state* $X = 0$ *of the system (1.12) is globally attractive with the finite attraction time* $\tau_a(X_0)$ *if and only if the preceding conditions under a) hold for* $\Delta = +\infty$*.*

The difference between attraction and attraction with finite attraction time is illustrated by Fig. 1.8.

Definition 1.12 can be generalised to sets as follows.

**Definition 1.13** *a) A set* $\mathcal{A}$ *is attractive with the finite attraction time* $\tau_a$, $\tau_a = \tau_a(X_0;\mathcal{A}) \in \Re_+$*, if and only if there is* $\Delta > 0$ *and for every* $X_0 \in \Re^n$, $\rho(X_0,\mathcal{A}) < \Delta$*, there is* $\sigma \in \Re_+$, $\sigma = \sigma(X_0,\mathcal{A})$*, such that* $\underline{X}(t;X_0) \in \mathcal{A}$ *for all* $t \in [\sigma_a, +\infty[$*, and* $\tau_a(X_0;\mathcal{A})$ *is the minimal* $\sigma_a(X_0;\mathcal{A})$ *for the chosen* $X_0$*.*

*b) The set* $\mathcal{A}$ *is globally attractive with the finite attraction time* $\tau_a$ *if and only if the conditions under a) hold for* $\Delta = +\infty$*.*

The finite attraction time can be also called *the finite reachability time.*

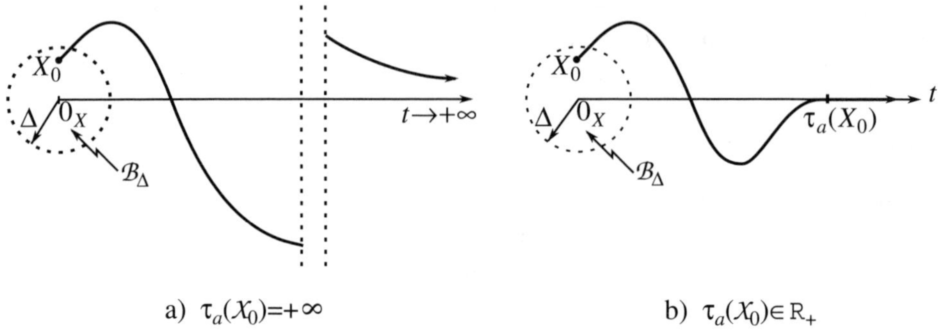


Figure 1.8: a) attraction of $X = 0$, b) attraction of $X = 0$ with the finite attraction time $\tau_a(X_0) \in \Re_+$.

### 1.1.7 Definitions of stability with finite attraction time

Definitions 1.6 and 1.12 lead to the following.

**Definition 1.14** *a) The state $X = 0$ of the system (1.12) is stable with the finite attraction time $\tau_a$ if and only if it is both stable and attractive with the finite attraction time $\tau_a$.*

*b) The state $X = 0$ of the system (1.12) is globally stable with the finite attraction time $\tau_a$ if and only if it is both stable and globally attractive with the finite attraction time $\tau_a$.*

*c) The state $X = 0$ of the system (1.12) is completely globally stable with the finite attraction time $\tau_a$ if and only if it is both globally stable and globally attractive with the finite attraction time $\tau_a$.*

Similarly, Definitions 1.7 and 1.13 yield the following notions.

**Definition 1.15** *a) A set $\mathcal{A}$ is stable with the finite attraction time $\tau_a$ if and only if it is both stable and attractive with the finite attraction time $\tau_a$.*

*b) The set $\mathcal{A}$ is globally stable with the finite attraction time $\tau_a$ if and only if it is both stable and globally attractive with the finite attraction time $\tau_a$.*

*c) The set $\mathcal{A}$ is completely globally stable with the finite attraction time $\tau_a$ if and only if it is both globally stable and globally attractive with the finite attraction time $\tau_a$.*

The preceding definitions are illustrated by Fig. 1.9.

### 1.1.8 Definitions of absolute stability with finite attraction time

In the framework of the Lurie systems (1.13) and (1.14) we can now introduce the following notions.

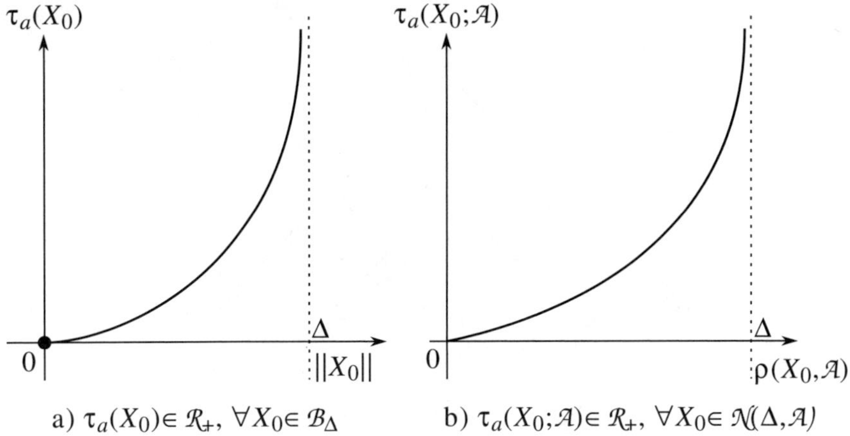


Figure 1.9: Dependence of $\tau_a$ on $X_0$. a) $X = 0$ is attractive with the finite attraction time $\tau_a$, b) the set $\mathcal{A}$ is attractive with the finite attraction time $\tau_a$. Although $\tau_a$ is finite for every $X_0 \in \mathcal{B}_\Delta$ in case a) and for every $X_0 \in \mathcal{N}(\Delta, \mathcal{A})$ in case b), in both cases $sup\,[\tau_a(X_0) : X_0 \in \mathcal{B}_\Delta] = +\infty$ and $sup\,[\tau_a(X_0; \mathcal{A}) : X_0 \in \mathcal{N}(\Delta, \mathcal{A})] = +\infty$.

**Definition 1.16** *a) The state $X = 0$ of the system (1.14) is absolutely stable on $\mathcal{N}_{(.)}(\mathcal{L})$ with the finite attraction time $\tau_a$ if and only if it is globally stable with the finite attraction time $\tau_a$, $\tau_a = \tau_a(X_0; f)$, for every $f \in \mathcal{N}_{(.)}(\mathcal{L})$.*

*b) The state $X = 0$ of the system (1.14) is absolutely stable with the finite attraction time $\tau_a$ on $\mathcal{N}_{(.)}(\mathcal{L})$ if and only if a) holds and $sup\,[\tau_a(X_0; f) : f \in \mathcal{N}_{(.)}(\mathcal{L})] < +\infty$ for every $X_0 \in \Re^n$.*

**Definition 1.17** *a) A set $\mathcal{A}$ is absolutely stable on $\mathcal{N}_{(.)}(\mathcal{L})$ with the finite attraction time $\tau_a$ if and only if it is globally stable with the finite attraction time $\tau_a$, $\tau_a = \tau_a(X_0; f)$, for every $f \in \mathcal{N}_{(.)}(\mathcal{L})$.*

*b) A set $\mathcal{A}$ is absolutely stable with the finite attraction time $\tau_a$ on $\mathcal{N}_{(.)}(\mathcal{L})$ if and only if a) holds and $sup\,[\tau_a(X_0; f) : f \in \mathcal{N}_{(.)}(\mathcal{L})] < +\infty$ for every $X_0 \in \Re^n$.*

## 1.2 Comments on the practical stability concept

### 1.2.1 Introductory comments

The practical stability concept aims at determining closeness in a practical, engineering sense, at clarifying the system behaviour either in the free or in a forced regime over a time interval that can be bounded and for initial states of practical, engineering significance. The closeness can be defined via a suitably accepted distance function and/or via sets. The latter appears more general and will be utilised

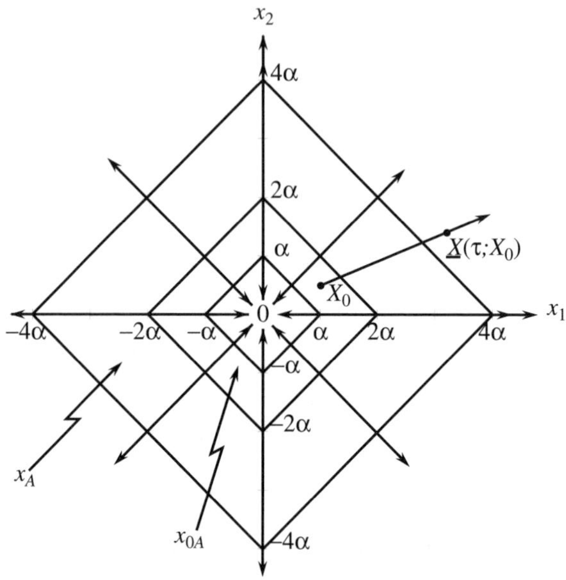


Figure 1.10: The state portrait of the system of Example 1.1 and the sets $\mathcal{X}_{0A}$ and $\mathcal{X}_A$.

in the sequel for the system (1.18):

$$\frac{\mathrm{d}X}{\mathrm{d}t} = f(X, \mathtt{i}), \quad f : \Re^n \times \Re^l \to \Re^n, \tag{1.18}$$

where $X : \Re \to \Re^n$ and $\mathtt{i} : \Re \to \Re^l$, $\mathtt{i}$ is an input (vector function).

Since the Lyapunov stability concept cannot meet the goals described above then Chetaev [29] emphasised the need for a non-Lyapunov stability concept. La Salle and Lefschetz [121] were probably the first to introduce a non-Lyapunov stability concept in the western literature.

### 1.2.2 Definition of practical stability

By referring to LaSalle and Lefschetz [121], Weiss and Infante [208], [209], Michel [144]–[146], Grujić [60]–[63] and Martynyuk [133] we shall accept the following practical stability definition in which $\Re_\tau = [0, \tau[$ and $\mathcal{I}$ is the family of inputs of our interest. Besides a set $\mathcal{X}_{(\cdot)}$ is assumed to have a non-empty interior $\mathring{\mathcal{X}}_{(\cdot)}$.

**Definition 1.18** *The system (1.18) is practically stable with respect to* $\{\tau,\ \mathcal{X}_{0A},\ \mathcal{X}_A, \mathcal{I}\}$ *if and only if its motions obey* $\underline{X}(t; X_0; \mathtt{i}) \in \mathcal{X}_A$ *for every* $(t, X_0, \mathtt{i}) \in \Re_\tau \times \mathcal{X}_{0A} \times \mathcal{I}$.

**Comment 1.16** *Let the system of Example 1.1 (Section 1.1.1) be reconsidered:*

$$\frac{\mathrm{d}X}{\mathrm{d}t} = (-\alpha + |x_1| + |x_2|)X.$$

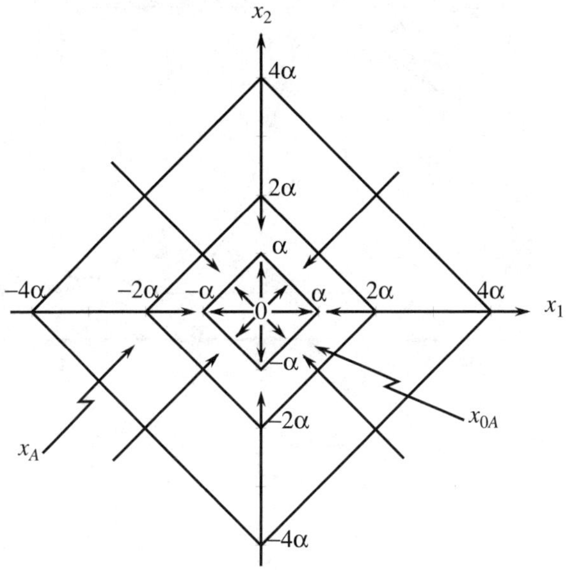


Figure 1.11: The state portrait of the system of Comment 1.17 and the sets $\mathcal{X}_{0A}$ and $\mathcal{X}_A$.

*If*

$$\mathcal{X}_{0A} = \{X : X \in \Re^2, \quad |x_1| + |x_2| \le 2\alpha\},$$
$$\mathcal{X}_A = \{X : X \in \Re^2, \quad |x_1| + |x_2| \le 4\alpha\}, \tag{1.19}$$

*and*

$$\mathcal{I} = \{0\},$$

*then for the system to be practically stable with respect to* $\{\tau, \mathcal{X}_{0A}, \mathcal{X}_A, \mathcal{I}\}$ *it is both necessary and sufficient that* $\tau$ *is sufficiently small,* $\tau \in \Re^+$*. If it is not satisfied, i.e. if* $\tau \ge \sigma$ *(Fig. 1.10) then the system is not practically stable with respect to* $\{\tau, \mathcal{X}_{0A}, \mathcal{X}_A, \mathcal{I}\}$*. Notice again that* $X = 0$ *is asymptotically stable.*

**Comment 1.17** *Let* $X \in \Re^2$ *and let the system (1.18) be specified in the following form:*

$$\frac{\mathrm{d}X}{\mathrm{d}t} = (\alpha - |x_1| - |x_2|)X, \quad \alpha \in \Re^+.$$

*Evidently,* $\mathcal{I} = \{0\}$*. If*

$$\mathcal{X}_{0A} = \{X : X \in \Re^2, \quad |x_1| + |x_2| \le 2\alpha\},$$

*and*

$$\mathcal{X}_A = \{X : X \in \Re^2, \quad |x_1| + |x_2| \le 4\alpha\},$$

*then the system is practically stable with respect to* $\{\tau, \mathcal{X}_{0A}, \mathcal{X}_A, I\}$ *for any* $\tau \in \Re^+$ *despite* $X = 0$ *is unstable, Fig. 1.11. The same holds if we redefine* $\mathcal{X}_A$ *and* $\mathcal{X}_{0A}$ *so that* $\mathcal{X}_A = \mathcal{X}_{0A} = \{X : X \in \Re^2, |x_1| + |x_2| \le \beta\}$ *for any* $\beta \in ]\alpha, +\infty[$.

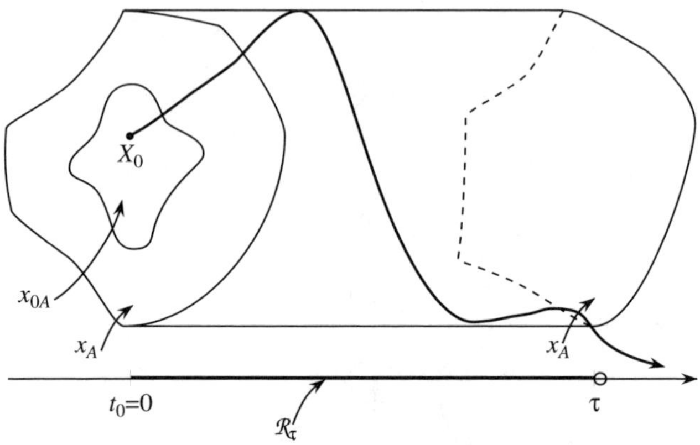


Figure 1.12: The system is practically stable with respect to $\{\tau, \mathcal{X}_{0A}, \mathcal{X}_A, \mathcal{I}\}$.

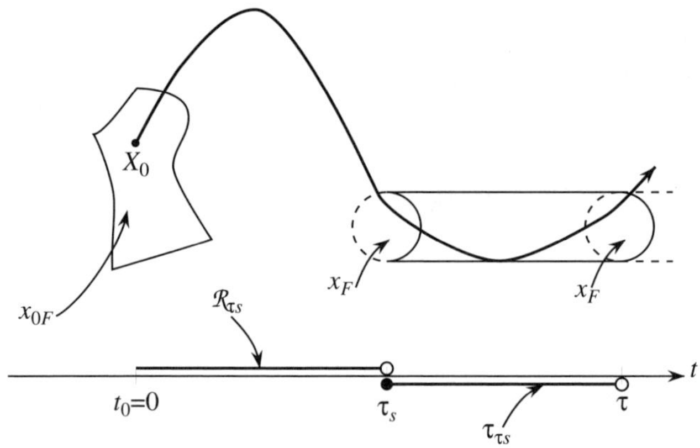


Figure 1.13: The system is practically contractive with the settling time $\tau_s$ with respect to $\{\tau, \mathcal{X}_{0F}, \mathcal{X}_F, \mathcal{I}\}$.

### 1.2.3 Definition of practical contraction with settling time

The notion of practical stability does not reflect any contraction property, which is expressed by the following in which $\mathcal{T}_{\tau_s} = [\tau_s, \tau[$.

**Definition 1.19** *Motions of the system (1.18) are practically contractive (or, for short, the system (1.18) is practically contractive) with settling time $\tau_s$ with respect to $\{\tau, \mathcal{X}_{0F}, \mathcal{X}_F, \mathcal{I}\}$ if and only if $\underline{X}(t; X_0; \mathtt{i}) \in \mathcal{X}_F$ for every $(t, X_0, \mathtt{i}) \in \mathcal{T}_{\tau s} \times \mathcal{X}_{0F} \times \mathcal{I}$.*

The difference between practical stability and practical contraction with settling time $\tau_s$ is illustrated by Fig. 1.12 and Fig. 1.13.

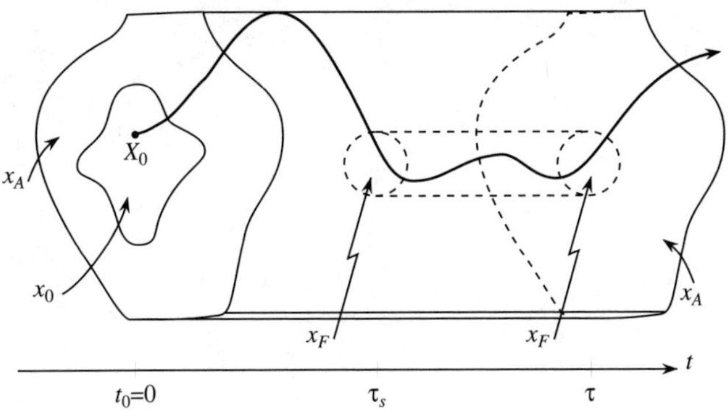


Figure 1.14: The system is practically stable with the settling time $\tau_s$ with respect to $\{\tau, \mathcal{X}_0, \mathcal{X}_A, \mathcal{X}_F, \mathcal{I}\}$.

### 1.2.4 Definition of practical stability with settling time

Definitions 1.18 and 1.19 combined imply the following notion.

**Definition 1.20** *The system (1.18) is practically contractively stable (for short, practically stable) with the settling time* $\tau_s$ *with respect to* $\{\tau, \mathcal{X}_0, \mathcal{X}_A, \mathcal{X}_F, \mathcal{I}\}$, $\mathcal{X}_F \subset \mathcal{X}_A$, *if and only if both*

*a)* $\underline{X}(t; X_0; \mathtt{i}) \in \mathcal{X}_A$ for every $t \in \mathfrak{R}_\tau$,

*and*

*b)* $\underline{X}(t; X_0; \mathtt{i}) \in \mathcal{X}_F$ for every $t \in \mathcal{T}_{\tau s}$,

*hold for every* $(X_0, \mathtt{i}) \in \mathcal{X}_0 \times \mathcal{I}$.

This definition is illustrated by Fig. 1.14.

# Chapter 2

# Stability domain concepts

## 2.1 Introductory comments

In order to get complete information about the causality between initial states and systems motions, concepts of domains of various stability properties were introduced. In the framework of the Lyapunov stability the notion of attraction domain of the origin was defined by Zubov [218] and Hahn [94], and notions of the stability domain and the asymptotic stability domain were defined by Grujić [64], [65], [68]–[70], [72], and used by Grujić *et al.* [88]–[90]. The concept of practical stability domains was introduced by Grujić [70].

In the literature (e.g. LaSalle and Lefschetz [121] and Zubov [218]) the notion of "region of asymptotic stability" has been used in the sense of the attraction domain. In what follows the difference between them will be clarified.

## 2.2 Domains of Lyapunov stability properties

### 2.2.1 The notion of domain

The term "domain" denotes a set that can be, but need not be, open or closed.

Domains of Lyapunov stability properties will be called for short "Lyapunov stability domains" in a general sense incorporating domains of stability, of attraction and of asymptotic stability. In the closer sense the notion "Lyapunov stability domain" will be used for the domain of stability (for short, the stability domain).

Lyapunov stability domains will be studied herein in the framework of time-invariant continuous-time nonlinear systems governed by

$$\frac{dX}{dt} = f(X), \tag{2.1}$$

with possibly certain specific features that will be described when they are needed.

In the literature (e.g. LaSalle and Lefschetz [121]) the notion "region" has been used for an open connected set. We do not wish *a priori* to impose such a restriction on the largest set of initial states appropriate for a corresponding stability property. Therefore, we shall use the term "domain" in general rather than "region". In case a domain is open and connected then we can also call it a "region".

### 2.2.2 Definitions of stability domains

The definitions of stability domains were introduced to comply with the definitions of stability of a state and of a set [64], [65], [68], [69], [70], [72].

**Definition 2.1** *(a) The state $X = 0$ of the system (2.1) has the stability domain denoted by $\mathcal{D}_s$ if and only if both*

*(i) for every $\varepsilon \in \Re^+$ there is a neighbourhood $\mathcal{D}_s(\varepsilon)$ of $X = 0$ such that $||\underline{X}(t; X_0)|| < \varepsilon$ for all $t \in \Re_+$ holds provided only that $X_0 \in \mathcal{D}_s(\varepsilon)$,*

*and*

*(ii) $\mathcal{D}_s = \cup [\mathcal{D}_s(\varepsilon) : \varepsilon \in \Re^+]$.*

*(b) The state $X = 0$ of the system (2.1) has the strict domain of stability (the strict stability domain) $\mathcal{D}_{sc}$ if and only if*

*(i) it has the domain $\mathcal{D}_s$ of stability,*

*(ii) $\mathcal{D}_{sc}(\varepsilon)$ is the largest connected neighbourhood of $X = 0$ which is subset of $\mathcal{D}_s(\varepsilon)$, $\mathcal{D}_{sc}(\varepsilon) \subseteq \mathcal{D}_s(\varepsilon)$, for every $\varepsilon \in \Re^+$,*

*(iii) $\mathcal{D}_{sc} = \cup [\mathcal{D}_{sc}(\varepsilon) : \varepsilon \in \Re^+]$.*

**Comment 2.1** *The state $X = 0$ of the system (2.1) has the stability domain iff it is stable due to Definition 2.1 and Definition 1.2. (Section 1.1.1). Moreover, it is globally stable iff its strict stability domain $\mathcal{D}_{sc}$ is its whole state space $\Re^n$: $\mathcal{D}_{sc} = \Re^n$. The significance of the latter is the following: for arbitrary $X_0 \in \Re^n$ there is (possibly sufficiently large) $\varepsilon \in \Re^+$ such that $||\underline{X}(t; X_0)|| < \varepsilon$ for all $t \in \Re_+$. In other words, system motions are bounded on $\Re_+$ for every initial state $X_0$.*

Since $\mathcal{D}_s(\varepsilon)$ can be disconnected, and $\mathcal{D}_{sc}(\varepsilon)$ cannot, then the latter rather than the former corresponds to $\mathcal{B}_{\delta(\varepsilon)}$ for every $\varepsilon \in \Re^+$. In other words, $\mathcal{D}_{sc}$ is the domain of stability in the strict Lyapunov sense, while $\mathcal{D}_s$ is the domain of stability in a wider, more practical Lyapunov sense.

In order to illustrate the preceding definition let us consider the following example [70].

**Example 2.1** Let a second order system (2.1) be in the following form:

$$\frac{dX}{dt} = (-\alpha + |x_1| + |x_2|)X, \ \alpha \in \Re^+.$$

It has the infinite set $\mathcal{S}_e$ of equilibrium states,

$$\mathcal{S}_e = \left\{X : X \in \Re^2, \ (X = 0) \ \text{or} \ (|x_1| + |x_2| = \alpha)\right\}.$$

We are now interested only in stability of $X = 0$. Let $\varepsilon^* = \alpha\sqrt{2}/2$. For any $\varepsilon \in ]0, \varepsilon^*]$, the maximal $\delta(\varepsilon)$ denoted by $\delta_M(\varepsilon)$ (Definition 1.2, Section 1.1.1) obeys $\delta_M(\varepsilon) = \varepsilon$. For any $\varepsilon \in [\varepsilon^*, +\infty[$ the maximal $\delta_M(\varepsilon) = \varepsilon^*$.

However

$$||\underline{X}(t, X_0)|| < \varepsilon, \quad \forall t \in \Re_+, \quad \text{iff } X_0 \in \begin{cases} \mathcal{B}_\varepsilon & , \varepsilon \in ]0, \varepsilon^*], \\ \mathcal{B}_\varepsilon \cap \mathcal{S}_\alpha & , \varepsilon \in ]\varepsilon^*, \alpha], \\ \mathcal{S}_\alpha & , \varepsilon \in ]\alpha, +\infty[, \end{cases}$$

where

$$\mathcal{S}_\alpha = \left\{X : X \in \Re^2, \ |x_1| + |x_2| \le \alpha\right\},$$

which is illustrated by Fig. 2.1. Evidently, $\partial\mathcal{S}_\alpha = \mathcal{S}_e - \mathcal{O}$.

It is now obvious (Fig. 2.1) that for $\varepsilon^* = \alpha\sqrt{2}/2$

$$\mathcal{D}_s(\varepsilon) = \begin{cases} \mathcal{B}_\varepsilon & , \varepsilon \in ]0, \varepsilon^*], \\ \mathcal{B}_\varepsilon \cap \mathcal{S}_\alpha & , \varepsilon \in ]\varepsilon^*, \alpha], \\ \mathcal{S}_\alpha & , \varepsilon \in ]\alpha, +\infty[, \end{cases}$$

so that

$$\mathcal{D}_s = \cup\left[\mathcal{D}_s(\varepsilon) : \varepsilon \in \Re^+\right] = \mathcal{S}_\alpha.$$

In this example the stability domain is the closed set $\mathcal{S}_\alpha$, i.e.

$$\mathcal{D}_s = \left\{X : X \in \Re^2, \ |x_1| + |x_2| \le \alpha\right\},$$

which is showed in Fig. 2.1f. The strict stability domain $\mathcal{D}_{sc}$ equals $\mathcal{D}_s$, $\mathcal{D}_{sc} = \mathcal{D}_s$.

**Definition 2.2** *(a) A set $\mathcal{A} \subset \Re^n$ (of states of the system (2.1)) has the stability domain denoted by $\mathcal{D}_s(\mathcal{A})$ if and only if both*

*(i) for every $\varepsilon \in \Re^+$ there is a neighbourhood $\mathcal{D}_s(\varepsilon; \mathcal{A})$ of $\mathcal{A}$ such that $\rho[\underline{X}(t; X_0), \mathcal{A}] < \varepsilon$ for all $t \in \Re_+$ provided only $X_0 \in \mathcal{D}_s(\varepsilon; \mathcal{A})$,*

*and*

*(ii) $\mathcal{D}_s(\mathcal{A}) = \cup[\mathcal{D}_s(\varepsilon; \mathcal{A}) : \varepsilon \in \Re^+]$.*

*(b) A set $\mathcal{A} \subset \Re^n$ (of the states of system (2.1)) has the strict domain of stability (the strict stability domain) denoted by $\mathcal{D}_{sc}(\mathcal{A})$ if and only if*

*(i) it has the stability domain $\mathcal{D}_s$,*

*(ii) $\mathcal{D}_{sc}(\varepsilon; \mathcal{A})$is the largest connected neighbourhood of $\mathcal{A}$ which is a subset of $\mathcal{D}_s(\varepsilon; \mathcal{A})$, $\mathcal{D}_{sc}(\varepsilon; \mathcal{A}) \subseteq \mathcal{D}_s(\varepsilon; \mathcal{A})$, for all $\varepsilon \in \Re^+$,*

*(iii) $\mathcal{D}_{sc}(\mathcal{A}) = \cup[\mathcal{D}_{sc}(\varepsilon; \mathcal{A}) : \varepsilon \in \Re^+]$.*

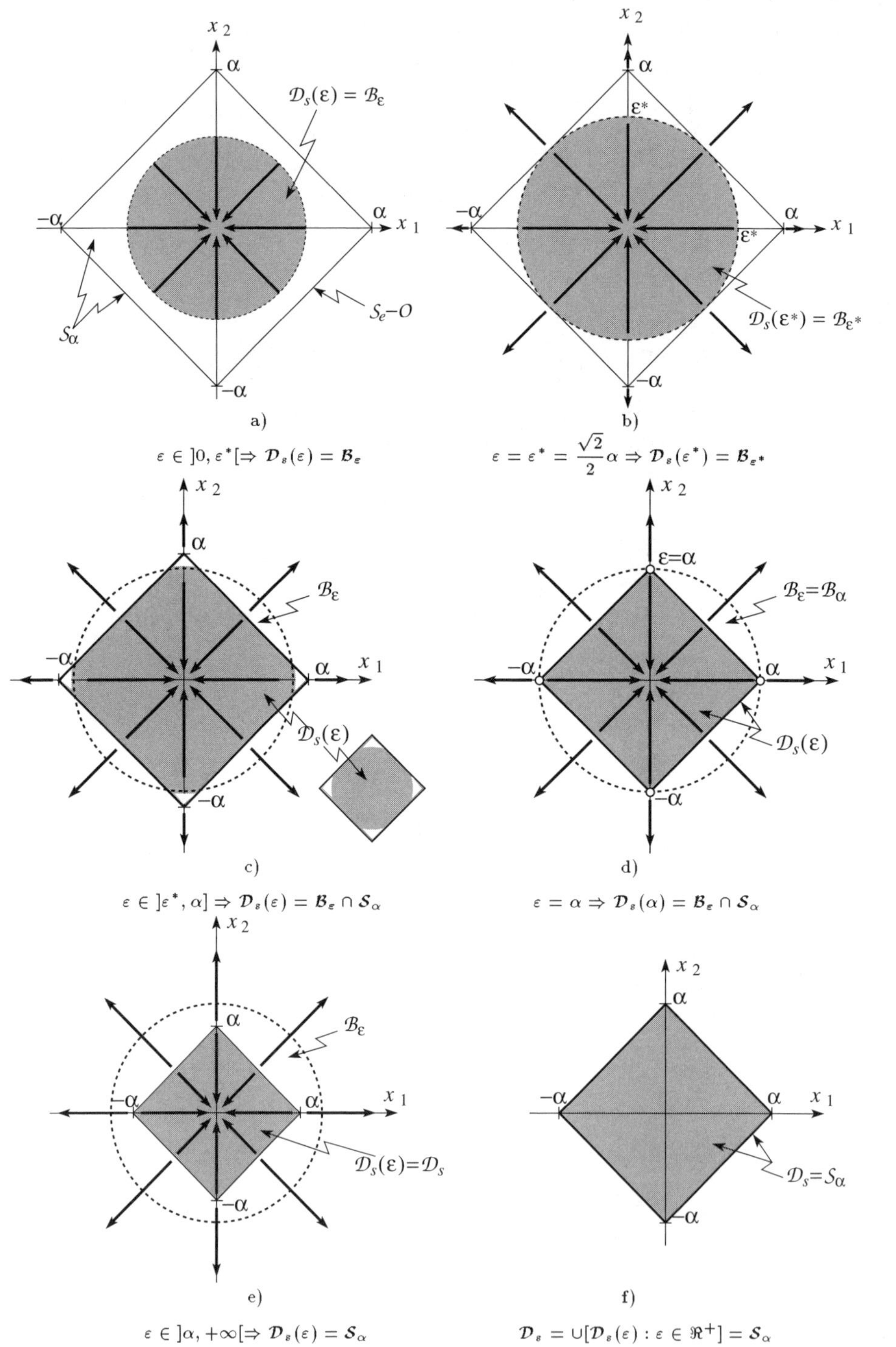


Figure 2.1: Dependence of $\mathcal{D}_s(\varepsilon)$ on $\varepsilon$.

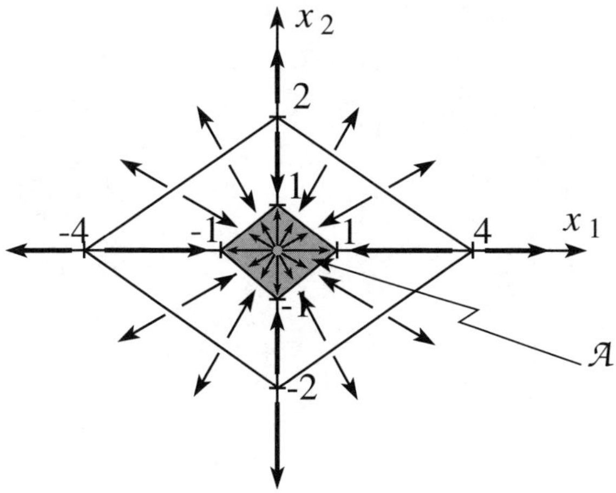


Figure 2.2: The state portrait of the system $\dot{X} = (1 - |x_1| - |x_2|)\,(4 - |x_1| - 2|x_2|)X$. The shaded area together with its boundary represents the set $\mathcal{A}$.

**Example 2.2** Let the second order system (2.1) be specified by

$$\frac{dX}{dt} = (1 - |x_1| - |x_2|)(4 - |x_1| - 2|x_2|)X.$$

We are interested in stability of the set $\mathcal{A}$,

$$\mathcal{A} = \left\{X : X \in \Re^2,\ \ |x_1| + |x_2| \leq 1\right\}.$$

The state portrait of the system is given in Fig. 2.2. The set $\mathcal{S}_e$ of the equilibrium states of the system is found as

$$\mathcal{S}_e = \left\{X : X \in \Re^2,\ (X = 0)\ \text{ or }\ (|x_1| + |x_2| = 1)\ \text{ or }\ (|x_1| + 2|x_2| = 4)\right\}.$$

The graphical analysis of dependence of $\mathcal{D}_s(\varepsilon; \mathcal{A})$ on $\varepsilon$ is presented in Fig. 2.3. In this example (Definition 1.3, Section 1.1.1)

$$\delta_M(\varepsilon) = \begin{cases} \varepsilon & ,\ \varepsilon \in \left]0, \dfrac{2\sqrt{5}}{5}\right], \\ \dfrac{2\sqrt{5}}{5} & ,\ \varepsilon \in \left[\dfrac{2\sqrt{5}}{5}, +\infty\right[. \end{cases}$$

However,

$$\|\underline{X}(t, X_0)\| < \varepsilon,\ \ \forall t \in \Re_+,\ \ \text{iff } X_0 \in \begin{cases} \mathcal{N}(\varepsilon, \mathcal{A}) & ,\ \varepsilon \in \left]0, \dfrac{2\sqrt{5}}{5}\right], \\ \mathcal{N}(\varepsilon, \mathcal{A}) \cap \mathcal{S}_4 & ,\ \varepsilon \in \left]\dfrac{2\sqrt{5}}{5}, 3\right], \\ \mathcal{S}_4 & ,\ \varepsilon \in\, ]3, +\infty[, \end{cases}$$

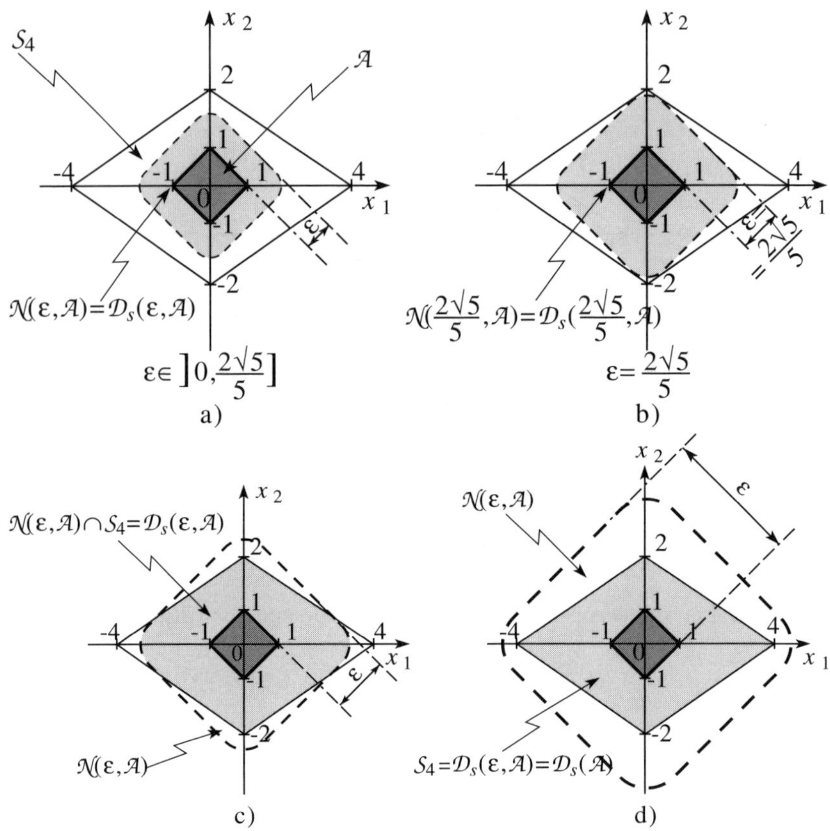


Figure 2.3: The dependence of the subset $\mathcal{D}_s(\epsilon, \mathcal{A})$ of $\mathcal{D}_s(\mathcal{A})$ on $\epsilon \in ]0, +\infty[$ (Example 2.2):

a) $\mathcal{D}_s(\epsilon, \mathcal{A}) = \mathcal{N}(\epsilon, \mathcal{A}) = \mathcal{N}(\epsilon, \mathcal{A}) \cap \mathcal{S}_4, \quad \epsilon \in ]0, \frac{2\sqrt{5}}{5}]$.

b) $\mathcal{D}_s(\frac{2\sqrt{5}}{5}, \mathcal{A}) = \mathcal{N}(\frac{2\sqrt{5}}{5}, \mathcal{A}) = \mathcal{N}(\frac{2\sqrt{5}}{5}, \mathcal{A}) \cap \mathcal{S}_4, \quad \epsilon = \frac{2\sqrt{5}}{5}$.

c) $\mathcal{D}_s(\epsilon, \mathcal{A}) = \mathcal{N}(\epsilon, \mathcal{A}) \cap \mathcal{S}_4, \quad \epsilon \in [\frac{2\sqrt{5}}{5}, 3]$.

d) $\mathcal{D}_s(\epsilon, \mathcal{A}) = \mathcal{S}_4, \quad \epsilon \in [3, +\infty[$.

Hence $\mathcal{D}_s(\mathcal{A}) = \cup [\mathcal{D}_s(\epsilon, \mathcal{A}) : \epsilon \in ]0, +\infty[\,] = \mathcal{S}_4$.

where

$$\mathcal{S}_4 = \left\{ X : X \in \Re^2, \ |x_1| + 2|x_2| \leq 4 \right\}.$$

This shows, which is illustrated by Fig. 2.3, that

$$\mathcal{D}_s(\varepsilon; \mathcal{A}) = \begin{cases} \mathcal{N}(\varepsilon, \mathcal{A}) & , \ \varepsilon \in \left]0, \dfrac{2\sqrt{5}}{5}\right], \\ \mathcal{N}(\varepsilon, \mathcal{A}) \cap \mathcal{S}_4 & , \ \varepsilon \in \left]\dfrac{2\sqrt{5}}{5}, 3\right], \\ \mathcal{S}_4 & , \ \varepsilon \in ]3, +\infty[, \end{cases}$$

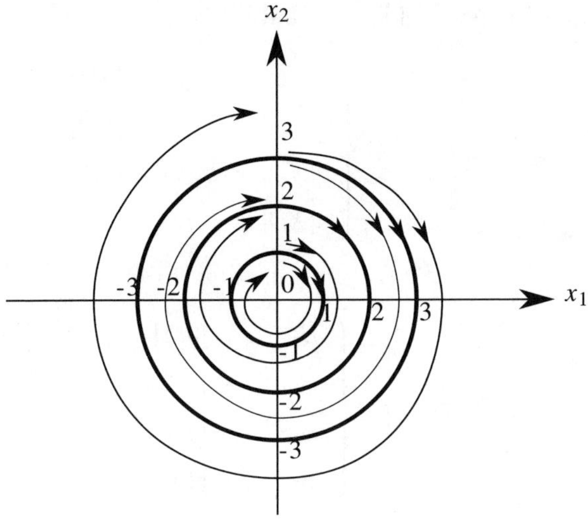


Figure 2.4: The state portrait of the system (Example 2.3) described via the polar coordinates $\rho = \|X\|$, $\theta = arctan\left(\frac{x_2}{x_1}\right)$ by $\dot{\rho} = -\rho(1-\rho^2)(4-\rho^2)(9-\rho^2)$, $\dot{\theta} = -1$.

so that the stability domain $\mathcal{D}_s(\mathcal{A})$ of the set $\mathcal{A}$ is found as $\mathcal{D}_s(\mathcal{A}) = \mathcal{S}_4$,

$$\mathcal{D}_s(\mathcal{A}) = \left\{X : X \in \Re^2, \ \ |x_1| + 2|x_2| \leq 4\right\}.$$

In this example $\mathcal{D}_{sc} = \mathcal{D}_s$.

**Note 2.1** In case $\mathcal{A} = \mathcal{O}$, $\mathcal{O} = \{0\} \subset \Re^n$, then Definition 2.2 reduces to Definition 2.1.

**Note 2.2** A nonlinear dynamical system can have both a stable limit cycle and the domain of stability of the set $\mathcal{O}$ (of $X = 0$). The next example illustrates this assertion.

**Example 2.3** A second order nonlinear system described by

$$\frac{dX}{dt} = \begin{bmatrix} -(1-x_1^2-x_2^2)(4-x_1^2-x_2^2)(9-x_1^2-x_2^2) & 1 \\ -1 & -(1-x_1^2-x_2^2)(4-x_1^2-x_2^2)(9-x_1^2-x_2^2) \end{bmatrix} X$$

has three limit cycles: $x_1^2 + x_2^2 = 1$, $x_1^2 + x_2^2 = 4$ and $x_1^2 + x_2^2 = 9$.

The zero state is the system unique equilibrium state. It is also stable. This is illustrated by Fig. 2.4.

It is now obvious in view of Fig. 2.4 that

$$||\underline{X}(t, X_0)|| < \varepsilon \quad \text{iff} \quad \begin{cases} ||X_0|| < \varepsilon, & \varepsilon \in ]0,1], \\ ||X_0|| \le 1, & \varepsilon \in ]1,2], \\ ||X_0|| < \varepsilon, & \varepsilon \in ]2,3], \\ ||X_0|| \le 3, & \varepsilon \in ]3,+\infty]. \end{cases}$$

This means

$$\mathcal{D}_s(\varepsilon) = \begin{cases} \mathcal{B}_\varepsilon, & \varepsilon \in ]0,1], \\ \bar{\mathcal{B}}_1, & \varepsilon \in ]1,2], \\ \mathcal{B}_\varepsilon, & \varepsilon \in ]2,3], \\ \bar{\mathcal{B}}_3, & \varepsilon \in ]3,+\infty], \end{cases}$$

which yields

$$\begin{aligned} \mathcal{D}_s &= \cup\,[\mathcal{D}_s(\varepsilon) : \varepsilon \in ]0,+\infty[] = [\cup\,(\mathcal{B}_\varepsilon : \varepsilon \in ]0,1])] \cup \bar{\mathcal{B}}_1 \cup [\mathcal{B}_\varepsilon : \varepsilon \in ]2,3]] \cup \bar{\mathcal{B}}_3 \\ &= \mathcal{B}_1 \cup \bar{\mathcal{B}}_1 \cup \mathcal{B}_3 \cup \bar{\mathcal{B}}_3 = \bar{\mathcal{B}}_3. \end{aligned}$$

Hence, the stability domain $\mathcal{D}_s$ of $X = 0$ of the system equals compact circle $\bar{\mathcal{B}}_3$,

$$\mathcal{D}_s = \bar{\mathcal{B}}_3 = \{X : X \in \Re^2,\ ||X|| \le 3\}.$$

Besides, the stability domain $\mathcal{D}_s$ equals the strict stability domain $\mathcal{D}_{sc}$ of $X = 0$, $\mathcal{D}_s = \mathcal{D}_{sc}$.

### 2.2.3 Definitions of attraction domains

The notion of the attraction domain has been widely used (cf. [94], [218]).

**Definition 2.3** *(a) The state $X = 0$ of the system (2.1) has the attraction domain denoted by $\mathcal{D}_a$ if and only if both (i) and (ii) hold:*

*(i) for every $\xi \in \Re^+$, there is $\tau \in \Re_+$, $\tau = \tau(X_0, \xi)$, such that $||\underline{X}(t; X_0)|| < \xi$ for all $t \in ]\tau, +\infty[$ holds provided only that $X_0 \in \mathcal{D}_a$,*

*and*

*(ii) $\mathcal{D}_a$ is a neighbourhood of $X = 0$.*

*(b) The state $X = 0$ of the system (2.1) has the strict domain of attraction (the strict attraction domain) denoted by $\mathcal{D}_{ac}$ if and only if $\mathcal{D}_{ac}$ is the largest connected neighbourhood of $X = 0$, which is subset of $\mathcal{D}_a$, $\mathcal{D}_{ac} \subseteq \mathcal{D}_a$.*

**Note 2.3** The condition *(i)* of the preceding definition can be expressed in an equivalent form *(i′)*,

*(i′) $lim\,[||\underline{X}(t; X_0)|| : t \to +\infty] = 0$ provided only that $X_0 \in \mathcal{D}_a$.*

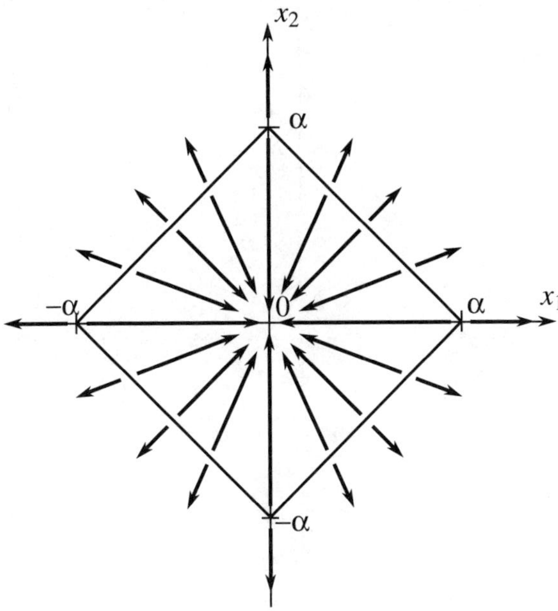


Figure 2.5: The state portrait of the system: $\dot{X} = (-\alpha + |x_1| + |x_2|)X$ (Example 2.4).

**Example 2.4** Let the system considered in Example 2.1 be reanalyzed,

$$\frac{dX}{dt} = (-\alpha + |x_1| + |x_2|)X, \ \alpha \in \Re^+.$$

Its state portrait is shown in Fig. 2.5.

The attraction domain $\mathcal{D}_a$ of $X = 0$ of the system is the interior

$$\mathring{\mathcal{S}}_\alpha = \left\{X : X \in \Re^2, \ \ |x_1| + |x_2| < \alpha\right\}$$

of the set

$$\mathcal{S}_\alpha = \left\{X : X \in \Re^2, \ \ |x_1| + |x_2| \leq \alpha\right\}, \ \mathcal{D}_a = \mathring{\mathcal{S}}_\alpha$$

$$\mathcal{D}_a = \left\{X : X \in \Re^2, \ \ |x_1| + |x_2| < \alpha\right\}.$$

The boundary

$$\partial\mathcal{S}_\alpha = \left\{X : X \in \Re^2, \ \ |x_1| + |x_2| = \alpha\right\}$$

is a subset of the set

$$\mathcal{S}_e = \left\{X : X \in \Re^2, \ \ (X = 0) \ \text{ or } \ (|x_1| + |x_2| = \alpha)\right\}$$

of all equilibrium states. Hence, $\underline{X}(t, ; X_0) \equiv X_0$ for every $X_0 \in \partial\mathcal{S}_\alpha$. Therefore, $\partial\mathcal{S}_\alpha \cap \mathcal{D}_a = \emptyset$ and $\mathcal{D}_a$ is open set. In this example, $\mathcal{D}_a = \mathcal{D}_{ac}$.

The domain $\mathcal{D}_a$ of attraction of $X = 0$ of the system is open (bounded connected) set, while its domain $\mathcal{D}_s = \mathcal{D}_{sc}$ of stability is closed (also bounded connected) set.

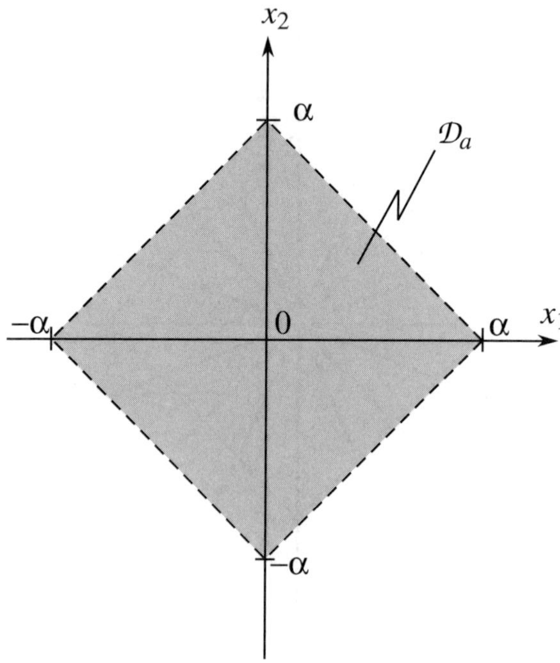


Figure 2.6: The attraction domain $\mathcal{D}_a$ of $X = 0$ of the system: $\dot{X} = (-\alpha + |x_1| + |x_2|)X$.

They are interrelated so that $\mathcal{D}_a$ is the interior of $\mathcal{D}_s$ and $\mathcal{D}_s$ is the closure of $\mathcal{D}_a$, hence, $\mathcal{D}_a$ is subset of $\mathcal{D}_s$,

$$\mathcal{D}_a = \mathring{\mathcal{D}}_s \ \ \mathcal{D}_s = \bar{\mathcal{D}}_a, \ \ \mathcal{D}_s \supset \mathcal{D}_a.$$

**Example 2.5** Let

$$\psi(||X_0||) = \begin{cases} 0 & \text{iff} \quad ||X_0|| \leq 1, \\ \dfrac{||X_0|| + 1}{||X_0|| - 1} & \text{iff} \quad ||X_0|| > 1, \end{cases}$$

and the system motions be defined by

$$\underline{X}(t; X_0) = exp\,(-t)\,[1 + \psi(X_0)t]\,X_0.$$

They satisfy the initial condition, $\underline{X}(0; X_0) \equiv X_0$, and are continuous in $t \in \Re_+$ for every $X_0 \in \Re^n$. However, they are not continuous in $X_0$ for $1 - \varepsilon < ||X_0|| < 1 + \varepsilon$ and arbitrarily small $\varepsilon \in \Re^+$.

Two cases of $||X_0||$ should be considered:

A) $||X_0|| \leq 1$ implies $||\underline{X}(t; X_0)|| = exp\,(-t)||X_0|| \leq ||X_0||$ for all $t \in \Re_+$. Hence, $max\,[||\underline{X}(t; X_0)|| : t \in \Re_+] = ||X_0||$, and $||\underline{X}(t; X_0)|| \to 0$ as $t \to +\infty$.

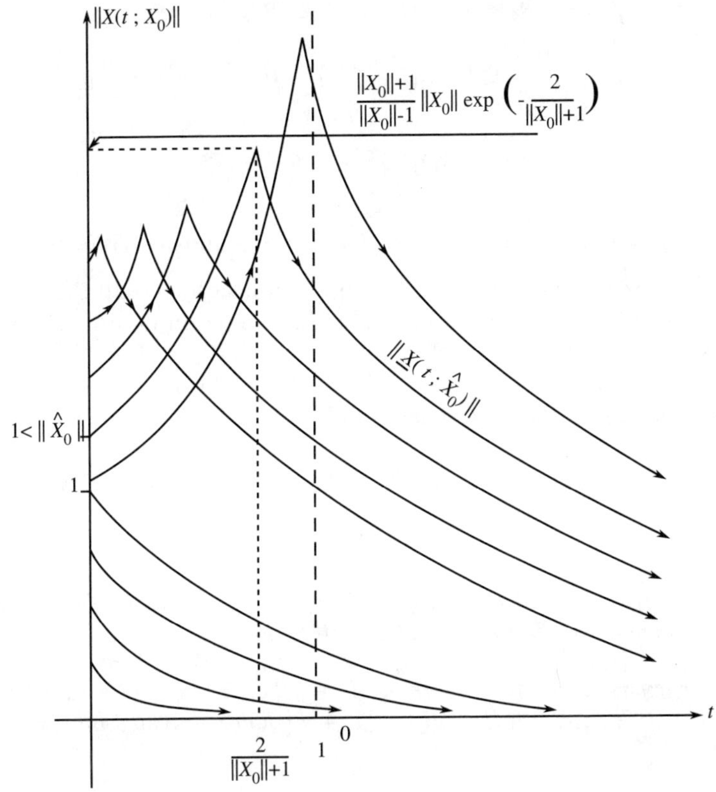


Figure 2.7: The norm of motions defined by $\underline{X}(t; X_0) = [1+\psi(X_0)t]X_0\, exp\,(-t)$, $\psi(X_0) = 0$ iff $||X_0|| \leq 1$ and $\psi(X_0) = \dfrac{||X_0||+1}{||X_0||-1}$ iff $||X_0|| > 1$ (Example 2.5).

B) $||X_0|| > 1$ yields $||\underline{X}(t; X_0)|| = exp\,(-t)\left(1 + \dfrac{||X_0||+1}{||X_0||-1}t\right)||X_0||$. In this case,

$$max\,[||\underline{X}(t; X_0)|| : \ t \in \Re_+] = \frac{||X_0||+1}{||X_0||-1}||X_0||\, exp\left(-\frac{2}{||X_0||+1}\right) \to +\infty$$

as $||X_0|| \to 1^+$ and

$$||\underline{X}(t; X_0)|| \to 0$$

as $t \to +\infty$ for every $X_0 \in \Re^n$, $||X_0|| > 1$.

Now, we can conclude that

$$\mathcal{D}_s(\varepsilon) \quad = \quad \begin{cases} \{X : X \in \Re^n, ||X|| < \varepsilon\} = \mathcal{B}_\varepsilon,\ \varepsilon \in \,]0,1] \\ \\ \bar{\mathcal{B}}_1 \cup \{X : X \in \Re^n, ||X|| > 1, \\ \\ \dfrac{||X||+1}{||X||-1}||X||\, exp\left(-\dfrac{2}{||X||+1}\right) < \varepsilon\} \cap \mathcal{B}_\varepsilon,\ \varepsilon \in \,]1, +\infty], \end{cases}$$

$$\mathcal{D}_{sc}(\varepsilon) = \begin{cases} \{X : X \in \Re^n, \ ||X|| < \varepsilon\} = \mathcal{B}_\varepsilon, & \varepsilon \in ]0,1] \\ \{X : X \in \Re^n, \ ||X|| \leq 1\} = \bar{\mathcal{B}}_1, & \varepsilon \in ]1,+\infty], \end{cases}$$

so that

$$\mathcal{D}_s = \cup[\mathcal{D}_s(\varepsilon) : \varepsilon \in \Re^+] = \Re^n$$

and

$$\mathcal{D}_{sc} = \cup[\mathcal{D}_{sc}(\varepsilon) : \varepsilon \in \Re^+] = (\cup[\mathcal{B}_\varepsilon : \varepsilon \in ]0,1]]) \cup \bar{\mathcal{B}}_1 = \bar{\mathcal{B}}_1$$

The strict stability domain $\mathcal{D}_{sc}$ of $X = 0$ is compact hyperball $\bar{\mathcal{B}}_1$. Furthermore, we conclude by examining the motions that the attraction domain $\mathcal{D}_a$ of $X = 0$ is the whole state space $\Re^n$, $\mathcal{D}_a = \Re^n$ and $\mathcal{D}_a = \mathcal{D}_{ac}$, $\mathcal{D}_a = \mathcal{D}_{ac} \supset \mathcal{D}_{sc}$, and $\mathcal{D}_a$ is open set, $\mathcal{D}_a = \mathcal{D}_{ac} = \Re^n$. This analysis is illustrated by Fig. 2.7.

This example shows that the strict stability domain $\mathcal{D}_{sc}$ can be a subset of the attraction domain $\mathcal{D}_a$.

Notice that the stability domain $\mathcal{D}_s = \mathcal{D}_a = \Re^n$, which illustrates $\mathcal{D}_s \neq \mathcal{D}_{sc}$, $\mathcal{D}_{sc} \subset \mathcal{D}_s$.

**Definition 2.4** *(a) A set $\mathcal{A} \subset \Re^n$ (of states of the system (2.1)) has the attraction domain denoted by $\mathcal{D}_a(\mathcal{A})$ if and only if both*

*(i) for every $\zeta \in \Re^+$, there is $T \in \Re_+$, $T = T(X_0, \zeta; \mathcal{A})$, such that $\rho[\underline{X}(t; X_0), \mathcal{A}] < \zeta$ for all $t \in ]T, +\infty[$ holds provided only $X_0 \in \mathcal{D}_a$,*

*and*

*(ii) $\mathcal{D}_a$ is a neighbourhood of the set $\mathcal{A}$.*

*(b) A set $\mathcal{A} \subset \Re^n$ (of states of the system (2.1)) has the strict domain of attraction (the strict attraction domain) denoted by $\mathcal{D}_{ac}(\mathcal{A})$ if and only if it has the attraction domain $\mathcal{D}_a(\mathcal{A})$ and $\mathcal{D}_{ac}(\mathcal{A})$ is the largest connected neighbourhood of $\mathcal{A}$ which is subset of $\mathcal{D}_a(\mathcal{A})$, $\mathcal{D}_{ac}(\mathcal{A}) \subseteq \mathcal{D}_a(\mathcal{A})$.*

**Note 2.4** The condition $(i)$ of Definition 2.4 can be expressed in an equivalent form $(i')$,

*$(i')$ $lim\{\rho[\underline{X}(t; X_0), \mathcal{A}] : t \to +\infty\} = 0$ provided only that $X_0 \in \mathcal{D}_a$.*

**Example 2.6** The attraction domain $\mathcal{D}_a(\mathcal{A})$ of the set $\mathcal{A}$,

$$\mathcal{A} = \{X : X \in \Re^2, \ |x_1| + |x_2| \leq 1\},$$

of states of the system described by (see Fig. 2.2)

$$\frac{dX}{dt} = (1 - |x_1| - |x_2|)(4 - |x_1| - 2|x_2|)X,$$

is found as (Fig. 2.8)

$$\mathcal{D}_a(\mathcal{A}) = \{X : X \in \Re^2, \ |x_1| + 2|x_2| < 4\}.$$

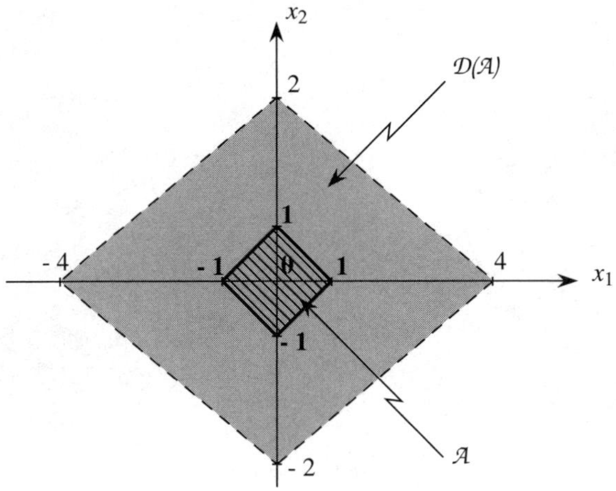


Figure 2.8: The attraction domain $\mathcal{D}_a(\mathcal{A})$ of the set $\mathcal{A} = \{X : X \in \Re^2,\ |x_1| + |x_2| \leq 1\}$ of states of the system $\dot{X} = (1 - |x_1| - |x_2|)(4 - |x_1| - 2|x_2|)X$.

The attraction domain $\mathcal{D}_a(\mathcal{A})$ of the set $\mathcal{A}$ is an open subset of its stability domain $\mathcal{D}_s(\mathcal{A})$ (Fig. 2.3d). In fact $\mathcal{D}_a(\mathcal{A}) = \mathcal{D}_{ac}(\mathcal{A}) = \overset{\circ}{\mathcal{D}}_s(\mathcal{A})$, $\bar{\mathcal{D}}_a(\mathcal{A}) = \mathcal{D}_s(\mathcal{A})$.

Since $\mathcal{D}_a(\mathcal{A})$ is bounded, then the set $\mathcal{A}$ is not globally attractive.

**Example 2.7** The set $\mathcal{A}$,

$$\mathcal{A} = \left\{X : X \in \Re^2,\ x_1^2 + x_2^2 = 4\right\},$$

represents a limit cycle of the system

$$\frac{dX}{dt} = \begin{bmatrix} -(1-x_1^2-x_2^2)(4-x_1^2-x_2^2)(9-x_1^2-x_2^2) & 1 \\ -1 & -(1-x_1^2-x_2^2)(4-x_1^2-x_2^2)(9-x_1^2-x_2^2) \end{bmatrix} X$$

The attraction domain $\mathcal{D}_a(\mathcal{A})$ of the set $\mathcal{A}$ is bounded, open and connected,

$$\mathcal{D}_a(\mathcal{A}) = \left\{X :\ 1 < x_1^2 + x_2^2 < 9\right\},\ \mathcal{D}_a(\mathcal{A}) = \mathcal{D}_{ac}(\mathcal{A}),$$

which is shown in Fig. 2.9a. Its stability domain $\mathcal{D}_s(\mathcal{A})$,

$$\mathcal{D}_s(\mathcal{A}) = \left\{X :\ 1 \leq x_1^2 + x_2^2 \leq 9\right\} = \overline{\mathcal{D}}_a(\mathcal{A}),$$

is shown in Fig. 2.9b. In this example $\mathcal{D}_s(\mathcal{A}) \supset \mathcal{D}_a(\mathcal{A})$.

### 2.2.4 Definitions of asymptotic stability domains

Asymptotic stability domains were defined in [64], [65], [68]–[70], [72] to correspond to the definitions of asymptotic stability of a state and of a set (Section 1.1.3).

**Definition 2.5** *(a) The state $X = 0$ of the system (2.1) has the asymptotic stability domain denoted by $\mathcal{D}$ if and only if both*

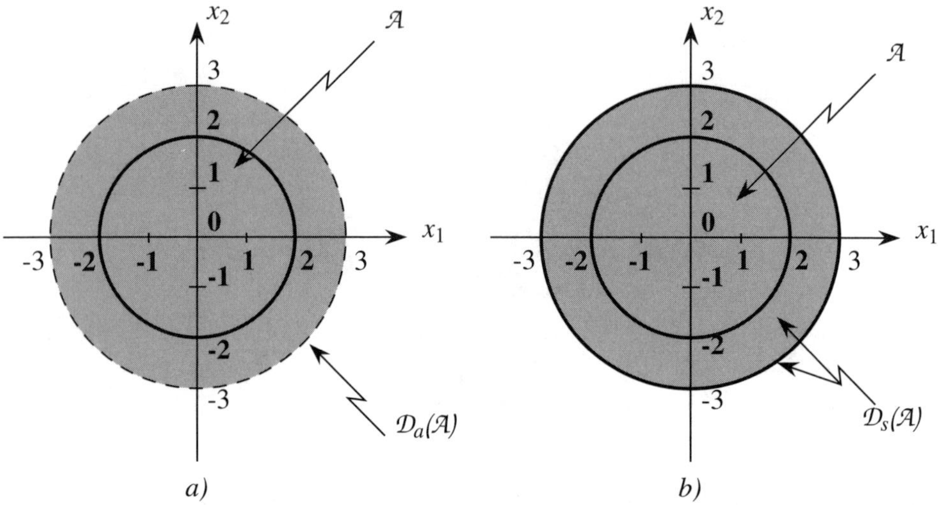


Figure 2.9: a) The domain $\mathcal{D}_a(\mathcal{A})$ of attraction of the set $\mathcal{A} = \{x : x \in \Re^2,\ x_1^2 + x_2^2 = 4\}$; b) The domain $\mathcal{D}_s(\mathcal{A})$ of stability of the set $\mathcal{A}$ (Example 2.7).

*(i) it has the stability domain $\mathcal{D}_s$ and the attraction domain $\mathcal{D}_a$,*

*and*

*(ii) $\mathcal{D} = \mathcal{D}_s \cap \mathcal{D}_a$.*

*(b) The state $X = 0$ of the system (2.1) has the strict domain of asymptotic stability (the strict asymptotic stability domain) denoted by $\mathcal{D}_c$ if and only if it has both the strict stability domain $\mathcal{D}_{sc}$ and the strict attraction domain $\mathcal{D}_{ac}$ and $\mathcal{D}_c = \mathcal{D}_{sc} \cap \mathcal{D}_{ac}$.*

**Comment 2.2** If the state $X = 0$ of the system (2.1) has both $\mathcal{D}_s$ – the stability domain and $\mathcal{D}_a$ – the attraction domain, then they are neighbourhoods of it. Hence, $\mathcal{D}$ – the asymptotic stability domain is also a neighbourhood of $X = 0$. It can be closed, can be open, which will be illustrated via examples. It need not be connected in general. However, $\mathcal{D}_c$ is a connected neighbourhood of $X = 0$ due to such property of $\mathcal{D}_{sc}$ and $\mathcal{D}_{ac}$, and $\mathcal{D}_c = \mathcal{D}_{ac} \cap \mathcal{D}_{sc}$.

**Example 2.8** Let the second order system analyzed in Example 2.1 and Example 2.4 be further considered,

$$\frac{dX}{dt} = (-\alpha + |x_1| + |x_2|)X,\ \alpha \in \Re^+.$$

It was shown that $X = 0$ of the system has both the stability domain $\mathcal{D}_s$,

$$\mathcal{D}_s = \left\{X : X \in \Re^2,\ |x_1| + |x_2| \leq \alpha\right\},\ \mathcal{D}_{sc} = \mathcal{D}_s,$$

and the attraction domain $\mathcal{D}_a$,

$$\mathcal{D}_a = \left\{X : X \in \Re^2,\ |x_1| + |x_2| < \alpha\right\} = \mathring{\mathcal{D}}_s,\ \mathcal{D}_{ac} = \mathcal{D}_a.$$

In view of Definition 2.5, the state $X = 0$ of the system has also the asymptotic stability domain $\mathcal{D}$,

$$\mathcal{D} = \mathcal{D}_s \cap \mathcal{D}_a = \left\{X : X \in \Re^2,\ |x_1| + |x_2| < \alpha\right\} = \mathcal{D}_a,\ \mathcal{D} = \mathcal{D}_c.$$

In this example $\mathcal{D}$ equals $\mathcal{D}_a$ and they are subsets of $\mathcal{D}_s$. They are exactly the interior $\overset{\circ}{\mathcal{D}}_s$ of $\mathcal{D}_s$, $\mathcal{D}_a = \mathcal{D} = \overset{\circ}{\mathcal{D}}_s$, in this case.

**Example 2.9** The zero state of the second order nonlinear system

$$\frac{dX}{dt} = \begin{bmatrix} -(1-\|X\|^2)(4-\|X\|^2)(9-\|X\|^2) & 1 \\ -1 & -(1-\|X\|^2)(4-\|X\|^2)(9-\|X\|^2) \end{bmatrix} X$$

has the stability domain $\mathcal{D}_s$ (Example 2.3),

$$\mathcal{D}_s = \bar{\mathcal{B}}_3 = \left\{X : X \in \Re^2,\ \|X\| \le 3\right\},\ \mathcal{D}_{sc} = \mathcal{D}_s,$$

and the attraction domain $\mathcal{D}_a$ (Fig. 2.4),

$$\mathcal{D}_a = \mathcal{B}_1 = \left\{X : X \in \Re^2,\ \|X\| < 1\right\},\ \mathcal{D}_{ac} = \mathcal{D}_a,$$

Hence, its asymptotic stability domain $\mathcal{D}$,

$$\mathcal{D} = \mathcal{D}_s \cap \mathcal{D}_a = \mathcal{B}_1 = \left\{X : X \in \Re^2,\ \|X\| < 1\right\},\ \mathcal{D}_c = \mathcal{D},$$

equals $\mathcal{D}_a$ and they are subsets of $\mathcal{D}_s$, but they are also proper subsets of the interior $\overset{\circ}{\mathcal{D}}_s$ of $\mathcal{D}_s$.

**Example 2.10** In the case of the system defined by its motions,

$$\underline{X}(t; X_0) = e^{-t}\left[1 + \psi(X_0)t\right] X_0,$$

$$\psi(X_0) = \begin{cases} 0 & ,\ \text{iff}\ \ \|X_0\| \le 1, \\ \dfrac{\|X_0\|+1}{\|X_0\|-1} & ,\ \text{iff}\ \ \|X_0\| = 1 \end{cases},$$

the zero state has both the strict stability domain $\mathcal{D}_{sc}$ (Example 2.5), $\mathcal{D}_{sc} = \overline{B}_1 = \{X : \|X\| \le 1\}$, $\mathcal{D}_s = \Re^n$, $\mathcal{D}_{sc} \subset \mathcal{D}_s$, $\mathcal{D}_{sc} \neq \mathcal{D}_s$, and the strict attraction domain $\mathcal{D}_{ac} = \Re^n$, $\mathcal{D}_a = \mathcal{D}_{ac}$.

Hence, it has also the strict asymptotic stability domain $\mathcal{D}_c$, $\mathcal{D}_c = \mathcal{D}_{sc} \cap \mathcal{D}_{ac} = \overline{B}_1 \cap \Re^n = \overline{B}_1 = \{X : \|X\| \le 1\}$, which is compact hyperball $\overline{B}_1$ and equal to $\mathcal{D}_{sc}$. They are, in this case, subsets of $\mathcal{D}_{ac}$, $\mathcal{D}_c \subset \mathcal{D}_{ac}$.

Notice that $\mathcal{D}_{ac}$ is open and $\mathcal{D}_c$ is closed in this example. Besides, $\mathcal{D}_s = \mathcal{D}_a = \mathcal{D} = \Re^n$, $\mathcal{D}_c \subset \mathcal{D}$, $\mathcal{D}_c \neq \mathcal{D}$.

Hence, the asymptotic stability domain $\mathcal{D}$ equals the whole state space $\Re^n$, but the strict asymptotic stability domain $\mathcal{D}_c$ does not. The state $X = 0$ is globally asymptotically stable, but it is not completely globally asymptotically stable.

**Definition 2.6** *(a) A set $\mathcal{A} \subset \Re^n$ [of states of the system (2.1)] has the asymptotic stability domain denoted by $\mathcal{D}(\mathcal{A})$ if and only if both*

*(i) it has the stability domain $\mathcal{D}_s(\mathcal{A})$ and the attraction domain $\mathcal{D}_a(\mathcal{A})$,*

*and*

*(ii) $\mathcal{D}(\mathcal{A}) = \mathcal{D}_s(\mathcal{A}) \cap \mathcal{D}_a(\mathcal{A})$.*

*(b) A set $\mathcal{A} \subset \Re^n$ [of states of the system (2.1)] has the strict domain of asymptotic stability (the strict asymptotic stability domain) denoted by $\mathcal{D}_c(\mathcal{A})$ if and only if it has both the strict stability domain $\mathcal{D}_{sc}(\mathcal{A})$ and the strict attraction domain $\mathcal{D}_{ac}(\mathcal{A})$, and $\mathcal{D}_c(\mathcal{A}) = \mathcal{D}_{sc}(\mathcal{A}) \cap \mathcal{D}_{ac}(\mathcal{A})$.*

**Example 2.11** The set $\mathcal{A}$,

$$\mathcal{A} = \{X : X \in \Re^2, \ \ |x_1| + |x_2| \leq 1\},$$

of states of the system

$$\frac{dX}{dt} = (1 - |x_1| - |x_2|)(4 - |x_1| - 2|x_2|)X,$$

has both the stability domain $\mathcal{D}_s(\mathcal{A})$ (Example 2.2),

$$\mathcal{D}_s(\mathcal{A}) = \{X : X \in \Re^2, \ \ |x_1| + 2|x_2| \leq 4\}, \quad \mathcal{D}_{sc}(\mathcal{A}) = \mathcal{D}_s(\mathcal{A}),$$

and the attraction domain $\mathcal{D}_a(\mathcal{A})$ (Example 2.6),

$$\mathcal{D}_a(\mathcal{A}) = \{X : X \in \Re^2, \ \ |x_1| + 2|x_2| < 4\} = \mathring{\mathcal{D}}_s(\mathcal{A}), \quad \mathcal{D}_{ac}(\mathcal{A}) = \mathcal{D}_a(\mathcal{A}).$$

Hence, its asymptotic stability domain $\mathcal{D}(\mathcal{A})$ equals $\mathcal{D}_a(\mathcal{A})$ that is the interior $\mathring{\mathcal{D}}_s(\mathcal{A})$ of $\mathcal{D}_s(\mathcal{A})$. Notice that $X = 0$ is unstable, thus it does not have the stability domain and the asymptotic stability domain.

**Comment 2.3** *If $X = 0$ is the unique equilibrium state of the system (2.1) and is stable then as soon as $X_0 \notin \mathcal{D}_s$ the motion $\underline{X}(t; X_0)$ is either not defined on the whole time axis $\Re_+$ or it is not bounded. However, such a statement cannot be applied either to $\mathcal{D}_a$ and $\mathcal{D}$ if $X = 0$ is also attractive as soon as $\mathcal{D}_a \subset \mathcal{D}_s$, or to $\mathcal{D}_{sc}$ if $\mathcal{D}_{sc} \neq \mathcal{D}_s$.*

### 2.2.5 Definitions of exponential stability domains

In order to explain the relative sense of the definitions of exponential stability domains we shall first consider the system of Example 2.1 once more.

**Example 2.12** The motions of the system

$$\frac{dX}{dt} = (-\alpha + |x_1| + |x_2|)X, \ \alpha \in \Re^+.$$

obey the estimate

$$||\underline{X}(t;X_0)|| \leq \beta||X_0||\, exp\,(-\gamma t), \quad \forall t \in \Re_+,$$

only for

$$X_0 \in \mathcal{S}_\zeta = \{X : X \in \Re^2, \quad |x_1| + |x_2| \leq \zeta\},$$

and for

$$\zeta \in ]0, \alpha[, \quad \gamma = \alpha - \zeta,$$

with

$$\beta = 1.$$

We can accept only $\zeta \in ]0, \alpha[$. Once $\zeta$ has been accepted then $\gamma$ is completely determined by $\gamma = \alpha - \zeta$. The bigger $\zeta$, the larger $\mathcal{S}_\zeta$ and the smaller $\gamma$. In other words, the larger set $\mathcal{S}_\zeta$ over which the motions obey the exponential estimate

$$||\underline{X}(t;X_0)|| \leq ||X_0||\, exp\,(-\gamma t), \quad \forall t \in \Re_+,$$

the smaller rate $\gamma$ of the exponential convergence of the motions to the origin. If we wish to find $\gamma > 0$ in the limiting case $\zeta = \alpha$, i.e.

$$\mathcal{S}_\zeta = \mathcal{S}_\alpha = \{X : X \in \Re^2, \quad |x_1| + |x_2| \leq \alpha\},$$

then we conclude that $\gamma > 0$ does not exist. The same holds for the interior $\mathring{\mathcal{S}}_\alpha = \{X : X \in \Re^2, \ |x_1| + |x_2| < \alpha\}$ of $\mathcal{S}_\alpha$. The consequence of this is that there does not exist the maximum set $\mathcal{S}_\zeta$ (or $\mathring{\mathcal{S}}_\zeta$) for which we can find $\gamma$ obeying the exponential estimate because $\zeta \in ]0, \alpha[$ and $max\,]0, \alpha[$ does not exist, although $sup\ ]0, \alpha[ = \alpha$ – exists. Therefore, we cannot speak of the largest (or the maximum) set of system states over which the exponential estimate holds for some $\beta \in [1, +\infty[$ and $\gamma \in ]0, +\infty[$. Instead, we can look for the largest set of system states over which the exponential estimate holds with respect to given $\beta \in [1, +\infty[$ and given $\gamma \in ]0, +\infty[$.

**Definition 2.7** *(a) The state $X = 0$ of the system (2.1) has the domain $\mathcal{D}_e(\beta, \gamma)$ of exponential stability with respect to $(\beta, \gamma)$ if and only if both*

*(i) $\mathcal{D}_e(\beta, \gamma)$ is a neighbourhood of $X = 0$,*

*and*

*(ii) the exponential estimate*

$$||\underline{X}(t;X_0)|| \leq \beta||X_0||\, exp\,(-\gamma t), \quad \forall t \in \Re_+,$$

*holds provided only that $X_0 \in \mathcal{D}_e(\beta, \gamma)$, where $\beta \in [1, +\infty[$ and $\gamma \in \Re^+$.*

*(b) The state $X = 0$ of the system (2.1) has the strict domain $\mathcal{D}_{ec}(\beta, \gamma)$ of exponential stability (the strict exponential stability domain) with respect to $(\beta, \gamma)$ if and only if $\mathcal{D}_{ec}(\beta, \gamma)$ is the largest connected neighbourhood of $X = 0$ which is subset of $\mathcal{D}_e(\beta, \gamma)$, $\mathcal{D}_{ec}(\beta, \gamma) \subseteq \mathcal{D}_e(\beta, \gamma)$.*

**Comment 2.4** *The state* $X = 0$ *of the system*

$$\frac{dX}{dt} = (-\alpha + |x_1| + |x_2|)X, \quad \alpha \in \Re^+.$$

*has* $\mathcal{D}_e(1,\gamma)$ *for any* $\gamma \in ]0,\alpha[$ *and it equals* $\mathcal{S}_\zeta$,

$$\mathcal{D}_e(1,\gamma) = \mathcal{S}_\zeta = \{X : X \in \Re^2, \ \ |x_1| + |x_2| \le \zeta\},$$

$$\zeta = \alpha - \gamma, \qquad \mathcal{D}_{ec}(1,\gamma) = \mathcal{D}_e(1,\gamma).$$

*However, for the set* $\mathcal{S}_\alpha$,

$$\mathring{\mathcal{S}}_\alpha = \{X : X \in \Re^2, \ \ |x_1| + |x_2| < \alpha\},$$

*we cannot find* $\beta \in ]1,+\infty[$ *and* $\gamma \in ]0,\alpha[$ *for which* $\mathring{\mathcal{S}}_\alpha$ *satisfies Definition 2.7, despite* $\mathring{\mathcal{S}}_\alpha$ *being the asymptotic stability domain of* $X = 0$, $\mathring{\mathcal{S}}_\alpha = \mathcal{D}$. *This is clear, because*

$$\mathring{\mathcal{S}}_\alpha = \cup[\mathcal{S}_\zeta : \zeta \in ]0,\alpha[],$$

*which means that we should find* $\gamma = min\,(\alpha - \zeta : \zeta \in ]0,\alpha[)$. *This minimum does not exist, and* $inf\ (\alpha - \zeta : \zeta \in ]0,\alpha[) = 0$. *The asymptotic stability domain* $\mathcal{D} = \mathring{\mathcal{S}}_\alpha$ *is not the exponential stability domain with respect to any* $(\beta,\gamma) \in [1,+\infty[\times]0,\alpha[$. *In fact*

$$\mathcal{D}_e(1,\gamma) \subset \mathcal{D}, \ \ \forall \gamma \in ]0,\alpha[,$$

*and*

$$\mathcal{D}_e(1,\gamma) = \mathring{\mathcal{S}}_\zeta = \{X : X \in \Re^2, \ \ |x_1| + |x_2| < \zeta\}, \quad \zeta = \alpha - \gamma.$$

**Definition 2.8** *(a) A set* $\mathcal{A} \subset \Re^n$ *(of states of the system (2.1)) has the domain* $\mathcal{D}_e(\mathcal{A};\beta,\gamma)$ *of exponential stability with respect to* $(\beta,\gamma)$ *if and only if both*

*(i)* $\mathcal{D}_e(\mathcal{A};\beta,\gamma)$ *is a neighbourhood of the set* $\mathcal{A}$,

*and*

*(ii) the exponential estimate*

$$\rho[\underline{X}(t;X_0),\mathcal{A}] \le \beta\rho(X_0,\mathcal{A})\,exp\,(-\gamma t) \ \ \textit{for all } t \in \Re_+$$

*holds provided only that* $X_0 \in \mathcal{D}_e(\mathcal{A};\beta,\gamma)$, *where* $\beta \in [1,+\infty[$ *and* $\gamma \in \Re^+$.

*(b) A set* $\mathcal{A} \subset \Re^n$ *(of states of the system (2.1)) has the strict domain* $\mathcal{D}_{ec}(\mathcal{A};\beta,\gamma)$ *of exponential stability (the strict exponential stability domain) with respect to* $(\beta,\gamma)$ *if and only if* $\mathcal{D}_{ec}(\mathcal{A};\beta,\gamma)$ *is the largest connected neighbourhood of* $\mathcal{A}$ *which is subset of* $\mathcal{D}_e(\mathcal{A};\beta,\gamma)$, $\mathcal{D}_{ec}(\mathcal{A};\beta,\gamma) \subseteq \mathcal{D}_e(\mathcal{A};\beta,\gamma)$.

### 2.2.6 Definitions of asymptotic stability domains on $\mathcal{N}_{(\cdot)}(\cdot\cdot)$

Let the system (2.1) be of the Lurie form (Section 1.1.5),

$$\frac{dX}{dt} = AX + Bf(w), \tag{2.2a}$$

$$w = CX + Df(w). \tag{2.2b}$$

In order to explain the need for the study of domains of asymptotic stability on $\mathcal{N}_i(\cdot)$ we present the following simple example.

**Example 2.13** Let $n = 1$ and

$$\frac{dX}{dt} = -sin\ X.$$

This system has infinitely many equilibrium points located at $X = k\pi$, where $k$ is any integer. Hence, $X = 0$ obviously is not asymptotically stable in the large. However, it is asymptotically stable with the domain of asymptotic stability $\mathcal{D} = ]-\pi, \pi[$. Over $\mathcal{S} = \mathcal{D}$ the nonlinearity $f$, $f(X) = -sin\ X$, belongs to the family $\mathcal{N}_1(\mathcal{L}, \mathcal{M}; \mathcal{S})$ for $\mathcal{L} = [0, 1]$ and $\mathcal{M} = [-1, 1[$. If the system is embedded into the class of Lurie systems (2.2), then we can speak only about asymptotic stability of $X = 0$ for a particular $f(\cdot)$ or for any $f \in \mathcal{N}_1(\mathcal{L}, \mathcal{M}; \mathcal{S})$, or for any $f \in \mathcal{N}_0(\mathcal{L}; \mathcal{S})$. This means that we can look only for the asymptotic stability domain for a particular $f$; e.g. $f(X) = -sin\ X$, or for every $f \in \mathcal{N}_1(\mathcal{L}, \mathcal{M}; \mathcal{S})$, or for every $f \in \mathcal{N}_0(\mathcal{L}; \mathcal{S})$.

Let $\mathcal{D}^f$ denote the asymptotic stability domain of $X = 0$ of the (Lurie) system (2.2) for a particular nonlinearity $f$.

**Definition 2.9** *The state $X = 0$ of the system (2.2) has the [strict] asymptotic stability domain on $\mathcal{N}_i(\mathcal{L}, \mathcal{M}; \mathcal{S})$, which is denoted by $\mathcal{D}_i(\mathcal{L}, \mathcal{M}; \mathcal{S})$ [$\mathcal{D}_{ic}(\mathcal{L}, \mathcal{M}; \mathcal{S})$], if and only if*

*a) it has the [strict] asymptotic stability domain $\mathcal{D}^f$ [$\mathcal{D}^f_c$] for every $f \in \mathcal{N}_i(\mathcal{L}, \mathcal{M}; \mathcal{S})$*

*and*

*b) $\mathcal{D}_i(\mathcal{L}, \mathcal{M}; \mathcal{S}) = \cap[\mathcal{D}^f : f \in \mathcal{N}_i(\mathcal{L}, \mathcal{M}; \mathcal{S})]$ is a neighbourhood of $X = 0$ [$\mathcal{D}_{ic}(\mathcal{L}, \mathcal{M}; \mathcal{S}) = \cap[\mathcal{D}^f_c : f \in \mathcal{N}_i(\mathcal{L}, \mathcal{M}; \mathcal{S})]$ is a connected neighbourhood of $X = 0$], respectively.*

This definition was introduced in [68], [69]. It can be extended to sets as follows:

**Definition 2.10** *A set $\mathcal{A} \subset \Re^n$ (of the states of the system (2.2)) has the [strict] asymptotic stability domain on $\mathcal{N}_i(\mathcal{L}, \mathcal{M}; \mathcal{S})$, which is denoted by $\mathcal{D}_i(\mathcal{L}, \mathcal{M}; \mathcal{A}; \mathcal{S})$ [$\mathcal{D}_{ic}(\mathcal{L}, \mathcal{M}; \mathcal{A}; \mathcal{S})$], if and only if both*

*a) it has the [strict] asymptotic stability domain $\mathcal{D}^f$ [$\mathcal{D}^f_c$] for every $f \in \mathcal{N}_i(\mathcal{L}, \mathcal{M}; \mathcal{S})$*

*and*

*b)* $\mathcal{D}_i(\mathcal{L}, \mathcal{M}; \mathcal{A}; \mathcal{S}) = \cap[\mathcal{D}^f : f \in \mathcal{N}_i(\mathcal{L}, \mathcal{M}; \mathcal{A}; \mathcal{S})]$ *is a neighbourhood of the set* $\mathcal{A}$ *[*$\mathcal{D}_{ic} = \cap[\mathcal{D}_c^f : f \in \mathcal{N}_i(\mathcal{L}, \mathcal{M}; \mathcal{A}; \mathcal{S})]$ *is a connected neighbourhood of the set* $\mathcal{A}$*], respectively.*

## 2.3 Domains of practical stability properties

### 2.3.1 Definitions of domains of practical stability

By following [70] and Section 1.2.2 we accept the following definition for the system (2.3):

$$\frac{dX}{dt} = f(X, \mathtt{i}),\ f : \Re^n \times \Re^l \rightarrow \Re^n. \tag{2.3}$$

**Definition 2.11** *The system (2.3) has the domain of practical stability with respect to* $\{\tau, \mathcal{X}_A, \mathcal{I}\}$*, which is denoted by* $\mathcal{D}_{ps}(\tau, \mathcal{X}_A, \mathcal{I})$*, if and only if both*

*a) its motions obey*

$$\underline{X}(t; X_0; \mathtt{i}) \in \mathcal{X}_A \text{ for every } (t, \mathtt{i}) \in \Re_\tau \times \mathcal{I}$$

*provided only that* $X_0 \in \mathcal{D}_{ps}(\tau, \mathcal{X}_A, \mathcal{I})$,

*and*

*b) the interior* $\overset{\circ}{\mathcal{D}}_{ps}(\tau, \mathcal{X}_A, \mathcal{I})$ *of* $\mathcal{D}_{ps}(\tau, \mathcal{X}_A, \mathcal{I})$ *is non-empty.*

When $\tau$, $\mathcal{X}_A$ and $\mathcal{I}$ are prespecified then we may replace $\mathcal{D}_{ps}(\tau, \mathcal{X}_A, \mathcal{I})$ by $\mathcal{D}_{ps}$.

**Definition 2.12** *A set* $\mathcal{A}$ *(of states of the system (2.3)) has the domain of practical stability with respect to* $\{\tau, \mathcal{X}_A, \mathcal{I}\}$*, which is denoted by* $\mathcal{D}_{ps}(\tau, \mathcal{X}_A, \mathcal{I}; \mathcal{A})$*, if and only if both*

*a) the system motions obey*

$$\underline{X}(t; X_0; \mathtt{i}) \in \mathcal{X}_A \text{ for every } (t, \mathtt{i}) \in \Re_\tau \times \mathcal{I}$$

*provided only that* $X_0 \in \mathcal{D}_{ps}(\tau, \mathcal{X}_A, \mathcal{I}; \mathcal{A})$,

*and*

*b)* $\mathcal{D}_{ps}(\tau, \mathcal{X}_A, \mathcal{I}; \mathcal{A})$ *is a neighbourhood of the set* $\mathcal{A}$.

When $\tau$, $\mathcal{X}_A$ and $\mathcal{I}$ are known then we may write $\mathcal{D}_{ps}(\mathcal{A})$ instead of $\mathcal{D}_{ps}(\tau, \mathcal{X}_A, \mathcal{I}; \mathcal{A})$.

**Comment 2.5** *If we are interested in the practical stability domain of a state* $X^*$ *then we can apply Definition 2.12 by settling* $\mathcal{A} = \{X^*\}$.

### 2.3.2 Definitions of domains of practical contraction with settling time

As for practical stability domains, we first introduce the notion of the domain of practical contraction with settling time for the system (2.3) (see Section 1.2.3).

**Definition 2.13** *The system (2.3) has the domain of practical contraction with the settling time* $\tau_s$ *with respect to* $\{\tau, \mathcal{X}_F, \mathcal{I}\}$, *which is denoted by* $\mathcal{D}_{pc}(\tau, \tau_s, \mathcal{X}_F, \mathcal{I})$, *if and only if both*

*a) its motions obey*

$$\underline{X}(t; X_0; \mathtt{i}) \in \mathcal{X}_F \quad \textit{for every} \quad (t, \mathtt{i}) \in \mathcal{T}_{\tau s} \times \mathcal{I}$$

*provided only that* $X_0 \in \mathcal{D}_{pc}(\tau, \tau_s, \mathcal{X}_F, \mathcal{I})$,

*and*

*b) the interior* $\mathring{\mathcal{D}}_{pc}(\tau, \tau_s, \mathcal{X}_F, \mathcal{I})$ *of* $\mathcal{D}_{pc}(\tau, \tau_s, \mathcal{X}_F, \mathcal{I})$ *is non-empty.*

When $\tau$, $\tau_s$, $\mathcal{X}_F$ and $\mathcal{I}$ are prespecified then we may write $\mathcal{D}_{pc}$ instead of $\mathcal{D}_{pc}(\tau, \tau_s, \mathcal{X}_F, \mathcal{I})$.

**Definition 2.14** *A set* $\mathcal{A}$ *(of states of the system (2.3)) has the domain of practical contraction with the settling time* $\tau_s$ *with respect to* $\{\tau, \mathcal{X}_F, \mathcal{I}\}$, *which is denoted by* $\mathcal{D}_{pc}(\tau, \tau_s, \mathcal{X}_F, \mathcal{I}; \mathcal{A})$, *if and only if both*

*a) the system motions obey*

$$\underline{X}(t; X_0; \mathtt{i}) \in \mathcal{X}_F \quad \textit{for every} \quad (t, \mathtt{i}) \in \mathcal{T}_{\tau s} \times \mathcal{I}$$

*provided only that* $X_0 \in \mathcal{D}_{pc}(\tau, \tau_s, \mathcal{X}_F, \mathcal{I}; \mathcal{A})$,

*and*

*b)* $\mathcal{D}_{pc}(\tau, \tau_s, \mathcal{X}_F, \mathcal{I}; \mathcal{A})$ *is a neighbourhood of the set* $\mathcal{A}$.

When $\tau$, $\tau_s$, $\mathcal{X}_F$ and $\mathcal{I}$ are known then we may replace $\mathcal{D}_{pc}(\tau, \tau_s, \mathcal{X}_F, \mathcal{I}; \mathcal{A})$ by $\mathcal{D}_{pc}(\mathcal{A})$.

**Comment 2.6** *If we are interested in the domain of practical contraction of a state* $X^*$ *then we may use Definition 2.14 with* $\mathcal{A} = \{X^*\}$.

### 2.3.3 Definitions of domains of practical stability with settling time

In view of the preceding definition and the notion of practical stability with settling time (Section 1.2.4) we accept the following definition:

**Definition 2.15** *The system (2.3) has the domain of practical (contractive) stability with the settling time $\tau_s$ with respect to $\{\tau, \mathcal{X}_A, \mathcal{X}_F, \mathcal{I}\}$, which is denoted by $\mathcal{D}_p(\tau, \tau_s, \mathcal{X}_A, \mathcal{X}_F, \mathcal{I})$, if and only if*

*a) both*

$$1) \quad \underline{X}(t; X_0; \mathtt{i}) \in \mathcal{X}_A \quad \textit{for every} \quad (t, \mathtt{i}) \in \Re_\tau \times \mathcal{I},$$

*and*

$$2) \quad \underline{X}(t; X_0; \mathtt{i}) \in \mathcal{X}_F \quad \textit{for every} \quad (t, \mathtt{i}) \in \mathcal{T}_{\tau s} \times \mathcal{I}$$

*hold provided only $X_0 \in \mathcal{D}_p(\tau, \tau_s, \mathcal{X}_A, \mathcal{X}_F, \mathcal{I})$,*

*and*

*b) the interior $\mathring{\mathcal{D}}_p(\tau, \tau_s, \mathcal{X}_A, \mathcal{X}_F, \mathcal{I})$ of $\mathcal{D}_p(\tau, \tau_s, \mathcal{X}_A, \mathcal{X}_F, \mathcal{I})$ is non-empty.*

When $\tau$, $\tau_s$, $\mathcal{X}_A$, $\mathcal{X}_F$ and $\mathcal{I}$ are given then we may write $\mathcal{D}_p$ instead of $\mathcal{D}_p(\tau, \tau_s, \mathcal{X}_A, \mathcal{X}_F, \mathcal{I})$.

For the set $\mathcal{A}$ we deduce from Definition 2.12 and Definition 2.14 the following [70]:

**Definition 2.16** *A set $\mathcal{A}$ (of states of the system (2.3)) has the domain of practical (contractive) stability with the settling time $\tau_s$ with respect to $\{\tau, \mathcal{X}_A, \mathcal{X}_F, \mathcal{I}\}$, which is denoted by $\mathcal{D}_p(\tau, \tau_s, \mathcal{X}_A, \mathcal{X}_F, \mathcal{I}; \mathcal{A})$, if and only if*

*a) both*

$$1) \quad \underline{X}(t; X_0; \mathtt{i}) \in \mathcal{X}_A \quad \textit{for every} \quad (t, \mathtt{i}) \in \Re_\tau \times \mathcal{I},$$

*and*

$$2) \quad \underline{X}(t; X_0; \mathtt{i}) \in \mathcal{X}_F \quad \textit{for every} \quad (t, \mathtt{i}) \in \Re_{\tau s} \times \mathcal{I}$$

*hold provided only that $X_0 \in \mathcal{D}_p(\tau, \tau_s, \mathcal{X}_A, \mathcal{X}_F, \mathcal{I}; \mathcal{A})$,*

*and*

*b) $\mathcal{D}_p(\tau, \tau_s, \mathcal{X}_A, \mathcal{X}_F, \mathcal{I}; \mathcal{A})$ is a neighbourhood of the set $\mathcal{A}$.*

When $\tau$, $\tau_s$, $\mathcal{X}_A$, $\mathcal{X}_F$ and $\mathcal{I}$ are known and fixed then we may write $\mathcal{D}_p(\mathcal{A})$ in the sense of $\mathcal{D}_p(\tau, \tau_s, \mathcal{X}_A, \mathcal{X}_F, \mathcal{I}; \mathcal{A})$.

# Chapter 3

# Qualitative features of stability domains properties

## 3.1 Introductory comments

### 3.1.1 Definition of a motion

If a (physical, technical) system is described by a first order vector differential equation in (3.1a)

$$\frac{dq}{dt} = r(q, \mathtt{W}), \quad r : \Re^n \times \Re^l \to \Re^n, \quad q \in \Re^n \tag{3.1a}$$

and by an algebraic vector equation in (3.1b)

$$y = g(q, \mathtt{W}), \quad g : \Re^n \times \Re^l \to \Re^m, \quad y \in \Re^m \tag{3.1b}$$

then the former describes its internal dynamics that via (3.1b) determines its output behaviour under the influence of an input (vector function) $\mathtt{W}$.

Let $y_d : \Re \to \Re^m$ denote a specific output response of the system (3.1), which is of a particular interest, which is aimed to and, therefore, called a *desired output (vector function)* of the system.

**Definition 3.1** *The system (3.1) is in the nominal (desired) regime with respect to $y_d$ if and only if $y(t) \equiv y_d(t)$. A pair $(q^*, \mathtt{W}^*)$ is nominal (desired) with respect to $y_d$ if and only if the system (3.1) is in the nominal regime with respect to the same $y_d$.*

**Theorem 3.1** In order for a pair $(q^*, \mathtt{W}^*)$ to be nominal denoted by $(q_N, \mathtt{W}_N)$ with respect to a desired output $y_d$ of the system (3.1a) it is both necessary and sufficient that it satisfies the following identities:

$$\frac{dq^*(t)}{dt} \equiv r[q^*(t), \mathtt{W}^*(t)], \tag{3.2a}$$

$$g[q^*(t), \mathtt{W}^*(t)] \equiv y_d(t). \tag{3.2b}$$

$\nabla\nabla\nabla$

**Proof.** The statement of Theorem 3.1 follows directly from (3.1) and Definition 3.1.

A nominal motion $q_N$ with respect to a desired output $y_d$ of the system (3.1) is an unperturbed motion in Lyapunov's terminology (Section 1.1.1).

**Assumption 3.1** A nominal pair $(q_N, \mathtt{W}_N)$ with respect to $y_d$ is known and it is time invariant.

From now on it is accepted that the Assumption 3.1 holds. This means that $(q_N, \mathtt{W}_N)$ is elementwise constant solution to (3.2).

Let the following change of variables (i.e. translation of coordinate systems) be defined:

$$X = q - q_N, \quad \mathtt{i} = \mathtt{W} - \mathtt{W}_N, \quad f(X, \mathtt{i}) = r(q_N + X, \mathtt{W}_N + \mathtt{i}) - r(q_N, \mathtt{W}_N). \tag{3.3}$$

It is easy to show that after subtracting (3.2a) from (3.1a) and using (3.3) we derive (3.4),

$$\frac{dX}{dt} = f(X, \mathtt{i}), \quad f : \mathcal{X} \times \mathcal{I} \to \Re^n. \tag{3.4}$$

Notice that the function $f$ obeys (3.5),

$$f(0, 0) = 0, \tag{3.5}$$

as soon as it is defined (as above) by (3.3).

In case $\mathtt{i}(t) \equiv 0$ then we simplify the notation as follows:

$$f(X, 0) \equiv f(X), \tag{3.6}$$

so that (3.4) reduces to (3.7) (see Section 1.1.1),

$$\frac{dX}{dt} = f(X), \quad f : \mathcal{X} \to \Re^n. \tag{3.7}$$

In the sequel we shall study either properties of systems described by (3.4) or by (3.7). Once we determine a solution $\underline{X}$ to (3.4) or to (3.7), for a given $\mathtt{i}$, i.e. for $\mathtt{i} = 0$, respectively, then we easily determine $q = q_N + X$, $\mathtt{W} = \mathtt{W}_N + \mathtt{i}$ and $y = g(q, \mathtt{W})$. This justifies the study of (3.4), or (3.7), since Assumption 3.1 holds. For its validity we have to solve the equations (3.2) for a given $y_d$.

**Definition 3.2** *A function $\underline{X} : \mathcal{I}_0 \times \mathcal{X}_0 \times \mathcal{I} \to \mathcal{X}$ is a solution (a motion) of the system (3.4) through $X_0$ at $t = 0$ if and only if $X_0 \in \mathcal{X}_0$ and $\underline{X}$ satisfy (i)–(iv):*

*(i)* $\underline{X}$ *is defined in* $\mathcal{X}$, $\underline{X}(t; X_0; \mathtt{i}) \in \mathcal{X}$, *if and only if* $t \in \mathcal{I}_0$ *for every* $\mathtt{i} \in \mathcal{I}$,

*(ii)* $\underline{X}$ *is continuous and differentiable in* $t \in \mathcal{I}_0$ *for all* $\mathtt{i} \in \mathcal{I}$,

*(iii)* $\underline{X}$ *identically satisfies the equation (3.4) on* $\mathcal{I}_0 \times \mathcal{I}$, *that is*

$$\frac{d}{dt}\underline{X}(t; X_0; \mathtt{i}) \equiv f[\underline{X}(t; X_0; \mathtt{i}), \mathtt{i}(t)],$$

*and*

*(iv)* $\underline{X}$ *fulfils the initial condition,*

$$\underline{X}(0; X_0; \mathtt{i}) \equiv X_0.$$

$\mathcal{I} = 0 = \{\mathtt{i} : \mathtt{i} = 0\}$ for the system (3.7). Hence, we shall use $\underline{X}(t; X_0)$ in the sense of $\underline{X}(t; X_0; 0) : \underline{X}(t; X_0) \equiv \underline{X}(t; X_0; 0)$.

In general $\mathcal{X}_0$ is a subset of $\Re^n$. In special cases $\mathcal{X}_0 = \Re^n$. The time interval $\mathcal{I}_0$, $\mathcal{I}_0 \subseteq \Re$, depends on $X_0 \in \mathcal{X}_0$:

$$\mathcal{I}_0 = \mathcal{I}_0(X_0) = (l^-, l^+) \subseteq \Re.$$

It is the maximal time interval over which $\underline{X}$ is defined (with respect to $\mathcal{X}$). Since $\mathcal{X}$ is the domain of definition of $f$ (3.7), or of $f$ (3.4) with respect to $X$, then the condition *(iii)* of Definition 3.2 can be satisfied only when $\underline{X}(t; X_0; \mathtt{i}) \in \mathcal{X}$.

With $\underline{X}(\mathcal{I}_0)$ we denote the set of all vectors $\underline{X}(t; X_0)$ over $t \in \mathcal{I}_0$: $\underline{X}(\mathcal{I}_0) = \{X : \exists t \in \mathcal{I}_0,\ \underline{X}(t; X_0) = X\}$, and $\underline{X}(\mathcal{T}; X_0)$ designates the set of all vectors $\underline{X}(t; X_0)$ over $t \in \mathcal{T}$ for $\mathcal{T} \subseteq \mathcal{I}_0$, $\mathcal{T} \neq \emptyset : \underline{X}(\mathcal{T}; X_0) = \{X : \exists t \in \mathcal{T}, \underline{X}(t; X_0) = X\}$. The set $\underline{X}(\mathcal{I}_0)$ is called a trajectory (of the system through $X_0$), and $\underline{X}(\mathcal{T}; X_0)$ is its arc (over $\mathcal{T}$).

### 3.1.2 Existence of motions

There are various theorems on existence of motions of the system (3.7). Their common feature is that they provide sufficient conditions (rather than necessary and sufficient conditions) for existence of motions. Here will be presented the classical results by referring to [157].

**Theorem 3.2** *In order for the system (3.7) to have a motion through* $X_0 \in \mathcal{X}$ *at* $t = 0$ *it is sufficient that there exists a compact (closed and bounded) neighbourhood* $\overline{\mathcal{N}}(X_0)$ *of* $X_0$, $\overline{\mathcal{N}}(X_0) \subset \mathcal{X}$, *such that the function* $f$ *is continuous on* $\overline{\mathcal{N}}(X_0)$ : $f(X) \in C[\overline{\mathcal{N}}(X_0)]$. *Then the motion is defined in the time interval* $]-\alpha, \alpha[$, *where*

$$\alpha = \frac{\rho[X_0, \partial\overline{\mathcal{N}}(X_0)]}{max\ [||f(X)|| : X \in \overline{\mathcal{N}}(X_0)]}$$

∇∇∇

Nemytskii and Stepanov [157, Theorem 1.21] established the following result by using the preceding theorem.

**Theorem 3.3** *If, as time increases, a given motion $\underline{X}$ remains in a compact nonempty subset $\mathcal{X}_c$ of the interior $\mathcal{N}(X_0)$ of the neighbourhood $\overline{\mathcal{N}}(X_0)$ on which the conditions of Theorem 3.2 are fulfilled, then the motion $\underline{X}$ may be continued for (is defined on) the whole interval $\overline{\Re}_+ = [0, +\infty]$.* ∇∇∇

This theorem is very important for discovering a link between stability of $X = 0$ and the existence of solutions, which is stated as follows:

**Theorem 3.4** *(a) If $X = 0$ is stable then $\underline{X}(t; X_0)$ exists on $\overline{\Re}_+$ for every $X_0 \in \mathcal{D}_s$.*

*(b) If $X = 0$ is attractive and $\underline{X}(t; X_0)$ is continuous in $t \in [0, \tau(X_0, ||X_0||)]$ for every $X_0 \in \mathcal{D}_a$ then $\underline{X}(t; X_0)$ exists on $\overline{\Re}_+$ for every $X_0 \in \mathcal{D}_a$.*

*(c) If $X = 0$ is asymptotically stable then $\underline{X}(t; X_0)$ exists on $\overline{\Re}_+$ for every $X_0 \in \mathcal{D}$.*

*(d) If $X = 0$ is exponentially stable then $\underline{X}(t; X_0)$ exists on $\overline{\Re}_+$ for every $X_0 \in \mathcal{D}_e(\beta, \gamma)$.* ∇∇∇

**Proof.** Theorem 3.3 and

(a) the definition of $\mathcal{D}_s$ (Definition 2.1, Section 2.2.2) imply directly the statement under (a).

(b) the definition of $\mathcal{D}_a$ (Definition 2.3, Section 2.2.3) yield directly the statement under (b), by noting that $\xi = ||X_0||$ in $\tau(X_0, \xi)$ in the same definition.

(c) the statement (c) follows from (a) and (b) due to the definition of $\mathcal{D}$ (Definition 2.5, Section 2.2.4).

(d) The statement under (d) is a direct corollary to Theorem 3.3 in view of the definition of $\mathcal{D}_e(\beta, \gamma)$ (Definition 2.7, Section 2.2.5).

Several other criteria for the existence of motion follow. The next one is due to Zubov [218, p. 14].

**Theorem 3.5** *If the function $f$ is defined and continuous on $\Re^n$: $f(X) \in C(\Re^n)$, with bounded norm on $\Re^n$,*

$$sup\,[||f(X]|| : X \in \Re^n] < +\infty \tag{3.8}$$

*then motions of the system (3.7) are defined on $\Re$ for every $X_0 \in \Re^n$.* ∇∇∇

In case the function $f$ is continuous on $\Re^n$ but does not obey the condition (3.8), then Zubov proposed [218, p. 15] a time (scale) transformation by introducing a new time variable $\theta$:

$$d\theta = dt[1 + ||f(X)||] \tag{3.9}$$

so that the system (3.7) takes a new form (3.10):

$$\frac{dX}{d\theta} = \frac{f(X)}{1 + ||f(X)||}. \tag{3.10}$$

Since the function $\frac{f}{1+||f||}$ obeys all the conditions of Theorem 3.5 for $f(X) \in C(\Re^n)$ then the motions of the system (3.10) are defined on $\Re$ for every $X_0 \in \Re^n$. Zubov stated [218, p. 15], that "geometrically the integral curves of the systems" (3.7) "and" (3.10) "will coincide", where "integral curve" is synonym to "motion" and to "solution".

### 3.1.3 Existence and uniqueness of motions

**Definition 3.3** *(a) A motion $\underline{X}$ of a system is backward-time unique through $X_0$ if and only if it and any other motion $\underline{X}^*$ of the same system through $X_0$ obey (3.11)*

$$\underline{X}^*(\sigma - t; \sigma, X_0) = \underline{X}(\sigma - t; \sigma, X_0) \tag{3.11}$$

*for all $\sigma$ in an interval over which both $\underline{X}^*(\sigma; X_0)$ and $\underline{X}(\sigma; X_0)$ are defined and for all $t \in \Re_+$ for which both $\underline{X}^*(\sigma - t; \sigma, X_0)$ and $\underline{X}(\sigma - t; \sigma, X_0)$ are defined.*

*(b) A motion $\underline{X}$ of a system is forward-time unique through $X_0$ if and only if it and any other motion $\underline{X}^*$ of the same system through $X_0$ obey (3.12)*

$$\underline{X}^*(\sigma + t; \sigma, X_0) = \underline{X}(\sigma + t; \sigma, X_0) \tag{3.12}$$

*for all $\sigma$ in an interval over which both $\underline{X}^*(\sigma; X_0)$ and $\underline{X}(\sigma; X_0)$ are defined and for all $t \in \Re_+$ for which both $\underline{X}^*(\sigma + t; \sigma, X_0)$ and $\underline{X}(\sigma + t; \sigma, X_0)$ are defined.*

*(c) A motion $\underline{X}$ of a system is unique through $X_0$ if and only if it is both backward-time unique through $X_0$ and forward-time unique through $X_0$.*

In order to illustrate the preceding definition several examples follow.

**Example 3.1** The first order system

$$\frac{dx}{dt} = x^{2/3}$$

has solutions $\underline{x}_\psi$ through $x_0 = 0$ at $t = 0$ determined by

$$\underline{x}_\psi(t; 0) = \begin{cases} \left[\frac{1}{3}(t+\xi)\right]^3 & , \ t \in ]-\infty, -\xi] \quad , \ \forall \xi \in \Re_+, \\ 0 & , \ t \in [-\xi, \zeta] \\ \left[\frac{1}{3}(t-\zeta)\right]^3 & , \ t \in [\zeta, +\infty[ \quad , \ \forall \zeta \in \Re_+, \quad \psi \in \{\xi, \zeta\} \end{cases}$$

and the trivial solution

$$\underline{x}(t; 0) = 0, \ \ \forall t \in \Re.$$

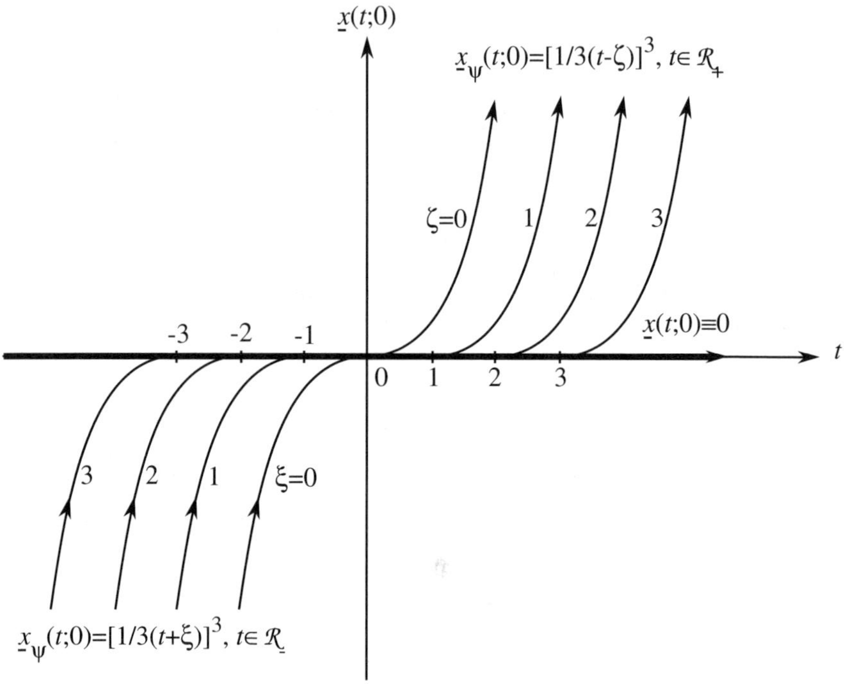


Figure 3.1: Solutions of $\dot{x} = x^{2/3}$.

The system has infinitely many solutions through $x_0 = 0$ at $t = 0$ in both the forward-time direction and in the backward-time direction. Altogether, it does not have a unique solution through $x_0 = 0$ at $t = 0$, Fig. 3.1.

**Example 3.2** The first order system

$$\frac{dx}{dt} = sgn\, x, \quad sgn\, x \begin{cases} = -1, & x < 0 \\ \in [-1, 1], & x = 0, \quad sgn\, 0 \text{ is fixed}, \\ = 1, & x > 0, \end{cases}$$

has solutions determined by

$$\underline{x}(t; x_0) \begin{cases} = x_0 + t\, sgn\, x_0\,, \quad t \in \,]-|x_0|, +\infty[\,, \quad (x_0 \neq 0) \in \Re, \\ \left.\begin{array}{l} = \quad 0, \quad \forall t \in \Re \quad , \text{ if } sgn\, 0 = 0 \\ \quad \text{does not exist} \quad , \text{ if } sgn\, 0 \neq 0 \end{array}\right\}, \qquad x_0 = 0. \end{cases}$$

Through every $(x_0 \neq 0) \in \Re$ at $t = 0$ there is the unique solution $\underline{x}(t; x_0) = x_0 + t\, sgn\, x_0$ defined on $]-|x_0|, +\infty[$. Through $x_0 = 0$ at $t = 0$ the system has the unique motion $\underline{x}(t; x_0) = 0$ defined on $\Re$ if $sgn\, 0 = 0$. However, if $sgn\, 0 \neq 0$ then the system does not have a solution through $x_0 = 0$ at $t = 0$, Fig. 3.2.

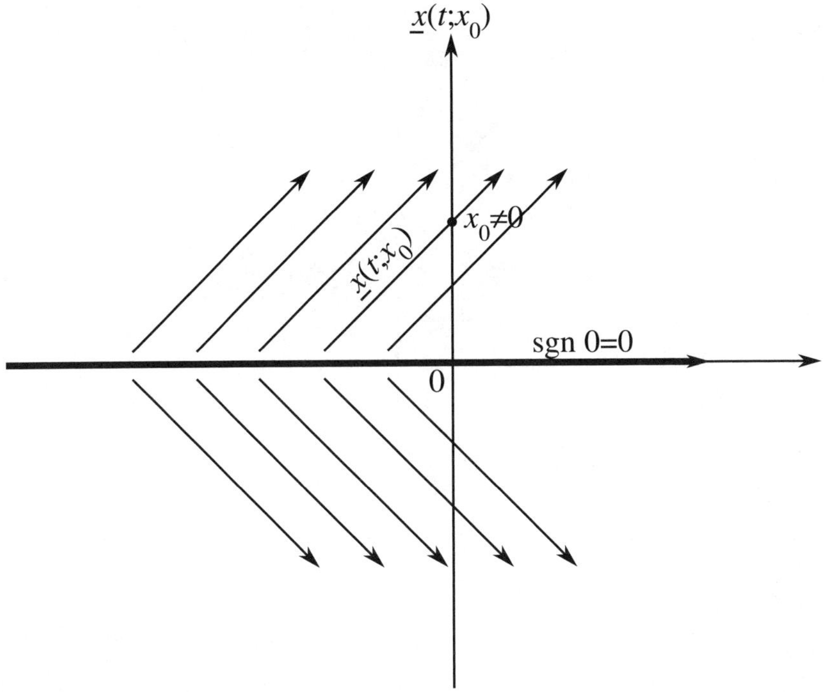


Figure 3.2: Solutions of $\dot{x} = sgn\, x$ (Example 3.2). If $sgn\, 0 \neq 0$ then the system does not have a solution through $x_0 = 0$ at $t = 0$.

**Example 3.3** Let

$$sign\, x = \begin{cases} sgn\, x & , \ x \neq 0 \\ 0 & , \ x = 0 \end{cases}$$

define the first order system

$$\frac{dx}{dt} = -sign\, x.$$

The motions are found in the form

$$\underline{x}(t; x_0) = \begin{cases} x_0 - t\ sign\, x_0 & , \ t \in ]-\infty, |x_0|[ & , \ (x_0 \neq 0) \in \Re \\ 0 & , \ t \in \Re & , \ x_0 = 0. \end{cases}$$

Through every $x_0 \in \Re$ at $t = 0$ the system has the unique motion. The motion $\underline{x}(t; x_0)$ is defined on $]-\infty, |x_0|[$ if $x_0 \neq 0$. The motion $\underline{x}(t; 0) \equiv 0$ is defined on $\Re$, Fig. 3.3. It is unique despite $f$, $f(x) = -sign\, x$, is not Lipschitzian on any neighbourhood of $x = 0$.

**Example 3.4** The first order system

$$\frac{dx}{dt} = -x^{1/2}$$

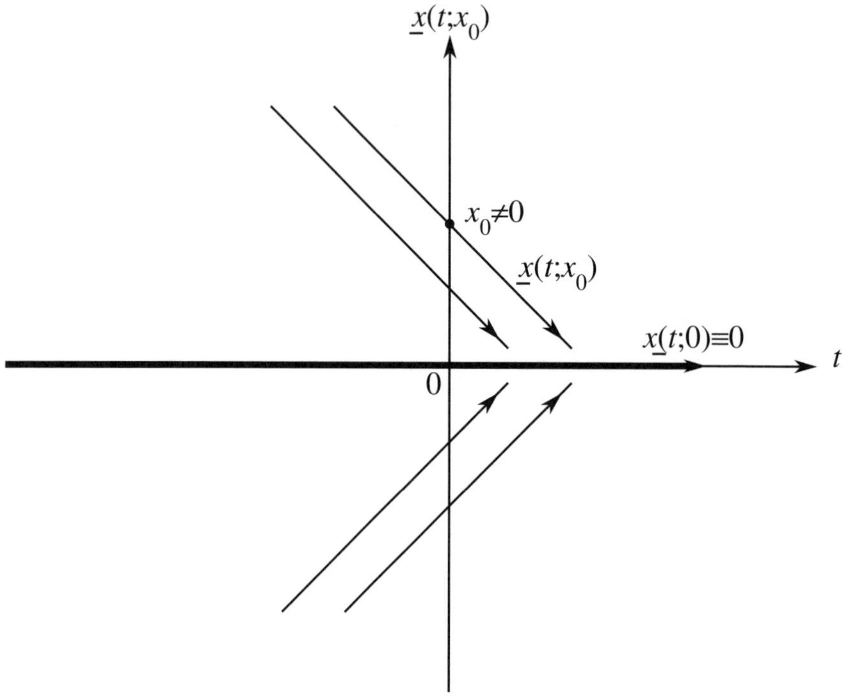


Figure 3.3: Motions of the system $\dot{x} = -sign\, x$.

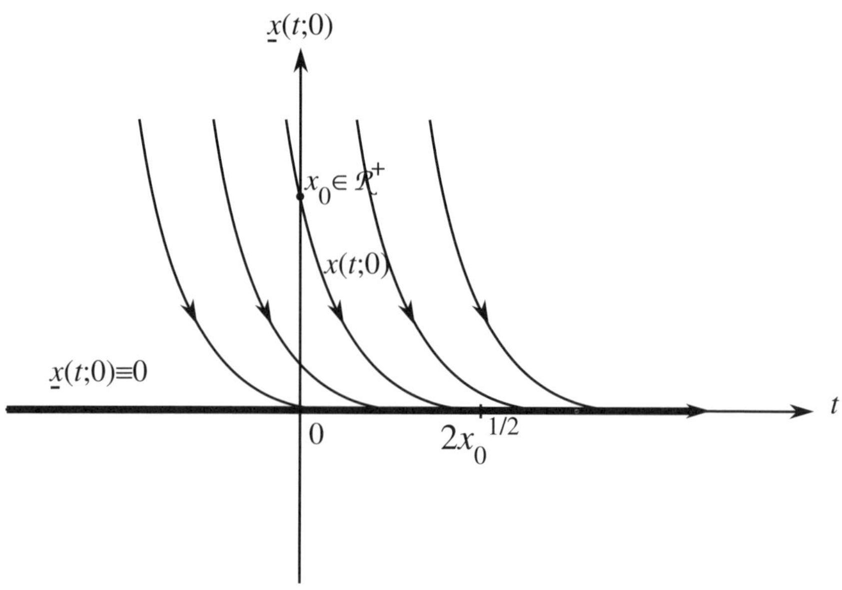


Figure 3.4: Motions of the system $\dot{x} = -x^{1/2}$.

has forward-time unique motions through every $x_0 \in \Re_+$, which are defined by

$$\underline{x}(t;x_0)\begin{cases} = \left[\sqrt{|x_0|}-\frac{1}{2}t\right]^2 & , t \in \left]-\infty, 2\sqrt{|x_0|}\right] \\ = \quad 0 & , t \in \left[2\sqrt{|x_0|}, +\infty\right[ \end{cases} \Bigg\} , x_0 \in \Re^+,$$
$$\quad = 0 \quad , t \in \Re \quad , x_0 = 0,$$
$$\quad \text{does not exist} \quad , x_0 \in \Re^-.$$

Motions are unique through every $x_0 \in \Re^+$, but not through $x_0 = 0$ because there is not a backward-time unique motion through $x_0 = 0$. They obey $\underline{x}[t;\sigma,\underline{x}(\sigma;0,0)] = \underline{x}(t;0,0) = 0$ for $\sigma \in \Re_+$ and $t \in [\sigma,+\infty[$, but not vice versa, Fig. 3.4.

**Example 3.5** The first order system

$$\frac{dx}{dt} = -x^3$$

has unique motions through every $x_0 \in \Re$ at $t = 0$, which are defined by

$$\underline{x}(t;x_0) = \begin{cases} 0 & , t \in \Re & , x_0 = 0, \\ \dfrac{x_0}{(1+2t|x_0|^2)^{1/2}} & , t \in \left]-\dfrac{1}{2|x_0|^2}, +\infty\right[ & , (x_0 \neq 0) \in \Re. \end{cases}$$

They exist on the intervals $\mathcal{I}_0 = \left]-\dfrac{1}{2x_0^2}, +\infty\right[$, Fig. 3.5. Notice that $\mathcal{I}_0(x_0) = \left]-\dfrac{1}{2|x_0|^2}, +\infty\right[ \to \Re$ as $|x_0| \to 0$, but for any $x_0 \in (\Re^+ \cup \Re^-)$ the motion is not defined on $\left]-\infty, -\dfrac{1}{2|x_0|^2}\right]$.

**Definition 3.4** *A function $f$ is a Lipschitz function on a compact set $\overline{\mathcal{S}}$ with a nonempty interior $\mathring{\mathcal{S}} = \mathcal{S}$ if and only if there is $L \in ]0,+\infty[$ that satisfies*

$$||f(X^1) - f(X^2)|| \leq L||X^1 - X^2||, \qquad X^1, X^2 \in \overline{\mathcal{S}}. \tag{3.13}$$

The number $L$ is called *a Lipschitz constant of the function $f$ on $\overline{\mathcal{S}}$*, the condition (3.13) is *the Lipschitz condition* and the family of all the functions $f$ obeying the Lipschitz condition on the set $\overline{\mathcal{S}}$ is denoted by $Lip\,(\overline{\mathcal{S}})$, $f \in Lip\,(\overline{\mathcal{S}})$.

If we wish to emphasize the mutual dependence between $L$ and $\overline{\mathcal{S}}$ in Definition 3.4 then we can write $\overline{\mathcal{S}}_L$ for $\overline{\mathcal{S}}$, $\overline{\mathcal{S}}_L = \overline{\mathcal{S}}$.

The next criterion for the existence and uniqueness of motions is classical.

**Theorem 3.6** *In order for the system (3.7) to have a unique motion $\underline{X}$ through $X_0$ at $t = 0$ it is sufficient that there exists a compact neighbourhood $\overline{\mathcal{N}}(X_0)$ of $X_0$ on which the function $f$ is Lipschitzian, $f \in Lip\,[\overline{\mathcal{N}}(X_0)]$.* $\nabla\nabla\nabla$

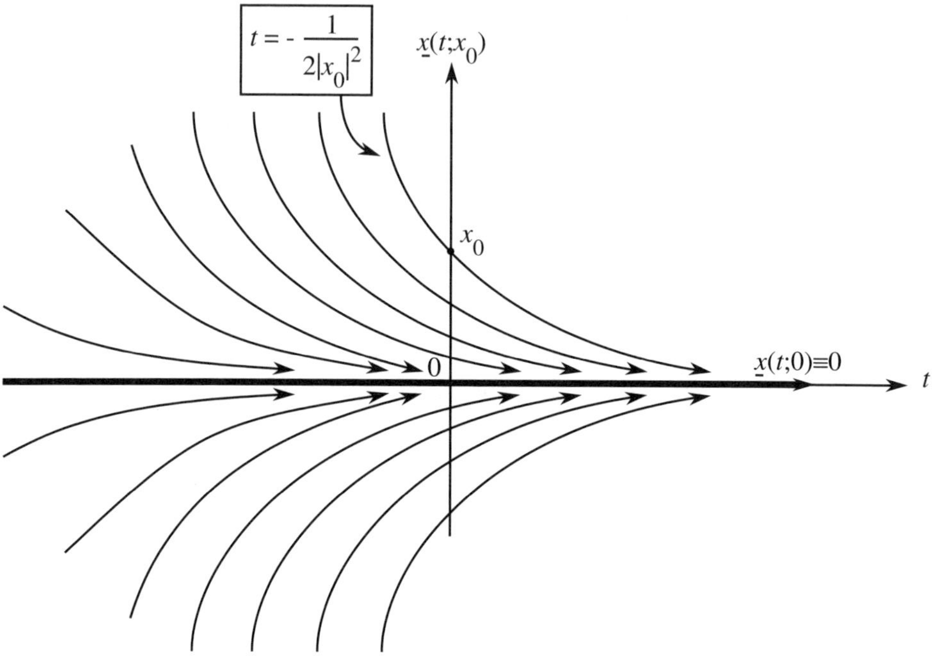


Figure 3.5: Motions of the system $\dot{x} = -x^3$.

**Comment 3.1** *If a function $f \in Lip(\overline{\mathcal{S}})$ then it is defined and continuous on $\overline{\mathcal{S}}$, but vice versa is not true. For example the function $f$, $f(x) = |x|^{1/2}$, is defined and continuous on $\Re$. However, there is not $\overline{\mathcal{N}}(\mathcal{O})$ such that $f(x) \in Lip[\mathcal{N}(\mathcal{O})]$, i.e. $f$ is not Lipschitzian at $x = 0$.*

**Comment 3.2** *If a function $f$ is defined, continuous and continuously differentiable in $X \in \overline{\mathcal{S}}$, where $\mathcal{S} = \mathring{\mathcal{S}}$ is nonempty, then $f(X) \in Lip(\overline{\mathcal{S}})$.*

Theorem 3.6 and comment 3.2 explain the following:

**Theorem 3.7** *If a function $f$ is continuously differentiable on a compact set $\overline{\mathcal{S}}$ with the nonempty interior $\mathring{\mathcal{S}} = \mathcal{S}$ then the system (3.7) has a unique motion $\underline{X}$ through every $X_0 \in \mathcal{S}$ at $t = 0$.* $\nabla\nabla\nabla$

**Comment 3.3** *The function $f$, $f(x) = -x^3$ is continuously differentiable on $\Re$, $f(x) \in C^{(1)}(\Re)$, Example 3.5. The system $\dot{x} = -x^3$ has a unique motion through every $x_0 \in \Re$ at $t = 0$, Fig. 3.5. However, for any $(x_0 \neq 0) \in \Re$ the motion $\underline{x}$ is not defined (does not exist) on $\left]-\infty, -\frac{1}{2|x_0|^2}\right]$.*

**Comment 3.4** *The function $f$, $f(x) = x^{1/2}$, is defined and continuous on $\Re_+$. It is Lipschitzian on every compact subset $\overline{\mathcal{S}}$ of $\Re^+$ with nonempty interior $\mathcal{S}$. Besides, it is continuously differentiable on $\Re^+$. However, it is neither differentiable nor*

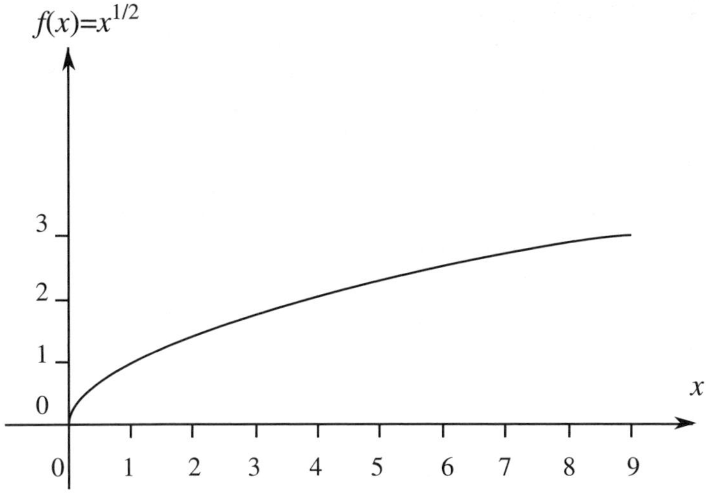


Figure 3.6: The graph of the function $f$, $f(x) = x^{1/2}$.

*Lipschitzian at $x = 0$, Fig. 3.6. Nevertheless, the system has the unique solution $\underline{x}(t;0) \equiv 0$ through $x_0 = 0$ at $t = 0$, Fig. 3.4. This illustrates that the Lipschitz condition is sufficient but not necessary for the existence and uniqueness of motions.*

### 3.1.4 Continuity of motions in initial conditions

The Lipschitzian property of the function $f$ is important also for continuity of motions $\underline{X}$ in the initial condition $X_0$, $X_0 = X(0)$. This is shown by what follows [157, Theorem 2.22], the proof of which can be found in [157, p. 14].

**Theorem 3.8** *Let $f(X) \in Lip[\overline{\mathcal{N}}(X_0)]$, where $\overline{\mathcal{N}}(X_0)$ is a compact neighbourhood of $X_0 \in \mathcal{X}$, $\overline{\mathcal{N}}(X_0) \subset \mathcal{X}$. If a motion $\underline{X}$ through $X_0$ at $t = 0$ is defined and in $\overline{\mathcal{N}}(X_0)$ for all $t \in [0,T]$, then for every $\varepsilon \in \Re^+$ there is $\delta \in \Re^+$ such that $||X^* - X_0|| < \delta$ implies that a motion through $X^*$ at $t = 0$ is also defined for $t \in [0,T]$ and $||\underline{X}(t;X^*) - \underline{X}(t;X_0)|| < \varepsilon$ for all $t \in [0,T]$.*

$\nabla\nabla\nabla$

**Comment 3.5** *The conditions of Theorem 3.8 are only sufficient for continuity of motions with respect to the initial conditions.*

### 3.1.5 Differentiability of motions in initial conditions

From Theorem 7.1 by Rouche and Mawhin [178, p. 111], we deduce the following:

**Theorem 3.9** *If $\mathcal{X}$ is open and $f(X) \in C^{(1)}(\mathcal{X})$ then the solutions of the system (3.4) $\underline{X}(t;X_0) \in C^{(2,1)}(\Re_+ \times \mathcal{X})$.*

$\nabla\nabla\nabla$

## 3.2 Generalised motions

### 3.2.1 Motivation

Let a simple first order unstable linear plant described by $y^{(1)} - y = u$ be controlled by locally stabilising nonlinear control $u$, $u = k\ sign\, e$, $k \in \Re^+$, $e = y_d - y$, where $y_d$ is a desired output accepted equal to zero, $y_d(t) \equiv 0$. Let $x = y$, $X = (x)$. Then the overall closed-loop feedback control system is described by

$$\frac{dx}{dt} = x - k\ sign\, x.$$

Its motions are found as

$$\underline{x}(t;x_0) = k\ sign\, x_0 + (x_0 - k\ sign\, x_0)\, exp\,(t), \begin{cases} t \in ]-\infty, \tau_0[, & \text{if } |x_0| \in ]0,k[, \\ t \in \Re, & \text{if } x_0 = 0 \text{ or } |x_0| \in [k, +\infty[, \end{cases}$$

where $\tau_0 = ln\,\dfrac{k\ sign\, x_0}{k\ sign\, x_0 - x_0}$. If $|x_0| \in ]0,k[$ then motions do not exist on $[\tau_0, +\infty[$ because $\underline{x}(t;x_0)$ is not differentiable at $t = \tau_0$ (Definition 3.2). However, the real system operates well for $t \in [\tau_0, +\infty[$ (see Example 3.11). This motivates us to generalise the notion of motion.

### 3.2.2 Dini derivatives

Let $t^* \in \overline{\Re}$, $\mathcal{T}_{t^*}^-$ be a set of a sequence $t_k$ converging to $t^*$ on the left as $k \to +\infty$ if $t^* > -\infty$, ($t^* = +\infty$ is permitted),

$$\mathcal{T}_{t^*}^- = \{t_k : t_k \in \Re, t_k < t^*,\ k = 0,1,2,\ldots, t_k \to t^* \text{ as } k \to +\infty\},\ t^* \in ]-\infty, +\infty],$$

and, analogously,

$$\mathcal{T}_{t^*}^+ = \{t_k : t_k \in \Re, t_k > t^*,\ k = 0,1,2,\ldots, t_k \to t^* \text{ as } k \to +\infty\},\ t^* \in [-\infty, +\infty[.$$

**Definition 3.5** *(i) A number $\alpha \in \Re$ is the partial limit of a function $x$, $x : \Re \to \Re$, over a sequence set $\mathcal{T}_{t^*}^-$ $[\mathcal{T}_{t^*}^+]$, if and only if for every $\varepsilon \in \Re^+$ there is an integer $N$ such that $k > N$ implies $|x(t_k) - \alpha| < \varepsilon$,*

*(ii) the symbol $\alpha = -\infty$ $[\alpha = +\infty]$ is the partial limit of a function $x : \Re \to \Re$ over a sequence set $\mathcal{T}_{t^*}^-$ $[\mathcal{T}_{t^*}^+]$, if and only if for every $\varepsilon \in ]0, +\infty[$ there is an integer $N$ such that, respectively, $k > N$ implies $x(t_k) < -\varepsilon^{-1}$ $[x(t_k) > \varepsilon^{-1}]$,*

*(iii) the greatest [the smallest] partial limit of a function $x : \Re \to \Re$ over all the sequence sets $\mathcal{T}_{t^*}^-$ is its left upper [left lower] limit at $t = t^*$, which is denoted by*

$$lim\, sup\,[x(t) : t \to t^{*-}] = \overline{lim}\,[x(t) : t \to t^{*-}],$$

$$[lim\, inf\,[x(t) : t \to t^{*-}] = \underline{lim}\,[x(t) : t \to t^{*-}]\,],$$

*respectively,*

(iv) *the greatest [the smallest] partial limit of a function* $x : \Re \to \Re$ *over all the sequence sets* $\mathcal{T}_{t^*}^{+}$ *is its right upper [right lower] limit at* $t = t^*$*, which is denoted by*

$$lim\,sup\,[x(t) : t \to t^{*+}] = \overline{lim}\,[x(t) : t \to t^{*+}],$$

$$[lim\,inf\,[x(t) : t \to t^{*+}] = \underline{lim}\,[x(t) : t \to t^{*+}]\,],$$

*respectively.*

If $x(t) = 0$ for $t = 0$ and $x(t) = sign[sin(t^{-2})]$ for $t \neq 0$ then $\overline{lim}\,[x(t) : t \to 0^-] = \overline{lim}\,[x(t) : t \to 0^+] = 1$ and $\underline{lim}\,[x(t) : t \to 0^-] = \underline{lim}\,[x(t) : t \to 0^+] = -1$.

Now, we can define Dini derivatives [143], [217]:

**Definition 3.6** *Let* $\mathcal{V}$ *be a continuous scalar function of* $X$ *on* $\mathcal{X}$*,* $\mathcal{V}(X) \in C(\mathcal{X})$ *and let the system (3.7) have a solution* $\underline{X}$ *through* $X \in \mathcal{X}$ *at* $t = 0$*. Then,*

(i) *the forward-time upper right Dini derivative of* $\mathcal{V}$ *along the solution* $\underline{X}$ *at* $X$ *is*

$$\mathcal{D}^+\mathcal{V}(X) = lim\,sup\left\{\frac{\mathcal{V}[\underline{X}(\theta; X)] - \mathcal{V}(X)}{\theta} : \theta \to 0^+\right\},$$

(ii) *the forward-time lower right Dini derivative of* $\mathcal{V}$ *along the solution* $\underline{X}$ *at* $X$ *is*

$$\mathcal{D}_+\mathcal{V}(X) = lim\,inf\left\{\frac{\mathcal{V}[\underline{X}(\theta; X)] - \mathcal{V}(X)}{\theta} : \theta \to 0^+\right\},$$

(iii) *the forward-time upper left Dini derivative of* $\mathcal{V}$ *along the solution* $\underline{X}$ *at* $X$ *is*

$$\mathcal{D}^-\mathcal{V}(X) = lim\,sup\left\{\frac{\mathcal{V}[\underline{X}(\theta; X)] - \mathcal{V}(X)}{\theta} : \theta \to 0^-\right\},$$

(iv) *the forward-time lower left Dini derivative of* $\mathcal{V}$ *along the solution* $\underline{X}$ *at* $X$ *is*

$$\mathcal{D}_-\mathcal{V}(X) = lim\,inf\left\{\frac{\mathcal{V}[\underline{X}(\theta; X)] - \mathcal{V}(X)}{\theta} : \theta \to 0^-\right\}.$$

(v) *The function* $\mathcal{V}$ *has the forward-time left Dini derivative along the motion* $\underline{X}$ *at* $X$*, which is denoted by* $\mathcal{D}_l\mathcal{V}(X)$*, if and only if* $\mathcal{D}^-\mathcal{V}(X) = \mathcal{D}_-\mathcal{V}(X)$ *and then* $\mathcal{D}_l\mathcal{V}(X) = \mathcal{D}^-\mathcal{V}(X) = \mathcal{D}_-\mathcal{V}(X)$.

(vi) *The function* $\mathcal{V}$ *has the forward-time right Dini derivative along the motion* $\underline{X}$ *at* $X$*, which is denoted by* $\mathcal{D}_r\mathcal{V}(X)$*, if and only if* $\mathcal{D}^+\mathcal{V}(X) = \mathcal{D}_+\mathcal{V}(X)$ *and then* $\mathcal{D}_r\mathcal{V}(X) = \mathcal{D}^+\mathcal{V}(X) = \mathcal{D}_+\mathcal{V}(X)$.

(vii) *The function* $\mathcal{V}$ *has the forward-time Eulerian derivative along the motion* $\underline{X}$ *at* $X$*, which is denoted by* $d_f[\mathcal{V}(X)]/dt$ *or by* $\dot{\mathcal{V}}_f(X)$*, if and only if it has both* $\mathcal{D}_l\mathcal{V}(X)$ *and* $\mathcal{D}_r\mathcal{V}(X)$ *and* $\mathcal{D}_l\mathcal{V}(X) = \mathcal{D}_r\mathcal{V}(X)$*, and then* $d_f[\mathcal{V}(X)]/dt = \dot{\mathcal{V}}_f(X) = \mathcal{D}_l\mathcal{V}(X) = \mathcal{D}_r\mathcal{V}(X)$.

$\mathcal{D}^*\mathcal{V}(X)$ means that both $\mathcal{D}^+\mathcal{V}(X)$ and $\mathcal{D}^-\mathcal{V}(X)$ may be used.

In order to clarify the difference between the forward-time derivatives and the backward-time derivatives we define the latter as follows:

**Definition 3.7** *Let $\mathcal{V}$ be a continuous scalar function of $X$ on $\mathcal{X}$, $\mathcal{V}(X) \in C(\mathcal{X})$ and let the system (3.7) have a solution $\underline{X}$ through $X \in \mathcal{X}$ at $t = 0$. Then,*

*(i) the backward-time upper right Dini derivative of $\mathcal{V}$ along the solution $\underline{X}$ at $X$ is*

$$^{+}\mathcal{D}\mathcal{V}(X) = lim\,sup\left\{\frac{\mathcal{V}[\underline{X}(-\theta;X)] - \mathcal{V}(X)}{\theta} : \theta \to 0^+\right\},$$

*(ii) the backward-time lower right Dini derivative of $\mathcal{V}$ along the solution $\underline{X}$ at $X$ is*

$$_{+}\mathcal{D}\mathcal{V}(X) = lim\,inf\left\{\frac{\mathcal{V}[\underline{X}(-\theta;X)] - \mathcal{V}(X)}{\theta} : \theta \to 0^+\right\},$$

*(iii) the backward-time upper left Dini derivative of $\mathcal{V}$ along the solution $\underline{X}$ at $X$ is*

$$^{-}\mathcal{D}\mathcal{V}(X) = lim\,sup\left\{\frac{\mathcal{V}[\underline{X}(-\theta;X)] - \mathcal{V}(X)}{\theta} : \theta \to 0^-\right\},$$

*(iv) the backward-time lower left Dini derivative of $\mathcal{V}$ along the solution $\underline{X}$ at $X$ is*

$$_{-}\mathcal{D}\mathcal{V}(X) = lim\,inf\left\{\frac{\mathcal{V}[\underline{X}(-\theta;X)] - \mathcal{V}(X)}{\theta} : \theta \to 0^-\right\}.$$

*(v) The function $\mathcal{V}$ has the backward-time left Dini derivative along the motion $\underline{X}$ at $X$, which is denoted by $_{l}\mathcal{D}\mathcal{V}(X)$, if and only if $^{-}\mathcal{D}\mathcal{V}(X) = {}_{-}\mathcal{D}\mathcal{V}(X)$ and then $_{l}\mathcal{D}\mathcal{V}(X) = {}^{-}\mathcal{D}\mathcal{V}(X) = {}_{-}\mathcal{D}\mathcal{V}(X)$.*

*(vi) The function $\mathcal{V}$ has the backward-time right Dini derivative along the motion $\underline{X}$ at $X$, which is denoted by $_{r}\mathcal{D}\mathcal{V}(X)$, if and only if $^{+}\mathcal{D}\mathcal{V}(X) = {}_{+}\mathcal{D}\mathcal{V}(X)$ and then $_{r}\mathcal{D}\mathcal{V}(X) = {}^{+}\mathcal{D}\mathcal{V}(X) = {}_{+}\mathcal{D}\mathcal{V}(X)$.*

*(vii) The function $\mathcal{V}$ has the backward-time Eulerian derivative along the motion $\underline{X}$ at $X$, which is denoted by $d_b[\mathcal{V}(X)]/dt$ or by $\dot{\mathcal{V}}_b(X)$, if and only if has both $_{l}\mathcal{D}\mathcal{V}(X)$ and $_{r}\mathcal{D}\mathcal{V}(X)$ and $_{l}\mathcal{D}\mathcal{V}(X) =_{r} \mathcal{D}\mathcal{V}(X)$, and then $d_f[\mathcal{V}(X)]/dt = \dot{\mathcal{V}}_b(X) =_{l} \mathcal{D}\mathcal{V}(X) = {}_{r}\mathcal{D}\mathcal{V}(X)$.*

Definition 3.6 and Definition 3.7 directly imply the next relationships among backward-time derivatives and forward-time derivatives:

$$^{-}\mathcal{D}\mathcal{V}(X) = -\mathcal{D}_{+}\mathcal{V}(X), \quad {}_{-}\mathcal{D}\mathcal{V}(X) = -\mathcal{D}^{+}\mathcal{V}(X),$$

$$^{+}\mathcal{D}\mathcal{V}(X) = -\mathcal{D}_{-}\mathcal{V}(X), \quad {}_{+}\mathcal{D}\mathcal{V}(X) = -\mathcal{D}^{-}\mathcal{V}(X),$$

$$_{l}\mathcal{D}\mathcal{V}(X) = -\mathcal{D}_{r}\mathcal{V}(X), \quad {}_{r}\mathcal{D}\mathcal{V}(X) = -\mathcal{D}_{l}\mathcal{V}(X),$$

and

$$\dot{\mathcal{V}}_b(X) = -\dot{\mathcal{V}}_f(X).$$

These equations and our interest in the forward-time system behaviour justify the usage of the forward-time derivatives only from now on. Hence, we shall use a simplified notation $d\mathcal{V}(X)/dt$, or $\dot{\mathcal{V}}(X)$, for the forward-time Eulerian derivative of $\mathcal{V}$ along $\underline{X}$ at $X$ instead of $d_f[\mathcal{V}(X)]/dt$, or $\dot{\mathcal{V}}_f(X)$, respectively. The preceding clarifications were necessary in order to define generalised motions.

### 3.2.3 Generalised motions

#### Definition

In the sequel "almost everywhere in $t \in \mathcal{I}_0$" means "for all $t \in \mathcal{I}_0$ except for $t$ belonging to a subset $\mathcal{M}_0$ of $\mathcal{I}_0$ with zero measure", which is denoted by "$\forall_a t \in \mathcal{I}_0$". Besides, "$t^-$" denotes "$t - 0 = lim\,[\alpha : \alpha \to t,\ \alpha < t]$", and "$t^+ = t + 0 = lim\,[\alpha : \alpha \to t,\ \alpha > t]$".

**Definition 3.8** *A function* $\underline{X} : \mathcal{I}_0 \times \mathcal{X}_0 \times \mathcal{I} \to \mathcal{X}$ *is a generalised solution (a generalised motion) of the system (3.4) through* $X_0$ *at* $t = 0$ *if and only if* $X_0 \in \mathcal{X}_0$ *and* $\underline{X}$ *satisfy (i)–(iv):*

*(i)* $\underline{X}$ *is defined in* $\mathcal{X}$, $\underline{X}(t; X_0; \mathtt{i}) \in \mathcal{X}$, *if and only if* $t \in \mathcal{I}_0$, $\mathcal{I}_0 = (l^-, l^+)$, *for every* $\mathtt{i} \in \mathcal{I}$,

*(ii) for every* $\mathtt{i} \in \mathcal{I}$ *and each* $X_0 \in \mathcal{X}_0$, $\underline{X}$ *is*

*(a) continuous in* $t \in \mathcal{I}_0$,

*(b) differentiable almost everywhere in* $t \in \mathcal{I}_0$.

*At* $t^* \in \mathcal{M}_0$, $\mathcal{M}_0$ *has the zero measure,* $\mathcal{M}_0 \subset \mathcal{I}_0$, *at which* $\underline{X}$ *is not differentiable in time* $t$ *the following holds*

*(c)* $\underline{X}$ *has both* $\mathcal{D}_l\underline{X}(t; X_0; \mathtt{i})$ *and* $\mathcal{D}_r\underline{X}(t; X_0; \mathtt{i})$ *if* $t^*$ *is an interior point of* $\mathcal{I}_0$, $t^* \in ]l^-, l^+[$,

*(d)* $\underline{X}$ *has* $\mathcal{D}_r\underline{X}(t; X_0; \mathtt{i})$ *if* $l^- \in \mathcal{I}_0$, $\mathcal{I}_0 = [l^-, l^+)$ *and* $t^* = l^-$,

*(e)* $\underline{X}$ *has* $\mathcal{D}_l\underline{X}(t; X_0; \mathtt{i})$ *if* $l^+ \in \mathcal{I}_0$, $\mathcal{I}_0 = (l^-, l^+]$ *and* $t^* = l^+$,

*(iii) for every* $\mathtt{i} \in \mathcal{I}$

*(a)* $\underline{X}$ *identically satisfies (3.4) almost everywhere on* $\mathcal{I}_0$,

$$\frac{d}{dt}\underline{X}(t; X_0; \mathtt{i}) = f[\underline{X}(t; X_0; \mathtt{i}), \mathtt{i}(t)],\ \ \forall_a t \in \mathcal{I}_0,$$

*(b) if* $t^* \in \mathcal{M}_0$ *is an internal point of* $\mathcal{I}_0$ *then*

$$\mathcal{D}_l \underline{X}(t; X_0; \mathtt{i}) = f[\underline{X}(t^-; X_0; \mathtt{i}), \mathtt{i}(t^-)],\ t = t^*,$$

$$\mathcal{D}_r \underline{X}(t; X_0; \mathtt{i}) = f[\underline{X}(t^+; X_0; \mathtt{i}), \mathtt{i}(t^+)],\ t = t^*,$$

*(c) if* $l^- \in \mathcal{I}_0$ *and* $t^* = l^-$ *then*

$$\mathcal{D}_r \underline{X}(t; X_0; \mathtt{i}) = f[\underline{X}(t^+; X_0; \mathtt{i}), \mathtt{i}(t^+)],\ t = t^* = l^-,$$

*(d) if* $l^+ \in \mathcal{I}_0$ *and* $t^* = l^+$ *then*

$$\mathcal{D}_l \underline{X}(t; X_0; \mathtt{i}) = f[\underline{X}(t^-; X_0; \mathtt{i}), \mathtt{i}(t^-)],\ t = t^* = l^+,$$

*and*

*(iv)* $\underline{X}$ *fulfils the initial condition,*

$$\underline{X}(0; X_0; \mathtt{i}) \equiv X_0.$$

The motivation for introducing the notion of generalised motions will be illustrated by reconsidering the systems $\dot{x} = sgn\, x$ (Example 3.2) and $\dot{x} = -sgn\, x$ (Example 3.3).

**Example 3.6** Generalised motions of the first-order system

$$\frac{dx}{dt} = sgn\, x$$

are determined by

$$\underline{x}(t; x_0) \begin{cases} \left.\begin{aligned} &= 0 &&, t \in ]-\infty, -|x_0|[, \\ &= x_0 + t\ sgn\, x_0 &&, t \in [-|x_0|, \infty[, \end{aligned}\right\} x_0 \in \Re,\ \text{if } sgn\, 0 = 0, \\[2ex] \text{does not exist},\ t \in \Re,\ x_0 = 0,\ (sgn\, 0 \neq 0) \in ]-1, +1[, \\[2ex] \left.\begin{aligned} &\text{does not exist} &&, t \in ]-\infty, 0[, \\ &= t\ sgn\, x_0 &&, t \in [0, \infty[, \end{aligned}\right\} x_0 = 0,\ \text{if } (sgn\, 0 \neq 0) \in \{-1, +1\}, \\[2ex] \left.\begin{aligned} &\left.\begin{aligned} &\text{does not exist} &&, t \in ]-\infty, -|x_0|[, \\ &= x_0 + t\ sgn\, x_0 &&, t \in [-|x_0|, \infty[, \end{aligned}\right\} \text{if } sgn\, x_0 = sgn\, 0, \\ &\left.\begin{aligned} &\text{does not exist} &&, t \in ]-\infty, -|x_0|], \\ &= x_0 + t\ sgn\, x_0 &&, t \in ]-|x_0|, \infty[, \end{aligned}\right\} \text{if } sgn\, x_0 \neq sgn\, 0, \end{aligned}\right\} \begin{gathered} (x_0 \neq 0) \in \Re \\ \text{and} \\ sgn\, 0 \neq 0. \end{gathered} \end{cases}$$

They are depicted in Fig. 3.7.

Now, we can compare motions with generalised motions as follows:

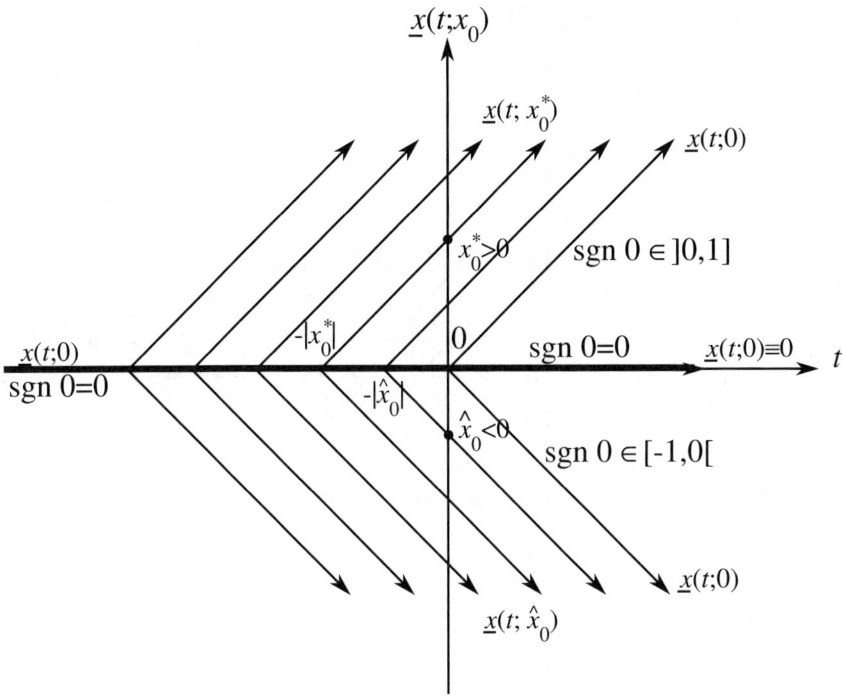


Figure 3.7: Generalised motions of the system $\dot{x} = sgn\, x$.

a) $(x_0 \neq 0) \in \Re$, $sgn\, 0 = \ 0$.
In this case the difference is in the time interval $\mathcal{I}_0$ over which motions and generalised motions are defined. $\mathcal{I}_0 = ]-|x_0| + \infty[$ is the interval of definition of motions, and $\mathcal{I}_0 = \Re$ is the time interval over which the system has defined generalised motions through every $(x_0 \neq 0) \in \Re$ since $sgn\, 0 = \ 0$. They are unique.

b) $x_0 = 0$, $sgn\, 0 = \ 0$.
System has unique both motion and generalised motion $\underline{x}(t; x_0) \equiv 0$.

c) $x_0 \in \ \Re$, $sgn\, 0 \neq \ 0$.
Since $\underline{\dot{x}}(t; x_0) \ = \ sgn\, \underline{x}(t; x_0)$, then $\underline{x}(t; x_0)$ should have the following form: $\underline{\dot{x}}(t; x_0) = x_0 + t \ sgn\, x_0$. They imply both $\underline{\dot{x}}(t; x_0) = sgn\, x_0$ and $\underline{\dot{x}}(t; x_0) = sgn\, \underline{x}(t; x_0) = sgn(x_0 + t \ sgn\, x_0)$. For them to be equal it is necessary and sufficient that: $sgn\, x_0 = sgn(x_0 + t \ sgn\, x_0)$ which requires: $t \in [0, \infty[$ if $(sgn\, 0 \neq 0) \in \{-1, +1\}$ and $x_0 = 0$; $t \in [-|x_0|, \infty[$ if $sgn\, x_0 = sgn\, 0$, $(x_0 \neq 0) \in \Re$ and $sgn\, 0 \neq 0$; $t \in ]-|x_0|, \infty[$ if $sgn\, x_0 \neq sgn\, 0$, $(x_0 \neq 0) \in \Re$ and $sgn\, 0 \neq 0$.

**Example 3.7** The first order system

$$\dot{x} = -sign\, x$$

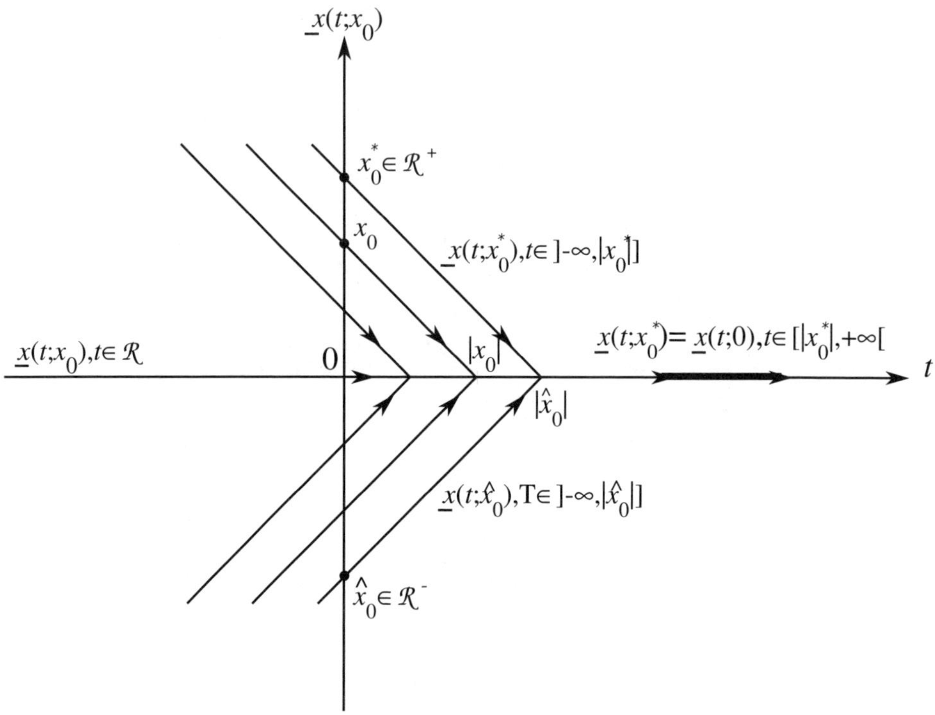


Figure 3.8: Generalised motions of the system $\dot{x} = -sign\, x$. Through $x_0 \neq 0$ there is not a backward-time unique generalised motion.

has generalised motions determined by

$$\underline{x}(t; x_0) = \begin{cases} x_0 - t\ sign\, x_0 & , t \in ]-\infty, |x_0|] \\ 0 & , t \in [|x_0|, +\infty[, \end{cases}$$

which are shown in Fig. 3.8. They are forward-time unique and defined on $\Re$ for every $x_0 \in \Re$. Besides, they coincide with the motion $\underline{x}(t; 0) \equiv 0$ on $[|x_0|, +\infty[$ for every $x_0 \in \Re$. They are not backward-time unique for $x_0 \neq 0$.

**Note 3.1** Generalised motions of the system $\dot{x} = -sign\, x$ obey $\underline{x}[t; \sigma, \underline{x}(\sigma; 0, 0)] = \underline{x}(t; 0, 0)$ for $t \in [\sigma, +\infty[$, $\sigma \in \Re$, i.e. in the forward-time direction. They do not obey this equation for $t \in ]-\infty, \sigma[$; i.e. in the backward-time direction.

**Note 3.2** All the definitions of Chapter 1 and Chapter 2 are valid also in case $\underline{X}$ therein represents a generalised motion. The same holds for Definition 3.6 and Definition 3.7 (Section 3.2.2).

### Existence of generalised motions

From the proof by Nemytskii and Stepanov [157, p. 3] of Theorem 3.2 [157, Theorem 1.11] and Definition 3.8 it follows:

**Theorem 3.10** *In order for the system (3.7) to have a generalised motion through $X_0 \in \mathcal{X}$ at $t = 0$, it is sufficient that there exists a compact neighbourhood $\overline{\mathcal{N}}(X_0)$ of $X_0$, $\overline{\mathcal{N}}(X_0) \subset \mathcal{X}$, such that the function $f$ is defined on $\overline{\mathcal{N}}(X_0)$ and continuous on $\overline{\mathcal{N}}(X_0)$ except possibly at $X_0$ : $f(X) \in C[\overline{\mathcal{N}}(X_0) - \{X_0\}]$. Then, the generalised motion is defined in the interval $]-\alpha, \alpha[$, where*

$$\alpha = \frac{\rho[X_0, \partial\overline{\mathcal{N}}(X_0)]}{max[||f(X)|| : X \in \overline{\mathcal{N}}(X_0)]}.$$

∇∇∇

This theorem directly implies the following result.

**Theorem 3.11** *If, as time increases, a given generalised motion $\underline{X}$ remains in a compact nonempty subset $\mathcal{X}_c$ of the interior $\mathcal{N}(X_0)$ of the neighbourhood $\overline{\mathcal{N}}(X_0)$ on which the conditions of Theorem 3.8 are fulfilled, then the generalised motion may be continued for the whole interval $\Re_+$.*

∇∇∇

**Note 3.3** Theorem 3.4 is valid also in the case where $\underline{X}$ represents a generalised motion of the system (3.7).

**Theorem 3.12** *If the function $f$ is defined on $\Re^n$ with a bounded norm on $\Re^n$, $sup\,[||f(X)|| : X \in \Re^n] < +\infty$, and continuous on $\Re^n$ except possibly at $\hat{X}_0 \in \Re^n$ : $f(X) \in C(\Re^n - \{\hat{X}_0\})$, then generalised motions of the system (3.7) are defined on $\Re$ for every $\hat{X}_0 \in \Re^n$.*

∇∇∇

### 3.2.4 Limit points and limit sets

In the sequel $\underline{X}(t; X_0)$ will be called a motion or generalised motion when $\underline{X}$ is a motion or generalised motion, respectively, and as soon as $t$ is arbitrary in $\mathcal{I}_0$ rather than fixed.

By following Nemytskii and Stepanov [157], Zubov [218, p. 21], Bhatia and Szegö [10, p. 28], and LaSalle [120, p. 27], and referring to Definition 3.8 we accept the following definition. Notice that the maximal interval on which $\underline{X}(t; X_0)$ exists was denoted by $\mathcal{I}_0 = \mathcal{I}_0(X_0; \underline{X}) = (l^-, l^+)$.

**Definition 3.9** *(a) A point $y$ is an $\alpha$-limit point of a motion $\underline{X}(t; X_0)$ if and only if there is a sequence $t_n \to l^-$ as $n \to +\infty$ such that $\underline{X}(t_n; X_0) \to y$ as $n \to +\infty$.*

*$y$ is also called a negative limit point of $\underline{X}(t; X_0)$.*

*(b) A point $z$ is an $\omega$-limit point of a motion $\underline{X}(t; X_0)$ if and only if there is a sequence $\tau_k \to l^+$ as $k \to +\infty$ such that $\underline{X}(\tau_k; X_0) \to z$ as $k \to +\infty$.*

*$z$ is also called a positive limit point of $\underline{X}(t; X_0)$.*

*(c) If $\underline{X}(t; X_0)$ is a generalised motion then (a) and (b) are still valid.*

**Comment 3.7** *Let*

a) $\dot{x} = x^{2/3}$, *Example 3.1, Fig. 3.1. Every motion* $\underline{x}(t;0)$ *has* $\alpha$*-limit point* $y = -\infty$ *for* $\xi \neq 0$*, and for* $\omega$*-limit point* $z = +\infty$ *for* $\zeta \neq 0$*. If* $\xi = \zeta = 0$ *then* $y = z = 0$.

b) $\dot{\underline{x}} = sgn\, x$, *Example 3.2, Fig. 3.2, Example 3.6, Fig. 3.7. The* $\alpha$*-limit point* $y$ *of* $\underline{x}(t;x_0)$ *is zero,* $y = 0$*, for every* $x_0 \in \Re$*. The* $\omega$*-limit point* $z$ *of* $\underline{x}(t;x_0)$ *is* $z = (+\infty\ sign\, x_0)$ *if* $(x_0 \neq 0) \in \Re$*, and* $z = 0$ *if* $x_0 = 0$ *and* $sgn\, 0 = 0$*. In all other cases in which the system has generalised motions, the* $\omega$*-limit point* $z$ *of* $\underline{x}(t;x_0)$ *is* $z = (+\infty\ sign\, x_0)$.

c) $\dot{x} = -sign\, x$, *Example 3.3, Fig. 3.3, Example 3.7, Fig. 3.8. If* $(x_0 \neq 0) \in \Re$ *then both the motion and the generalised motion have the* $\alpha$*-limit point* $y = (+\infty\ sgn\, 0)$ *and the* $\omega$*-limit point* $z = 0$*. The motion* $\underline{x}(t;0) \equiv 0$ *has* $y = z = 0$.

d) $\dot{x} = -x^{1/2}$, *Example 3.4, Fig. 3.4. If* $x_0 \in \Re^+$ *then* $y = +\infty$ *and* $z = 0$*. If* $x_0 = 0$ *then* $y = z = 0$.

e) $\dot{x} = -x^3$, *Example 3.5, Fig. 3.5. If* $(x_0 \neq 0) \in \Re^+$ *then* $y = (+\infty\ sign\, x_0)$ *and* $z = 0$*. If* $x_0 = 0$ *then* $y = z = 0$.

**Note 3.4** The preceding analysis illustrates that limit points depend on a (generalised) motion $\underline{X}$ and an initial $X_0$,

$$y = y(X_0; \underline{X}), \quad z = z(X_0; \underline{X}).$$

In the case where a system has a unique motion through $X_0 \in \mathcal{X}_0$ at $t = 0$ then the limit points depend on the initial $X_0$ only,

$$y = y(X_0), \quad z = z(X_0).$$

**Definition 3.10** *(a) The set of all negative* ($\alpha-$) *limit points* $y$ *of* $\underline{X}(t;X_0)$ *is its negative limit set denoted by* $\mathcal{L}^-$,

$$\mathcal{L}^- = \mathcal{L}^-(X_0; \underline{X}).$$

*(b) The set of all positive* ($\omega-$) *limit points* $z$ *of* $\underline{X}(t;X_0)$ *is its positive limit set denoted by* $\mathcal{L}^+$,

$$\mathcal{L}^+ = \mathcal{L}^+(X_0; \underline{X}).$$

**Comment 3.8** *If* $\underline{X}(t;X_0)$ *is unique through* $X_0$ *at* $t = 0$ *then*

$$\mathcal{L}^-(X_0; \underline{X}) = \mathcal{L}^-(X_0), \quad \mathcal{L}^+(X_0; \underline{X}) = \mathcal{L}^+(X_0).$$

**Comment 3.9** *Let*

a) $\dot{x} = x^{2/3}$, *Example 3.1, Fig. 3.1.:* $\mathcal{L}^-(0; \underline{x}_\zeta) = \{-\infty\}$ *and* $\mathcal{L}^+\{0; \underline{x}_\zeta\} = \{+\infty\}$, $\forall \zeta \in \Re_+$*. However,* $\mathcal{L}^-(0;\underline{x}) = \mathcal{L}^+(0;\underline{x}) = \{0\}$*. The origin* $x = 0$ *is not either positively or negatively Lagrange stable, or equivalently, the family* $\mathcal{F}_0(0)$ *is neither positively or negatively precompact.*

*b)* $\dot{x} = sgn\, x$, *Example 3.2, Fig. 3.2, motions:* $\mathcal{L}^-(x_0) = \{0\}, \forall x_0 \in \Re$, $\mathcal{L}^+\{x_0\} = \{+\infty\}$, $\forall (x_0 \neq 0) \in \Re$ *and* $\mathcal{L}^+(0) = \{0\}$ *in case* $sgn\, 0 = 0$; *Example 3.6, Fig. 3.7, generalised motions:* $\mathcal{L}^-(x_0) = \{0\}$, $\forall x_0 \in \Re$ *for which generalised motions exist,* $\mathcal{L}^+\{x_0\} = \{+\infty\ sgn\, x_0\}$, $\forall x_0 \in \Re$ *for which generalised motions exist.*

*c)* $\dot{x} = -sign\, x$, *Example 3.3, Fig. 3.3 and Example 3.7, Fig. 3.8, motions and generalised motions:* $\mathcal{L}^-(x_0) = \{+\infty\ sgn\, x_0\}$, $\forall (x_0 \neq 0) \in \Re$, $\mathcal{L}^-(0) = \mathcal{L}^+(0) = \{0\}$ *and* $\mathcal{L}^+(x_0) = \{0\}$, $\forall x_0 \in \Re$.

*d)* $\dot{x} = -x^{1/2}$, *Example 3.4, Fig. 3.4, motions (coincide with generalised motions):* $\mathcal{L}^-(x_0) = \{+\infty\}$, $\forall x_0 \in \Re^+$, $\mathcal{L}^-(0) = \mathcal{L}^+(0) = \{0\}$, $\mathcal{L}^+(x_0) = \{0\}$, $\forall x_0 \in \Re_+$.

*e)* $\dot{x} = -x^3$, *Example 3.5, Fig. 3.5, motions:* $\mathcal{L}^-(x_0) = \{+\infty\ sign\, x_0\}$, $\forall (x_0 \neq 0) \in \Re$, $\mathcal{L}^-(0) = \mathcal{L}^+(0) = \{0\}$, $\mathcal{L}^+(x_0) = \{0\}$, $\forall x_0 \in \Re$.

**Example 3.8 (Bhatia and Szegö [10, p. 30])** The second order system,

$$X = (x_1\ \ x_2)^T, \quad f = (f_1\ \ f_2)^T,$$

$$\frac{dX}{dt} = f(X),$$

is determined by

$$f_1(X) = \frac{x_1(1 - x_1^2) - x_2(1 - x_2^2)^2}{(1 + x_1^2)(1 + \|X\|)},$$

$$f_2(X) = \begin{cases} \left(\dfrac{x_1}{1 - x_1^2} + x_2\right) \dfrac{1}{1 + \|X\|} &, \ |x_1| \neq 1, \\ sign\, x_1 &, \ |x_1| = 1. \end{cases}$$

System trajectories are shown in Fig. 3.9.

The positive limit set is found as follows for every $X_0 \in \Re$, $X_0 \neq 0, |x_{10}| < 1$:

$$\mathcal{L}^+(X_0) = \{X : X \in \Re^2,\ |x_1| = 1\}.$$

This equation shows that the vertical lines through $x_1 = -1$ and $x_1 = 1$ constitute the positive limit set $\mathcal{L}^+(X_0)$ of every motion $\underline{X}(t; X_0)$ for any $X_0$ in the strip between the lines and out of the origin.

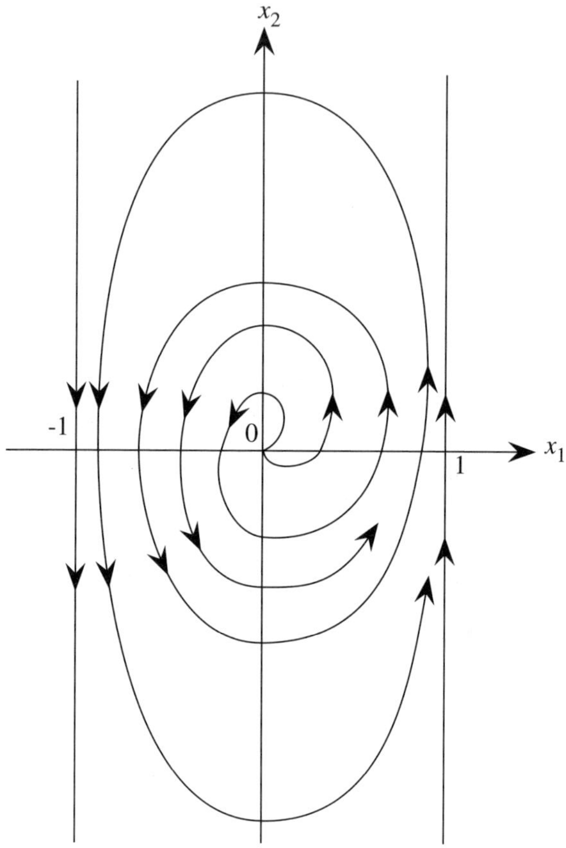


Figure 3.9: Trajectories of the system of Example 3.8.

### 3.2.5 Limit sets, Lagrange stability, precompactness and stability domains

At first we present the definition of Lagrange stability [157], [10, p. 46], [152]:

**Definition 3.11** *(a) A point $x_0 \in \mathcal{X}_0$ is positively [negatively] Lagrange stable relative to $\mathcal{X}$ and relative to system (3.7) if and only if the closure $\overline{\underline{X}([0, l^+(X_0; \underline{X})[; X_0)}$ of every arc $\underline{X}([0, l^+(X_0; \underline{X})[; X_0)$ [the closure $\overline{\underline{X}(]l^-(X_0; \underline{X}), 0]; X_0)}$ of every arc $\underline{X}(]l^-(X_0; \underline{X}), 0]; X_0)$], respectively, is compact and in $\mathcal{X}$.*

*(b) A set $\mathcal{A}_0 \subseteq \mathcal{X}_0$ is positively [negatively] Lagrange stable relative to $\mathcal{X}$ and relative to the system (3.7) if and only if every point $X_0 \in \mathcal{A}_0$ is positively [negatively] Lagrange stable relative to $\mathcal{X}$ and relative to the system (3.7).*

(c) *The expression "relative to X" is omitted if and only if*

$$\bigcap_{\underline{X}} \mathcal{I}_0(X_0; \underline{X}) = \bigcap_{\underline{X}} ]l^-(X_0; \underline{X}), l^+(X_0; \underline{X})[ = \Re.$$

(d) *The expression "and relative to the system (3.7)" is omitted if and only if the system is prespecified.*

**Note 3.5** Definition 3.11 is general. It holds even if the system (3.7) has multiple (generalised) motions through $X_0$ at $t = 0$. If the system has a unique (generalised) motion $\underline{X}(t; X_0)$ then $l^{\pm}(X_0; \underline{X}) = l^{\pm}(X_0)$ and there is a unique arc $\underline{X}([0, l^+(X_0)[; X_0)$ $[\underline{X}(]l^-(X_0), 0]; X_0)]$. In such a case the intersection in (c) taken over all motions $\underline{X}$ passing through $X_0$ at $t = 0$ reduces to $\mathcal{I}_0(X_0; \underline{X})$, that is,

$$\mathcal{I}_0(X_0) = ]l^-(X_0), l^+(X_0)[ = \Re.$$

Besides, Lagrange stability of $X_0 \in \mathcal{X}_0$ guarantees boundedness of all motions through $X_0 \in \mathcal{X}_0$.

In view of Definition 3.11 we slightly generalise the definition of motion precompactness by LaSalle [120, p. 28, Definition 3.1].

**Definition 3.12** *(a) A family $\underline{\mathcal{F}}_0 = \{\underline{X} : \underline{X}(0; X_0) = X_0\}$, $\underline{\mathcal{F}}_0 = \underline{\mathcal{F}}_0(X_0)$, of (generalised) motions of the system (3.7) is positively [negatively] precompact relative to $\mathcal{X}$ if and only if every $\underline{X} \in \underline{\mathcal{F}}_0$ is bounded for all $t \in [0, l^+(X_0; \underline{X})[$ $[t \in ]l^-(X_0; \underline{X}), 0]]$ and has no positive [negative] limit point on the boundary $\partial\mathcal{X}$ of $\mathcal{X}$,*

$$\mathcal{L}^+(X_0; \underline{X}) \cap \partial\mathcal{X} = \emptyset [\mathcal{L}^-(X_0; \underline{X}) \cap \partial\mathcal{X} = \emptyset], \quad \forall \underline{X} \in \mathcal{F}_0,$$

*respectively.*

(b) *A set $\mathcal{A}_0 \subseteq \mathcal{X}_0$ is positively [negatively] precompact relative to $\mathcal{X}$ and relative to the system (3.7) if and only if the family $\underline{\mathcal{F}}_0(X_0)$ of (generalised) motions of the system is positively [negatively] precompact relative to $\mathcal{X}$ for every $X_0 \in \mathcal{A}_0$ respectively.*

(c) *The expression "relative to X" is omitted if and only if $\mathcal{X} = \Re^n$.*

(d) *The expression "relative to the system (3.7)" is omitted if and only if the system is prespecified.*

(e) *The expressions "A family $\underline{\mathcal{F}}_0 = \{\underline{X} : \underline{X}(0; X_0) = X_0\}$, $\underline{\mathcal{F}}_0 = \underline{\mathcal{F}}_0(X_0)$, of (generalised) motions" and "the family $\underline{\mathcal{F}}_0(X_0)$ of (generalised) motions" are replaced by "A (generalised) motion $\underline{X}(t; X_0)$" and "the (generalised) motion $\underline{X}(t; X_0)$", respectively, if and only if the (generalised) motion $\underline{X}(t; X_0)$ is unique.*

**Note 3.6 (LaSalle [120, p. 28])** If $\underline{X}(t; X_0)$ is positively precompact relative to $\mathcal{X}$ then $l^+(X_0; \underline{X}) = +\infty$. Analogously, if $\underline{X}(t; X_0)$ is negatively precompact relative to $\mathcal{X}$ then $l^-(X_0; \underline{X}) = -\infty$. Hence, if $\underline{X}(t; X_0)$ is precompact relative to $\mathcal{X}$ then $\mathcal{I}_0(X_0; \underline{X}) = \Re$.

**Note 3.7** Definition 3.11 and Definition 3.12 are equivalent as pointed out by La Salle for the case of unique motions [120, p. 28].

Definition 3.11, Definition 3.12 and the definitions of the stability domains (Definition 3.1 – Definition 3.6, Section 2.2) directly imply the following result that concerns the system (3.7) by noting that (generalised) motions are continuous in $t \in \mathcal{I}_0$ (Definition 3.2 and Definition 3.8).

**Theorem 3.13** *(a-1) If $X = 0$ is stable then its domain $\mathcal{D}_s$ of stability is positively precompact (positively Lagrange stable).*

*(a-2) If a set $\mathcal{A}$ is stable then its domain $\mathcal{D}_s(\mathcal{A})$ of stability is positively precompact (positively Lagrange stable).*

*(b-1) If $X = 0$ is attractive then its domain $\mathcal{D}_a$ of attraction is positively precompact (positively Lagrange stable).*

*(b-2) If a set $\mathcal{A}$ is attractive then its domain $\mathcal{D}_a(\mathcal{A})$ of attraction is positively precompact (positively Lagrange stable).*

*(c-1) If $X = 0$ is asymptotically stable then $\mathcal{D} = \mathcal{D}_a \cup \mathcal{D}_s$ is positively precompact (positively Lagrange stable).*

*(c-2) If a set $\mathcal{A}$ is asymptotically stable then $\mathcal{D}(\mathcal{A}) = \mathcal{D}_a(\mathcal{A}) \cup \mathcal{D}_s(\mathcal{A})$ is positively precompact (positively Lagrange stable).*

**Example 3.9** $x = 0$ of the system $\dot{x} = -x^3$ (Example 3.5, Fig. 3.5) is completely globally asymptotically stable:

$$\mathcal{D}_s = \mathcal{D}_a = \mathcal{D} = \Re.$$

The whole space $\Re$ is positively precompact (positively Lagrange stable).

**Example 3.10** The set $\mathcal{A}_0 = \{X : X \in \Re^2, |x_1| \leq 1\}$ is not positively precompact (positively Lagrange stable) relative to the system of Example 3.8.

**Example 3.11** The first order system

$$\frac{dx}{dt} = x - k\ sign\, x, \quad k \in \Re^+,$$

has generalised motions determined by

$$\underline{x}(t; x_0) = \begin{cases} k\ sign\, x_0 + (x_0 - k\ sign x_0)\, exp\,(t)\ , & t \in \Re,\ (x_0 = 0 \text{ or } |x_0| \in [k, +\infty[), \\ \left.\begin{array}{ll} k\ sign\, x_0 + (x_0 - k\ sign x_0)\ exp\,(t) & ,\ t \in [0, t_0], \\ 0 & ,\ t \in [t_0, +\infty[, \end{array}\right\} & ,\ |x_0| \in ]0, k[, \end{cases}$$

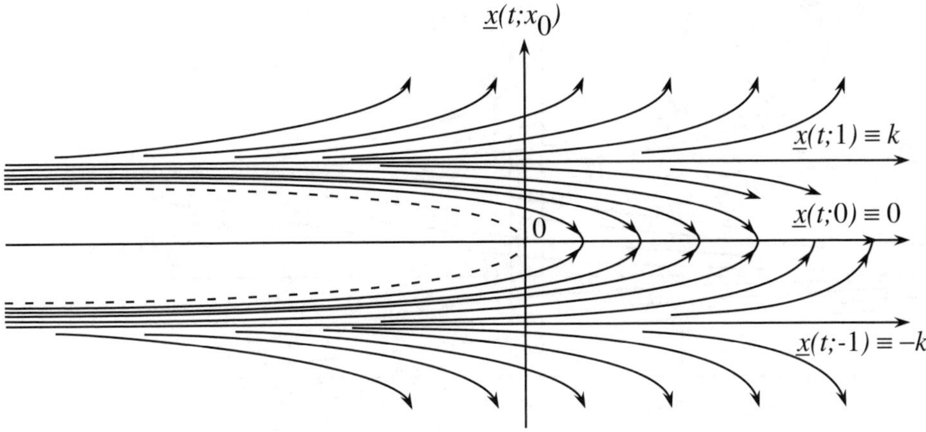


Figure 3.10: Generalised motions of the system $\dot{x} = x - k\, sign\, x$, $k \in \Re^+$, Example 3.11

where $t_0 = ln\,\dfrac{k}{k - |x_0|}$. The generalised motions are shown in Fig. 3.10. In this connection see Section 3.2.1 for system motions.

$X = 0$ is asymptotically stable with stability domains

$$\mathcal{D}_s = [-k, k], \quad \mathcal{D}_a = ]-k, k[, \quad \mathcal{D} = ]-k, k[.$$

They are positively precompact (positively Lagrange stable). They are also negatively precompact (negatively Lagrange stable).

**Example 3.12** The first order system

$$\frac{dx}{dt} = x - x^2 sign\, x = x(1 - |x|)$$

has generalised motions determined by

$$\underline{x}(t; x_0) = \frac{x_0}{|x_0| + (1 - |x_0|)\, exp\,(-t)}, \quad \forall x_0 \in \Re.$$

They are shown in Fig. 3.11. Let

$$\mathcal{A}_0 = \{x : x \in \Re, \ \ |x| \leq 1\}.$$

The set $\mathcal{A}_0$ is completely asymptotically stable in the large. Its stability domains are

$$\mathcal{D}_s(\mathcal{A}) = \mathcal{D}_a(\mathcal{A}) = \mathcal{D}(\mathcal{A}) = \Re.$$

They are positively Lagrange stable (positively precompact). However, they are not negatively Lagrange stable (negatively precompact) because

$$|\underline{x}(t; x_0)| \to +\infty \text{ as } t \to t_0 = ln\,\frac{|x_0| - 1}{|x_0|}, \quad |x_0| \in ]1, +\infty[,$$

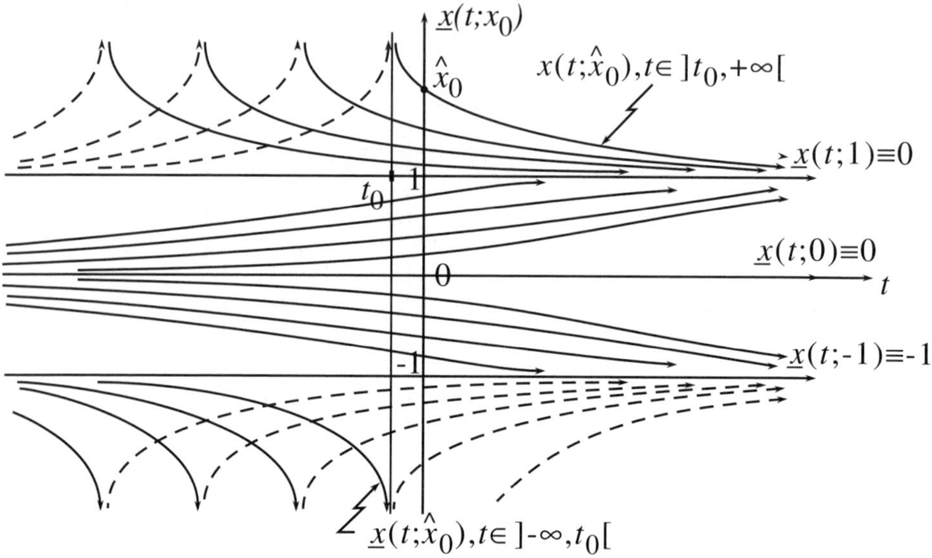


Figure 3.11: Generalised motions and motions of the system $\dot{x} = x - x^2 \; sign \, x$.

so that

$$\mathcal{L}^-(x_0) = \{+\infty \; sign \, x_0\}, \quad |x_0| \in ]1, +\infty[.$$

Notice that

$$\mathcal{L}^+(x_0) = \{sign \, x_0\}, \quad |x_0| \in \Re^+,$$

$$\mathcal{L}^-(0) = \mathcal{L}^+(0) = \{0\},$$

and

$$\mathcal{L}^-\{x_0\} = \mathcal{L}^+\{x_0\} = \{sign \, x_0\}, \quad |x_0| = 1.$$

### 3.2.6 Invariance properties of sets

By following LaSalle [120, p. 28, Definition 4.1], we define invariance properties of sets at follows:

**Definition 3.13** *(a) A set $\mathcal{A}$, $\mathcal{A} \subset \Re^n$, is positively [negatively] weakly invariant relative to the system (3.7) if and only if $X_0 \in \mathcal{A} \cap \mathcal{X}_0$ implies that all system generalised motions $\underline{X}(t; X_0) \in \mathcal{A}$ for all $t \in [0, l^+(X_0; \underline{X})[$ $[t \in ]l^-(X_0; \underline{X}), 0]]$, respectively. The adverb "weakly" is omitted if and only if $l^+(X_0; \underline{X}) = +\infty$ $[l^-(X_0; \underline{X}) = -\infty]$, respectively, for all $X_0 \in \mathcal{A} \cap \mathcal{X}_0$ and for all system generalised motions $\underline{X}(t; X_0)$ through $X_0$ at $t = 0$.*

(b) *A set $\mathcal{A}$, $\mathcal{A} \subset \Re^n$, is weakly invariant if and only if it is both positively weakly invariant and negatively weakly invariant. The adverb "weakly" is omitted if and only if $\mathcal{I}_0(X_0; \underline{X}) = \Re$ for all $X_0 \in \mathcal{A} \cap \mathcal{X}_0$ and for all system generalised motions $\underline{X}(t; X_0)$ through $X_0$ at $t = 0$.*

(c) *The expression "relative to the system (3.7)" is omitted if and only if the system is prespecified.*

This definition incorporates the classical definition of invariance properties of sets [157, p. 330], [10, p. 20], [120, p. 28].

**Definition 3.14** *(a) A set $\mathcal{A}$, $\mathcal{A} \subset \Re^n$, is positively [negatively] invariant relative to the system (3.7) if and only if $X_0 \in \mathcal{A}$ implies that all system generalised motions $\underline{X}(t; X_0) \in \mathcal{A}$ for all $t \in \Re_+$ [$t \in \Re_-$], respectively.*

(b) *A set $\mathcal{A}$, $\mathcal{A} \subset \Re^n$, is invariant relative to the system (3.7) if and only if it is both positively and negatively invariant relative to the system.*

(c) *The expression "relative to the system (3.7)" is omitted if and only if the system is prespecified.*

**Example 3.13** Let

a) $\dot{x} = x^{2/3}$, Example 3.1, Fig. 3.1. The set $\mathcal{O} = \{0\}$ is neither positively nor negatively invariant.

b) $\dot{x} = -x^{1/2}$, Example 3.4, Fig. 3.4. The set $\mathcal{O} = \{0\}$ is invariant.

c) $\dot{x} = sgn\, x$, Example 3.6, Fig. 3.7. The set $\mathcal{O} = \{0\}$ is invariant iff $sgn\, 0 = 0$.

d) $\dot{x} = x - k\ sgn\, x$, $k \in \Re^+$ Example 3.11, Fig. 3.10. The sets $\mathcal{O} = \{0\}$, $\mathcal{A}_1 = \{x : x \in ]-k, k[\}$, $\mathcal{A}_2 = \{x : x = k\}$, $\mathcal{A}_3 = \{x : x = -k\}$, $\mathcal{A}_4 = \{x : |x| \leq k\}$, $\mathcal{A}_5 = \{x : x \in [k, +\infty[\}$, $\mathcal{A}_6 = \{x : x \in ]k, +\infty[\}$, $\mathcal{A}_7 = \{x : x \in [-k, -\infty[\}$, $\mathcal{A}_8 = \{x : x \in ]-k, -\infty[\}$, and $\Re$ are invariant. Notice that

$$\overline{\mathcal{A}_1} = \mathcal{A}_4, \;\; \mathring{\mathcal{A}}_4 = \mathcal{A}_1, \;\; \mathcal{A}_2 \cup \mathcal{A}_3 = \partial\mathcal{A}_4.$$

e) $\dot{x} = x(1 - |x|)$, Example 3.12, Fig. 3.11. The invariant sets are $\mathcal{O} = \{0\}$, $\mathcal{A}_1 = \{x : -1 < x < 1\}$, $\mathcal{A}_2 = \{1\}$, $\mathcal{A}_3 = \{-1\}$, $\mathcal{A}_4 = \{x : -1 \leq x \leq 1\}$, $\mathcal{A}_5 = \{x : x \in [1, +\infty[\}$, $\mathcal{A}_6 = \{x : x \in ]1, +\infty[\}$, $\mathcal{A}_7 = \{x : x \in [-1, -\infty[\}$, $\mathcal{A}_8 = \{x : x \in ]-1, -\infty[\}$ and $\Re$.

**Note 3.8 (LaSalle [120, p. 28])** If a set $\mathcal{A}$ is precompact relative to $\mathcal{X}$ and weakly invariant then it is invariant.

**Definition 3.15** *(a) Generalised motions $\underline{X}(t; X_0)$ of the system (3.7) are continuous in $X_0 \in \mathcal{X}$ for any $t \in \mathcal{I}_0 = \mathcal{I}_0(X_0; \underline{X})$, which is denoted by $\underline{X}(t; X_0) \in C_0(\mathcal{I}_0 \times \mathcal{X}_0)$, if and only if for every $(\varepsilon, t) \in \Re^+ \times \mathcal{I}_0$ there is $\delta \in \Re^+$, $\delta = \delta(\varepsilon; t; X_0; \underline{X})$, such that $||X^* - X_0||$ implies $||\underline{X}(t; X^*) - \underline{X}(t; X_0)|| < \varepsilon$ for all motions $\underline{X}(t; X_0)$; or equivalently, for any sequence $X_n$ converging to $X_0$, $X_n \to X_0$ as $n \to +\infty$, and every $(\varepsilon, t) \in \Re^+ \times \mathcal{I}_0$ there is a natural number $m$ such that $n > m$ implies $||\underline{X}(t; X_n) - \underline{X}(t; X_0)|| < \varepsilon$ for all motions $\underline{X}(t; X_0)$, which is denoted by $\underline{X}(t; X_n) \to \underline{X}(t; X_0)$ as $X_n \to X_0$.*

(b) *Generalised motions* $\underline{X}(t; X_0)$ *of the system (3.7) are continuous in* $t \in \mathcal{I}_0(X_0; \underline{X}) = \mathcal{I}_0$ *for every* $X_0 \in \mathcal{X}$, *which is denoted by* $\underline{X}(t; X_0) \in C_t(\mathcal{I}_0 \times \mathcal{X}_0)$, *if and only if for every motion* $\underline{X}(t; X_0)$ *and every* $(\varepsilon, t, X_0) \in \Re^+ \times \mathcal{I} \times \mathcal{X}_0$ *there is* $\delta \in \Re^+$, $\delta = \delta(\varepsilon; t; X_0; \underline{X})$, *such that* $|t^* - t| < \delta$ *implies* $||\underline{X}(t^*; X_0) - \underline{X}(t; X_0)|| < \varepsilon$; *or equivalently, for every motion* $\underline{X}(t; X_0)$ *and every sequence* $t_n$ *converging to* $t$, $t_n \to t$ *as* $n \to +\infty$, *there is a natural number* $m$ *such that* $n > m$ *implies* $||\underline{X}(t_n; X_0) - \underline{X}(t; X_0)|| < \varepsilon$, *which is denoted by* $\underline{X}(t_n; X_0) \to \underline{X}(t; X_0)$ *as* $t_n \to t$.

**Note 3.9** $C(\mathcal{I}_0 \times \mathcal{X}_0) \subset [C_0(\mathcal{I}_0 \times \mathcal{X}_0) \cap C_t(\mathcal{I}_0 \times \mathcal{X}_0)]$.

**Theorem 3.14** *If the system (3.7) has generalised motions* $\underline{X}(t; X_0)$ *continuous in* $X_0 \in \mathcal{X}_0$ *for every* $t \in \mathcal{I}_0$, $\underline{X}(t; X_0) \in C_0(\mathcal{I}_0 \times \mathcal{X}_0)$, *then the system has a unique generalised motion* $\underline{X}(t; X_0)$ *through every* $X_0 \in \mathcal{X}_0$ *at* $t = 0$. ∇∇∇

**Proof.** Let $\underline{X}(t; X_0) \in C_0(\mathcal{I}_0 \times \mathcal{X}_0)$ and $(X_0, \varepsilon, t) \in \mathcal{X}_0 \times \Re^+ \times \mathcal{I}_0$ be arbitrarily chosen. Let $\underline{X}_1$ and $\underline{X}_2$ be two generalised motions of the system through $X_0$ at $t = 0$. Let $X^* = X_0$. Hence, $||X^* - X_0|| = 0 < \delta$, $\delta = min\ \{\delta(\varepsilon; t; X_0; \underline{X}_i) : i = 1, 2\}$ (Definition 3.15a). Since $\underline{X}(t; X_0) \in C_0(\mathcal{I}_0 \times \mathcal{X}_0)$ then (Definition 3.15a) $||\underline{X}_i(t; X^*) - \underline{X}_k(t; X_0)|| = ||\underline{X}_i(t; X_0) - \underline{X}_k(t; X_0)|| < \varepsilon$, $i, k \in \{1, 2\}$. Letting $\varepsilon \to 0$ it follows that $||\underline{X}_i(t; X_0) - \underline{X}_k(t; X_0)|| = 0$ for every $(t; X_0) \in \mathcal{I}_0 \times \mathcal{X}_0$ and $i, k \in \{1, 2\}$, which proves the uniqueness of $\underline{X}(t; X_0)$ through $X_0$ at $t = 0$.

**Lemma 3.1** *If the system (3.7) has generalised motions* $\underline{X}(t; X_0)$ *continuous in* $X_0 \in \mathcal{X}_0$ *for every* $t \in \mathcal{I}_0(X_0; \underline{X})$, $\underline{X}(t; X_0) \in C_0(\mathcal{I}_0 \times \mathcal{X}_0)$, *then the motions obey the group property (3.14),*

$$\underline{X}[t; \underline{X}(\sigma; X_0)] = \underline{X}(t + \sigma; X_0), \qquad \forall t \in \mathcal{I}_0(X_0; \underline{X}),\ \forall \sigma \in \mathcal{I}_0(X_0; \underline{X}),$$

$$(\sigma + t) \in \mathcal{I}_0(X_0; \underline{X}). \qquad (3.14)$$

∇∇∇

**Proof.** Let $\underline{X}(t; X_0) \in C_0(\mathcal{I}_0 \times \mathcal{X}_0)$. Hence $\underline{X}(t; X_0)$ is unique for every $X_0 \in \mathcal{X}_0$ (Theorem 3.14) so that $\mathcal{I}_0(X_0; \underline{X}) = \mathcal{I}_0(X_0)$ ($= \mathcal{I}_0$, when $X_0$ is fixed). Let $X_0 \in \mathcal{X}_0$, $t \in \mathcal{I}_0$ and $\sigma \in \Re$ be arbitrarily chosen provided $(\sigma + t) \in \mathcal{I}_0$. The theorem can be proved in two ways.

(a) Let at first the equation (3.7) be integrated from 0 to $\sigma$,

$$\underline{X}(\sigma; X_0) = X_0 + \int_0^{\sigma} f[\underline{X}(\tau; X_0)]d\tau = \underline{X}(\sigma; 0, X_0), \qquad (3.15)$$

and now from 0 to $\sigma + t$,

$$\underline{X}(\sigma + t; X_0) = X_0 + \int_0^{\sigma+t} f[\underline{X}(\tau; X_0)]d\tau = \underline{X}(\sigma + t; 0, X_0), \qquad (3.16)$$

We can present (3.16) as follows due to additivity of the integral and (3.15),

$$\underline{X}(\sigma + t; 0, X_0) = \underline{X}(\sigma; 0, X_0) + \int_{\sigma}^{\sigma+t} f[\underline{X}(\tau; 0, X_0)]d\tau, \qquad (3.17)$$

Since $\underline{X}(t;0,X_0) = \underline{X}[t;\sigma,\underline{X}(\sigma;X_0)]$ then

$$\int_\sigma^{\sigma+t} f[\underline{X}(\tau;0,X_0)]d\tau = \int_\sigma^{\sigma+t} f\{\underline{X}[\tau;\sigma,\underline{X}(\sigma;X_0)]\}d\tau. \tag{3.18}$$

In this equation we replace $\tau$ by $u+\sigma$,

$$\int_\sigma^{\sigma+t} f\{\underline{X}[\tau;\sigma,\underline{X}(\sigma;X_0)]\}d\tau = \int_0^{t} f\{\underline{X}[u+\sigma;\sigma,\underline{X}(\sigma;X_0)]\}du. \tag{3.19}$$

The system (3.7) is time invariant so that

$$\underline{X}[u+\sigma;\sigma,\underline{X}(\sigma;X_0)] = \underline{X}[u;0;\underline{X}(\sigma;0,X_0)],$$

which with (3.17) through (3.19) yields (3.20),

$$\underline{X}(\sigma+t;0,X_0) = \underline{X}[t;0;\underline{X}(\sigma;0,X_0)], \tag{3.20}$$

and proves the theorem.

(b) The statement of the theorem results from uniqueness of generalised motions.

The following results slightly generalise those by Nemytskii and Stepanov [157, p. 331], Bhatia and Szegö [10], and LaSalle [120, p. 29].

**Theorem 3.15** *If the system (3.7) has generalised motions $\underline{X}(t;X_0)$ continuous in $X_0 \in \mathcal{X}_0$ for every $t \in \mathcal{I}_0(X_0;\underline{X})$, $\underline{X}(t;X_0) \in C_0(\mathcal{I}_0 \times \mathcal{X}_0)$, then*

*(a) The closure $\overline{\mathcal{A}}$ of a positively [negatively] weakly invariant set $\mathcal{A}$ is positively [negatively] weakly invariant, respectively.*

*(b) The closure $\overline{\mathcal{A}}$ of a weakly invariant set $\mathcal{A}$ is weakly invariant.*

*(c) If a set $\mathcal{A}$ is invariant and precompact relative to $\mathcal{X}$ then its closure $\overline{\mathcal{A}}$ is invariant.*

*(d) If a set $\mathcal{A}$ is invariant and $\overline{\mathcal{A}} \subset \mathcal{X}$ then its closure $\overline{\mathcal{A}}$ is invariant.* ∇∇∇

**Proof.** Let the system have generalised motions $\underline{X}(t;X_0) \in C_0(\mathcal{I}_0 \times \mathcal{X}_0)$.

(a) Let $\mathcal{A}$ be positively [negatively] weakly invariant. If $X_0 \in \mathcal{A}$ then $\underline{X}(t;X_0) \in \overline{\mathcal{A}}$ for all $t \in (\Re_+ \cap \mathcal{I}_0)$ $[t \in (\Re_- \cap \mathcal{I}_0)]$, due to positive [negative] invariance of $\mathcal{A}$ and $\mathcal{A} \subseteq \overline{\mathcal{A}}$. Let $X_0 \in \partial\mathcal{A}$ and $X_n \in \mathcal{A}$, $X_n \to X_0$ as $n \to +\infty$. Then, $\underline{X}(t;X_0) \in C_0(\mathcal{I}_0 \times \mathcal{X}_0)$ implies $\underline{X}(t;X_n) \to \underline{X}(t;X_0)$ for every $t \in \mathcal{I}_0$. This and $\underline{X}(t;X_n) \in \mathcal{A}$ for every $t \in (\Re_+ \cap \mathcal{I}_0)$ $[t \in (\Re_- \cap \mathcal{I}_0)]$, due to positive [negative] invariance, imply $\underline{X}(t;X_0) \in \mathcal{A}$ for every $t \in (\Re_+ \cap \mathcal{I}_0)$ $[t \in (\Re_- \cap \mathcal{I}_0)]$, respectively.

(b) The statement under (b) results from (a) in view of Definition 3.14a.

(c) Let the set $\mathcal{A}$ be invariant and precompact relative to $\mathcal{X}$. Then for any $X_n \in \mathcal{A}$ both $\mathcal{I}_0(X_n) = \Re$ and $[\mathcal{L}^-(X_n) \cup \mathcal{L}^+(X_n)] \cap \partial\mathcal{X} = \emptyset$. Let $X_0 \in \partial\mathcal{A}$ and $X_n$ be defined as in (a) so that for every $t \in \Re$, $\underline{X}(t; X_n) \to \underline{X}(t; X_0)$ as $n \to +\infty$, which implies $\mathcal{L}^{\mp}(X_n) \to \mathcal{L}^{\mp}(X_0)$. Therefore, $[\mathcal{L}^-(X_0) \cup \mathcal{L}^+(X_0)] \cap \mathcal{X} = \emptyset$. This shows that $\underline{X}(t; X_0)$ is precompact relative to $\mathcal{X}$, which together with (b) proves that $\underline{X}(t; X_0) \in \overline{\mathcal{A}}$ for all $t \in \Re$.

(d) Let the set $\mathcal{A}$ be invariant and $\overline{\mathcal{A}} \subset \mathcal{X}$. Since $\mathcal{X}$ is open, then $\overline{\mathcal{A}}$ is compact, which with invariance of $\mathcal{A}$ and $\mathcal{A} \subseteq \overline{\mathcal{A}}$ prove precompactness of $\underline{X}(t; X_0)$ for every $X_0 \in \mathcal{A}$. This and (c) imply invariance of $\overline{\mathcal{A}}$.

What follows broadens slightly a theorem by Bhatia and Szegö [10, p. 22, Theorem 1.2.35].

**Theorem 3.16** *Let $\overline{\mathcal{A}}_i \subset \mathcal{X}$, $i = 1, 2, \ldots, m$, and let the system (3.7) have generalised motion $\underline{X}(t; X_0) \in C_0(\mathcal{I}_0 \times \mathcal{X}_0)$. Then,*

*(a) if $\mathcal{A}_i$ is positively [negatively] weakly invariant, $i = 1, 2, \ldots, m$, then $\cup[\mathcal{A}_i : i = 1, 2, \ldots, m]$ is positively [negatively] weakly invariant, respectively,*

*(b) if $\cap[\mathcal{A}_i : i = 1, 2, \ldots, m] \neq \emptyset$ and $\mathcal{A}_i$ is positively [negatively] weakly invariant, then $\cap[\mathcal{A}_i : i = 1, 2, \ldots, m]$ is positively [negatively] weakly invariant, respectively,*

*(c) if $\mathcal{A}_i \subset \mathcal{A}_j$, $\mathcal{A}_j - \mathcal{A}_i \neq \emptyset$, and $\mathcal{A}_i$ and $\mathcal{A}_j$ are positively [negatively] weakly invariant, then $(\mathcal{A}_j - \mathcal{A}_i)$ is also positively [negatively] weakly invariant, respectively.*

*(d) All the statements (a)–(c) hold also if "weakly" is omitted in each case.*
$\nabla\nabla\nabla$

**Proof.** The proof repeats that by Bhatia and Szegö [10, Theorem 1.2.35]. Notice that the generalised system motions are unique (Theorem 3.14).

(a) The statement (a) is obvious due to $\overline{\mathcal{A}}_i \subset \cup\left[\overline{\mathcal{A}}_i : i = 1, 2, \ldots, m\right]$.

(b) Let $\mathcal{P} = \cap[\mathcal{A}_i : i = 1, 2, \ldots, m] \neq \emptyset$ and $\mathcal{A}_i$ be positively [negatively] weakly invariant so that $X_0 \in \mathcal{P}$ implies $\underline{X}(t; X_0) \in \mathcal{A}_i$, $t \in (\Re_+ \cap \mathcal{I}_0)$ $[t \in (\Re_- \cap \mathcal{I}_0)]$, because $\mathcal{P} \subseteq \mathcal{A}_i$, $i = 1, 2, \ldots, m$. Hence, $\underline{X}(t; X_0) \in \mathcal{P}$, $t \in (\Re_+ \cap \mathcal{I}_0)$ $[t \in (\Re_- \cap \mathcal{I}_0)]$, respectively.

(c) Let $\mathcal{A}_i \subset \mathcal{A}_j$, $\mathcal{A}_j - \mathcal{A}_i \neq \emptyset$ and both be positively [negatively] weakly invariant. If the statement were wrong then there would exist $X_0 \in (\mathcal{A}_j - \mathcal{A}_i)$ implying $\underline{X}(t; X_0) \notin (\mathcal{A}_j - \mathcal{A}_i)$ at some $\sigma \in (\Re_+ \cap \mathcal{I}_0)$ $[\sigma \in (\Re_- \cap \mathcal{I}_0)]$. Since $X_0 \in \mathcal{A}_j$ yields $\underline{X}(t; X_0) \in \mathcal{A}_j$, for all $t \in (\Re_+ \cap \mathcal{I}_0)$ $[t \in (\Re_- \cap \mathcal{I}_0)]$, then, altogether, $\underline{X}(t; X_0) \in \mathcal{A}_i$, for all $t \in ([\sigma, +\infty[ \cap \mathcal{I}_0)$ $[t \in (]-\infty, \sigma] \cap \mathcal{I}_0)]$, which contradicts the invariance property of $\mathcal{A}_i$ due to $X_0 \notin \mathcal{A}_i$. Hence, such $X_0 \in (\mathcal{A}_j - \mathcal{A}_i)$ does not exist. The statement is proved.

(d) Since all $\overline{\mathcal{A}}_i \subset \mathcal{X}$ then, under the conditions of (a) or (b) or (c), $\mathcal{A}$ is positively [negatively] precompact, hence, positively [negatively] invariant, which holds also for $\cup[\mathcal{A}_i : i = 1,\ 2, \ldots,\ m]$, for $\mathcal{P}$ and for $\mathcal{A}_j - \mathcal{A}_i$. Therefore, the proofs of (a)–(c) are true for all $t \in \Re_+$ $[t \in \Re_-]$, respectively, which implies the statement (d).

This theorem is the key-stone for the next one.

**Theorem 3.17** *Let $\mathcal{A}$ be an open set, $\overline{\mathcal{A}} \subset \mathcal{X}$ and generalised motions of the system (3.7) $\underline{X}(t; X_0) \in C_0(\mathcal{I}_0 \times \mathcal{X})$.*

*(a) If $\mathcal{A}$ is positively [negatively] weakly invariant then its boundary $\partial\mathcal{A}$ is positively [negatively] weakly invariant.*

*(b) If $\mathcal{A}$ is positively [negatively] invariant then its boundary $\partial\mathcal{A}$ is also positively [negatively] invariant.*

*(c) If $\mathcal{A}$ is invariant then its boundary $\partial\mathcal{A}$ is also invariant.* ∇∇∇

**Proof.** Let $\mathcal{A}$ be open, $\mathcal{A} \subset \mathcal{X}$ and $\underline{X}(t; X_0) \in C(\mathcal{I}_0 \times \mathcal{X}_0)$.

(a) Let $\mathcal{A}$ be positively [negatively] weakly invariant. Now, (a) of Theorem 3.15 and (c) of Theorem 3.16 prove positive [negative] weak invariance of $\partial\mathcal{A}$ because $\partial\mathcal{A} = \overline{\mathcal{A}} - \mathcal{A}$.

(b) The statement results from (d) of Theorem 3.15 and from (c) and (d) of Theorem 3.16.

(c) The statement (b) and Definition 3.14b imply the statement under (c).

### 3.2.7 Invariance properties of limit sets

At first we broaden Theorem 5.1 by LaSalle [120] to the system (3.7) with generalised motions.

**Theorem 3.18** *Let generalised motions $\underline{X}(t; X_0)$ of the system (3.7) be continuous in $X_0 \in \mathcal{X}_0$ for every $t \in \mathcal{I}_0(X_0; \underline{X})$, $\underline{X}(t; X_0) \in C_0(\mathcal{I}_0 \times \mathcal{X}_0)$. For every $X_0 \in \mathcal{X}$ the positive limit set $\mathcal{L}^+(X_0; \underline{X})$ is closed and weakly invariant.* ∇∇∇

**Proof.** Let $\underline{X}(t; X_0) \in C_0(\mathcal{I}_0 \times \mathcal{X}_0)$. Then, $\underline{X}(t; X_0)$ is unique, $\forall X_0 \in \mathcal{X}_0$, (Theorem 3.14), so that $\mathcal{L}^+(X_0; \underline{X}) = \mathcal{L}^+(X_0)$.

Let us first prove closeness, for which the proof of (i) of 1.3.13 Theorem by Bhatia and Szegö in [10] is repeated except one detail which will be explained in the sequel. Let $y \in \overline{\mathcal{L}^+(X_0)}$ for arbitrarily chosen $X_0$, let $y_k \in \mathcal{L}^+(X_0)$, $y_k \to y$ as $k \to +\infty$. It must be shown that $y \in \mathcal{L}^+(X_0)$. For each $k$ there exists a sequence $t_{kn} \to l^+(X_0)$ as $n \to +\infty$, and $\underline{X}(t_{kn}; X_0) \to y_k$. We may assume without loss of generality that $\rho[\underline{X}(t_{kn}; X_0), y_k] < 1/k$ holds for each $k$. In (the proof of i) of) Theorem 1.3.13 in [10], Bhatia and Szegö request, in addition, $t_{kk} > k$, which is not possible in this framework as soon as $l^+(X_0) < +\infty$. Then considering the sequence $t_n$, $t_n = t_{nn}$, we see that $t_n \to l^+(X_0)$ and $\underline{X}(t_n; X_0) \to y$ as $n \to +\infty$, since

$$\rho[\underline{X}(t_n; X_0), y] \le \rho[\underline{X}(t_n; X_0), y_n] + \rho(y_n, y) < \frac{1}{n} + \rho(y_n, y) \to 0.$$

From $(1/n) + \rho(y_n, y)] \to 0$ as $n \to +\infty$ it now follows $\rho[\underline{X}(t_n; X_0), y] \to 0$ as $n \to +\infty$.

Hence $y \in \mathcal{L}^+(X_0)$ that proves closeness of $\mathcal{L}^+(X_0)$.

In order to prove weak invariance we repeat the proof of Theorem 5.1 of Chapter 2 by LaSalle in [120]. Let $y \in \mathcal{L}^+(X_0) \cap \mathcal{X}$, and let $t \in \mathcal{I}_0(y)$. Since $\mathcal{L}^+(X_0) \cap \mathcal{X}$ is nonempty then $l^+(X_0) = +\infty$, and there is a sequence $t_n$ such that $t_n \to +\infty$ and $\underline{X}(t_n; X_0) \to y$ as $n \to +\infty$. From $\underline{X}(t; X_0) \in C_0(\mathcal{I} \times \mathcal{X}_0)$ it follows for $t \in \mathcal{I}_0[\underline{X}(t_n; X_0)]$ that (Lemma 3.1) $\underline{X}[t; \underline{X}(t_n; X_0)] = \underline{X}(t + t_n; X_0) \to \underline{X}(t; y)$ as $n \to +\infty$. Therefore, $\underline{X}(t; y) \in \mathcal{L}^+(X_0)$ and $\mathcal{L}^+(X_0)$ is weakly invariant.

**Theorem 3.19** *Let generalised motions $\underline{X}(t; X_0)$ of the system (3.7) be continuous in $X_0 \in \mathcal{X}_0$ for every $t \in \mathcal{I}_0(X_0; \underline{X})$, $\underline{X}(t; X_0) \in C_0(\mathcal{I}_0 \times \mathcal{X}_0)$. In order for $\mathcal{L}^+(X_0)$, $X_0 \in \mathcal{X}_0$ is arbitrary, to be nonempty, in $\mathcal{X}$ and compact it is both necessary and sufficient that $\underline{X}(t; X_0)$ is positively precompact relative to $\mathcal{X}$.* ∇∇∇

**Proof.** Let $\underline{X}(t; X_0) \in C_0(\mathcal{I}_0 \times \mathcal{X}_0)$. Then $\underline{X}(t; X_0)$ is unique for any $X_0 \in \mathcal{X}_0$ (Theorem 3.14).

*Necessity.* Let $\mathcal{L}^+(X_0)$ be nonempty, in $\mathcal{X}$ and compact, $X_0 \in \mathcal{X}_0$ is arbitrary. Since $\underline{X}(t; X_0) \to \mathcal{L}^+(X_0)$ as $t \to +\infty$ then $\underline{X}(t; X_0)$ is bounded because $\underline{X}(t; X_0) \in [C_0(\mathcal{I}_0 \times \mathcal{X}_0) \cap C_t(\mathcal{I}_0 \times \mathcal{X}_0)]$ and $\mathcal{L}^+(X_0)$ is bounded, and $\underline{X}(t; X_0)$ does not have a positive limit point on the boundary of $\mathcal{X}$ because $\mathcal{L}^+(X_0)$ is in $\mathcal{X}$. Hence, $\underline{X}(t; X_0)$ is precompact relative to $\mathcal{X}$ (Definition 3.12).

*Sufficiency.* Let $\underline{X}(t; X_0)$ be positively precompact relative to $\mathcal{X}$. Then, it is bounded and does not have a positive limit point on the boundary of $\mathcal{X}$ (Definition 3.12). Hence, $\mathcal{L}^+(X_0)$ is in $\mathcal{X}$, nonempty and bounded. It is also closed (Theorem 3.18), thus it is compact.

**Note 3.10** Theorem 3.18, Theorem 3.19 and the next theorem hold for $\mathcal{L}^-(X_0; \underline{X})$ provided "positive" and "positively" are replaced by "negative" and "negatively", respectively.

**Theorem 3.20** *Let generalised motions $\underline{X}(t; X_0)$ of the system (3.7) be continuous in $X_0 \in \mathcal{X}_0$ for every $t \in \mathcal{I}_0(X_0; \underline{X})$, $\underline{X}(t; X_0) \in C_0(\mathcal{I}_0 \times \mathcal{X}_0)$. If $\mathcal{X}_0$ is positively precompact relative to $\mathcal{X}$ then $\mathcal{L}^+(X_0)$ is connected for every $X_0 \in \mathcal{X}_0$.* ∇∇∇

**Proof.** Let $\underline{X}(t; X_0) \in C_0(\mathcal{I}_0 \times \mathcal{X}_0)$. Hence, $\underline{X}(t; X_0)$ is unique $\forall X_0 \in \mathcal{X}_0$. $\mathcal{L}^+(X_0)$ is nonempty, in $\mathcal{X}$ and compact (Theorem 3.19). If $\mathcal{L}^+(X_0)$ is assumed non-connected, then there are two disjoint subsets $\mathcal{L}_i^+(X_0) \subset \mathcal{L}^+(X_0)$, $i = 1, 2$,

$\mathcal{L}_1^+(X_0) \cup \mathcal{L}_2^+(X_0) = \mathcal{L}^+(X_0)$. The subsets are also nonempty and compact. There are $\alpha, \beta \in \Re^+$ such that $\mathcal{N}[\alpha, \mathcal{L}_1^+(X_0)] \cap \mathcal{N}[\beta, \mathcal{L}_2^+(X_0)] = \emptyset$. In view of Definition 3.10 there are two sequences $t_n$ and $\tau_n$, $y_1 \in \mathcal{L}_1^+(X_0)$ and $y_2 \in \mathcal{L}_2^+(X_0)$ such that $\underline{X}(t_n; X_0) \to y_1$ and $\underline{X}(\tau_n; X_0) \to y_2$ as $n \to +\infty$. From this and $\underline{X}(t; X_0) \in C_t(\mathcal{I}_0 \times \mathcal{X}_0)$ it follows that there is a sequence $\sigma_n$, $t_n < \sigma_n < \tau_n$ such that $\underline{X}(\sigma_n; X_0) \in \{(\mathcal{N}[\alpha, \mathcal{L}_1^+(X_0)] \cup \mathcal{N}[\beta, \mathcal{L}_2^+(X_0)]) - (\mathcal{L}_1^+(X_0) \cup \mathcal{L}_2^+(X_0))\}$. There exist sequences $\zeta_n$ and $\xi_n$ such that $\underline{X}(\sigma_m + \zeta_n; X_0) \to y_1$ and $\underline{X}(\sigma_m + \xi_n; X_0) \to y_2$ as $n \to +\infty$, for arbitrary natural number $m$. In view of Lemma 3.1, $\underline{X}(\sigma_m + \zeta_n; X_0) = \underline{X}[\zeta_n; \underline{X}(\sigma_m; X_0)] \to y_1$ as $n \to +\infty$ and $\underline{X}(\sigma_m + \xi_n; X_0) = \underline{X}[\xi_n; \underline{X}(\sigma_m; X_0)] \to y_2$ as $n \to +\infty$. This and $\underline{X}(t; X_0) \in C_t(\mathcal{I}_0 \times \mathcal{X}_0)$ imply existence of a sequence $\psi_n$, $t_n < \psi_n < \tau_n$, and $w \in \{(\mathcal{N}[\alpha, \mathcal{L}_1^+(X_0)] \cup \mathcal{N}[\beta, \mathcal{L}_2^+(X_0)]) - (\mathcal{L}_1^+(X_0) \cup \mathcal{L}_2^+(X_0))\}$ such that $\underline{X}(\sigma_m + \psi_n; X_0) = \underline{X}[\psi_n; \underline{X}(\sigma_m; X_0)] \to w$. Hence, $w \in \mathcal{L}^+(X_0)$ which contradicts $w \in \{(\mathcal{N}[\alpha, \mathcal{L}_1^+(X_0)] \cup \mathcal{N}[\beta, \mathcal{L}_2^+(X_0)]) - (\mathcal{L}_1^+(X_0) \cup \mathcal{L}_2^+(X_0))\}$ because $\mathcal{L}_1^+(X_0) \cup \mathcal{L}_2^+(X_0) = \mathcal{L}^+(X_0)$. Hence, $\mathcal{L}_1^+(X_0)$ and $\mathcal{L}_2^+(X_0)$ are not disjoint, i.e. $\mathcal{L}^+(X_0)$ is connected.

## 3.3 System regimes

### 3.3.1 Forced regimes and the free regime

**Definition 3.16** *(a) The system (3.4) is in a forced regime if and only if there is $t \in \Re_+$ such that $\mathtt{i}(t) \neq 0$.*

*(b) The system (3.4) is in a free regime if and only if $\mathtt{i}(t) = 0$ for all $t \in \Re_+$.*

The system (3.4) can be either in a forced or in a free regime. However, the system (3.7) is in a free regime.

### 3.3.2 Periodic regimes

**Definition 3.17** *The system (3.4) is in a periodic regime if and only if there is $\sigma \in \Re^+$ such that its generalised motions obey*

$$\underline{X}(t + \sigma; X_0; \mathtt{i}) \equiv \underline{X}(t; X_0; \mathtt{i}).$$

*The minimal $\sigma$ is called period. It is denoted by $T$. The generalised motion $\underline{X}(t; X_0; \mathtt{i})$ is called periodic (with period $T$).*

In literature it can be found that $\sigma = 0$, i.e. $T = 0$, is permitted, which means that time-invariant (elementwise constant) $\underline{X}(t; X_0; \mathtt{i}) \equiv const$ is considered as a periodic motion.

### 3.3.3 Stationary regimes and stationary points

**Definition 3.18** *The system (3.4) is in a stationary regime if and only if its generalised motions through $X_0$ obey*

$$\underline{X}(t; X_0; \mathtt{i}) = X_0, \quad \forall t \in \Re_+.$$

*The point $X_0$ is called a stationary point (stationary state, steady state), and it is denoted by $X_s$, $X_0 = X_s$.*

**Theorem 3.21** *In order for $X^* \in \mathcal{X}_0$ to be a stationary point of the system (3.4) it is necessary and sufficient that both*

*(a) the system has forward-time unique generalised motion $\underline{X}(t; X^*; \mathtt{i})$ through $X^*$,*

*and*

*(b) $f[X^*, \mathtt{i}(t)] = 0, \quad \forall t \in \Re_+$.* ∇∇∇

**Proof.** *Necessity.* Let $X^* = X_s$ be a stationary point of the system (3.4). Then, in view of Definition 3.18, any two generalised motions $\underline{X}_1(t; X^*; \mathtt{i})$ and $\underline{X}_2(t; X^*; \mathtt{i})$ obey

$$\underline{X}_k(t; X^*; \mathtt{i}) = X^*, \quad \forall t \in \Re_+, \quad k = 1, 2.$$

Hence, the system (3.4) has the forward-time unique generalised motion $\underline{X}(t; X^*; \mathtt{i}) = X^*$ through $X^*$, which obeys

$$\frac{d}{dt}\underline{X}(t; X^*; \mathtt{i}) = 0, \quad \forall t \in \Re_+.$$

This and (3.4) imply (b).

*Sufficiency.* Let (a) and (b) hold and $\underline{\psi}(t) = X^*$ for all $t \in \Re_+$. Hence,

$$\underline{\psi}(0) = X^*,$$

and

$$\frac{d}{dt}\underline{\psi}(t) = 0 = f[\underline{\psi}(t), \mathtt{i}(t)], \quad \forall t \in \Re_+.$$

The last result shows that $\underline{\psi}(\cdot)$ is a generalised motion of the system (3.4), $\underline{\psi}(t) = \underline{X}(t; 0; X^*; \mathtt{i}) = X^*$, for all $t \in \Re_+$. Since $\underline{X}(t; 0; X^*; \mathtt{i}) \equiv X^*$ is unique then $X^*$ is a stationary point.

### 3.3.4 Equilibrium regimes and equilibrium points

**Definition 3.19** *The system (3.4) is in an equilibrium regime if and only if its generalised motions in the free regime obey*

$$\underline{X}(t; X_0; 0) = X_0, \quad \forall t \in \Re_+.$$

*The point $X_0$ is called an equilibrium point (an equilibrium state), and it is denoted by $X_e$, $X_0 = X_e$.*

**Theorem 3.22** *In order for $X^* \in \mathcal{X}_0$ to be an equilibrium point (an equilibrium state) of the system (3.4) it is necessary and sufficient that both*

*(a) the system has forward-time unique generalised motion $\underline{X}(t; X^*; \mathtt{i})$ through $X^*$ in the free regime,*

*and*

*(b) $f(X^*, 0) = 0$.* ∇∇∇

**Proof.** Theorem 3.22 results directly from Theorem 3.21 due to $\mathtt{i}(t) \equiv 0$ (Definition 3.19).

A direct consequence of Theorem 3.22 is the following result due to (3.5)

**Corollary 3.1** *In order for $X^* = 0$ to be an equilibrium point of the system (3.4) it is necessary and sufficient that the system has forward-time unique generalised motion $\underline{X}(t; 0; 0) \equiv 0$ through $X^* = 0$ in the free regime.*

**Note 3.11** Theorem 3.21 and Theorem 3.22 directly apply to the system (3.7).

**Note 3.12** When we study equilibrium states and their properties it is the same to study them for the system (3.4) or for the system (3.7) due to (3.6) and Definition 3.19.

The next two results generalise Theorem 2.10 and Corollary 2.11 by Nemytskii and Stepanov [157, p. 332].

**Theorem 3.23** *If, for any $\varepsilon \in \Re^+$, there exists a point $y \in \mathcal{N}(\varepsilon, X^*)$, such that the right half-trajectory $\underline{X}(\Re_+; y; 0) \subset \mathcal{N}(\varepsilon, X^*)$, then $X^*$ is an equilibrium point of the system (3.4) as soon as system generalised motions $\underline{X}(t; X_0; 0) \in C_0(\Re_0 \times \mathcal{X}_0)$.* ∇∇∇

**Proof.** Let generalised motions of the system (3.4) $\underline{X}(t; X_0; 0) \in C_0(\Re_+ \times \mathcal{X}_0)$. Let for any $\varepsilon \in \Re^+$ there exist a point $y \in \mathcal{N}(\varepsilon, X^*)$ such that the right half-trajectory $\underline{X}(\Re_+; y; 0) \subset \mathcal{N}(\varepsilon, X^*)$. Let it be assumed that $X^*$ is not an equilibrium state of the system. Hence, for some $\sigma \in \Re^+$, $\underline{X}(\sigma; X^*; 0) \neq X^*$. Let $\rho[X^*, \underline{X}(\sigma; X^*; 0)]$

$= d > 0$. From $\underline{X}(t; X_0; 0) \in C_0(\Re_+ \times \mathcal{X}_0)$ it follows that there is $\varepsilon \in \left]0, \frac{d}{2}\right[$ such that $t \in [0, \sigma]$ and $\rho(z, X^*) < \varepsilon < \frac{d}{2}$ imply

$$\rho[\underline{X}(t; z; 0), \underline{X}(t; X^*; 0)] < \frac{d}{2}.$$

By the condition of the theorem there is $y \in \mathcal{N}(\varepsilon, X^*)$ such that

$$\underline{X}(\Re_+; y; 0) \subset \mathcal{N}(\varepsilon, X^*).$$

The triangle property of a distance function yields

$$\rho[X^*, \underline{X}(\sigma; X^*; 0)] < \rho[X^*, \underline{X}(\sigma; y; 0)] + \rho[\underline{X}(\sigma; y; 0), \underline{X}(\sigma; X^*; 0)],$$

which with the above results (in which we set $z = y$) leads to

$$\rho[X^*, \underline{X}(\sigma; y; 0)] + \rho[\underline{X}(\sigma; y; 0), \underline{X}(\sigma; X^*; 0)] < \varepsilon + \frac{d}{2} < d,$$

so that

$$\rho[X, \underline{X}(\sigma; X^*; 0)] < d.$$

This contradicts existence of $\sigma \in \Re^+$ as assumed and proves the theorem.

The requirement for $\underline{X}(t; X_0; 0) \in C_0(\Re_+ \times \mathcal{X}_0)$ can be omitted in case a point $X^*$ is stable. This is explained by the following statement.

**Theorem 3.24** *If $X^* \in \mathcal{X}_0$ is stable point of the system (3.4) then it is an equilibrium point of the system* $\nabla\nabla\nabla$

**Proof.** Let $X^* \in \mathcal{X}_0$ be stable point of the system (3.4). Then (Definition 1.1 of the Section 1.1.1 and Note 1.2), for any $\varepsilon \in \Re^+$ there is $\delta = \delta(\varepsilon) \in ]0, \varepsilon]$ such that $\rho(X_0, X^*) < \delta$ implies

$$\rho[\underline{X}(t; X_0; 0), X^*] < \varepsilon \text{ for all } t \in \Re_+$$

and for all the generalised motions of the system. Let $\varepsilon_n \in \Re_+$ be a sequence converging to zero as $n \to +\infty$. Then, $\delta(\varepsilon_n) \to 0$ as $n \to +\infty$. Since $\rho(X^*, X^*) < \delta(\varepsilon_n)$ for every $n \in \{1, 2, \ldots\}$ then for any generalised motion $\underline{X}(t; X^*; 0)$ holds

$$\rho[\underline{X}(t; X^*; 0), X^*] < \varepsilon_n \text{ for all } t \in \Re_+.$$

As $n \to +\infty$, $\varepsilon_n \to 0$ so that

$$\rho[\underline{X}(t; X^*; 0), X^*] = 0 \text{ for all } t \in \Re_+$$

and for every generalised motion $\underline{X}(t; X^*; 0)$. Equivalently, every generalised motion $\underline{X}(t; X^*; 0)$ obeys

$$\underline{X}(t; X^*; 0) = X^* \text{ for all } t \in \Re_+$$

which proves that $X^*$ is an equilibrium state of the system (Definition 3.19).

**Theorem 3.25** *Let generalised motions of the system (3.4)* $\underline{X}(t; X_0; 0) \in C_0(\Re_+ \times \mathcal{X}_0)$. *If for any* $\varepsilon \in \Re^+$, *there exists a point* $y \in \mathcal{N}(\varepsilon, X^*)$ *such that*

$$lim\,[\underline{X}(t; y; 0) : t \to +\infty] = X^* \in \Re^n$$

*then* $X^*$ *is an equilibrium point of the system.* ∇∇∇

**Proof.** Let the conditions of the theorem hold. From the definition of a limit, for any $\varepsilon \in \Re_+$, there is $\sigma \in \Re_+$ such that

$$\rho[\underline{X}(t; y; 0), X^*] < \varepsilon \quad \text{for all} \quad t \in [\sigma, +\infty[.$$

Let $z = \underline{X}(\sigma; y; 0)$, $z \in \mathcal{N}(\varepsilon, X^*)$, so that

$$\underline{X}(\Re_+; z; 0) \subset \mathcal{N}(\varepsilon, X^*), \tag{3.21}$$

in view of $\underline{X}[t; \underline{X}(\sigma; y; 0); 0] \equiv \underline{X}(t + \sigma; y; 0)$ (Lemma 3.1). The conditions of Theorem 3.23 are satisfied due to (3.21), which proves that $X$ is an equilibrium state of the system.

Theorem 3.25 is general. It can be linked with attraction of $X^* \in \mathcal{X}_0$ as follows.

**Theorem 3.26** *Let generalised motions of the system (3.4)* $\underline{X}(t; X_0; 0) \in C_0(\Re_+ \times \mathcal{X}_0)$. *If a point* $X^* \in \mathcal{X}_0$ *is attractive then it is an equilibrium point of the system.* ∇∇∇

**Proof.** Let $\underline{X}(t; X_0; 0) \in C_0(\Re_+ \times \mathcal{X}_0)$ and $X^* \in \mathcal{X}_0$ be attractive. Then, there is $\Delta > 0$ such that $\rho(y, X^*) < \Delta$ implies

$$lim\,[\underline{X}(t; y; 0) : t \to +\infty] = X^*. \tag{3.22}$$

Let $\varepsilon \in \Re^+$ be arbitrarily chosen and $\rho(y, X^*) < min(\Delta, \varepsilon)$. Then, (3.22) holds. Hence, all the conditions of Theorem 3.25 are satisfied, which completes the proof.

**Note 3.13** Theorems 3.23–3.26 hold obviously for the system (3.7).

## 3.4 Invariance properties of sets of equilibrium states

The set of all the equilibrium states denoted by $\mathcal{S}_e$ has interesting invariance properties.

**Theorem 3.27** *Let* $\mathcal{S}$ *be any nonempty subset of* $\mathcal{S}_e$ *– the set of the equilibrium states of the system (3.4),* $\mathcal{S} \subseteq \mathcal{S}_e$. *Then the set* $\mathcal{S}$ *is positively invariant.* ∇∇∇

**Proof.** Let $(\mathcal{S} \neq \emptyset) \subseteq \mathcal{S}_e$. Let $X_0 \in \mathcal{S}$ be arbitrary. Then $X_0 = X_e$, i.e.

$$\underline{X}(t; X_0) = X_0, \quad \forall t \in \Re_+,$$

due to Definition 3.19. Hence,

$$\underline{X}(t; X_0) \in \mathcal{S}, \quad \forall t \in \Re_+.$$

**Note 3.14** Theorem 3.27 is valid also for the system (3.7) due to (3.6) and Definition 3.19.

**Theorem 3.28** *If the system (3.7) has generalised motions $\underline{X}(t; X_0)$ continuous in $X_0 \in \mathcal{X}_0$ for every $t \in \Re$, $\underline{X}(t; X_0) \in C_0(\Re \times \mathcal{X}_0)$, with $\mathcal{I}_0(X_0) = \Re$ for all $X_0 \in \mathcal{X}_0$, then every nonempty subset $\mathcal{S}$ of the set $\mathcal{S}_e$ of its equilibrium points is invariant.* $\nabla\nabla\nabla$

**Proof.** Let $\mathcal{I}_0(X_0) = \Re$, $\forall X_0 \in \mathcal{X}_0$ and $\underline{X}(t; X_0) \in C_0(\Re \times \mathcal{X}_0)$. Then, Lemma 3.1 yields

$$\underline{X}[t; \underline{X}(\sigma; X_0)] = \underline{X}(t + \sigma; X_0), \quad \forall t \in \Re, \quad \forall \sigma \in \Re, \quad \forall X_0 \in \mathcal{X}_0. \tag{3.23}$$

Let $(\mathcal{S} \neq \emptyset) \subseteq \mathcal{S}_e$ be arbitrary. Let $X_0 \in \mathcal{S}_e$ be also arbitrary. Then $X_0 = X_e$ so that (Definition 3.19)

$$\underline{X}(t; X_0) = X_0, \quad \forall t \in \Re_+. \tag{3.24}$$

Let us assume that there is $\sigma \in \Re^-$ such that

$$\underline{X}(\sigma; X_0) \neq X_0. \tag{3.25}$$

Since (3.23) is true for any $t \in \Re$ then we may take $\tau \in \Re_+$ such that $\tau + \sigma \in \Re_+$, which with (3.24) implies

$$\underline{X}(\tau + \sigma; X_0) = X_0. \tag{3.26}$$

(3.23) and (3.26) show that

$$\underline{X}(\tau + \sigma; X_0) = \underline{X}[\tau; \underline{X}(\sigma; X_0)] = X_0, \quad \tau \in \Re_+, \quad (\tau + \sigma) \in \Re_+. \tag{3.27}$$

Since $\tau \in \Re_+$ and the system has unique generalised motions (Theorem 3.14) then Definition 3.19, (3.24) and (3.27) imply

$$\underline{X}(\sigma; X_0) = X_0,$$

which disproves (3.25) and shows that

$$\underline{X}(\sigma; X_0) \in \mathcal{S}, \quad \forall \sigma \in \Re^-.$$

This, (3.24) and $X_0 \in \mathcal{S}$ complete the proof.

This theorem directly implies the next result.

**Corollary 3.2** *If the system (3.7) has generalised motions $\underline{X}(t; X_0)$ continuous in $X_0 \in \mathcal{X}_0$ for every $t \in \Re$, $\underline{X}(t; X_0) \in C_0(\Re \times \mathcal{X}_0)$, with $\mathcal{I}_0(X_0) = \Re$ for all $X_0 \in \mathcal{X}_0$ then the set $\mathcal{S}_e$ of the equilibrium states cannot be reached in finite time by any generalised motion $\underline{X}(t; X_0)$ as soon as $X_0 \in (\mathcal{X}_0 - \mathcal{S}_e)$.*

**Note 3.15** If we reconsider the system $\dot{x} = -sign\, x$ (Example 3.7, Fig. 3.8), then we conclude that $\mathcal{S}_e = \{0\}$ is the set of its equilibrium states and it is singleton (i.e. $x_e = 0$ is its unique equilibrium state). For any $x_0 \in \Re$ the generalised motion $\underline{x}(t; x_0)$ reaches $\mathcal{S}_e = \{0\}$ in finite time that equals $|x_0|$. This does not contradict Corollary 3.2 because the system generalised motions are not continuous in $x_0 \in ]-\delta, \delta[$ for any $\delta \in \Re^+$ and for every $t \in \Re$. The same is valid for the system of Example 3.1.

# 3.5 Dynamical and generalised dynamical systems

We shall present at first the classical definition of dynamical systems [10], [120], [157], [218]. Then, it will be discussed from the point of view of the preceding results.

## 3.5.1 Definition and properties of dynamical systems

### Definition of dynamical systems

In the next definition "dynamical system" means "time-invariant continuous-time dynamical system".

**Definition 3.20** *A system with motions (or generalised motions) $\underline{X}$ is a dynamical system if and only if its (generalised) motions have the properties $P_1$ through $P_3$:*

$P_1$ *Existence: Every system (generalised) motion $\underline{X}$ is defined in $\Re^n$ on $\Re \times \Re^n$,*

$$\underline{X}(t; X_0) \in \Re^n, \quad \forall (t, X_0) \in \Re \times \Re^n,$$

*and satisfies the initial conditions,*

$$\underline{X}(0; X_0) = X_0, \quad \forall X_0 \in \Re^n.$$

$P_2$ *Continuous Dependence: Every (generalised) motion $\underline{X}$ is continuous in $(t, X_0) \in \Re \times \Re^n$,*

$$\underline{X}(t; X_0) \in C(\Re \times \Re^n).$$

$P_3$ *Group Property: for every $(s, t, X_0) \in \Re \times \Re \times \Re^n$ every (generalised) motion $\underline{X}$ satisfies*

$$\underline{X}[t; \underline{X}(s; X_0)] = \underline{X}(t + s; X_0).$$

**Comment 3.10** *Property $P_1$ means that $\mathcal{I}_0(X_0) = ]l^-, l^+[$ equals $\Re$, i.e. $l^- = -\infty$ and $l^+ = +\infty$ for every $X_0 \in \Re^n$.*

**Properties of dynamical systems**

**Comment 3.11** *What follows is due to Nemytskii and Stepanov [157, p. 327], Zubov [218, p. 20], and La Salle [120, p. 28]. Property $P_2$ can be interpreted in either of the following two equivalent ways:*

*a) If $(t_n, X_n) \in \Re \times \Re^n$ and it is a sequence converging to $(t, X) \in \Re \times \Re^n$, $(t_n, X_n) \to (t, X)$, then $\underline{X}(t_n; X_n)$ converges to $\underline{X}(t; X)$, $\underline{X}(t_n; X_n) \to \underline{X}(t; X)$.*

*b) For arbitrary $(s, y, \varepsilon) \in \Re \times \Re^n \times \Re^+$ there exists $\delta = \delta(\varepsilon; s, y) \in \Re^+$ such that $max\,\{\rho(X, y), |t - s|\} < \delta$ guarantees $\rho[\underline{X}(s; y), \underline{X}(t; X)] < \varepsilon$.*

This was used by Nemytskii and Stepanov [157, II′ on p. 327], to prove the next statement on motions, which is broadened to generalised motions.

**Theorem 3.29** *For any $(X, \sigma, \varepsilon) \in \Re^n \times \Re^+ \times \Re^+$, no matter how large $\sigma$ is and how small $\varepsilon$ is, there exists $\delta = \delta(\varepsilon; \sigma, X) \in \Re^+$ such that $\rho(X, y) < \delta$ guarantees $\rho[\underline{X}(t; X), \underline{X}(t; y)] < \varepsilon$ for all $t \in [-\sigma, \sigma]$ and for all (generalised) motions $\underline{X}$ of a dynamical system.* $\nabla\nabla\nabla$

**Comment 3.12** *Property $P_2$ concerns two different (generalised) motions through two different initial points $X$ and $y$ at two different instants $s$ and $t$, $\underline{X}(s; y)$ and $\underline{X}(t; X)$. However, Theorem 3.29 concerns such two (generalised) motions at the same time $t \in \Re$. It in fact proves Note 3.9, that is that $\underline{X}(t; X_0) \in C(\mathcal{I}_0 \times \mathcal{X}_0)$ implies both $\underline{X}(t; X_0) \in C_t(\mathcal{I}_0 \times \mathcal{X}_0)$ and $\underline{X}(t; X_0) \in C_0(\mathcal{I}_0 \times \mathcal{X}_0)$.*

**Note 3.16** In view of Theorem 3.29 and Comment 3.12, we conclude that Theorems 3.15–3.20, 3.23, 3.25 and 3.28 and Corollary 3.2 hold for dynamical systems. Theorems 3.21, 3.22, 3.24, 3.26 and 3.27 are valid for dynamical systems because those theorems are general.

From Property $P_2$ we deduce uniqueness of motions.

**Theorem 3.30** *A dynamical system has a unique (generalised) motion $\underline{X}$ through every $X_0 \in \Re^n$ at $t = 0$.* $\nabla\nabla\nabla$

**Proof.** The theorem can be proved in two ways. a) Let a system be dynamical. Definition 3.9 is valid. In view of Property $P_2$, Theorem 3.13 holds. Let it be assumed that the statement of Theorem 3.30 is false. Then, there exists $(\theta, X_0) \in \Re \times \Re^n$ such that through $X_0$ at $t = 0$ the system has two different (generalised) motions $\underline{X}_1$ and $\underline{X}_2$, $\underline{X}_k(0; X_0) = X_0$, $k = 1, 2$, that is $\underline{X}_1(\theta; X_0) \neq \underline{X}_2(\theta; X_0)$. Hence $\mu = \rho[\underline{X}_1(\theta; X_0), \underline{X}_2(\theta; X_0)] > 0$. Let $(\sigma, \varepsilon) \in ]\theta, +\infty[\times]0, \mu[$ and let $\delta = \delta(\varepsilon; \sigma, X_0)$ obey Theorem 3.13. Since $\delta > 0$ and $\rho(X_0, X_0) = 0$ then $\rho(X_0, X_0) < \delta$ so that $\rho[\underline{X}_1(\theta; X_0), \underline{X}_2(\theta; X_0)] < \varepsilon$, due to Theorem 3.13. The last inequality contradicts $\rho[\underline{X}_1(\theta; X_0), \underline{X}_2(\theta; X_0)] = \mu > \varepsilon$, which is the consequence of the assumption that $\mu > 0$. Hence, $\mu = 0$, or $\underline{X}_1(\theta; X_0) = \underline{X}_2(\theta; X_0)$. This further contradicts the existence of $\theta \in \Re$ at which the (generalised) motions $\underline{X}_1$ and $\underline{X}_2$ differ. Therefore, $\underline{X}_1(t; X_0) = \underline{X}_2(t; X_0) = \underline{X}(t; X_0)$ for all $t \in \Re$ and every $X_0 \in \Re^n$.

b) The theorem results from Definition 3.15a (see Note 3.6) and Theorem 3.14 (see Note 3.9 and Comment 3.12).

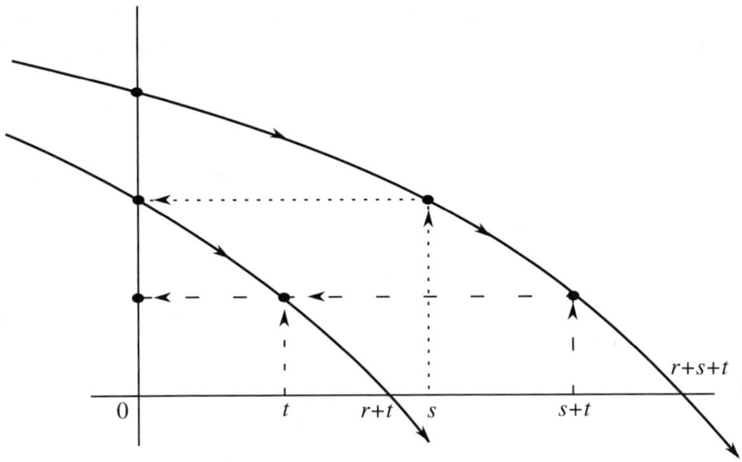


Figure 3.12: A graphical explanation of Group Property $P_3$: $x^* = \underline{x}(s; x_0)$, $\underline{x}(t; x^*) = \underline{x}[t; \underline{x}(s; x_0)] = \underline{x}(s+t; x_0)$. The same holds for any $r, s, t \in \Re$ and any $x_0 \in \Re$.

**Comment 3.13** *Bhatia and Szegö [10, p. 11] proved that every motion $\underline{X}$ of a dynamical system is a commutative group characterised by the following axioms, the first three of which are axioms of a group:*

(*i*) $\underline{X}[t; \underline{X}(0; X_0)] \equiv \underline{X}(t+0; X_0) \equiv \underline{X}(t; X_0)$.

(*ii*) $\underline{X}[t; \underline{X}(-t; X_0)] \equiv \underline{X}(t-t; X_0) \equiv \underline{X}(0; X_0)$.

(*iii*) $\underline{X}(t; \underline{X}[s; \underline{X}(r; X_0)]) \equiv \underline{X}(t; \underline{X}[s+r; X_0]) \equiv \underline{X}(t+s+r; X_0)$.

(*iv*) $\underline{X}[t; \underline{X}(s; X_0)] \equiv \underline{X}(t+s; X_0) \equiv \underline{X}(s+t; X_0) \equiv \underline{X}[s; \underline{X}(t; X_0)]$.

**Comment 3.14** *Group Property $P_3$ means that the (generalised) motion at time $(t+s)$, which passes through $X_0$ at $t=0$, equals the motion at time $t$, which passes through $X^*$ at $t=0$, provided for $X^*$ is taken $\underline{X}(s; X_0)$ achieved at time $s$ from $X_0$ at $t=0$, Fig. 3.12.*

**Comment 3.15** *Let us consider the systems of the preceding examples.*

*a)* $\dfrac{dX}{dt} = X^{2/3}$ *was analysed in Example 3.1, Fig. 3.1. It is not a dynamical system because it does not have a unique motion (or generalised motion) through $X_0 = 0$ at $t = 0$, i.e. not through every $X_0 \in \Re$ at $t = 0$ (Theorem 3.14).*

*b)* $\dfrac{dX}{dt} = sgn\, X$ *was considered in Example 3.2, Fig. 3.2, and Example 3.6, Fig. 3.7. Its motions through any $(X_0 \neq 0) \in \Re$ at $t = 0$ are not defined on $\Re$, and its generalised motions do not have Group Property $P_3$ which will be shown as follows: let $t = 3|X_0| \in \Re^+$, $s = -2|X_0| \in \Re^-$ and $sgn\, 0 \in \{-1, 0, +1\}$. Then, for $X_0 \in \Re^+$, the generalised motion is determined by*

$\underline{X}(s;X_0) = 0$ *and* $\underline{X}(t;X_0) = X_0 + t \; sgn \, X_0 = X_0 + 3|X_0| \, sgn \, X_0 = 4X_0$. *Further,* $\underline{X}(s+t;X_0) = \underline{X}(-2|X_0|+3|X_0|;X_0) = X_0+|X_0| \, sgn \, X_0 = 2X_0$, *and* $\underline{X}[t;\underline{X}(s;X_0)] = \underline{X}(t;0) = \underline{X}(3|X_0|;0) = X_0 + 3|X_0| \, sgn \, X_0 = 4X_0$ *Hence,* $\underline{X}[t;\underline{X}(s;X_0)] \neq \underline{X}(s+t;X_0)$. *The system is not a dynamical system.*

c) $\dfrac{dX}{dt} = -sgn \, X$, *Example 3.3, Fig. 3.3, and Example 3.7. Fig. 3.8. Let* $X_0 \in \Re^+$, $s \in \,]-\infty, -2X_0[$, $t = -\dfrac{s}{2}$. *Then,* $\underline{X}(t;X_0) = 0$, $\underline{X}[s,\underline{X}(t;X_0)] = \underline{X}(s;0) = 0$, $\underline{X}(t+s;X_0) = \underline{X}\left(-\dfrac{s}{2}+s;X_0\right) = \underline{X}\left(\dfrac{s}{2};X_0\right) = X_0 - \dfrac{s}{2} \, sign \, X_0$ $= X_0 - \dfrac{s}{2} > 2X_0$. *Hence,* $\underline{X}[s;\underline{X}(t;X_0)] \neq \underline{X}(t+s;X_0)$. *The Property* $P_3$ *is not realised. The system is not a dynamical system.*

d) $\dfrac{dX}{dt} = -X^{1/2}$ *of Example 3.4, Fig. 3.4, does not obey Property* $P_1$. *It is not a dynamical system.*

(e) $\dfrac{dX}{dt} = X^3$ *is considered in Example 3.5, Fig. 3.5. It does not satisfy Property* $P_1$ *because motions (which coincide with generalised motions) do not exist on* $\left]-\infty, -\dfrac{1}{2X_0^2}\right]$ *for every* $(X_0 \neq 0) \in \Re$. *The system is not a dynamical system.*

**Example 3.14** The first order system, $n = 1$

$$\frac{dX}{dt} = - \, sat \, X, \quad sat \, X = \begin{cases} X & , \; |X| \le 1, \\ sign \, X & , \; |X| \in \Re_+, \; |X| > 1 \end{cases}$$

has motions defined by

$$\underline{X}(t;X_0) = \begin{cases} \left.\begin{array}{ll} (1 + ln \, |X_0| - t) \; sign \, X_0 \, , & t \in \,]-\infty, ln \, |X_0|] \\ exp \, (-t) \; X_0 & , \; t \in [\, ln \, |X_0|, +\infty[ \end{array}\right\} & , \; |X_0| \in \,]0,1[, \\ \quad 0 \qquad , \; t \in \Re & , \; X_0 = 0, \\ \left.\begin{array}{ll} (|X_0| - t) \; sign \, X_0 & , \; t \in \,]-\infty, |X_0| - 1] \\ exp \, (-t+|X_0|-1) \; sign \, X_0 \, , & t \in [|X_0| - 1, +\infty[ \end{array}\right\} & , \; |X_0| \in [1, +\infty[. \end{cases}$$

The motions are defined and continuous in $(t, X_0) \in \Re \times \Re$, and differentiable in $t \in \mathcal{I}_0$ for all $X_0 \in \Re$. The system has a unique motion through every $X_0 \in \Re$ at $t = 0$, Fig. 3.13. Hence, the system motions possess Property $P_1$ and Property $P_2$. They do not possess Property $P_3$, which can be verified as follows. Let for example $X_0 \in \,]0,1[$, $s \in [\, ln \, |X_0|, +\infty[$ and $t \in [(ln \, |X_0| - s), +\infty[$. Then,

$$\underline{X}(s;X_0) = exp \, (-s) \; X_0 \Rightarrow |\underline{X}(s;X_0)| \le 1,$$

$$t + s \in [\, ln \, |X_0|, +\infty[,$$

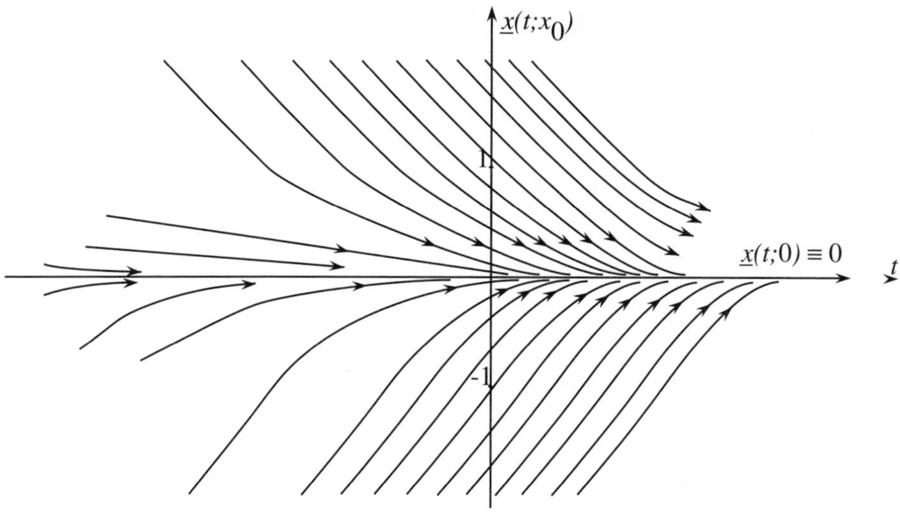


Figure 3.13: Motions of the system $\dot{x} = -sat\, x$ (Example 3.8).

$$\underline{X}(t+s;X_0) = exp\,(-s-t)\;X_0,$$

$$s \in [\,ln\,|X_0|, +\infty[ \Rightarrow t \in \Re,$$

$$\underline{X}[t;\underline{X}(s;X_0)] \;=\; exp\,(-t)\;\underline{X}(s;X_0) = exp\,(-t)\,exp\,(-s)\;X_0 = exp\,(-t-s)\;X_0,$$

$$\text{for} \quad t \in [\,ln\,|X_0|, +\infty[,$$

but

$$\underline{X}[t;\underline{X}(s;X_0)] = (1 + ln\,|\underline{X}(s;X_0)| - t)sign\,\underline{X}(s;X_0) = (1 + ln\,|X_0| - t - s)sign\,X_0,$$

$$\text{for} \quad t \in [-\infty, ln\,|X_0|[.$$

Hence,

$$\underline{X}[t,\underline{X}(s;X_0)] \neq \underline{X}(t+s;X_0), \quad t \in [-\infty, ln\,|X_0|[.$$

This illustrates that the motions do not possess Property 3. Hence, the system is not a dynamical system (Definition 3.9).

**Comment 3.16** *The function $f$ of the system $\dot{X} = -sat\,X$, $f(X) = -sat\,X$, is differentiable in $X \in (\Re - \{-1,1\})$. However, the motions are differentiable in time $t$ for all $t \in \Re$ and every $X_0 \in \Re$. This means they are differentiable also at time $t^*$ when $|\underline{X}(t^*;X_0)| = 1$, which further means $sat\,\underline{X}(t^*;X_0) = sign\,X_0$, $(X_0 \neq 0) \in \Re$. However, at that moment $f$ is not differentiable with respect to $X = \underline{X}(t^*;X_0)$. This illustrates that differentiability of $f$ on $\Re^n$ is not necessary for differentiability of motions in $t \in \Re$ for all $X_0 \in \Re^n$ in general.*

### 3.5.2 Definition and properties of generalised dynamical systems

#### Definition of generalised dynamical systems

Theorems 3.14–3.28 suggest generalisation of Definition 3.20.

**Definition 3.21** *A system is a generalised dynamical system if and only if its generalised motions $\underline{X}$ have the properties $P_1$ (Definition 3.20), $P_2'$ and $P_3$ (Definition 3.20), where*

$P_2'$ *Continuous Dependence: Every (generalised) motion $\underline{X}$ is continuous in $t \in \Re$ for every $X_0 \in \Re^n$ and in $X_0 \in \Re^n$ for every $t \in \Re$, that is*

$$\underline{X}(t; X_0) \in [C_t(\Re \times \Re^n) \cap C_0(\Re \times \Re^n)].$$

#### Properties of generalised dynamical systems

**Note 3.17** Theorems 3.14–3.28 hold for generalised dynamical systems provided $\mathcal{I}_0 \times \mathcal{X}_0 = \Re \times \Re^n$.

**Note 3.18** The system considered in Example 2.5 (Section 2.2.3) is not generalised dynamical system because $\underline{X}$ is not continuous in every $X_0 \in \Re^n$ for every $t \in \Re$ : $\underline{X}(t; X_0) \notin C_0(\Re \times \Re^n)$.

## 3.6 Stability properties and invariance properties of sets

At first we generalise Theorem 1.5.24 by Bhatia and Szegö [10, p. 54] to systems that need not be dynamical systems (in the sense of the classical mathematical Definition 3.20).

In what follows we consider the system (3.7) that has generalised motions $\underline{X}(t; X_0)$.

**Theorem 3.31** *If a closed set $\mathcal{A}$, $\mathcal{A} \subset \Re^n$ is stable then it is positively invariant.*

$\nabla\nabla\nabla$

**Proof.** (a) We essentially repeat the proof of Theorem 1.5.24 by Bhatia and Szegö [10, p. 54]. Let a closed set $\mathcal{A}$, $\mathcal{A} \subset \Re^n$, be stable. Since it is closed then

$$\cap [\mathcal{N}(\varepsilon, \mathcal{A}) : \varepsilon \in \Re^+] = \mathcal{A} \tag{3.28}$$

Since it is stable then $X_0 \in \mathcal{A}$ implies that every generalised motion $\underline{X}(t; X_0)$ of the system (3.7) obeys (3.29),

$$\underline{X}(t; X_0) \in \mathcal{N}(\varepsilon, \mathcal{A}) \quad \text{for all } t \in \Re \text{ and every } \varepsilon \in \Re^+, \tag{3.29}$$

because $\rho(X_0, \mathcal{A}) = 0 < \delta_M(\varepsilon)$ for every $\varepsilon \in \Re^+$. From (3.28) and (3.29) it follows that every generalised motion through $X_0 \in \mathcal{A}$ satisfies (3.30),

$$\underline{X}(t; X_0) \in \cap[\mathcal{N}(\varepsilon, \mathcal{A}) : \varepsilon \in \Re^+] = \mathcal{A} \quad \text{for all } t \in \Re_+, \tag{3.30}$$

which proves the theorem.

(b) The theorem can be proved also in the following way. Let a closed set $\mathcal{A}$, $\mathcal{A} \subset \Re^n$, be stable. Since it is closed, then

$$\rho(X, \mathcal{A}) = 0 \Leftrightarrow X \in \mathcal{A}. \tag{3.31}$$

Since it is stable then $X_0 \in \mathcal{A}$ guarantees that every generalised motion $\underline{X}(t; X_0)$ of the system (3.7) fulfils (3.32),

$$\rho[\underline{X}(t; X_0), \mathcal{A}] < \varepsilon \quad \text{for all } t \in \Re_+ \text{ and every } \varepsilon \in \Re^+, \tag{3.32}$$

because $\rho(X_0, \mathcal{A}) = 0 < \delta_M(\varepsilon)$. Let $\varepsilon \to 0^+$ which with (3.32) and (3.31) yield

$$\underline{X}(t; X_0) \in \mathcal{A} \text{ for all } t \in \Re_+.$$

Theorem 2.6.8 by Bhatia and Szegö [10, p. 137] is broadened to hold also for non-dynamical systems.

**Theorem 3.32** *Let $\mathcal{A}$, $\mathcal{A} \subset \Re^n$, be a compact set. If $\mathcal{A} = \cup[\mathcal{A}_i : i \in \mathcal{I}]$, where $\mathcal{I}$ is a finite index set and subsets $\mathcal{A}_i$ of $\mathcal{A}$, $\mathcal{A}_i \subset \mathcal{A}$, are connected and pairwise disjoint then for the set $\mathcal{A}$ to be stable it is necessary and sufficient that every $\mathcal{A}_i$ is stable.*
$\nabla\nabla\nabla$

**Proof.** Let $\mathcal{A}$, $\mathcal{A} \subset \Re^n$, be compact set and $\mathcal{A} = \cup[\mathcal{A}_i : i \in \mathcal{I}]$ with all $\mathcal{A}_i$'s connected, pairwise disjoint and with finite index set $\mathcal{I}$. Then, every $\mathcal{A}_i$ is compact and

$$\mathcal{N}(\varepsilon, \mathcal{A}) = \cup[\mathcal{N}(\varepsilon, \mathcal{A}_i) : i \in \mathcal{I}], \ \forall \varepsilon \in \Re^+. \tag{3.33}$$

*Necessity.* Let $\mathcal{A}$ be stable. Let $j \in \mathcal{I}$ be arbitrarily chosen. Let $\varepsilon^* \in \Re^+$ be such that

$$\mathcal{N}(\varepsilon^*, \mathcal{A}_j) \cap \mathcal{N}(\varepsilon^*, \mathcal{A}_k) = \emptyset, \ \forall (k \neq j) \in \mathcal{I}. \tag{3.34}$$

Such $\varepsilon^*$ exists because all $\mathcal{A}_i$ are compact and $\mathcal{I}$ is finite. Let $\varepsilon \in ]0, \varepsilon^*]$ be arbitrarily chosen. Then, $X_0 \in \mathcal{N}[\delta_M(\varepsilon; \mathcal{A}), \mathcal{A}]$ implies (Definition 1.3, Section 1.1.1.):

$$\underline{X}(t; X_0) \in \mathcal{N}(\varepsilon, \mathcal{A}), \ \ \forall t \in \Re_+. \tag{3.35}$$

Let $X^* \in \mathcal{N}[\delta_M(\varepsilon; \mathcal{A}), \mathcal{A}_j]$ be arbitrarily chosen. Since $\delta_M(\varepsilon; \mathcal{A}) \leq \varepsilon \leq \varepsilon^*$ then (3.33)–(3.35) imply

$$\underline{X}(t; X^*) \in \mathcal{N}(\varepsilon, \mathcal{A}_j), \ \ \forall t \in \Re_+.$$

For $\varepsilon \in ]\varepsilon^*, +\infty[$ it suffices to accept $\delta(\varepsilon; \mathcal{A}_j) = \delta_M(\varepsilon^*; \mathcal{A})$. Hence, for every $\varepsilon \in \Re^+$ there is $\delta(\varepsilon; \mathcal{A}_j) \in \Re^+$ obeying

$$\rho(X_0, \mathcal{A}_j) < \delta(\varepsilon; \mathcal{A}_j) \Rightarrow \rho[\underline{X}(t; X_0), \mathcal{A}_j] < \varepsilon, \ \ \forall t \in \Re_+.$$

This means stability of $\mathcal{A}_j$ because $\underline{X}(t; X_0)$ is also arbitrary generalised motion of the system (3.7).

*Sufficiency.* Let $\mathcal{A}_i$ be stable for any $i \in \mathcal{I}$. Let $\varepsilon \in \Re^+$ be arbitrarily chosen and

$$\delta(\varepsilon; \mathcal{A}) = min\ [\delta_M(\varepsilon; \mathcal{A}_i) : i \in \mathcal{I}].$$

Since $\delta_M(\varepsilon; \mathcal{A}_i) \in \Re^+$ for every $i \in \mathcal{I}$, and $\mathcal{I}$ is finite, then $\delta(\varepsilon; \mathcal{A}) \in \Re^+$. Let $X_0 \in \mathcal{N}[\delta(\varepsilon; \mathcal{A}), \mathcal{A}]$. Since $\mathcal{A}_i$-sets are pairwise disjoint and $\mathcal{A} = \cup[\mathcal{A}_i : i \in \mathcal{I}]$, then there is $j \in \mathcal{I}$ such that $X_0 \in \mathcal{N}[\delta(\varepsilon; \mathcal{A}), \mathcal{A}_j]$. The definition of $\delta(\varepsilon; \mathcal{A})$ yields $\mathcal{N}[\delta(\varepsilon; \mathcal{A}), \mathcal{A}_j] \subseteq \mathcal{N}[\delta_M(\varepsilon; \mathcal{A}_j), \mathcal{A}_j]$. Hence, $X_0 \in \mathcal{N}[\delta_M(\varepsilon; \mathcal{A}_j), \mathcal{A}_j]$ which guarantees $\underline{X}(t; X_0) \in \mathcal{N}(\varepsilon, \mathcal{A}_j)$, $\forall t \in \Re_+$, due to stability of $\mathcal{A}_j$. This result and (3.33) imply further $\underline{X}(t; X_0) \in \mathcal{N}(\varepsilon, \mathcal{A})$, $\forall t \in \Re_+$, which means stability of $\mathcal{A}$ because $\underline{X}(t; X_0)$ is arbitrary generalised motion of the system (3.7).

## 3.7 Invariance features of stability domains properties

Invariance properties of stability domains of sets hold also for stability domains of equilibrium states, which is obvious if we define a set $\mathcal{A} = \{X_e\}$. With this in mind it suffices to consider only invariance properties of stability domains of sets.

**Theorem 3.33** *Let the system (3.7) be a generalised dynamical system. Let a set $\mathcal{A}$, $\mathcal{A} \subset \Re^n$, be stable. Then, the sets $\mathcal{D}_s(\varepsilon; \mathcal{A})$ and $\mathcal{D}_{sc}(\varepsilon; \mathcal{A})$ for any $\varepsilon \in \Re^+$, the domain of stability $\mathcal{D}_s(\mathcal{A})$ and the strict domain of stability $\mathcal{D}_{sc}(\mathcal{A})$ of the set $\mathcal{A}$ are all positively invariant.* ∇∇∇

**Proof.** Let the system (3.7) be a generalised dynamical system. Then it has unique motions obeying the group property $P_3$ (Definition 3.20 and Definition 3.21). Let $\varepsilon \in \Re^+$ be arbitrary. Let $X_0 \in \mathcal{D}_s(\varepsilon; \mathcal{A})$ that implies $\underline{X}(t; X_0) \in \mathcal{N}(\varepsilon; \mathcal{A})$ for all $t \in \Re_+$. Let $\tau$ and $\sigma \in \Re_+$ be arbitrary and $y = \underline{X}(\sigma; X_0)$. Evidently, $(\tau + \sigma) \in \Re_+$ yields $\underline{X}(\tau + \sigma; X_0) \in \mathcal{N}(\varepsilon; \mathcal{A})$. Since (the Property $P_3$) $\underline{X}(\tau + \sigma; X_0) = \underline{X}[\tau; \underline{X}(\sigma; X_0)] = \underline{X}(\tau; y)$ then $\underline{X}(\tau; y) \in \mathcal{N}(\varepsilon; \mathcal{A})$ for all $\tau \in \Re_+$. This yields $y \in \mathcal{D}_s(\varepsilon; \mathcal{A})$, i.e. $\underline{X}(\sigma; X_0) \in \mathcal{D}_s(\varepsilon; \mathcal{A})$ for all $\sigma \in \Re_+$, which proves the positive invariance of $\mathcal{D}_s(\varepsilon; \mathcal{A})$. Let now $\hat{X}_0 \in \mathcal{D}_{sc}(\varepsilon; \mathcal{A})$ so that $\underline{X}(t; \hat{X}_0) \in \mathcal{N}(\varepsilon; \mathcal{A})$ for all $t \in \Re_+$. From $\underline{X}(t; X_0) \in C_t[\Re_+ \times \mathcal{D}_s(\mathcal{A})]$ which holds for a generalised dynamical system with stable set $\mathcal{A}$, it follows that $\hat{y} = \underline{X}(\sigma; \hat{X}_0)$ for any $\sigma \in \Re_+$, is connected with $\hat{X}_0$. Using this and $\underline{X}(\tau + \sigma; \hat{X}_0) = \underline{X}(\tau; \hat{y}) \in \mathcal{N}(\varepsilon; \mathcal{A})$ for all $\tau \in \Re_+$ we conclude that $\hat{y} = \underline{X}(\sigma; \hat{X}_0) \in \mathcal{D}_{sc}(\varepsilon; \mathcal{A})$ for any $\sigma \in \Re_+$, which proves positive invariance of $\mathcal{D}_{sc}(\varepsilon; \mathcal{A})$. Let $\tilde{X}_0 \in \mathcal{D}_s(\mathcal{A})$ be arbitrary. There is (Definition 2.2, Section 2.2.2) $\tilde{\varepsilon} \in \Re^+$ such that $\underline{X}(t; \tilde{X}_0) \in \mathcal{N}(\tilde{\varepsilon}; \mathcal{A})$ for all $t \in \Re_+$. Hence, $\tilde{X}_0 \in \mathcal{D}_s(\tilde{\varepsilon}; \mathcal{A})$ implies $\underline{X}(t; \tilde{X}_0) \in \mathcal{D}_s(\tilde{\varepsilon}; \mathcal{A}) \subseteq \mathcal{D}_s(\mathcal{A})$ for all $t \in \Re_+$ and proves positive invariance of $\mathcal{D}_s(\mathcal{A})$. Analogously, let $X_0^* \in \mathcal{D}_{sc}(\mathcal{A})$ be arbitrary.

Then there is $\varepsilon^* \in \Re^+$ such that $\underline{X}(t; X_0^*) \in \mathcal{N}(\varepsilon^*; \mathcal{A})$ for all $t \in \Re_+$. Hence, $X_0^* \in \mathcal{D}_{sc}(\varepsilon^*; \mathcal{A})$ yields $\underline{X}(t; X_0^*) \in \mathcal{D}_{sc}(\varepsilon^*; \mathcal{A}) \subseteq \mathcal{D}_{sc}(\mathcal{A})$ for all $t \in \Re_+$ and proves positive invariance of $\mathcal{D}_{sc}(\mathcal{A})$.

The next results broaden the validity of the classical theorems on properties of the domain of attraction from the framework of dynamical systems to generalised dynamical systems.

**Theorem 3.34** *Let the system (3.7) be a generalised dynamical system. Let a set $\mathcal{A}$, $\mathcal{A} \subset \Re^n$, be attractive. Then its domain of attraction $\mathcal{D}_a(\mathcal{A})$ is open and connected: $\mathcal{D}_a(\mathcal{A}) = \mathcal{D}_{ac}(\mathcal{A})$. Besides, $\mathcal{D}_a(\mathcal{A})$, its closure $\overline{\mathcal{D}}_a(\mathcal{A})$ and its boundary $\partial\mathcal{D}_a(\mathcal{A})$, if it is non-empty, are all invariant.* ∇∇∇

**Proof.** Let the system (3.7) be a generalised dynamical system. Let a set $\mathcal{A}$, $\mathcal{A} \subset \Re^n$, be attractive. Then it has the domain of attraction $\mathcal{D}_a(\mathcal{A})$ (Definition 1.5, Section 1.1.2; and Definition 2.4, Section 2.2.3). Let $X_0 \in \mathcal{D}_a(\mathcal{A})$ be arbitrarily chosen and $\zeta \in ]0, ||X_0||[$ be also arbitrarily chosen. There is $\tau = \tau(X_0, \zeta) \in \Re_+$ (Definition 2.3, Section 2.2.3) such that $\rho[\underline{X}(t; X_0), \mathcal{A}] < \zeta$ for all $t \in ]\tau, +\infty[$. Let $\psi \in \Re^+$ be such that $\mathcal{N}(\psi, \mathcal{A}) \subset \mathcal{D}_a(\mathcal{A})$. Such $\psi$ exists because both $\mathcal{D}_a(\mathcal{A})$ and $\mathcal{N}(\psi, \mathcal{A})$ are neighbourhoods of $\mathcal{A}$. Let $\varepsilon = m^{-1}\psi \in \Re^+$ for any $m \in \{4, 5, \ldots\}$. Since the system (3.7) is a generalised dynamical system, then (Definition 3.21, Property $P_2'$) there is $\delta = \delta(\varepsilon, \sigma) \in \Re^+$ for $\sigma = \tau(X_0, \varepsilon)$ such that $X^* \in \mathcal{N}(\delta, X_0)$ guarantees $\underline{X}(\sigma; X^*) \in \mathcal{N}(\varepsilon, y)$ for $y = \underline{X}(\sigma; X_0)$. Now, $\mathcal{N}(\psi, \mathcal{A}) \subset \mathcal{D}_a(\mathcal{A})$, $\varepsilon = m^{-1}\psi$, $m \in \{4, 5, \ldots\}$ and $y = \underline{X}(\sigma; X_0) \in \mathcal{N}(\varepsilon, \mathcal{A})$ imply $\mathcal{N}(\varepsilon, \mathcal{A}) \subset \mathcal{N}(\psi, \mathcal{A})$ and $\mathcal{N}(\varepsilon, \mathcal{A}) \subset \mathcal{D}_a(\mathcal{A})$. Hence, $\underline{X}(\sigma; X^*) \in \mathcal{N}(\psi, \mathcal{A})$ and $\underline{X}(\sigma; X^*) \in \mathcal{D}_a(\mathcal{A})$. Let $z = \underline{X}(\sigma; X^*)$, $z \in \mathcal{D}_a(\mathcal{A})$ so that for any $\zeta \in \Re^+$ there is $\tau(z, \zeta) \in \Re_+$ obeying $\rho[\underline{X}(t; X_0), \mathcal{A}] < \zeta$ for all $t \in ]\tau(z, \zeta), +\infty[$. Since $\underline{X}(t; z) = \underline{X}[t; \underline{X}(\sigma; X^*)] = \underline{X}(t + \sigma; X^*)$ (Definition 3.20 and Definition 3.21, Property $P_3$) then $X^* \in \mathcal{N}(\delta, X_0)$ ensures for any $\zeta \in \Re^+$ the existence of $\tau(X^*, \zeta) \in \Re_+$ such that $\rho[\underline{X}(t; X^*), \mathcal{A}] < \zeta$ for all $t \in ]\tau(X^*, \zeta), +\infty[$. One possible choice for $\tau(X^*, \zeta)$ is $\tau(X^*, \zeta) = [\sigma + \tau(z, \zeta)] \in \Re_+$. Hence, any $X^* \in \mathcal{N}(\delta, X_0)$ is in $\mathcal{D}_a(\mathcal{A})$. $\mathcal{D}_a(\mathcal{A})$ is open. Its connectedness is due to (a) of Definition 2.4 (Section 2.2.3) and the properties $P_1$ and $P_2'$ of generalised dynamical systems (Definition 3.21).

The system (3.7) has unique generalised motions (Definition 3.21 and Property $2'$; and Theorem 3.14), which possess Property 3, $\underline{X}(t + \alpha; X_0) = \underline{X}[t; \underline{X}(\alpha; X_0)]$, $\forall (t, \alpha) \in \Re \times \Re$, $\forall X_0 \in \Re^n$. If $\mathcal{D}_a(\mathcal{A})$ were not invariant then there would exist $(\beta, w) \in \Re \times \mathcal{D}_a(\mathcal{A})$ such that $\underline{X}(\beta; w) \notin \mathcal{D}_a(\mathcal{A})$. However $\underline{X}(\gamma + \beta; w) = \underline{X}[\gamma; \underline{X}(\beta; w)]$ for every $\gamma \in \Re$. Let $v = \underline{X}(\beta; w)$. From $w \in \mathcal{D}_a(\mathcal{A})$ it follows $\underline{X}(t; w) \in \mathcal{N}(\zeta, \mathcal{A})$ for all $t \in ]\tau(\zeta, w), +\infty[$ so that $\underline{X}(\gamma + \beta; w) = \underline{X}(\gamma; v) \in \mathcal{N}(\zeta, \mathcal{A})$ for all $\gamma \in ]\tau(\zeta, w) - \beta, +\infty[$ and for any $\zeta \in \Re^+$. Therefore $v = \underline{X}(\beta; w) \in \mathcal{D}_a(\mathcal{A})$ that disproves $\underline{X}(\beta; w) \notin \mathcal{D}_a(\mathcal{A})$ and proves invariance of $\mathcal{D}_a(\mathcal{A})$.

Since $\mathcal{D}_a(\mathcal{A})$ is open and invariant then $\overline{\mathcal{D}}_a(\mathcal{A})$ and $\partial\mathcal{D}_a(\mathcal{A})$, if it is nonempty, are invariant (Theorem 3.17).

**Comment 3.17.** *Since $\partial\mathcal{D}_a(\mathcal{A})$ of $\mathcal{D}_a(\mathcal{A})$ is invariant when the system (3.7) is a generalised dynamical system then obviously*

$$lim\,\{\rho[\underline{X}(t; X_0), \mathcal{A}] : t \to +\infty\} > 0 \quad \text{for } X_0 \in \partial\mathcal{D}_a.$$

There is an interesting relationship among domains of stability properties [81]:

**Theorem 3.35** *Let a set $\mathcal{A}$, $\mathcal{A} \subset \Re^n$, be asymptotically stable for the system (3.7) with generalised motions $\underline{X}(t; X_0) \in [C_t(\Re_+ \times \mathcal{X}_0) \cap C_0(\Re_+ \times \mathcal{X}_0)]$, where $[\mathcal{D}_s(\mathcal{A}) \cup \mathcal{D}_a(\mathcal{A})] \subseteq \mathcal{X}_0$. Then,*

$$\mathcal{D}_a(\mathcal{A}) = \mathcal{D}_{ac}(\mathcal{A}) \subseteq \mathcal{D}_{sc}(\mathcal{A}) \subseteq \mathcal{D}_s(\mathcal{A})$$

*and*

$$\mathcal{D}(\mathcal{A}) = \mathcal{D}_c(\mathcal{A}) = \mathcal{D}_a(\mathcal{A}) = \mathcal{D}_{ac}(\mathcal{A}). \qquad \nabla\nabla\nabla$$

**Proof.** Let a set $\mathcal{A}$, $\mathcal{A} \subset \Re^n$, be asymptotically stable for the system (3.7) whose generalised motions $\underline{X}(t; X_0) \in [C_t(\Re_+ \times \mathcal{X}_0) \times C_0(\Re_+ \times \mathcal{X}_0)]$ and $\mathcal{X}_0 \supseteq [\mathcal{D}_s(\mathcal{A}) \cup \mathcal{D}_a(\mathcal{A})]$. Then, $\underline{X}(t; X_0)$ is unique, $X_0 \in \mathcal{X}_0$. Let $X_0 \in \mathcal{D}_a(\mathcal{A})$ be arbitrarily chosen, which is possible because the set $\mathcal{A}$ has $\mathcal{D}_a(\mathcal{A})$, $\mathcal{D}_s(\mathcal{A})$ and $\mathcal{D}(\mathcal{A})$ due to its asymptotic stability. Hence, $lim\, \{\rho[\underline{X}(t; X_0), \mathcal{A}] : t \to +\infty\} = 0$. This, $\underline{X}(t; X_0) \in C_t(\Re_+ \times \mathcal{X}_0)$ and $\mathcal{D}_a(\mathcal{A}) \subseteq \mathcal{X}_0$ imply $max\, \{\rho[\underline{X}(t; X_0), \mathcal{A}] : t \in \Re_+\} = \mu \in \Re^+$. Let $\varepsilon = 2\mu \in \Re^+$. Hence, $\underline{X}(t; X_0) \in \mathcal{N}(\varepsilon; \mathcal{A})$, $\forall t \in \Re_+$, which implies $X_0 \in \mathcal{D}_s(\varepsilon; \mathcal{A})$ and $X_0 \in \mathcal{D}_s$ (Definition 2.2, Section 2.2.2). Hence, $\mathcal{D}_a(\mathcal{A}) \subseteq \mathcal{D}_s(\mathcal{A})$ so that $\mathcal{D}(\mathcal{A}) = \mathcal{D}_a(\mathcal{A})$ due to $\mathcal{D}(\mathcal{A}) = \mathcal{D}_s(\mathcal{A}) \cap \mathcal{D}_a(\mathcal{A})$. From $\underline{X}(t; X_0) \in C_t(\Re_+ \times \mathcal{X}_0)$ and $\mathcal{D}_a(\mathcal{A}) \subseteq \mathcal{X}_0$ it follows that for any $X_0 \in \mathcal{D}_a(\mathcal{A})$ and any $t \in \Re_+$ the point $y = \underline{X}(t; X_0)$ is connected with $X_0$. This and existence of $\tau \in \Re_+$ such that $\underline{X}(\tau; X_0) \in \mathcal{D}_{ac}(\mathcal{A})$ due to $lim\, \{\rho[\underline{X}(t; X_0), \mathcal{A}] : t \to +\infty\} = 0$ imply that $X_0$ is connected with $\underline{X}(\tau; X_0)$. Hence, $X_0 \in \mathcal{D}_{ac}(\mathcal{A})$ and $\mathcal{D}_a(\mathcal{A}) \subseteq \mathcal{D}_{ac}(\mathcal{A})$. By definition, $\mathcal{D}_{ac}(\mathcal{A}) \subseteq \mathcal{D}_a(\mathcal{A})$. Therefore $\mathcal{D}_a(\mathcal{A}) = \mathcal{D}_{ac}(\mathcal{A})$. Let $X_0 \in \mathcal{D}_{ac}(\mathcal{A}) = \mathcal{D}_a(\mathcal{A})$ and $\varepsilon = 2\mu \in \Re^+$ be defined as above so that for all $t \in \Re_+$, $\underline{X}(t; X_0) \in \mathcal{N}(\varepsilon; \mathcal{A})$, which together with $lim\, \{\rho[\underline{X}(t; X_0), \mathcal{A}] : t \to +\infty\} = 0$ and $\underline{X}(t; X_0) \in C_t[\Re_+ \times \mathcal{D}_a(\mathcal{A})]$ imply existence of $\tau \in \Re_+$ such that $\underline{X}(\tau; X_0) \in \mathcal{D}_{sc}(\varepsilon; \mathcal{A})$, connectedness of $\underline{X}(t; X_0)$ at any $t \in \Re_+$ with $\underline{X}(\tau; X_0)$ and $\underline{X}(t; X_0) \in \mathcal{D}_{sc}(\varepsilon; \mathcal{A}) \subseteq \mathcal{D}_{sc}(\mathcal{A})$ for all $t \in \Re_+$. Hence, $X_0 \in \mathcal{D}_a(\mathcal{A}) = \mathcal{D}_{ac}(\mathcal{A})$ implies $X_0 \in \mathcal{D}_{sc}(\mathcal{A})$ and $\mathcal{D}_a(\mathcal{A}) = \mathcal{D}_{ac}(\mathcal{A}) \subseteq \mathcal{D}_{sc}$. By the definitions $\mathcal{D}_{sc}(\mathcal{A}) \subseteq \mathcal{D}_s(\mathcal{A})$, $\mathcal{D}_c(\mathcal{A}) = \mathcal{D}_{sc}(\mathcal{A}) \cap \mathcal{D}_{ac}(\mathcal{A})$ and $\mathcal{D}(\mathcal{A}) = \mathcal{D}_s(\mathcal{A}) \cap \mathcal{D}_a(\mathcal{A})$. Therefore, $\mathcal{D}(\mathcal{A}) = \mathcal{D}_a(\mathcal{A}) = \mathcal{D}_{ac}(\mathcal{A}) = \mathcal{D}_c(\mathcal{A}) \subseteq \mathcal{D}_{sc}(\mathcal{A}) \subseteq \mathcal{D}_s(\mathcal{A})$.

**Comment 3.18.** *This theorem explains why many authors use the term "region of attraction" for the "domain of asymptotic stability". It also justifies the use of "the asymptotic stability domain" for "the strict asymptotic stability domain".*

**Comment 3.19.** *Various other qualitative features of domains of stability properties were discovered and described by Bhatia and Szegö [10], Genesio et al. [53] and Chiang and Thorp [30].*

We shall particularly consider stability of equilibrium states on the boundaries of domains of stability properties.

## 3.8 Features of equilibrium states on boundaries of domains of stability properties

At first we establish a theorem discovering that the boundary of attraction domain cannot contain either a stable or an attractive equilibrium state.

**Theorem 3.36** *If a set* $\mathcal{A} \subset \Re^n$ *is attractive and the boundary* $\partial\mathcal{D}_a(\mathcal{A})$ *of its domain of attraction contains at least one equilibrium state of a given system then every such an equilibrium state is unstable and non-attractive.* ∇∇∇

**Proof.** Let a set $\mathcal{A} \subset \Re^n$ be attractive and let $X_e \in \partial\mathcal{D}_a(\mathcal{A})$ be an equilibrium state.

$X_e$ *is unstable.* If $X_e$ were stable then for every $\varepsilon \in \Re^+$ there would be $\delta = \delta(\varepsilon; X_e) \in \Re^+$ such that $\rho(X_0, X_e) < \delta$ guarantees $\rho[\underline{X}(t; X_0), X_e] < \varepsilon$ for all $t \in \Re_+$. Let $\rho(X_e, \mathcal{A}) = d$ and $\varepsilon^* \in \left]0, \frac{d}{2}\right[$. Let $X_0 \in [\mathcal{N}(\delta^*, X_e) \cap \mathcal{D}_a(\mathcal{A})]$, $\delta^* = \delta(\varepsilon^*; X_e) \in \Re^+$. Hence, $\rho[\underline{X}(t; X_0), X_e] < \varepsilon^* < \frac{d}{2}$ for all $t \in \Re_+$, or, $\rho[\underline{X}(t; X_0), \mathcal{A}] \geq \frac{d}{2} > 0$ for all $t \in \Re_+$ that contradicts $X_0 \in \mathcal{D}_a(\mathcal{A})$. Therefore, $X_e$ cannot be stable.

$X_e$ *cannot be attractive.* If $X_e$ were attractive then there would exist $\Delta > 0$ such that $\rho(X_0, X_e) < \Delta$ implies $lim\{\rho[\underline{X}(t; X_0), X_e] : t \to +\infty\} = 0$. Let $X_0 \in [\mathcal{N}(\Delta, X_e) \cap \mathcal{D}_a(\mathcal{A})]$. Then $lim\{\rho[\underline{X}(t; X_0), \mathcal{A}] : t \to +\infty\} > 0$ that contradicts $X_0 \in \mathcal{D}_a(\mathcal{A})$. Hence $X_e \in \partial\mathcal{D}_a(\mathcal{A})$ cannot be attractive.

This theorem yields the following:

**Theorem 3.37** *If a set* $\mathcal{A} \subset \Re^n$ *is asymptotically stable and its domains of stability* $[\mathcal{D}_s(\mathcal{A})]$, *of attraction* $[\mathcal{D}_a(\mathcal{A})]$ *and of asymptotic stability* $[\mathcal{D}(\mathcal{A})]$ *are interrelated by*

$$\mathcal{D}(\mathcal{A}) = \mathcal{D}_a(\mathcal{A}) \subseteq \mathcal{D}_s(\mathcal{A})$$

*then every equilibrium state* $X_e$ *of a given system, which is in the boundary* $\partial\mathcal{D}(\mathcal{A})$, $X_e \in \partial\mathcal{D}(\mathcal{A})$, *is unstable and non-attractive.* ∇∇∇

**Proof.** Since $\mathcal{D}(\mathcal{A}) = \mathcal{D}_a(\mathcal{A})$ and they exist as requested then the statement of the theorem results from Theorem 3.36.

# Chapter 4

# Foundations of the Lyapunov method

## 4.1 Introductory comment

Definitions of stability properties and of their domains are stated via (generalised) motions of systems. A test of a stability feature via its definition demands knowledge of system (generalised) motions (for infinitely many different initial points $X_0$ from a neighbourhood of an equilibrium state or a set). Determination of the (generalised) motions further requires solving the system's mathematical model, which is rarely possible analytically in the closed form for nonlinear systems described by nonlinear first-order vector differential equation (4.1):

$$\frac{dX}{dt} = f(X). \tag{4.1}$$

Papers by Poincaré [165], [166] inspired Lyapunov [127] to pose the problem of testing stability properties by using the equation (4.1) directly rather than by solving it. If $f(X) = AX$ the system is linear and stability properties are tested via real parts of the eigenvalues of the matrix $A$, and eventually also via their multiplicities. However, when $f$ is nonlinear, then it was necessary to discover another, essentially different, method. It was discovered and established in 1892 by Lyapunov [127] and has been well-known as *the direct method of Lyapunov*, or *the second method of Lyapunov*, or simply as *the Lyapunov method.* It is based on the concept of *sign* definite functions $v : \Re^n \to \Re$ and their behaviour $v[\underline{X}(t; X_0)]$ along (generalised) motions $\underline{X}(t; X_0)$, which is to be tested without solving the equation (4.1) for $\underline{X}(t; X_0)$. In order to achieve this goal, Lyapunov used the total time derivative of a function $v$ assumed to be positive definite (that will be defined in the next section) along an arbitrary (generalised) motion of the system. Lyapunov supposed that the function $v$ is differentiable on a hyperball $\mathcal{B}_\alpha$, $v(X) \in C^{(1)}[\mathcal{B}_\alpha]$, $\alpha \in \Re^+$. Then its total time derivative $dv(X)/dt$ along system motion at a point $X = \underline{X}(t; X_0)$ is its Eulerian derivative that can be expressed via $f(X)$,

$$\dot{v}(X) = \frac{d}{dt}v(X) = \frac{d}{dt}v[\underline{X}(t; X_0)] = [grad\, v(X)]^T \frac{dX}{dt} = [grad\, v(X)]^T f(X). \tag{4.2}$$

The behaviour of $v$ along $\underline{X}(t; X_0)$ is now studied via the *sign* of $\dot{v}(X)$ at every $X \in \mathcal{B}_\alpha$.

This immediately poses questions of the sense, origin and construction of a function $v$ for a given system (4.1). These questions will be mathematically clarified in the next chapter. Their excellent physical clarification in the framework of conservative systems can be found in the book by Rouche and Mawhin [178, pp. 3–4].

In what follows the classic Lyapunov method and the related results will be briefly synthesised. For extended studies see [1]–[10], [14]–[17], [22]–[33], [36]–[38], [40]–[42], [45]–[47], [49], [51], [53]–[56], [59], [64], [66]–[70], [72], [73], [80], [87]–[90], [94]–[98], [100]–[102], [104], [106]–[108], [112]–[124], [126], [128]–[132], [134]–[140], [147]–[149], [151]–[155], [158]–[160], [167], [170], [171], [174]–[193], [196]–[204], [210]–[218].

## 4.2 Sign definite functions

### 4.2.1 Sign semi-definite functions

**Definition 4.1** *A function $v : \Re^n \to \Re$ is:*

a) *positive semi-definite with respect to a set $\mathcal{A}$, $\mathcal{A} \subset \Re^n$, if and only if there is a neighbourhood $\mathcal{N}(\mathcal{A})$ of the set $\mathcal{A}$ such that*

1) *the function $v$ is continuous on $\mathcal{N}(\mathcal{A})$:*

$$v(X) \in C[\mathcal{N}(\mathcal{A})],$$

2) *the function $v$ is non-positive on the interior $\mathring{\mathcal{A}}$ of the set $\mathcal{A}$:*

$$v(X) \leq 0, \quad \forall X \in \mathring{\mathcal{A}},$$

3) *the function $v$ vanishes on the boundary $\partial\mathcal{A}$ of the set $\mathcal{A}$:*

$$v(X) = 0, \quad \forall X \in \partial\mathcal{A},$$

4) *the function $v$ is non-negative on $\mathcal{N}(\mathcal{A})$,*

$$v(X) \geq 0, \quad \forall X \in [\mathcal{N}(\mathcal{A}) - \mathring{\mathcal{A}}],$$

*and*

5) *there exists $y \in \mathcal{N}(\mathcal{A})$ at which the function $v$ has a positive real value*

$$\exists y \in \mathcal{N}(\mathcal{A}), \quad v(y) \in \Re^+.$$

b) *positive semi-definite with respect to a set $\mathcal{A}$, $\mathcal{A} \subset \Re^n$, on a set $\mathcal{S}$ if and only if the conditions under a) are satisfied for $\mathcal{N}(\mathcal{A}) = \mathcal{S}$.*

*c) positive semi-definite with respect to a set $\mathcal{A}$, $\mathcal{A} \subset \Re^n$, in the whole (or in the large or globally positive semi-definite with respect to a set $\mathcal{A}$) if and only if the conditions under a) are satisfied for $\mathcal{N}(\mathcal{A}) = \Re^n$.*

*d) The expression "with respect to a set $\mathcal{A}$, $\mathcal{A} \subset \Re^n$", is omitted if and only if $\mathcal{A} = \mathcal{O} = \{0\}$.*

**Example 4.1** Let a function $v$ be defined by

1) $v(X) = \mathrm{a}^T X$. It is continuous on $\Re^n$ and vanishes at the origin. It is not positive semi-definite because there is not a neighbourhood of $X = 0$ on which it is non-negative.

2) $v(x) = sign\,|x|$. It is non-negative on $\Re$ and vanishes at $x = 0$. However, it is not positive semi-definite because it is not continuous at $x = 0$.

3) $v(x) = cos\, x$. It is non-negative and continuous on $\left[-\frac{\pi}{2}, \frac{\pi}{2}\right]$. However, it does not vanish at $x = 0:\ cos\,0 = 1 \neq 0$. Thus it is not positive semi-definite.

4) $v(X) = \left|\sum_{i=1}^{n} \alpha_i x_i\right|$ is positive semi-definite in the whole.

5) $v(x) = sin\,|x|$ is positive semi-definite on $[-\pi, \pi]$.

6) $v(x) = \begin{cases} 0 & , |x| \leq 1, \\ |x| - 1 & , |x| \geq 1, \end{cases}$ is positive semi-definite with respect to $\mathcal{A} = \{x : x \in \Re,\ |x| \leq 1\}$ in the whole.

7) $v(x) = \begin{cases} 0 & , |x| \leq 1, \\ (|x| - 1)(4 - |x|) & , |x| \geq 1, \end{cases}$, is positive semi-definite with respect to $\mathcal{A} = \mathcal{B}_1$ on the set $\mathcal{S} = \{x : x \in \Re,\ |x| \leq 4\}$, but not in the large.

8) $v(x) \equiv 0$ is not positive semi-definite because there are not a neighbourhood $\mathcal{N}$ of $x = 0$ and $y \in \mathcal{N}$ such that $v(y) > 0$.

**Definition 4.2** *A function $v : \Re^n \to \Re$ is (globally) negative semi-definite [with respect to a set $\mathcal{A}$, $\mathcal{A} \subset \Re^n$] $\{$on a set $\mathcal{S}\}$ if and only if the function $-v$ is (globally) positive semi-definite [with respect to the set $\mathcal{A}$] $\{$on the set $\mathcal{S}\}$, respectively.*

**Definition 4.3** *(a) A matrix $H \in \Re^{n \times n}$ is positive semi-definite if and only if its quadratic form $X^T HX$ is positive semi-definite.*

*(b) A matrix $H \in \Re^{n \times n}$ is negative semi-definite if and only if its quadratic form $X^T HX$ is negative semi-definite.*

The next theorem is a slight modification [88, p. 22], of the classical criteria for semi-definite properties [50, Vol. 1, p. 307). Let $H = (h_{ij}) \in \Re^{n\times n}$. Its principal minors $H^{j_1 j_2 \ldots j_r}_{i_1 i_2 \ldots i_r}$ are defined by

$$H^{j_1 j_2 \ldots j_r}_{i_1 i_2 \ldots i_r} = \begin{vmatrix} h_{i_1 j_1} & h_{i_1 j_2} & \cdots & h_{i_1 j_r} \\ h_{i_2 j_1} & h_{i_2 j_2} & \cdots & h_{i_2 j_r} \\ - & - & \cdots & - \\ h_{i_r j_1} & h_{i_r j_2} & \cdots & h_{i_r j_r} \end{vmatrix}, \quad \begin{array}{l} \forall r = 1, 2, \ldots, n, \\ \forall i_k \in \{1, 2, \ldots, n\}, \\ \forall j_k \in \{1, 2, \ldots, n\}, \\ \forall k = 1, 2, \ldots, r. \end{array}$$

**Theorem 4.1** *In order for a matrix $H = H^T \in \Re^{n\times n}$ to be positive semi-definite, it is necessary and sufficient that*

*1) all its principal minors $H^{j_1 j_2 \ldots j_r}_{i_1 i_2 \ldots i_r}$ are non-negative,*

$$H^{j_1 j_2 \ldots j_r}_{i_1 i_2 \ldots i_r} \geq 0, \quad \begin{array}{l} \forall r = 1, 2, \ldots, n, \\ \forall i_k \in \{1, 2, \ldots, n\}, \\ \forall j_k \in \{1, 2, \ldots, n\}, \\ \forall k = 1, 2, \ldots, r. \end{array}$$

*and*

*2) at least one of its principal minors is positive,*

$$\exists r = \{1, 2, \ldots, n\ \} \ ; \ \exists i_k \in \{1, 2, \ldots, n\}, \ \exists j_k \in \{1, 2, \ldots, n\}, \ \exists k = 1, 2, \ldots, r;$$
$$H^{j_1 j_2 \ldots j_r}_{i_1 i_2 \ldots i_r} > 0.$$

$\nabla\nabla\nabla$

**Note 4.1** The necessary and sufficient conditions for negative semi-definitness of a matrix $H \in \Re^{n\times n}$ reduce to the necessary and sufficient conditions for positive semi-definiteness of the matrix $(-H)$ (Definition 4.3b).

### 4.2.2 Sign definite functions

#### Definitions of sign definite functions

**Definition 4.4** *A function $v : \Re^n \to \Re$ is:*

*a) positive definite with respect to a set $\mathcal{A}$, $\mathcal{A} \subset \Re^n$, if and only if there is a neighbourhood $\mathcal{N}(\mathcal{A})$ of the set $\mathcal{A}$ such that*

*1) the function $v$ is continuous on $\mathcal{N}(\mathcal{A})$:*

$$v(X) \in C[\mathcal{N}(\mathcal{A})],$$

*2) the function $v$ is non-positive on the interior $\mathring{\mathcal{A}}$ of the set $\mathcal{A}$:*

$$v(X) \leq 0, \quad \forall X \in \mathring{\mathcal{A}},$$

*3) the function $v$ vanishes on the boundary $\partial\mathcal{A}$ of the set $\mathcal{A}$,*

$$v(X) = 0, \quad \forall X \in \partial\mathcal{A},$$

*4) the function $v$ has positive values on $\mathcal{N}(\mathcal{A})$ out of the closure $\overline{\mathcal{A}}$ of the set $\mathcal{A}$,*

$$v(X) > 0, \quad \forall X \in [\mathcal{N}(\mathcal{A}) - \overline{\mathcal{A}}],$$

*b) positive definite with respect to a set $\mathcal{A}$, $\mathcal{A} \subset \Re^n$, on a set $\mathcal{S}$ if and only if the conditions under a) are satisfied for $\mathcal{N}(\mathcal{A}) = \mathcal{S}$.*

*c) positive definite with respect to a set $\mathcal{A}$, $\mathcal{A} \subset \Re^n$, in the whole (or in the large, or globally positive definite with respect to a set $\mathcal{A}$) if and only if the conditions under a) are satisfied for $\mathcal{N}(\mathcal{A}) = \Re^n$.*

*d) The expression "with respect to a set $\mathcal{A}$, $\mathcal{A} \subset \Re^n$", is omitted if and only if $\mathcal{A} = \mathcal{O} = \{0\}$.*

**Note 4.2** A necessary condition for (global) positive definiteness of a function $v$ with respect to a set $\mathcal{A}$ (on a set $\mathcal{S}$) is its (global) positive semi-definiteness with respect to the set $\mathcal{A}$ (on the set $\mathcal{S}$), respectively.

**Example 4.2** Let a function $v$ be defined by

1) $v(X) = \left|\sum_{i=1}^{n} \alpha_i x_i\right|$. It is not positive definite because it vanishes on the hyperplane $\alpha_1 x_1 + \alpha_2 x_2 + \ldots + \alpha_n x_n = 0$, which means that there is not a neighbourhood $\mathcal{N}$ of $X = 0$ such that $v(X) > 0$ for all $(X \neq 0) \in \mathcal{N}$.

2) $v(x) = sin\,|x|$ is positive definite on $]-\pi, \pi[$. It is not positive definite on $[-\pi, \pi]$ because $v(x) = 0$ for $|x| = \pi$.

3) $v(x) = \begin{cases} 0 & , |x| \leq 1 \\ |x| - 1 & , |x| \geq 1 \end{cases}$, is globally positive definite with respect to the set $\mathcal{A} = \{x : x \in \Re,\ |x| \leq 1\}$. However, it is not positive definite because $v(x) = 0$ on $\mathcal{A}$ that means there is not a neighbourhood $\mathcal{N}$ of $x = 0$ on which $v(x) > 0$ out of the origin.

4) $v(x) = \begin{cases} 0 & , |x| \leq 1, \\ (|x| - 1)(4 - |x|) & , |x| \geq 1, \end{cases}$ is positive definite with respect to the set $\mathcal{A} = \{x : x \in \Re^1,\ |x| \leq 1\} = [-1, 1]$ on the set $\mathcal{S} = \{x : x \in \Re^1,\ |x| < 4\} = ]-4, 4[$. It is not positive definite on $\overline{\mathcal{S}} = [-4, 4]$ because $v(x) = 0$ for $|x| = 4$.

5) $v(X) = \sum_{i=1}^{k} \alpha_i |x_i|$ is positive definite iff $k = n$ and $\alpha_i \in \Re^+$ for every $i = 1, 2, \ldots, k = n$. Under these conditions it is globally positive definite, too.

**Definition 4.5** *A function $v : \Re^n \to \Re$ is (globally) negative definite [with respect to a set $\mathcal{A}$, $\mathcal{A} \subset \Re^n$] {on a set $\mathcal{S}$} if and only if the function $-v$ is (globally) positive definite [with respect to the set $\mathcal{A}$] {on the set $\mathcal{S}$}, respectively.*

**Definition 4.6** *(a) A matrix $H \in \Re^{n \times n}$ is positive definite if and only if its quadratic form $X^T HX$ is positive definite.*

*(b) A matrix $H \in \Re^{n \times n}$ is negative definite if and only if its quadratic form is negative definite.*

The criterion for positive definiteness of a square symmetric matrix has the following simple form [50, Vol. 1, p. 306]:

**Theorem 4.2 (Positive definiteness criterion)** *In order for a matrix $H = (h_{ij}) = H^T \in \Re^{n \times n}$ to be positive definite it is necessary and sufficient that all its leading principal minors $H^{1\,2 \ldots k}_{1\,2 \ldots k}$ are positive,*

$$H^{1\,2 \ldots k}_{1\,2 \ldots k} = \begin{vmatrix} h_{11} & h_{12} & \ldots & h_{1k} \\ h_{21} & h_{22} & \ldots & h_{2k} \\ - & - & \ldots & - \\ h_{k1} & h_{k2} & \ldots & h_{kk} \end{vmatrix} > 0, \ \forall k \in 1, 2, \ldots, n.$$

∇∇∇

**Note 4.3** The necessary and sufficient conditions for negative definiteness of a matrix $H = H^T$ reduce to the necessary and sufficient conditions for positive definiteness of the matrix $(-H)$ (Definition 4.6). More precisely:

**Theorem 4.3** *In order for a matrix $H = (h_{ij}) = H^T \in \Re^{n \times n}$ to be negative definite it is necessary and sufficient that*

$$(-1)^k \begin{vmatrix} h_{11} & h_{12} & \ldots & h_{1k} \\ h_{21} & h_{22} & \ldots & h_{2k} \\ - & - & \ldots & - \\ h_{k1} & h_{k2} & \ldots & h_{kk} \end{vmatrix} > 0, \ \forall k \in 1, 2, \ldots, n.$$

∇∇∇

**Note 4.4** If a matrix $H \in \Re^{n \times n}$ is not symmetric, $H \neq H^T$, then it is positive (negative) definite iff its symmetric part $H_s = {}^1/_2(H + H^T)$ is positive (negative) definite, respectively. This follows from the fact that

$$X^T HX = X^T \left( H + \frac{1}{2} H^T - \frac{1}{2} H^T \right) X = X^T H_s X + X^T H_{as} X = X^T H_s X$$

because

$$X^T H_{as} X = \frac{1}{2} X^T \left(H - H^T\right) X = \frac{1}{2}\left[X^T H X - (X^T H^T X)^T\right] \equiv 0.$$

**Properties of sign definite functions**

**Definition 4.7** *(a) A set $\mathcal{V}_\zeta(\mathcal{A})$ is the largest connected neighbourhood of a set $\mathcal{A} \subset \Re^n$ such that*

*1) $v(X) < \zeta$ for all $X \in \mathcal{V}_\zeta(\mathcal{A})$,*

*and*

*2) $v(X) = \zeta$ for all $X \in \partial\mathcal{V}_\zeta(\mathcal{A})$*

*(b) $\mathcal{V}_\zeta(\mathcal{A}) = \mathcal{V}_\zeta$ if and only if $\mathcal{A} = \mathcal{O} = \{0\}$.*

**Lemma 4.1** *If $v(X) \in C(\mathcal{S})$ and $\mathcal{S}$ is an open connected set, then $X \in \mathcal{S}$ determines uniquely the value $v(X)$ of $v$.* ∇∇∇

**Proof.** Let $\mathcal{S}$ be an open and connected set and $v(X) \in C(\mathcal{S})$. Let it be assumed that some $y \in \mathcal{S}$ does not determine uniquely $v(y)$, i.e. that there are $\zeta_1$ and $\zeta_2 \in \Re$, $\zeta_1 \neq \zeta_2$, such that $v(y) = \zeta_1$ and $v(y) = \zeta_2$. Let $\varepsilon = {}^1\!/_4\,|\zeta_1 - \zeta_2|$. From the conditions of the lemma follows the existence of $\delta = \delta(\varepsilon; y) \in ]0, \varepsilon]$ ensuring that $||X - y|| < \delta$ implies $|v(X) - v(y)| < \varepsilon$, i.e. both $|v(X) - \zeta_1| < \varepsilon$ and $|v(X) - \zeta_2| < \varepsilon$. Since $X = y$ obeys $||X - y|| < \delta$ then $|v(X) - \zeta_1| = |\zeta_1 - \zeta_2| < \varepsilon = {}^1\!/_4\,|\zeta_1 - \zeta_2|$, which is a contradiction because $\zeta_1 \neq \zeta_2$ was assumed. Hence, $\zeta_1 = \zeta_2$, or in other words there is not $y \in \mathcal{S}$ as supposed.

A direct consequence of this lemma is the following

**Lemma 4.2** *If $v(X) \in C(\mathcal{S})$, $\zeta_1$ and $\zeta_2 \in \Re$ are such that $\overline{\mathcal{V}}_{\zeta_i}(\mathcal{A}) \subset \mathcal{S}$, and $\mathcal{S}$ is open connected set, $\mathcal{S} \subseteq \Re^n$, then $\zeta_1 \neq \zeta_2$ implies $\partial\mathcal{V}_{\zeta_1}(\mathcal{A}) \cap \partial\mathcal{V}_{\zeta_1}(\mathcal{A}) = \emptyset$.* ∇∇∇

**Proof.** Let $v(X) \in C(\mathcal{S})$, $\zeta_1$ and $\zeta_2 \in \Re$, $\zeta_1 \neq \zeta_2$, $\overline{\mathcal{V}}_{\zeta_i}(\mathcal{A}) \subset \mathcal{S}$, $\mathcal{S} = \mathring{\mathcal{S}}$ be connected. Let it be assumed that $\partial\mathcal{V}_{\zeta_1}(\mathcal{A}) \cap \partial\mathcal{V}_{\zeta_2}(\mathcal{A}) \neq \emptyset$. Then there is $y \in [\partial\mathcal{V}_{\zeta_1}(\mathcal{A}) \cap \partial\mathcal{V}_{\zeta_2}(\mathcal{A})]$, i.e. $y \in \mathcal{S}$, such that $v(y) = \zeta_1$ and $v(y) = \zeta_2$, which contradicts Lemma 4.1. Hence, $\partial\mathcal{V}_{\zeta_1}(\mathcal{A}) \cap \partial\mathcal{V}_{\zeta_1}(\mathcal{A}) = \emptyset$.

Rouche and Mawhin established [178, p. 9] the following:

**Lemma 4.3** *Let $v : \overline{\mathcal{B}}_\sigma \to \Re$ be a positive definite function, and $\zeta_0 = inf\,[v(X) : ||X|| = \sigma]$. For every $\zeta \in ]0, \zeta_0[$, there exists an open connected neighbourhood $\mathcal{N}_\zeta$ of the origin such that $v(X) < \zeta$ on $\mathcal{N}_\zeta$ and $v(X) = \zeta$ on $\partial\mathcal{N}_\zeta$. As $\zeta \to 0^+$, the sets $\mathcal{N}_\zeta$ enclose each other and their diameters tend to $0$.* ∇∇∇

**Note 4.5** In Lemma 4.3 the statement "the sets $\mathcal{N}_\zeta$ enclose each other" means $\mathcal{N}_{\zeta_1} \subset \mathcal{N}_{\zeta_2}$ and $\partial\mathcal{N}_{\zeta_1} \cap \partial\mathcal{N}_{\zeta_2} = \emptyset$ for $0 < \zeta_1 < \zeta_2 < \zeta_0$.

**Note 4.6** From Definition 4.7 and Lemma 4.3 it follows that $\mathcal{N}_\zeta = \mathcal{V}_\zeta$ in Lemma 4.3 for $\zeta \in ]0, \zeta_0[$.

### 4.2.3 Comparison functions

What follows on comparison functions is due to Hahn [94, p. 7 and p. 95].

#### Definitions of comparison functions

**Definition 4.8** *(a) A function $\varphi : [0,\alpha[\rightarrow \Re_+$ is a comparison function of the class $\mathcal{K}_{[0,\alpha[}$ if and only if*

*1) $\varphi$ is continuous on $[0,\alpha[$: $\varphi(\zeta) \in C([0,\alpha[)$,*

*2) $\varphi$ vanishes at the origin: $\varphi(0) = 0$,*

*3) $\varphi$ is strictly monotonously increasing on $[0,\alpha[$: $0 \leq \zeta_1 < \zeta_2 < \alpha$ imply $0 \leq \varphi(\zeta_1) < \varphi(\zeta_2)$.*

*(b) $\mathcal{K}_{[0,\alpha[} = \mathcal{K}$ if and only if $\alpha = +\infty$.*

**Example 4.3** Let a function $\psi$ be defined by

a)

$$\psi(\zeta) = \begin{cases} 0 & , \zeta = 0 \\ |\zeta|\left(1 + \left|cos\frac{1}{\zeta}\right|\right) & , \zeta \in \Re^+. \end{cases}$$

It is not strictly monotonously increasing on $[0,\alpha[$ for any $\alpha > 0$. Notice that $\psi(\zeta) \in C(\Re_+)$ and $\psi(0) = 0$.

b) $\psi(\zeta) = sin\,\zeta$, $\zeta \in \Re_+$. It is a comparison function of the class $\mathcal{K}_{[0,\alpha[}$ for any $\alpha \in \left]0, \frac{\pi}{2}\right]$.

c) $\psi(\zeta) = \zeta^k$, $\zeta \in \Re_+$, $k \in \Re^+$. It is a comparison function of the class $\mathcal{K}$.

**Definition 4.9** *A function $\varphi : \Re_+ \rightarrow \Re_+$ is a comparison function of the class $\mathcal{K}\Re$ if and only if $\varphi \in \mathcal{K}$ and $lim\,[\varphi(\zeta) : \zeta \rightarrow +\infty] = +\infty$.*

#### Properties of comparison functions

$\varphi^I$ denotes the inverse function to $\varphi$,

$$\varphi^I[\varphi(\zeta)] \equiv \zeta,$$

$$\varphi[\varphi^I(\xi)] \equiv \xi.$$

If $\varphi(\zeta) = \xi$ then $\zeta = \varphi^I(\xi)$.

$\varphi_2 \circ \varphi_1$ denotes the function composition, that is

$$\varphi_2 \circ \varphi_1(\zeta) = \varphi_2[\varphi_1(\zeta)].$$

**Lemma 4.4** *(a) If $\varphi_i \in \mathcal{K}$, $i = 1, 2$, then $\varphi_2 \circ \varphi_1 \in \mathcal{K}$.*

*(b) If $\varphi \in \mathcal{K}_{[0,\alpha[}$ and $\varphi(\alpha) = \beta$ then the inverse function $\varphi^I$ is defined at least on $[0, \beta]$ and $\varphi^I \in \mathcal{K}_{[0,\beta[}$.*

*If, in addition, $\varphi$ is defined for all $\zeta \geq 0$ and if $lim\,[\varphi(\zeta) : \zeta \to +\infty] = \beta$ then $\varphi^I$ is not defined on $]\beta, +\infty]$.*

*(c) If $\varphi_i \in \mathcal{K}_{[0,\alpha]}$, $i = 1, 2$, $\varphi_1(\zeta) > \varphi_2(\zeta)$ on $[0, \alpha]$ and $\varphi_1(\alpha) = \beta$ then $\varphi_1^I(\xi) < \varphi_2^I(\xi)$ on $[0, \beta]$.*

*(d) If $\varphi \in \mathcal{K}\Re$, hence $lim\,[\varphi(\zeta) : \zeta \to +\infty] = +\infty$, then $\varphi^I \in \mathcal{K}\Re$.* ∇∇∇

### 4.2.4 Positive definite functions and comparison functions

Definitions 4.4 and 3.6 Criterion by Rouche and Mawhin [10, p. 10] directly yield the following result.

**Lemma 4.5** *In order for a function $v : \overline{\mathcal{S}} \to \Re$, $\overline{\mathcal{S}} \subset \Re^n$, to be positive definite with respect to a set $\mathcal{A}$ (on the set $\overline{\mathcal{S}}$) it is necessary and sufficient that*

*1) there is a neighbourhood $\mathcal{N}(\mathcal{A}) \subseteq \mathcal{S}$, $(\mathcal{N}(\mathcal{A}) = \overline{\mathcal{S}})$, such that $v(X) \in C[\mathcal{N}(\mathcal{A})]$,*

*2) $v(X) \leq 0$ for all $x \in \overline{\mathcal{A}}$, and $v(X) = 0$ for every $X \in \partial\mathcal{A}$,*

*3) there are $\alpha \in \Re^+$ obeying $\mathcal{N}(\alpha, \mathcal{A}) \subseteq \mathcal{S}$ $(\mathcal{N}(\alpha, \mathcal{A}) \supseteq \overline{\mathcal{S}})$ and $\varphi \in \mathcal{K}_{[0,\alpha[}$ such that $v(X) \geq \varphi[\rho(X, \mathcal{A})]$ for all $X \in \mathcal{N}(\alpha, \mathcal{A})$ $(\varphi \in \mathcal{K}_{[0,\alpha]}$ such that $v(X) \geq \varphi[\rho(X, \mathcal{A})]$ for all $X \in \overline{\mathcal{S}})$, respectively* ∇∇∇

**Example 4.4** Let

1) $v(x) = sin\,|x|$ and $\mathcal{S} = ]-\alpha, \alpha[$, $\alpha \in ]0, \pi[$. Then, $\varphi(\zeta) = \left[\dfrac{sin\,\alpha}{\pi}\zeta\right] \in \mathcal{K}$ is a comparison function satisfying $v(x) \geq \varphi[\rho(x, \mathcal{O})] = \varphi(|x|) = \dfrac{sin\,\alpha}{\pi}|x|$ because $sin\,|x| \geq \dfrac{sin\,\alpha}{\pi}|x|$ for all $x \in ]-\alpha, \alpha[$. Notice that there is not a comparison function $\varphi \in \mathcal{K}_{[0,\pi]}$ for $sin\,|x|$ because $sin\,|x| \to 0$ as $|x| \to \pi$ that requires $\varphi(\pi) = 0$, which is impossible for $\varphi \in \mathcal{K}_{[0,\pi]}$. However, $v$ is positive definite on $]-\pi, \pi[$.

2) $v(X) = (||X|| - 1)\,[1 - sign\,(1 - ||X||)]$. It is globally positive definite with respect to the set $\mathcal{A} = \overline{\mathcal{B}}_1$. The function $\varphi(\zeta) = \zeta \in \mathcal{K}\Re$ obeys $v(X) \geq \varphi[\rho(X, \mathcal{A})]$ for all $X \in \Re^n$ because $\varphi[\rho(X, \mathcal{A})] = \rho(X, \mathcal{A}) = (||X|| - 1)\,[1 - sign\,(1 - ||X||)] = v(X)$ for all $X \in \Re^n$.

3) $v(X) = (||X|| - 1)\,[1 - sign\,(1 - ||X||)]\,exp\,(1 - ||X||)$. It is globally positive definite with respect to the set $\mathcal{A} = \overline{\mathcal{B}}_1$. However, there is no comparison function $\varphi \in \mathcal{K}$ that obeys $v(X) \geq \varphi[\rho(X, \mathcal{A})]$ for all $X \in \Re^n$ because $v(X) \to 0$ as $||X|| \to +\infty$, which demands $lim\,[\varphi(\zeta) : \zeta \to +\infty] = 0$.

No $\varphi \in \mathcal{K}$ can satisfy it. However, if we take $\alpha \in \Re^+$ arbitrary large then there is $\varphi_\alpha \in \mathcal{K}_{[0,\alpha]}$ obeying $v(X) = (||X|| - 1)\,[1 - sign\,(1 - ||X||)]\,exp\,(1 - ||X||) \geq \varphi_\alpha[\rho(X, \mathcal{A})]$. Such a comparison function $\varphi_\alpha \in \mathcal{K}$ is defined by $\varphi_\alpha[\rho(X, \mathcal{A})] = \beta\rho(X, \mathcal{A})\,exp\,(-\alpha)$ where $\beta \in \,]0, 1]$ can be arbitrarily accepted and $\rho(X, \mathcal{A}) = {}^1/_2(||X|| - 1)\,[1 - sign\,(1 - ||X||)]$. This explains why the closure $\overline{\mathcal{S}}$ of the set $\mathcal{S}$ is used in Lemma 4.5.

### 4.2.5 Radially unbounded and radially increasing positive definite functions

**Definition 4.10** *A positive definite function $v : \Re^n$ (with respect to a set $\mathcal{A} \subset \Re^n$) is radially unbounded (with respect to $\mathcal{A}$) if and only if $||X|| \to +\infty$ ($\rho(X, \mathcal{A}) \to +\infty$), respectively, implies $v(X) \to +\infty$.*

By following Hahn [94, p. 99], we present the relationship between radially unbounded functions of the class $\mathcal{K}\Re$ and positive definite functions.

**Note 4.7** A comparison function $\varphi \in \mathcal{K}\Re$ is radially unbounded because $\varphi(\zeta) \to +\infty$ as $\zeta \to +\infty$ (Definition 4.10).

**Lemma 4.6** *In order for a globally positive definite function $v : \Re^n \to \Re_+$ (with respect to a set $\mathcal{A} \subset \Re^n$) to be radially unbounded, it is necessary and sufficient that there exists a comparison function $\varphi \in \mathcal{K}\Re$ obeying*

$$v(X) \geq \varphi(||X||) \quad (v(X) \geq \varphi[\rho(X, \mathcal{A})]) \quad \textit{for all} \;\; X \in \Re^n,$$

*respectively.* ∇∇∇

**Example 4.5** Let

1) $v(X) = \big(||X|| - 1\big)\,\big[1 - sign\,\big(1 - ||X||\big)\big]$. It is a radially unbounded globally positive definite function with respect to $\mathcal{A} = \{X : \; ||X|| \leq 1\}$. A corresponding radially unbounded comparison function $\varphi \in \mathcal{K}\Re$ is given by $\varphi(\zeta) = 2\zeta$ because $v(X) = \big(||X|| - 1\big)\,\big[1 - sign\,\big(1 - ||X||\big)\big] \geq 2\rho(X, \mathcal{A})$ for all $X \in \Re^n$, where $\rho(X, \mathcal{A}) = {}^1/_2\,\big(||X|| - 1\big)\,\big[1 - sign\,\big(1 - ||X||\big)\big]$.

2) $v(X) = \big(||X|| - 1\big)\,\big[1 - sign\,\big(1 - ||X||\big)\big]\,exp\,\big(1 - ||X||\big)$. It is not radially unbounded because $v(X) \to 0$ as $\rho(X, \mathcal{A}) = {}^1/_2\,\big(||X|| - 1\big)\,\big[1 - sign\,\big(1 - ||X||\big)\big] \to +\infty$ although it is globally positive definite with respect to the set $\mathcal{A} = \overline{\mathcal{B}}_1$.

By following [73], [88] we introduce:

**Definition 4.11** *A function $v : \Re^n \to \Re$ is:*

*a) radially increasing on the boundary $\partial\mathcal{S}$ of a set $\mathcal{S} \subset \Re^n$ if and only if there is $\varepsilon \in \Re^+$ such that $v(\Lambda_1 X) < v(X) < v(\Lambda_2 X)$ for every $(\Lambda_1, \Lambda_2, X) \in \,]1 - \varepsilon, 1[\, \times \,]1, 1 + \varepsilon[\, \times \,\partial\mathcal{S}$.*

b) *radially increasing on the boundary* $\partial\mathcal{S}$ *of a set* $\mathcal{S} \subset \Re^n$ *with respect to a set* $\mathcal{A} \subset \Re^n$ *if and only if there are* $\varepsilon \in \Re^+$ *and a point* $y$ *in the interior* $\mathring{\mathcal{A}}$ *of the set* $\mathcal{A}$ *such that* $v[\lambda_1(X-y)] < v(X-y) < v[\lambda_2(X-y)]$ *for every* $(\lambda_1, \lambda_2, X) \in ]1-\varepsilon, 1[ \times ]1, 1+\varepsilon[ \times \partial\mathcal{S}$.

c) *radially increasing on a neighbourhood* $\mathcal{N}$ *of* $X=0$ *if and only if* $v(\lambda_1 X) < v(\lambda_2 X)$ *for all* $\lambda_1, \lambda_2 \in \Re_+$, $\lambda_1 < \lambda_2$, *and every* $(X \neq 0) \in \mathcal{N}$ *such that* $(\lambda_i X) \in \mathcal{N}$, $i = 1, 2$.

   *The expression "on a neighbourhood* $\mathcal{N}$ *of* $X = 0$*" is omitted if and only if* $\mathcal{N} = \Re^n$.

d) *radially increasing with respect to a set* $\mathcal{A} \subset \Re^n$ *on its neighbourhood* $\mathcal{N}(\mathcal{A})$ *if and only if there is* $y \in \mathring{\mathcal{A}}$ *such that* $v[\lambda_1(X-y)] < v(X-y) < v[\lambda_2(X-y)]$ *for all* $\lambda_1, \lambda_2 \in \Re_+$, $\lambda_1 < 1 < \lambda_2$, *and every* $(X \neq y) \in \mathcal{N}(\mathcal{A})$ *such that* $[\lambda_i(X-y)] \in [\mathcal{N}(\mathcal{A}) - \mathcal{A}]$.

   *The expression "on its neighbourhood* $\mathcal{N}(\mathcal{A})$*" is omitted if and only if* $\mathcal{N}(\mathcal{A}) = \Re^n$.

**Example 4.6** Let

1) $v(X) = (||X|| - 1)\left[1 - sign\left(1 - ||X||\right)\right]\left[1 - exp\left(1 - ||X||\right)\right]$. It is radially increasing with respect to the set $\mathcal{A} = \overline{\mathcal{B}}_1$. However, it is not radially increasing on any neighbourhood $\mathcal{N}$ of $X = 0$. It is radially unbounded globally positive definite with respect to $\overline{\mathcal{B}}_1$.

2) $v(X) = \left[1 - sign\left(1 - ||X||\right)\right]\left[1 - exp\left(1 - ||X||\right)\right]$. It is radially increasing globally positive definite with respect to $\mathcal{A} = \overline{\mathcal{B}}_1$. But it is neither radially unbounded nor radially increasing on any neighbourhood $\mathcal{N}$ of $X = 0$.

3)
$$v(X) = \begin{cases} 0 & , X = 0 \\ ||X||^2\left[\varepsilon + sin^2\left(\dfrac{1}{||X||}\right)\right] & , \varepsilon \in \Re^+,\ (X \neq 0) \in \Re^n. \end{cases}$$

It is not radially increasing on any neighbourhood $\mathcal{N}$ of $X = 0$, although it is radially unbounded globally positive definite.

By following the proof of Theorem 24.1 by Hahn [94, p. 99], we generalise it to sets as follows.

**Lemma 4.7** *Let* $v : \mathcal{S} \to \Re$ *be positive definite with respect to a compact set* $\mathcal{A} \subset \Re^n$ *on a set* $\mathcal{S} \subseteq \Re^n$, $\sigma \in \Re^+$ *be the largest number such that* $\mathcal{N}(\sigma, \mathcal{A}) \subseteq \mathcal{S}$, *let* $\varphi \in \mathcal{K}_{[0,\sigma]}$ *obey* $v(X) \geq \varphi[\rho(X, \mathcal{A})]$ *for all* $X \in \mathcal{N}(\sigma, \mathcal{A})$, $\alpha = \varphi(\sigma)$, *and let* $\mathcal{V}_\zeta$ *denote the largest connected neighbourhood of the set* $\mathcal{A}$ *such that* $v(X) < \zeta$ *for all* $X \in \mathcal{V}_\zeta$ *and* $v(X) = \zeta$ *for* $X \in \partial\mathcal{V}_\zeta$. *Then* $v(X) = \zeta$ *defines a closed hypersurface enclosing the set* $\mathcal{A}$ *for every* $\zeta \in ]0, \alpha[$. ∇∇∇

**Proof.** Let $\psi = min\,[\rho(X, \mathcal{A}) : X \in \partial\mathcal{V}_\alpha]$. Let $y \in \partial\mathcal{N}(\sigma, \mathcal{A})$. Then, $v(\beta y) = w(\beta)$ considered as a function of $\beta$ with respect to the fixed $y$ varies between zero and $\gamma \geq \alpha$ for $\beta$-variation between zero and one. Thus, there exists at least one $\theta \in ]0, 1[$ such that $v(\theta y) = \zeta \in ]0, \alpha[$. Let $\theta_m$ be minimal such $\theta$. Then, the set $\{X : X = \theta_m y, y \in \partial\mathcal{N}(\sigma, \mathcal{A})\}$ determines a closed hypersurface $v(X) = \zeta$ enclosing the set $\mathcal{A}$.

**Note 4.8** The hypersurface $v(X) = \zeta$ is closed for any $\zeta \in \Re^+$ if $v(X) \geq \varphi[\rho(X, \mathcal{A})]$ holds for all $X \in \Re^n$ and for radially unbounded $\varphi \in \mathcal{K}\Re$.

**Example 4.7** Let, by following Hahn [94, p. 99], $v(X) = \dfrac{\rho^2(X, \mathcal{A})}{\rho^2(X, \mathcal{A}) + 1}$, $X \in \Re^2$, and $\mathcal{A}$ be a compact set in $\Re^2$. Then $v(X) = \zeta > 1$ is not closed. In fact such a $v$ does not exist because $v(X) = \zeta$ yields $\rho^2(X, \mathcal{A}) = -\dfrac{\zeta}{\zeta - 1} < 0$, which is not possible.

**Lemma 4.8** *Let $v : \mathcal{S} \to \Re$ be positive definite on a set $\mathcal{S} \subset \Re^n$ with respect to a compact set $\mathcal{A} \subset \Re^n$, $\alpha \in \Re^+$ determine $\mathcal{V}_\alpha(\mathcal{A})$ as the largest connected neighbourhood of the set $\mathcal{A}$ such that $\overline{\mathcal{V}}_\alpha(\mathcal{A}) \subseteq \mathcal{S}$, $v(X) < \alpha$ for all $X \in \mathcal{V}_\alpha(\mathcal{A})$, $v(X) = \alpha$ for $X \in \partial\mathcal{V}_\alpha(\mathcal{A})$, and let arbitrary $\underline{Z} : \Re_+ \to \Re^n$, $\underline{Z}(t) \in C[\mathcal{V}_\alpha(\mathcal{A})]$ with $\underline{Z}(0) = Z_0 \in \mathcal{V}_\alpha(\mathcal{A})$. Then, for $\underline{Z}(t) \to \mathcal{A}$ as $t \to +\infty$ to imply strict monotonous decreasing of $v[\underline{Z}(t)]$ it is necessary for the hypersurface $v(X) = \zeta_1$ to enclose the hypersurface $v(X) = \zeta_2$, $0 \leq \zeta_2 < \zeta_1 < \alpha$, and that $v(X) = \zeta$ is closed for $\zeta \in ]0, \alpha[$.*
$\nabla\nabla\nabla$

**Proof.** Let all the conditions of the lemma hold. Then there is $\sigma \in \Re_+$ such that $\underline{Z}(t) \in \mathcal{V}_\alpha(\mathcal{A})$ for all $t \in [\sigma, +\infty]$ and $\underline{Z}(t) \to \mathcal{A}$ as $t \to +\infty$ implies the strict monotonous decreasing of $v[\underline{Z}(t)]$. Let $t_1$, $t_2 \in [\sigma, +\infty]$ be arbitrary and $t_2 > t_1$. Let $v[\underline{Z}(t_i)] = \zeta_i$, $i = 1, 2$. Hence, $\zeta_2 < \zeta_1$. Now, Lemma 4.1, Lemma 4.2 and Lemma 4.7 imply that the hypersurfaces are closed and do not intersect, which with the monotonous decreasing of $v[\underline{Z}(t)]$ prove that $v(X) = \zeta_1$ encloses $v(X) = \zeta_2$. If $t_2 = +\infty$ then $\underline{Z}(t_2) \in \partial\mathcal{A}$ so that $v[\underline{Z}(t_2)] = \zeta_2 = 0$ due to positive definiteness of $v$ on $\mathcal{S}$. Thus, the statement of the lemma is true for $0 \leq \zeta_2 < \zeta_1 < \alpha$.

**Lemma 4.9** *Let $v : \mathcal{S} \to \Re$ be radially increasing positive definite function on a set $\mathcal{S} \subset \Re^n$ with respect to a compact set $\mathcal{A} \subset \mathcal{S}$. Let $\alpha \in \Re^+$ and $\mathcal{V}_\alpha(\mathcal{A})$ be defined as in Lemma 4.8. Then the hypersurface $v(X) = \zeta$ is closed for every $\zeta \in ]0, \alpha[$ and $v(X) = \zeta_1$ encloses $v(X) = \zeta_2$ for $0 < \zeta_2 < \zeta_1 < \alpha$.* $\nabla\nabla\nabla$

**Proof.** Let all the conditions of the lemma hold. Then $v(X) = \zeta$ is closed for $\zeta \in ]0, \alpha[$ due to Lemma 4.7. Radial increasing of $v$ on $[\mathcal{V}_\alpha(\mathcal{A}) - \mathring{\mathcal{A}}]$ and Lemma 4.1 prove now that $v(X) = \zeta_1$ encloses $v(X) = \zeta_2$ for $0 < \zeta_2 < \zeta_1 < \alpha$ because if it were otherwise then there would exist $X \in \partial\mathcal{V}_{\zeta_2}$ and $\lambda \in ]0, 1[$ such that $[\lambda(X - y)] \in \partial\mathcal{V}_{\zeta_1}$, that would mean $v[\lambda(X - y)] < v(X - y)$ and would contradict radial increasing of $v$ (Definition 4.11d).

## 4.3 Uniquely bounded sets

### 4.3.1 Definition of uniquely bounded sets

Uniquely bounded sets and $\mathcal{O}$-uniquely bounded sets were discovered in [67] and further studied in [66], [80], [84].

**Definition 4.12** *(a) A set $\mathcal{S} \subset \Re^n$ is uniquely bounded, which is denoted by $\mathcal{U}$ : $\mathcal{S} = \mathcal{U}$, if and only if it is bounded and there are an open set $\mathcal{N}$ and a point $Z \in \Re^n$ called the centre of unique boundedness of the set $\mathcal{U}$ such that*

*1) $\overline{\mathcal{U}} \subseteq \overline{\mathcal{N}}$,*

*2) $Z$ is in the interior $\overset{\circ}{\mathcal{U}}$ of $\mathcal{U}$: $Z \in \overset{\circ}{\mathcal{U}}$,*

*3) for every $(X \neq Z) \in \overline{\mathcal{N}}$ and any $\alpha \in \Re^+$ there is exactly one positive number $\lambda$, $\lambda = \lambda(\alpha X; \mathcal{U}; Z) = \lambda(\alpha(X - Z); \mathcal{U}) \in \Re^+$, which obeys $\lambda(\alpha X; \mathcal{U}; Z)\,[\alpha(X - Z)] \in \partial\mathcal{U}$,*

*(b) A set $\mathcal{S} \subset \Re^n$ is $\mathcal{O}$-uniquely bounded, which is denoted by $\mathcal{U}_0$: $\mathcal{S} = \mathcal{U}_0$, if and only if it is a uniquely bounded set with its centre of unique boundedness at the origin, that is $Z = 0$. Then and only then $\lambda = \lambda(\alpha X; \mathcal{U}; 0) = \lambda(\alpha X; \mathcal{U})$ for any $\alpha \in \Re^+$.*

*(c) $\lambda = \lambda(\alpha X; \mathcal{U}; Z) = \lambda(\alpha(X - Z); \mathcal{U})$ will be denoted by $\lambda(X; Z)$: $\lambda(X; \mathcal{U}; Z) \equiv \lambda(X; Z)$ if and only if $\mathcal{U}$ and $\alpha$ are fixed.*

**Note 4.9** From $\lambda(\alpha X; \mathcal{U}; Z)[\alpha(X - Z)] \in \partial\mathcal{U}$ (Definition 4.12-3) it follows:

$$\lambda(\alpha X; \mathcal{U}; Z) \equiv \lambda[\alpha(X - Z); \mathcal{U}],$$

and

$$\lambda(\alpha X; Z) \equiv \lambda[\alpha(X - Z)]$$

in case $\mathcal{U}$ is fixed. If $\alpha = 1$ is also fixed then it is omitted.

Definition 4.12 will be illustrated by several examples taken from the reference [66].

**Example 4.8** Starlike sets with respect to 0 defined by Spivak [189] seem at first glance equivalent to $\mathcal{O}$-uniquely bounded sets. However, they are different. The set $\mathcal{S}$ defined by

$$\mathcal{S} = \{X : X \in \Re^2, (\|X\|^2 < 4, x_1 < 0) \vee (\|X\|^2 < 4, x_1 > 0, x_2 \in ]-2, 1[) \vee \\ \vee \ (x_1 = 0, x_2 \in ]-2, 1[)\}$$

is starlike with respect to 0 but it is not $\mathcal{O}$-uniquely bounded because the ray through $(0, 0)$ and $(0, 2)$ has infinitely many common points with the boundary of the set $\mathcal{S}$.

**Example 4.9** At first glance the family of uniquely bounded sets seems to be the family of starlike sets defined by Hocking and Young [103]. To show that the former is a proper subfamily of the latter let the following starlike set be considered,

$$\mathcal{S} = \left\{X : X \in \Re^2,\ \left[(x_1-2)^2 + x_2^2 \le 4\right] \vee \left[(x_1+2)^2 + x_2^2 \le 4\right]\right\}.$$

It is not uniquely bounded. There is not a point $Z \in \mathring{\mathcal{S}}$ that satisfies Definition 4.12.

**Example 4.10** Let $\mathcal{S} = \left\{X : X \in \Re^n,\ a \in \Re^n,\ ||X-a||^2 \le ||a||^2\right\}$ be uniquely bounded, with $Z = a$ as one choice for $Z$. However, the set $\mathcal{S}$ is not $\mathcal{O}$-uniquely bounded because $Z = 0$ cannot be chosen since $(X = 0) \in \partial\mathcal{S}$, that is $(X = 0) \notin \mathring{\mathcal{S}}$.

**Example 4.11** Let $\mathcal{S} = \left\{X : X \in \Re^n,\ a \in \Re^n,\ ||X-a||^2 \le 4||a||^2\right\}$. It is $\mathcal{O}$-uniquely bounded.

### 4.3.2 Properties of uniquely bounded sets

**Property UB.** *Let $\mathcal{U} \subset \Re^n$ be a bounded set and $\mathcal{N} \subset \Re^n$ be an open set satisfying $\overline{\mathcal{U}} \subseteq \overline{\mathcal{N}}$. There are a function $u : \Re^n \to \Re$ and a point $Z \in \mathring{\mathcal{U}}$ which satisfy the following:*

*1) $u(X-Z) = 0$ for $(X-Z) \in \overline{\mathcal{N}}$ iff $X = Z$,*

*2) if $\mathcal{N} = \Re^n$ then $u(X-Z) \to +\infty$ as $||X-Z|| \to +\infty$,*

*3) $u$ is continuous on $\overline{\mathcal{N}}$: $u(X-Z) \in C(\overline{\mathcal{N}})$,*

*4) $u$ is radially increasing on $\mathcal{N}$ with respect to $Z$, that is*

$$u[\alpha_1(X-Z)] < u[\alpha_2(X-Z)],$$

$$0 < \alpha_1 < \alpha_2 < +\infty,\quad \alpha_i(X-Z) \in \overline{\mathcal{N}},\quad i = 1, 2,$$

*5) there is $\xi \in \Re^+$ such that both 1) and 2) hold:*

$$1)\quad u(X-Z) \le \xi \ \text{ for } \ (X-Z) \in \overline{\mathcal{N}} \ \text{ iff } \ (X-Z) \in \overline{\mathcal{U}},$$

$$2)\quad u(X-Z) = \xi \ \text{ for } \ (X-Z) \in \overline{\mathcal{N}} \ \text{ iff } \ (X-Z) \in \partial\mathcal{U}.$$

**Lemma 4.10** *(a) For a bounded subset $\mathcal{U}$ of an open set $\mathcal{N}$, $\overline{\mathcal{U}} \subseteq \overline{\mathcal{N}}$, to be uniquely bounded it is both necessary and sufficient that it possesses the Property UB.*

*(b) For a bounded subset $\mathcal{U}$ of an open set $\mathcal{N}$, $\overline{\mathcal{U}} \subseteq \overline{\mathcal{N}}$, to be $\mathcal{O}$-uniquely bounded it is both necessary and sufficient that it possesses the Property UB with $Z = 0$.*

$\nabla\nabla\nabla$

**Proof.** *Necessity.* (a) Let a bounded subset $\mathcal{U}$ of an open set $\mathcal{N}$, $\overline{\mathcal{U}} \subseteq \overline{\mathcal{N}}$, be uniquely bounded. Definition 4.12 may be used. Let $u(X-Z) = [\lambda(X;\mathcal{U};Z)]^{-1} = [\lambda(X-Z;\mathcal{U})]^{-1}$.

1) Definition 4.12 yields $\lambda(Z;\mathcal{U};Z) = +\infty$. Hence, $u(Z-Z) = 0$.

2) $\lambda(X;\mathcal{U};Z) \to 0$ as $\|X\| \to +\infty$ in case $\mathcal{N} = \Re^n$ (Definition 4.12), which implies $u(X-Z) \to +\infty$ as $\|X-Z\| \to +\infty$ in case $\mathcal{N} = \Re^n$.

3) Let $(X_0 - Z) \in \overline{\mathcal{N}}$, $X_0 \neq Z$, be arbitrary. Let $X_n$ represent a sequence converging to $X_0$, $(X_n - Z) \in \overline{\mathcal{N}}$ and $X_n \to X_0$ as $n \to +\infty$. Since $(X_0 - Z) \in \overline{\mathcal{N}}$ and $(X_n - Z) \in \overline{\mathcal{N}}$ then (Definition 4.12) $u[\lambda(X_n;\mathcal{U};Z)(X_n - Z)] \equiv u[\lambda(X_0;\mathcal{U};Z)(X_0 - Z)] \equiv 1 = \xi$ because $[\lambda(X_n;\mathcal{U};Z)(X_n - Z)] \in \partial\mathcal{U}$ and $[\lambda(X_0;\mathcal{U};Z)(X_0 - Z)] \in \partial\mathcal{U}$. The preceding identities are possible when $n \to +\infty$ iff $\lambda(X_n;\mathcal{U};Z) \to \lambda(X_0;\mathcal{U};Z)$ as $n \to +\infty$ due to the definition of $\lambda$ (Definition 4.12). This proves $\lambda(X;\mathcal{U};Z) \in C(\overline{\mathcal{N}} - \{Z\})$. Hence, $u(X-Z) \in C(\overline{\mathcal{N}} - \{Z\})$ because $\lambda(X;\mathcal{U};Z) > 0, \forall (X-Z) \in \overline{\mathcal{N}}, X \neq Z$. Since $u(X-Z) \to 0$ as $X \to Z$ and $u(Z-Z) = 0$ then, altogether, $u(X-Z) \in C(\overline{\mathcal{N}})$.

4) Let $(X-Z) \in \mathcal{N}$ be arbitrary, $X \neq Z$. Let $\alpha_i \in \Re^+$, $i = 1, 2$, $\alpha_1 < \alpha_2$, be such that $\alpha_i(X-Z) \in \mathcal{N}$. Hence, $\lambda(\alpha_1 X;\mathcal{U};Z) > \lambda(\alpha_2 X;\mathcal{U};Z)$, where $\lambda(\alpha_i X;\mathcal{U};Z)\,[\alpha_i(X-Z)] \in \partial\mathcal{U}$, $i = 1, 2$. The last inequality implies $u[\alpha_1(X-Z)] < u[\alpha_2(X-Z)]$. This and $u(Z-Z) = 0$ prove radial increasing of $u$ on $\mathcal{N}$ with respect to $Z$.

5) Since $\lambda(X;\mathcal{U};Z) = 1$ iff $(X-Z) \in \partial\mathcal{U}$, it now immediately results (due to 1) and 4)) $u(X-Z) \leq 1$ for $(X-Z) \in \overline{\mathcal{N}}$ iff $(X-Z) \in \overline{\mathcal{U}}$ and $u(X-Z) = 1$ for $(X-Z) \in \overline{\mathcal{N}}$ iff $(X-Z) \in \partial\mathcal{U}$, which completes the proof if we set $\xi = 1$.

(b) If $\mathcal{U}$ is $\mathcal{O}$-uniquely bounded then $Z = 0$ (Definition 4.12).

*Sufficiency.* (a) Let all the conditions of the lemma hold. Let $\lambda(X;\mathcal{U};Z) = \xi[u(X-Z)]^{-1} = \lambda(X-Z;\mathcal{U})$. Radial increasing of $u$ on $\mathcal{N}$ from $u = 0$ with respect to $Z$ implies radial decreasing of $\lambda$ on $\mathcal{N}$ from $\lambda(Z;\mathcal{U};Z) = +\infty$, that is $\lambda(\alpha_1 X;\mathcal{U};Z) > \lambda(\alpha_2 X;\mathcal{U};Z)$, $\alpha_i \in \Re^+$, $\alpha_1 < \alpha_2$, $\alpha_i(X-Z) \in \mathcal{N}$, $i = 1, 2$. It now follows that $\lambda(y;\mathcal{U};Z) = 1$ for $(y-Z) \in \mathcal{N}$ iff $(y-Z) \in \partial\mathcal{U}$ due to 5.2) and $[\alpha_i(X-Z)] \in \partial\mathcal{U}$, $i = 1, 2$, iff $\alpha_1 = \alpha_2 = \lambda(X;\mathcal{U};Z)$, $\forall (X-Z) \in \overline{\mathcal{N}}$, $X \neq Z$. Hence, $\mathcal{U}$ is uniquely bounded (Definition 4.12).

(b) The conditions under (a) hold for $Z = 0$. Therefore, $\mathcal{U}$ is $\mathcal{O}$-uniquely bounded (Definition 4.12).

This lemma justifies the following definition [67], [66], [80].

**Definition 4.13** *(a) A function $u : \Re^n \to \Re$ is a generating function of a uniquely bounded set $\mathcal{U} \subset \Re^n$ on a set $\mathcal{S} \subseteq \Re^n$ if and only if the function $u$ and the set $\mathcal{U}$ satisfy Property UB for $\mathcal{S} = \mathcal{N}$.*

*(b) A function $u : \Re^n \to \Re$ is a generating function of an $\mathcal{O}$-uniquely bounded set $\mathcal{U} \subset \Re^n$ on a set $\mathcal{S} \subseteq \Re^n$ if and only if the function $u$ and the set $\mathcal{U}$ satisfy Property UB for $\mathcal{S} = \mathcal{N}$ and $Z = 0$.*

*(c) The expression "on a set $\mathcal{S}$" is omitted if and only if $\mathcal{S} = \mathcal{N} = \Re^n$.*

*(d) $u(X-Z)$ will be denoted by $u(X)$, $u(X) \equiv u(X-Z)$ if and only if $Z$ is fixed.*

### 4.3.3 $\mathcal{O}$-uniquely bounded sets and positive definite functions

**Lemma 4.11** *Let $u : \Re^n \to \Re$ be a generating function of an $\mathcal{O}$-uniquely bounded set $\mathcal{U}$ on a set $\mathcal{S}$. Then,*

1) *the function $u$ is positive definite on the set $\mathcal{S}$,*

2) *the set $\mathcal{U}_\zeta = \{X : X \in \Re^n, u(X) < \zeta\}$ is a bounded, connected and open neighbourhood of $X = 0$ with the boundary $\partial\mathcal{U}_\zeta = \{X : X \in \Re^n, u(X) = \zeta\}$ for every $\zeta \in \Re^+$ for which $\mathcal{U}_\zeta \subseteq \mathcal{S}$,*

3) *$\mathcal{U}_{\zeta_1} \subset \mathcal{U}_{\zeta_2}$ and $\partial\mathcal{U}_{\zeta_1} \cap \partial\mathcal{U}_{\zeta_2} = \emptyset$ for all $\zeta_i \in \Re^+$ for which $\mathcal{U}_{\zeta_i} \subseteq \mathcal{S}$, $i = 1, 2$, $\zeta_1 < \zeta_2$,*

4) *the hypersurface $u(X) = \zeta$ is closed for every $\zeta \in \Re^+$ for which $\mathcal{U}_\zeta \subseteq \mathcal{S}$,*

5) *if $\mathcal{S} = \Re^n$ then $u$ is radially both unbounded and increasing, and globally positive definite.* $\nabla\nabla\nabla$

**Proof.** Let $u : \Re^n \to \Re$ be the generating function of an $\mathcal{O}$-uniquely bounded set $\mathcal{U}$ on a set $\mathcal{S}$. Then, Definition 4.12b, Property UB and Lemma 4.10b hold due to Definition 4.13b.

1) Positive definiteness of $u$ on $\mathcal{S}$ results from 1), 3), and 4) of Property UB and $\mathcal{S} = \mathcal{N}$ (Definition 4.13b).

2) If $\mathcal{U}_\zeta \subseteq \mathcal{S}$ then it is bounded since $\zeta \in \Re^+$ and due to 1), 3)–5) of Property UB, Lemma 4.10b and Definition 4.10b. Connectedness follows from Definition 4.12b because for any $X$ and $y \in \mathcal{U}_\zeta$, $(\alpha X) \in \mathcal{U}_\zeta$ and $(\beta y) \in \mathcal{U}_\zeta$, as soon as $\alpha \in [0, 1]$ and $\beta \in [0, 1]$. Hence, $X$ and $y$ are linked via semi-lines $(\psi x)$ and $(\xi y)$ for $\psi$ and $\xi \in \Re_+$.

3) The statement under 3) results from 3)–5) of Property UB, Lemma 4.10b and Definition 4.13b.

4) The hypersurface $u(X) = \zeta$ is closed for $\zeta \in \Re^+$ for which $\mathcal{U}_\zeta \subseteq \mathcal{S}$ because of 3) and 4) of Property UB, Lemma 4.10b and Definition 4.13b.

5) $\mathcal{S} = \mathcal{N} = \Re^n$ and 2) and 4) of Property UB prove the radial both unboundedness and increasing of the function $u$ and its global positive definiteness.

### 4.3.4 Definition of uniquely bounded neighbourhoods of sets

**Definition 4.14** *(a) A set $\mathcal{S} \subset \Re^n$ is a uniquely bounded neighbourhood of a set $\mathcal{A} \subset \Re^n$, which is denoted by $\mathcal{N}_u(\mathcal{A})$, that is $\mathcal{S} = \mathcal{N}_u(\mathcal{A})$, if and only if there exists a point $w \in \Re^n$ called the centre of uniquely bounded neighbourhood $\mathcal{N}_u(\mathcal{A})$ such that*

1) $\mathcal{S} = \mathcal{N}_u(\mathcal{A})$ *is a bounded neighbourhood of the set* $\mathcal{A}$,

2) *the point* $w$ *is in the interior* $\mathring{\mathcal{A}}$ *of the set* $\mathcal{A}$: $w \in \mathring{\mathcal{A}}$,

3) *for every* $(X - w) \in (\Re^n - \mathring{\mathcal{A}})$ *there is exactly one* $\psi \in \Re^+$, $\psi = \psi(X; \mathcal{A}, \mathcal{S}; w)$ *which obeys* $[\psi(X - w)] \in \partial\mathcal{A}$,

4) *for every* $(X - w) \in (\Re^n - \mathring{\mathcal{A}})$ *there is exactly one* $\Psi \in \Re^+$, $\Psi = \Psi(X; \mathcal{A}, \mathcal{S}; w)$ *which satisfies* $[\Psi(X - w)] \in \partial\mathcal{S}$.

(b) *A set* $\mathcal{S} \subset \Re^n$ *is an* $\mathcal{O}$*-uniquely bounded neighbourhood of a set* $\mathcal{A} \subset \Re^n$, *which is denoted by* $\mathcal{N}_0(\mathcal{A})$, *that is* $\mathcal{S} = \mathcal{N}_0(\mathcal{A})$, *if and only if it is uniquely bounded neighbourhood of the set* $\mathcal{A}$ *with its centre of uniquely bounded neighbourhoodness at the origin:* $w = 0$. *Then and only then* $\psi = \psi(X; \mathcal{A}, \mathcal{S}; 0) = \psi(X; \mathcal{A}; \mathcal{N}_0)$ *and* $\Psi = \Psi(X; \mathcal{A}, \mathcal{S}; 0) = \Psi(X; \mathcal{A}; \mathcal{N}_0)$.

**Example 4.12** Let $\mathcal{A} = \{X : X \in \Re^2, [(x_1 - 2)^2 + x_2^2 \leq 4] \vee [(x_1 + 2)^2 + x_2^2 \leq 4]\}$ and $\mathcal{S} = \{X : X \in \Re^2, x_1^2 + (x_2 - 6)^2 \leq 100\}$. The set $\mathcal{S}$ is uniquely bounded with any $Z \in \mathring{\mathcal{S}}$, hence $\mathcal{O}$-uniquely bounded, too. However, it is not a uniquely bounded neighbourhood of $\mathcal{A}$, although it is a neighbourhood of $\mathcal{A}$, because the set $\mathcal{A}$ does not obey the conditions 2) and 3) of Definition 4.14.

**Example 4.13** The set $\mathcal{S} = \{X : X \in \Re^2, x_1^2 + (x_2 - 6)^2 \leq 100\}$ is a uniquely bounded neighbourhood of the set $\mathcal{A} = \{X : X \in \Re^2, (x_1 + 2)^2 + x_2^2 \leq 4\}$ (Definition 4.14).

### 4.3.5 Properties of uniquely bounded neighbourhoods of sets

**Lemma 4.12** *(a) In order for a set* $\mathcal{S} \subset \Re^n$ *to be a uniquely bounded neighbourhood, with the centre of uniquely bounded neighbourhoodness at* $w$, *of a set* $\mathcal{A} \subset \Re^n$ *it is necessary for the set* $\mathcal{A}$ *to be uniquely bounded with its centre* $Z$ *of unique boundedness at* $w$ : $Z = w$ *and* $\psi(X; \mathcal{A}; \mathcal{S}; w) = \lambda(X; \mathcal{A}; w)$.

(b) *In order for a set* $\mathcal{S} \subset \Re^n$ *to be an* $\mathcal{O}$*-uniquely bounded neighbourhood of a set* $\mathcal{A} \subset \Re^n$ *it is necessary for the set* $\mathcal{A}$ *to be* $\mathcal{O}$*-uniquely bounded.* ∇∇∇

**Proof.** (a) Let a set $\mathcal{S} \subset \Re^n$ be a uniquely bounded neighbourhood of a set $\mathcal{A} \subset \Re^n$. Since it is a bounded neighbourhood of the set $\mathcal{A}$ then $\mathcal{A}$ is also bounded. Let $X \in \Re^n$ be arbitrary. Two cases are possible:

1) $(X - w) \in (\Re^n - \mathring{\mathcal{A}})$. Then there is exactly one $\psi \in \Re^+$ such that $\psi(X - w) \in \partial\mathcal{A}$ (Definition 4.13). Hence, we may set $Z = w \in \mathring{\mathcal{A}}$ and $\lambda(X; \mathcal{A}; Z) \equiv \lambda(X; \mathcal{A}; w) \equiv \psi(X; \mathcal{A}, \mathcal{S}; w)$ (Definition 4.12).

2) $(X - w) \in \mathring{\mathcal{A}}$. Since $\mathcal{S} = \mathcal{N}_u(\mathcal{A})$ and $\mathring{\mathcal{A}} \subseteq \mathcal{A} \subset \mathcal{N}_u(\mathcal{A})$ then for any $V \in \Re^n$ obeying both $(V - w) \notin \mathring{\mathcal{A}}$ and $(V - w) = l(X - w)$ for some $l \in \Re^+$, $V = V(X, w)$, the value of $l$ is uniquely determined by $(X, V, w)$, $l = l(X, V, w)$. Since $\psi(lX; \mathcal{A}, \mathcal{S}; w)l(X, V, w)(X - w) = \psi(V; \mathcal{A}, \mathcal{S}; w)(V - w) \in \partial\mathcal{A}$ (Definition 4.13) then $\lambda(X - w) \in \partial\mathcal{A}$ for $\lambda = \lambda(X; \mathcal{A}; Z) \in \Re^+$ determined by $\lambda(X; \mathcal{A}; Z) = \psi[V(X, w); \mathcal{A}, \mathcal{S}; w]l(X, V, w)$ and by $Z = w$.

Altogether, it now follows that for every $(X - Z) \in \overline{\mathcal{S}}$, $X \neq Z$, there is exactly one $\lambda = \lambda(X; \mathcal{A}; Z) \in \Re^{+}$ such that $\lambda(X - Z) \in \partial\mathcal{A}$ with $Z = w$.

(b) The statement under (b) results directly from (a) and Definition 4.14b.

Lemma 4.12 is illustrated by Examples 4.12 and 4.13.

**Lemma 4.13** *Let a set $\mathcal{A} \subset \Re^n$ be uniquely bounded, and $\mathcal{N}_u(\mathcal{A})$ be its uniquely bounded neighbourhood with the common centre of unique boundedness $w \in \mathring{\mathcal{A}}$. Let for any $(X - w) \in \overline{\mathcal{N}}_u(\mathcal{A})$, $X \neq w$, $[\lambda(X; w)(X - w)] \in \partial\mathcal{A}$ and $[\Lambda(X; w)(X - w)] \in \partial\mathcal{N}_u(\mathcal{A})$. Then a function $v : \Re^n \to \Re$ defined by*

$$v(X) = \left\{ \begin{array}{cc} 0 & , X = w, \\ \dfrac{1 - \lambda(X; w)}{1 - \lambda[\Lambda(X; w)X; w]} \cdot \dfrac{\rho(X - w, \mathcal{A})}{\rho[\Lambda(X; w)(X - w), \mathcal{A}]} & , (X \neq w) \in \Re^n \end{array} \right\} =$$

$$= v(X - w)$$

*a) is globally positive definite with respect to the set $\mathcal{A}$,*

*b) is radially increasing with respect to the set $\mathcal{A}$,*

*and*

*c) $v(X) = 1$ for all $(X - w) \in \partial\mathcal{N}_u(\mathcal{A})$.* $\nabla\nabla\nabla$

**Proof.** a) From the proof of a-3 of Lemma 4.10 it follows that $\lambda(X; w) \in [\Re^n - \{w\}]$ and $\Lambda(X; w) \in C[\Re^n - \{w\}]$. Since $\lambda(X, w) = 1$ for all $(X - w) \in \partial\mathcal{A}$, $\lambda[\Lambda(X; w)X; \Lambda(X; w)w] \in ]0, 1[$ for all $X \in [\Re^n - \{w\}]$ because

$$[\Lambda(X; w)X - \Lambda(X; w)w] = [\Lambda(X; w)(X - w)] \in \partial\mathcal{N}_u(\mathcal{A})$$

for all $(X \neq w) \in \Re^n$ and $\rho(X - w, \mathcal{A}) \in C(\Re^n)$, then

1) $v(X) \in C(\Re^n - \mathcal{A})$,

2) $v(X) > 0$ for all $(X - w) \in (\Re^n - \overline{\mathcal{A}})$ and $v(X) \to 0$ as $(X - w) \to \partial\mathcal{A}$, $(X - w) \in (\Re^n - \overline{\mathcal{A}})$.

From $\rho(X - w, \mathcal{A}) = 0$ for $(X - w) \in \overline{\mathcal{A}}$ it results

3) $v(X) = 0$ for all $(X - w) \in \overline{\mathcal{A}}$.

Hence,

4) $v(X) \in C(\Re^n)$.

Now, 2) through 4) prove global positive definiteness of $v$ with respect to $\mathcal{A}$.

b) Notice that $\lambda(X-w;w) \equiv \lambda(X;w)$, $\Lambda(X-w;w) \equiv \Lambda(X;w)$, and $\lambda[\Lambda(X;w)X;w] \equiv \lambda[\Lambda(X-w)(X-w)]$ (Definition 4.12 and Note 4.9). Hence, $v(X) = v(X-w)$. Since $\lambda(\alpha_1 X;w) > \lambda(\alpha_2 X;w)$, since unique boundedness of $\mathcal{A}$ guarantees $\rho[\alpha_1(X-w),\mathcal{A}] < \rho[\alpha_2(X-w),\mathcal{A}]$, $\lambda[\Lambda(\alpha_1 X;w)\alpha_1 X;w] \equiv \lambda[\Lambda(\alpha_2 X;w)\alpha_2 X;w]$ and $\rho[\Lambda(\alpha_1 X;w)\alpha_1(X-w),\mathcal{A}] \equiv \rho[\Lambda(\alpha_2 X;w)\alpha_2(X-w),\mathcal{A}]$ because $[\Lambda(\alpha_1 X;w)\alpha_1(X-w)] \equiv [\Lambda(\alpha_2 X;w)\alpha_2(X-w)] \in \partial\mathcal{N}_u(\mathcal{A})$ for every $\alpha_i \in \Re^+$, $i = 1,2$, $\alpha_1 < \alpha_2$, and for all $(X \neq w) \in \Re^n$, then d) of Definition 4.11 is satisfied for $y = w$. Hence, the function $v$ is radially increasing with respect to the set $\mathcal{A}$ on $\mathcal{N}_u(\mathcal{A})$.

c) Let $(X-w) \in \partial\mathcal{N}_u(\mathcal{A})$. Then $\Lambda(X;w) = 1$ and $sign\,\rho(X,\mathcal{A}) = 1$. Hence,

$$v(X) = \frac{1-\lambda(X;w)}{1-\lambda(X;w)} \cdot \frac{1}{1} = 1,$$

because $\lambda(X;w) \in ]0,1[$.

In view of the proof of sufficiency of the conditions of Lemma 4.10, we get the following corollary to Lemma 4.13. In Corollary 4.1 the notation is in the sense of Definition 4.13(d).

**Corollary 4.1** *Let a set $\mathcal{A} \subset \Re^n$ be uniquely bounded with a generating function $u_1$ obeying $u_1(X) = \xi_1$ on $\partial\mathcal{A}$ and let $\mathcal{N}_u(\mathcal{A})$ be its uniquely bounded neighbourhood with the common centre $w$ of unique boundedness, $w \in \mathring{\mathcal{A}}$, and with a generating function $u_2$ satisfying $u_2(X) = \xi_2$ on $\partial\mathcal{N}_u(\mathcal{A})$, $\xi_i \in \Re^+$, $i = 1,2$. Then a function $v : \Re^n \to \Re$ defined by*

$$v(X) = \begin{cases} 0 & , \quad X = w, \\ \dfrac{u_1(X)-\xi_1}{u_1[\xi_2 u_2^{-1}(X)X]-\xi_1} \cdot \dfrac{u_1[\xi_2 u_2^{-1}(X)X]}{u_1(X)} \cdot \\ \quad \cdot \dfrac{\rho(X-w,\mathcal{A})}{\rho[\xi_2 u_2^{-1}(X)(X-w),\mathcal{A}]} & , \quad (X \neq w) \in \Re^n \end{cases}$$

*a) is globally positive definite with respect to the set $\mathcal{A}$,*

*b) is radially increasing with respect to the set $\mathcal{A}$,*

*c) $v(X) = 1$ for all $(X-w) \in \partial\mathcal{N}_u(\mathcal{A})$.*

**Example 4.14** Let $\mathcal{A} = \{X : X \in \Re^n, ||X-\mathbf{1}|| \leq \sqrt{n}\}$ and $\mathcal{S} = \{X : X \in \Re^n, \mathbf{1}^T|X-\mathbf{1}| \leq 4\sqrt{n}\}$, $\mathbf{1} = (1\,1\ldots 1)^T \in \Re^n$. The sets $\mathcal{A}$ and $\mathcal{S}$ are uniquely bounded with a common centre of unique boundedness $W = \mathbf{1}$. The set $\mathcal{S}$ is a neighbourhood of the set $\mathcal{A}$. Moreover, it is a uniquely bounded neighbourhood of the set $\mathcal{A}$. Their generating functions are given by $u_1(X) = ||X-\mathbf{1}|| = u_1(X-\mathbf{1})$ for the set $\mathcal{A}$ with $\xi_1 = \sqrt{n}$, and $u_2(X) = (\mathbf{1}^T|X-\mathbf{1}|) = u_2(X-\mathbf{1})$ for the set

$\mathcal{S} = \mathcal{N}_u(\mathcal{A})$ with $\xi_2 = 4\sqrt{n} = 4\xi_1$. The function $v$ defined in Lemma 4.13 has the following form due to $\lambda = \sqrt{n}u_1^{-1}$ and $\Lambda = 4\sqrt{n}u_2^{-1}$:

$$v(X) = \begin{cases} 0 & , X = \mathbf{1}, \\ \dfrac{1 - \dfrac{\sqrt{n}}{||X - \mathbf{1}||}}{1 - \dfrac{\sqrt{n}}{\left\| \dfrac{4\sqrt{n}}{\mathbf{1}^T|X - \mathbf{1}|}(X - \mathbf{1}) \right\|}} \cdot \dfrac{\rho(X - \mathbf{1}, \mathcal{A})}{\rho\left[\dfrac{4\sqrt{n}}{\mathbf{1}^T|X - \mathbf{1}|}(X - \mathbf{1}), \mathcal{A}\right]} & , (X \neq \mathbf{1}) \in \Re^n, \end{cases}$$

or,

$$v(X) = \begin{cases} 0 & , X = \mathbf{1}, \\ \dfrac{4(||X - \mathbf{1}|| - \sqrt{n})}{(4||X - \mathbf{1}|| - \mathbf{1}^T|X - \mathbf{1}|)} \cdot \dfrac{\rho(X - \mathbf{1}, \mathcal{A})}{\rho\left[\dfrac{4\sqrt{n}}{\mathbf{1}^T|X - \mathbf{1}|}(X - \mathbf{1}), \mathcal{A}\right]} & , (X \neq \mathbf{1}) \in \Re^n, \end{cases}$$

It has the following properties:

1) $v(X) = 0$ for all $(X - \mathbf{1}) \in \overline{\mathcal{A}} = \mathcal{A}$ because then $\rho(X - \mathbf{1}, \mathcal{A}) = 0$,

2) $v(X) \in C(\Re^n)$,

3) $v(X) > 0$ for all $(X - \mathbf{1}) \in (\Re^n - \overline{\mathcal{A}})$. This can be verified as follows:

$$\frac{4\sqrt{n}}{\mathbf{1}^T|X - \mathbf{1}|}(X - \mathbf{1}) \in \partial\mathcal{S}$$

due to $\mathbf{1}^T \left| \dfrac{4\sqrt{n}}{\mathbf{1}^T|X - \mathbf{1}|}(X - \mathbf{1}) \right| = 4\sqrt{n} = \xi_2$, so that

$$1 - \frac{\sqrt{n}}{\left\| \dfrac{4\sqrt{n}}{\mathbf{1}^T|X - \mathbf{1}|}(X - \mathbf{1}) \right\|} > 0 \quad \text{for} \quad (X - \mathbf{1}) \in (\Re^n - \overline{\mathcal{A}}).$$

For $(X - \mathbf{1}) \in (\Re^n - \overline{\mathcal{A}})$, $||X - \mathbf{1}|| > \sqrt{n}$. Hence, $1 - \dfrac{\sqrt{n}}{||X - \mathbf{1}||} > 0$. Besides, $\rho(X - \mathbf{1}, \mathcal{A}) > 0$ and $\rho\left[\dfrac{4\sqrt{n}}{\mathbf{1}^T|X - \mathbf{1}|}(X - \mathbf{1}), \mathcal{A}\right] > 0$ for all $(X - \mathbf{1}) \in (\Re^n - \overline{\mathcal{A}})$.

Now, we may conclude that $v$ is globally positive definite with respect to the set $\mathcal{A}$. Furthermore, for any $(X - \mathbf{1}) \in (\Re^n - \overline{\mathcal{A}})$ and any $\alpha \in \Re^+$,

$$v(\alpha X) \equiv v[\alpha(X - \mathbf{1})] =$$

$$= \begin{cases} 0 & , X = w, \\ \dfrac{4(\alpha\|X - \mathbf{1}\| - \sqrt{n})\rho[\alpha(X - \mathbf{1}), \mathcal{A}]}{(4\alpha\|X - \mathbf{1}\| - \alpha\mathbf{1}^T|X - \mathbf{1}|)\rho\left[\dfrac{4\sqrt{n}}{\alpha\mathbf{1}^T|X - \mathbf{1}|}\alpha(X - \mathbf{1}), \mathcal{A}\right]} & , X \neq w. \end{cases}$$

Unique boundedness of $\mathcal{A}$ ensures that $\rho$ is increasing in $\alpha$ for $X \neq \mathbf{1}$. This and

$$v(\alpha X) =$$

$$= \begin{cases} 0 & , X = \mathbf{1}, \\ \dfrac{4\left(\|X - \mathbf{1}\| - \dfrac{\sqrt{n}}{\alpha}\right)\rho[\alpha(X - \mathbf{1}), \mathcal{A}]}{(4\|X - \mathbf{1}\| - \mathbf{1}^T|X - \mathbf{1}|)\rho\left[\dfrac{4\sqrt{n}}{\mathbf{1}^T|X - \mathbf{1}|}(X - \mathbf{1}), \mathcal{A}\right]} & , (X \neq \mathbf{1}) \in \Re^n, \end{cases}$$

imply $v(\alpha_1 X) < v(\alpha_2 X)$ for all $\alpha_i \in \Re^+$, $\alpha_1 < \alpha_2$, and every $(X - \mathbf{1}) \in (\Re^n - \overline{\mathcal{A}})$. This shows that $v$ is radially increasing with respect to the set $\mathcal{A}$.

If $(X - \mathbf{1}) \in \partial\mathcal{S}$ then $\mathbf{1}^T|X - \mathbf{1}| = 4\sqrt{n}$ and $\rho(X - \mathbf{1}, \mathcal{A}) > 0$ so that

$$v(X) = \frac{4(\|X - \mathbf{1}\| - \sqrt{n})}{(4\|X - \mathbf{1}\| - 4\sqrt{n})} \cdot \frac{\rho(X - \mathbf{1}, \mathcal{A})}{\rho\left(\dfrac{4\sqrt{n}}{4\sqrt{n}}(X - \mathbf{1}), \mathcal{A}\right)} = 1 \ \text{ for all } \ (X - \mathbf{1}) \in \partial\mathcal{S}.$$

This verifies the last property of the function $v$ as specified in Lemma 4.13-c.

## 4.4 Dini derivatives and the Lyapunov method

### 4.4.1 Fundamental lemmae on Dini derivatives

Dini derivatives are defined in Section 3.2.2. For the following result see McShane [143, p. 200] and LaSalle [120, p. 30].

**Lemma 4.14** *Let* $\nu : [a, b] \rightarrow \Re$. *If*

*(i)* $\nu$ *is left lower semicontinuous on* $[a, b[$ *(i.e.* $\liminf\,[\nu(t) : t \rightarrow \tau^-] \geq \nu(\tau)$ *for each* $\tau \in [a, b[$*)*

*and*

*(ii)* $\mathcal{D}_+\nu(t) \leq 0$ *for all* $t \in [a, b[$ *except at the most countable number of points,*

*then* $\nu$ *is nonincreasing on* $[a, b[$*, hence differentiable almost everywhere, and*

$$\int_a^t \mathcal{D}_+\nu(\sigma)d\sigma \geq \nu(t) - \nu(a), \ \ \forall t \in [a, b[.$$

$\nabla\nabla\nabla$

This lemma leads to the following result.

**Lemma 4.15** *Let the system (4.1) have generalised motions* $\underline{X}(t; X_0) \in C_0(\mathcal{I}_0 \times \mathcal{X}_0)$, $\mathcal{X}_0 \subseteq \mathcal{X}$, $\mathcal{X} \subseteq \Re^n$, *and* $\nu(t) = v[\underline{X}(t; X_0)]$ *for arbitrary fixed* $X_0 \in \mathcal{X}_0$ *and for* $v(X) \in C(\mathcal{S})$, *where* $\mathcal{S}$, $\mathcal{S} \subseteq \mathcal{X}$, *is an open connected set satisfying either (i),*

*(i)* $\mathcal{D}_+ v(X) \leq 0$ *for all* $X \in \mathcal{S}$,

*or (ii),*

*(ii)* $\mathcal{D}^+ v(X) \leq 0$ *for all* $X \in \mathcal{S}$.

*Then,*

*(a)* $\nu$ *is nonincreasing on* $[0, b_0[$ *where* $b_0$ *is the first time when* $\underline{X}(t; X_0)$ *is not in* $\mathcal{S}$: $\underline{X}(t; X_0) \in \mathcal{S}$ *for all* $t \in [0, b_0[$ *and* $\underline{X}(b_0; X_0) \notin \mathcal{S}$,

*(b)* $\nu$ *is differentiable almost everywhere on* $[0, b_0[$,

*and*

*(c) either*

$$(i) \quad \int_0^t \mathcal{D}_+ v[\underline{X}(\sigma; X_0)] \geq v[\underline{X}(t; X_0)] - v(X_0), \quad \forall t \in [0, b_0[,$$

*or*

$$(ii) \quad \int_0^t \mathcal{D}^+ v[\underline{X}(\sigma; X_0)] \geq v[\underline{X}(t; X_0)] - v(X_0), \quad \forall t \in [0, b_0[,$$

*respectively.* $\nabla\nabla\nabla$

**Proof.** Let the system (4.1) have generalised motions $\underline{X}(t; X_0)$ for each $X_0 \in \mathcal{X}_0$. Hence, $\underline{X}(t; X_0) \in C_t(\mathcal{I}_0 \times \mathcal{X}_0)$ (Definition 3.8, Section 3.2.3). Let $v$, $\nu$ and $\mathcal{S}$ obey the conditions of the lemma. Altogether, $\nu$ is continuous on $[0, b_0]$. Lemma 4.14 and (i) or (ii) guarantee (a) and (b). The condition (i) implies (c-i) due to Lemma 4.15. The condition (ii) implies (c-ii) due to the same lemma because $\mathcal{D}^+ v(X) \geq \mathcal{D}_+ v(X)$ by definition and $\mathcal{D}^+ v(X) \leq 0$ on $\mathcal{S}$ in view of (ii).

**Note 4.10** It is important to note that Lemma 4.15

1) does not guarantee positive invariance of the set $\mathcal{S}$,

and

2) does not guarantee validity of the result (c) for $t \geq b_0$.

Therefore, it is important to find conditions for positive invariance of the set $\mathcal{S}$. The following result by Yoshizawa [217, p. 3], simplifies determination of Dini derivatives.

**Lemma 4.16** *Let the system (4.1) have generalised motions* $\underline{X}(t; X_0) \in C_0(\mathcal{I}_0 \times \mathcal{X}_0)$, $\mathcal{X}_0 \subseteq \mathcal{X} \subseteq \Re^n$, *and* $\mathcal{S}$, $\mathcal{S} \subseteq \mathcal{X}$, *be an open connected set. Let* $v : \mathcal{S} \to \Re$ *be Lipschitzian on every compact subset of* $\mathcal{S}$. *Then,*

$$1. \qquad \mathcal{D}^+ v(X) = lim\,sup \left\{ \frac{v[X + \theta f(X)] - v(X)}{\theta} : \theta \to 0^+ \right\}$$

*and*

$$2. \qquad \mathcal{D}_+ v(X) = lim\,inf \left\{ \frac{v[X + \theta f(X)] - v(X)}{\theta} : \theta \to 0^+ \right\}. \qquad \nabla\nabla\nabla$$

Applications of Lemma 4.16 will be illustrated by an example in which we shall make use of a function $\omega : \Re \times \Re \to \Re$ defined by

$$\omega(x, \dot{x}) = \begin{cases} -1 & , (x \in \Re^- \text{ and } \dot{x} \in \Re) \text{ or } (x = 0 \text{ and } \dot{x} \in \Re^-), \\ 0 & , (x = 0 \text{ and } \dot{x} = 0), \\ +1 & , (x \in \Re^+ \text{ and } \dot{x} \in \Re) \text{ or } (x = 0 \text{ and } \dot{x} \in \Re^+). \end{cases}$$

With the function $\omega$ we associate $\Omega : \Re^n \times \Re^n \to \Re^{n \times n}$ defined by

$$\Omega(x, \dot{x}) = diag\,\{\omega(x_1, \dot{x}_1)\ \ \omega(x_2, \dot{x}_2) \ldots \omega(x_n, \dot{x}_n)\}.$$

**Example 4.15** Let $v : \Re^1 \to \Re^1$, $v(x) = |x|$, and $n = 1$ so that the system (4.1) takes the scalar form

$$\frac{dx}{dt} = f(x).$$

Let the system have generalised motion $\underline{x}(t; x_0) \in C_0(\mathcal{I}_0 \times \mathcal{X}_0)$, $\mathcal{X}_0 \subseteq \mathcal{X} \subseteq \Re^1$. The function $v$ is globally Lipschitzian. We may apply Lemma 4.16,

$$\mathcal{D}^+ v(X) = lim\,sup \left\{ \frac{|x + \theta f(x)| - |x|}{\theta} : \theta \to 0^+ \right\} =$$

$$= \begin{cases} lim\,sup \left\{ \dfrac{-[x + \theta f(x)] - (-x)}{\theta} : \theta \to 0^+ \right\} , \\ \qquad (x \in \Re^-) \text{ or } (x = 0,\ f(x) \in \Re^-), \\ lim\,sup \left\{ \dfrac{x + \theta f(x) - x}{\theta} : \theta \to 0^+ \right\} , \\ \qquad (x = 0,\ f(x) = 0), \\ lim\,sup \left\{ \dfrac{x + \theta f(x) - x}{\theta} : \theta \to 0^+ \right\} , \\ \qquad (x \in \Re^+) \text{ or } (x = 0,\ f(x) \in \Re^+). \end{cases}$$

Hence, after cancelling $x$ and $(-x)$, and using $\dot{x} = f(x)$, we get

$$\mathcal{D}^+|x| = \begin{cases} -\dot{x} & , (x \in \Re^-) \text{ or } (x = 0,\ \dot{x} \in \Re^-), \\ 0 & , (x = 0,\ \dot{x} = 0), \\ \dot{x} & , (x \in \Re^+) \text{ or } (x = 0,\ \dot{x} \in \Re^+). \end{cases}$$

Now we can make use of the function $\omega$ so that

$$\mathcal{D}^+|x| = \omega(x, \dot{x})\dot{x} = \omega(x, \dot{x})f(x) = \omega[x, f(x)]f(x).$$

Obviously, $\mathcal{D}^+|x| = \mathcal{D}_+|x|$.

**Example 4.16** Let $v : \Re^n \to \Re^1$ be given in the natural form

$$v(X) = \mathsf{b}^T|X| = b_1|x_1| + b_2|x_2| + \ldots + b_n|x_n|,$$

with $\mathsf{b} = (b_1\ b_2 \ldots b_n)^T > 0$ elementwise (i.e. $b_i > 0$, $\forall i = 1, 2, \ldots, n$).
Now, by referring to Example 4.15 we derive

$$\mathcal{D}^+v(X) = \mathsf{b}^T\mathcal{D}^+|X| = \sum_{i=1}^{n} b_i\mathcal{D}^+|x_i| = \sum_{i=1}^{n} b_i\omega(x_i, \dot{x}_i)\dot{x}_i = \mathsf{b}^T\Omega(X, \dot{X})\dot{X}.$$

Altogether, along generalised motions of the system (4.1),

$$\mathcal{D}^+(\mathsf{b}^T|X|) = \mathsf{b}^T\Omega(X, \dot{X})\dot{X} = \mathsf{b}^T\Omega(X, \dot{X})f(X) = \mathsf{b}^T\Omega[X, f(X)]f(X),$$

and $\mathcal{D}_+(\mathsf{b}^T|X|) = \mathcal{D}^+(\mathsf{b}^T|X|)$.

Let $\mathcal{V}^{-1}(\zeta) = \{X : X \in \Re^n, v(X) = \zeta\}$.

### 4.4.2 LaSalle principle

LaSalle established the invariance principle [120, p. 30], that will be called the LaSalle principle.

**Theorem 4.4 (LaSalle principle)** *Let*

*1)* $v(X) \in C(\mathcal{X})$, $\mathcal{X} \subseteq \Re^n$,

*2)* $\mathcal{D}^+v(X) \leq 0$ *[or* $\mathcal{D}_+v(X) \leq 0$*] for all* $X \in \mathcal{X}$,

*and*

*3) the system (4.1) have generalised motions* $\underline{X}(t; X_0) \in C_0(\mathcal{I}_0 \times \mathcal{X}_0)$, $\mathcal{X}_0 \subseteq \mathcal{X}$.

*Then, for some $\zeta$, $\mathcal{L}^+(X_0) \cap \mathcal{X} \subset \mathcal{M}^* \cap \mathcal{V}^{-1}(\zeta)$, where $\mathcal{M}^*$ is the largest weakly invariant set in $\mathcal{E}_0 = \{X : X \in \mathcal{X}, \mathcal{D}^+v(X) = 0\}$ [or $\mathcal{E}_0 = \{X : X \in \mathcal{X}, \mathcal{D}_+v(X) = 0\}$], respectively.*

*If, in addition to 1)–3)*

*4) $\underline{X}(t; X_0)$ is precompact relative to $\mathcal{X}$,*

*then $\underline{X}(t; X_0) \rightarrow \mathcal{M} \cap \mathcal{V}^{-1}(\zeta)$ as $t \rightarrow +\infty$, where $\mathcal{M}$ is the largest invariant set in $\mathcal{E}_0$.* ∇∇∇

**Proof [120].** Let at first the conditions 1) through 3) hold. Let it be assumed that $y \in \mathcal{L}^+(X_0) \cap \mathcal{X}$. Then $l^+(X_0) = +\infty$, and there is a sequence $t_n$ such that $\underline{X}(t_n; X_0) \rightarrow y$ and $t_n \rightarrow +\infty$ as $n \rightarrow +\infty$. Now, $v[\underline{X}(t_n; X_0)] \rightarrow v(y)$ as $n \rightarrow +\infty$ (due to 1)); since $v[\underline{X}(t; X_0)] = \nu(t)$ is nonincreasing with respect to $t$ due to 2) (Lemma 4.15) then $v[\underline{X}(t; X_0)] \geq v(y)$ for all $t \in \Re_+$ and there is $\zeta$ such that $v[\underline{X}(t; X_0)] \rightarrow \zeta$ as $t \rightarrow +\infty$. Hence, for such $\zeta$, $v(y) = \zeta$ for each $y \in \mathcal{L}^+(X_0)$. Now, $\mathcal{L}^+(X_0)$ is weakly invariant, and therefore $\mathcal{L}^+(X_0) \subset \mathcal{E}_0$, and hence in $\mathcal{M}^*$. This proves $\mathcal{L}^+(X_0) \cap \mathcal{X} \subset \mathcal{M}^* \cap \mathcal{V}^{-1}(\zeta)$. If, in addition to 1)–3), the condition 4) also holds, then $\mathcal{L}^+(X_0)$ is invariant and $\mathcal{L}^+(X_0) \subset \mathcal{M} \cap \mathcal{V}^{-1}(\zeta)$. Since $\underline{X}(t; X_0) \rightarrow \mathcal{L}^+(X_0)$ as $t \rightarrow +\infty$, then $\underline{X}(t; X_0) \rightarrow \mathcal{M} \cap \mathcal{V}^{-1}(\zeta)$ as $t \rightarrow +\infty$.

LaSalle deduced the following results from the invariance principle (Theorem 4.4) [116, pp. 31, 33]:

**Corollary 4.2 [120]** *Let*

*1) $v(X) \in C(\mathcal{X})$, $\mathcal{X} \subseteq \Re^n$,*

*2) $\mathcal{D}^+v(X) \leq 0$ [or $\mathcal{D}_+v(X) \leq 0$] for all $X \in \mathcal{X}$,*

*3) the system (4.1) have generalised motions $\underline{X}(t; X_0) \in C_0(\mathcal{I}_0 \times \mathcal{X}_0)$, $\mathcal{X}_0 \subseteq \mathcal{X} \subseteq \Re^n$.*

*4) $\underline{X}(t; X_0)$ be a precompact generalised motion that remains in $\mathcal{X}$ for all $t \geq 0$,*

*and*

*5) the points of intersection of $\mathcal{M}$ (or, $\mathcal{E}_0$) with $\mathcal{V}^{-1}(\zeta)$ are isolated for each $\zeta$. Then, $\underline{X}(t; X_0)$ approaches an equilibrium point of the system (4.1) as $t \rightarrow +\infty$*

**Note 4.11** Since the system (4.1) has generalised motion $\underline{X}(t; X_0) \in C_0(\mathcal{I}_0 \times \mathcal{X}_0)$ then it is unique (Theorem 4.14, Section 3.2.6).

**Lemma 4.17 [120].** *Let*

*1) $\mathcal{S}$ be a positively invariant open set in $\mathcal{X}$ with the property that each solution starting in $\mathcal{S}$ is bounded and has no positive limit points on the boundary of $\mathcal{S}$,*

*2) $v(X) \in C(\mathcal{S})$,*

*3) $\mathcal{D}^+v(X) \leq 0$ [or, $\mathcal{D}_+v(X) \leq 0$] for all $X \in \mathcal{S}$,*

*4) the system (4.1) have generalised motions $\underline{X}(t; X_0) \in C_0(\mathcal{I}_0 \times \mathcal{S})$,*

*5) $\mathcal{M}^0 = \overline{\mathcal{M} \cap \mathcal{S}} \subset \mathcal{S}$,*

*and*

*6) $\mathcal{M}^0$ is precompact.*

*Then, $\mathcal{M}^0$ is attractive set and $\mathcal{S} \subseteq \mathcal{D}_a(\mathcal{M}^0)$.*

*If, in addition to 1)–5),*

*6) $v$ is constant on the boundary of $\mathcal{M}^0$,*

*then $\mathcal{M}^0$ is asymptotically stable set.* ∇∇∇

**Comment 4.1** *Effective applications of the preceding results by LaSalle need*

*1) adequate choice, construction of the function $v$,*

*2) effective test of precompactness,*

*and*

*3) effective test of positive invariance.*

### 4.4.3 Dini derivatives, positive definiteness, positive invariance and precompactness

**Lemma 4.18** *Let the system (4.1) have generalised motions $\underline{X}(t; X_0) \in C_0(\mathcal{I}_0 \times \mathcal{X}_0)$, $\mathcal{X}_0 \subseteq \mathcal{X} \subseteq \Re^n$, and $\mathcal{A}$, $\mathcal{A} \subset \mathcal{X}_0$, be a compact set. Then, the existence of a neighbourhood $\mathcal{N}(\mathcal{A})$ of $\mathcal{A}$ and of a function $v : \Re^n \to \Re$ obeying*

*1) $v$ is positive definite with respect to the set $\mathcal{A}$ on $\mathcal{N}(\mathcal{A})$,*

*2) $\mathcal{D}^+v(X) \leq 0$ [or, $\mathcal{D}_+v(X) \leq 0$] for all $X \in \mathcal{N}(\mathcal{A})$*

*is sufficient for any bounded set $\mathcal{V}_\zeta(\mathcal{A})$, $\zeta \in \Re^+$ and $\mathcal{V}_\zeta(\mathcal{A}) \subseteq \mathcal{N}(\mathcal{A})$, to be positively invariant and positively precompact relative to itself.* ∇∇∇

**Proof.** Let the system (4.1) have generalised motions $\underline{X}(t; X_0) \in C_0(\mathcal{I}_0 \times \mathcal{X}_0)$, $\mathcal{X}_0 \subseteq \mathcal{X} \subseteq \Re^n$, and $\mathcal{A} \subset \mathcal{X}_0$, be compact. Let there exist $\mathcal{N}(\mathcal{A})$ and $v$ satisfying the conditions 1) and 2). Let $\zeta \in \Re^+$ be such that $\mathcal{V}_\zeta(\mathcal{A})$ is bounded and $\mathcal{V}_\zeta(\mathcal{A}) \subseteq \mathcal{N}(\mathcal{A})$. Hence, the hypersurface $v(X) = \zeta$ is closed, which holds also for $v(X) = \xi$ that bounds $\mathcal{V}_\xi(\mathcal{A})$ as soon as $\xi \in \,]0, \zeta]$ due to Lemma 4.2. Let it be assumed that $\mathcal{V}_\zeta(\mathcal{A})$ is not positively invariant. Then, there are $X_0 \in \mathcal{V}_\zeta(\mathcal{A})$ and $\tau \in \mathcal{I}_0$ for which $\underline{X}(\tau; X_0) \in \partial\mathcal{V}_\zeta(\mathcal{A})$. From $\underline{X}(t; X_0) \in [C_t(\mathcal{I}_0 \times \mathcal{X}_0) \cap C_0(\mathcal{I}_0 \times \mathcal{X}_0)]$ (Definition 3.8, Section 3.2.3), $X_0 \in \mathcal{V}_\zeta(\mathcal{A})$, $\mathcal{V}_\zeta(\mathcal{A}) = \mathring{\mathcal{V}}_\zeta(\mathcal{A})$ and $\underline{X}(\tau; X_0) \in \partial\mathcal{V}_\zeta(\mathcal{A})$

follows the existence of a sequence $t_n$, $t_n \in [0, \tau[$, $t_n \to \tau$ as $n \to +\infty$, such that $v[\underline{X}(t_{n-1}; X_0)] < v[\underline{X}(t_n; X_0)]$. This inequality contradicts the conditions 1) and 2) due to Lemma 4.15 and $\underline{X}(t_k; X_0) \in \mathcal{V}_\zeta(\mathcal{A})$, $k = n-1, n$. The contradiction is a consequence of the assumption that $\mathcal{V}_\zeta(\mathcal{A})$ is not positively invariant, which proves its positive invariance. This further implies $v[\underline{X}(t; X_0)] \leq v(X_0) < \zeta$ for all $t \in \Re_+$ (Lemma 4.15). Hence, $\mathcal{L}^+(X_0) \subset \mathcal{V}_\zeta(\mathcal{A})$, i.e. $\mathcal{L}^+(X_0) \cap \partial\mathcal{V}_\zeta(\mathcal{A}) = \emptyset$ that now proves positive precompactness of $\mathcal{V}_\zeta(\mathcal{A})$ relative to itself.

**Definition 4.15** *A function $v$ obeying Lemma 4.18 will be called Lyapunov function.*

Lemma 4.18 can be linked with Lemmae 4.5 and 4.7 to yield the following result.

**Lemma 4.19** *Let the system (4.1) have generalised motions $\underline{X}(t; X_0) \in C_0(\mathcal{I}_0 \times \mathcal{X}_0)$, $\mathcal{X}_0 \subseteq \mathcal{X} \subseteq \Re^n$, and $\mathcal{A}$, $\mathcal{A} \subset \mathcal{X}_0$, be compact set. Let*

*1) $v$ be positive definite function with respect to $\mathcal{A}$ on a neighbourhood $\mathcal{N}(\mathcal{A})$ of $\mathcal{A}$,*

*2) $\sigma \in \Re^+$ be the greatest number for which $\sigma$-neighbourhood $\mathcal{N}(\sigma, \mathcal{A})$ is in $\mathcal{N}(\mathcal{A})$: $\mathcal{N}(\sigma, \mathcal{A}) \subseteq \mathcal{N}(\mathcal{A})$,*

*3) $\varphi \in \mathcal{K}_{[0,\sigma]}$ be such that $v(X) \geq \varphi[\rho(X, \mathcal{A})]$ for all $X \in \mathcal{N}(\sigma, \mathcal{A})$,*

*4) $\alpha = \varphi(\sigma) \in \Re^+$,*

*and*

*5) $\mathcal{D}^+ v(X) \leq 0$ [or, $\mathcal{D}_+ v(X) \leq 0$] for all $X \in \mathcal{N}(\mathcal{A})$.*

*Then the set $\mathcal{V}_\zeta(\mathcal{A})$ is positively invariant and positively precompact relative to itself for each $\zeta \in ]0, \alpha[$.* ∇∇∇

**Proof.** Let all the conditions of Lemma 4.19 hold, where the condition 3) and 4) result from the condition 2) due to Lemma 4.5. The set $\mathcal{V}_\zeta(\mathcal{A})$ is bounded for every $\zeta \in ]0, \alpha[$ in view of Lemma 4.7. Hence, all the conditions of Lemma 4.18 are fulfilled, which proves Lemma 4.19.

## 4.5 Stability theorems

Following Lyapunov [127], Halanay [95], Yoshizawa [217], Bhatia and Szegö [10] and Hahn [94] we shall outline synthesised results on stability conditions.

### 4.5.1 Stability of a set

**Theorem 4.5** *Let the system (4.1) have generalised motions* $\underline{X}(t;X_0) \in C_0(\mathcal{I}_0 \times \mathcal{X}_0)$, $\mathcal{X}_0 \subseteq \mathcal{X} \subseteq \Re^n$. *In order for a compact connected set* $\mathcal{A}$, $\mathcal{A} \subset \mathcal{X}_0$, *to be stable it is necessary and sufficient that there exist a connected neighbourhood* $\mathcal{N}(\mathcal{A})$ *of* $\mathcal{A}$ *and a function* $v : \mathcal{N}(\mathcal{A}) \to \Re$ *satisfying both 1) and 2),*

1) *$v$ is positive definite with respect to $\mathcal{A}$ on $\mathcal{N}(\mathcal{A})$,*

2) $\mathcal{D}^+ v(X) \leq 0$ *for all* $X \in \mathcal{N}(\mathcal{A})$. $\nabla\nabla\nabla$

**Proof.** Let the system (4.1) have generalised motions $\underline{X}(t;X_0) \in C_0(\mathcal{I}_0 \times \mathcal{X}_0)$, $\mathcal{X}_0 \subseteq \mathcal{X} \subseteq \Re^n$ and $\mathcal{A}$, $\mathcal{A} \subset \mathcal{X}_0$, be compact.

*Necessity.* Let $\mathcal{A}$ be stable. Let $\varepsilon \in \Re^+$ be arbitrary and let $\delta = \delta(\varepsilon;\mathcal{A}) \in \Re^+$ obey Definition 1.3, Section 1.1.1. Then, $\mathcal{N}(\delta,\mathcal{A}) \subset \mathcal{X}_0$ and $\mathcal{I}_0 = \Re_+$ for every $X_0 \in \mathcal{N}(\delta,\mathcal{A})$. Let $\mathcal{N}(\mathcal{A}) = \mathcal{N}(\delta,\mathcal{A})$. $\mathcal{N}(\mathcal{A})$ is connected neighbourhood of $\mathcal{A}$. Let $\varphi \in \mathcal{K}_{[0,\varepsilon]}$ and

$$v(X) = sup\,(\varphi\{\rho[\underline{X}(t+\sigma;t,X),\mathcal{A}]\} : \sigma \in \Re_+).$$

The right hand side of this equation is independent of $t$ because $\underline{X}(t+\sigma;t,X) \equiv \underline{X}(\sigma;0,X) \equiv \underline{X}(\sigma;X)$ due to time-invariance of the system (4.1), that is

$$v(X) = sup\,(\varphi\{\rho[\underline{X}(\sigma;X),\mathcal{A}]\} : \sigma \in \Re_+).$$

From $\underline{X}(t;X_0) \in C_t[\Re_+ \times \mathcal{N}(\mathcal{A})] \cap C_0[\Re_+ \times \mathcal{N}(\mathcal{A})]$ due to the conditions of the theorem and $\mathcal{N}(\mathcal{A}) \subseteq \mathcal{X}_0$ it now follows

(i) $v(X) \in C[\mathcal{N}(\mathcal{A})]$.

Stability of $\mathcal{A}$ implies (Theorem 3.31, Section 3.6) $\rho[\underline{X}(t;X),\mathcal{A}] = 0$ for all $t \in \Re_+$ and each $X \in \mathcal{A}$. Hence,

(ii) $v(X) = 0$ for all $X \in \mathcal{A}$.

From the definition of $v$ it follows

(iii) $v(X) \geq \varphi[\rho(X,\mathcal{A})]$ for all $X \in \mathcal{N}(\mathcal{A})$.

The results (i)–(iii) prove positive definiteness of $v$ with respect to $\mathcal{A}$ on $\mathcal{N}(\mathcal{A})$. The condition 1) is necessary. Let $\theta \in \Re^+$ be arbitrarily small and $X^* = \underline{X}(t+\theta;t,X) = \underline{X}(\theta;0,X) = \underline{X}(\theta;X)$. Now,

$$v(X^*) = sup\,(\varphi\{\rho[\underline{X}(t+\theta+\sigma;t+\theta,X^*),\mathcal{A}]\} : \sigma \in \Re_+).$$

From $\underline{X}(t;X_0) \in C_t[\Re_+ \times \mathcal{N}(\mathcal{A})] \cap C_0[\Re_+ \times \mathcal{N}(\mathcal{A})]$, Theorem 3.14 (Section 3.2.6), time invariance of the system (4.1) and the definition of $X^*$ follows $\underline{X}(t+\theta+\sigma;$ $t+\theta,X^*) = \underline{X}[t+\theta+\sigma;t+\theta,\underline{X}(\theta;X)] \equiv \underline{X}(t+\theta+\sigma;t,X) \equiv \underline{X}(\theta+\sigma;0,X) = \underline{X}(\theta+\sigma;X)$, so that for $\sigma_1 = \theta+\sigma$, $\sigma_1 \in [\theta,+\infty[$ due to $\sigma \in \Re_+$,

$$v(X^*) = v[\underline{X}(\theta;X)] = sup\,(\varphi\{\rho[\underline{X}(\sigma_1;X),\mathcal{A}]\} : \sigma_1 \in [\theta,+\infty[) \leq$$
$$\leq sup\,(\varphi\{\rho[\underline{X}(\sigma_1;X),\mathcal{A}]\} : \sigma_1 \in \Re_+) = v(X),\ \forall X \in \mathcal{N}(\mathcal{A})$$

where the inequality holds because $\theta \in \Re^+$. Hence,

$$\mathcal{D}^+ v(X) = lim\,sup \left\{ \frac{v[\underline{X}(\theta; X)] - v(X)}{\theta} : \theta \to 0^+ \right\} \leq 0,\ \forall X \in \mathcal{N}(\mathcal{A}).$$

The condition 2) is also necessary.

*Sufficiency.* Let $\varepsilon \in \Re^+$ be arbitrary, $\sigma \in \Re^+$ be such that $\mathcal{N}(\sigma, \mathcal{A}) \subseteq \mathcal{N}(\mathcal{A})$ and $\varphi \in \mathcal{K}_{[0,\sigma]}$ obey $v(X) \geq \varphi[\rho(X, \mathcal{A})]$, $\forall X \in \mathcal{N}(\sigma, \mathcal{A})$ (Lemma 4.7). Let $\alpha = \varphi(\sigma)$, $\delta = \delta(\varepsilon) \in \Re^+$ and $\zeta \in ]0, \alpha[$ obey $\mathcal{N}(\delta, \mathcal{A}) \subseteq \mathcal{V}_\zeta(\mathcal{A}) \subseteq \mathcal{N}(\varepsilon, \mathcal{A})$. Now Lemma 4.19, all the conditions of which are satisfied, guarantees

$$\underline{X}(t; X_0) \in \mathcal{V}_\zeta(\mathcal{A}) \Rightarrow \underline{X}(t; X_0) \in \mathcal{N}(\varepsilon, \mathcal{A}),\ \ \forall t \in \Re_+,\ \forall X_0 \in \mathcal{N}(\delta, \mathcal{A}).$$

This proves the stability of $\mathcal{A}$ (Definition 1.3, Section 1.1.1).

### 4.5.2 Stability of $X = 0$

Let $\mathcal{A} = \mathcal{O} = \{0\}$ in Theorem 4.5 that takes the following form.

**Theorem 4.6** *Let the system (4.1) have generalised motions* $\underline{X}(t; X_0) \in C_0(\mathcal{I}_0 \times \mathcal{X}_0)$, $\mathcal{O} \subset \mathring{\mathcal{X}}_0 \subseteq \mathcal{X}_0 \subseteq \mathcal{X} \subseteq \Re^n$. *In order for* $X = 0$ *to be stable it is necessary and sufficient that there exist a connected neighbourhood* $\mathcal{N}$ *of* $X = 0$ *and a function* $v : \mathcal{N} \to \Re$ *satisfying both 1) and 2):*

*1) v is positive definite on* $\mathcal{N}$,

*and*

*2)* $\mathcal{D}^+ v(X) \leq 0$ *for all* $X \in \mathcal{N}$. ∇∇∇

### 4.5.3 Comment

Theorem 4.5 and Theorem 4.6 present necessary and sufficient stability conditions in terms of the existence of a function $v$ with specified properties. However, they do not provide an algorithm to construct the function $v$ for a given system (4.1).

## 4.6 Asymptotic stability theorems

We shall synthesise results by Lyapunov [127], Malkin [130], Halanay [95], Yoshizawa [217], Bhatia and Szegö [10] and Hahn [94] on asymptotic stability via the LaSalle principle.

### 4.6.1 Asymptotic stability of a set

**Theorem 4.7** *Let the system (4.1) have generalised motions* $\underline{X}(t;X_0) \in C_0(\mathcal{I}_0 \times \mathcal{X}_0)$, $\mathcal{X}_0 \subseteq \mathcal{X} \subseteq \Re^n$. *In order for a compact connected set* $\mathcal{A}$, $\mathcal{A} \subset \mathcal{X}_0$, *to be asymptotically stable it is necessary and sufficient that there exist a connected neighbourhood* $\mathcal{N}(\mathcal{A})$ *of* $\mathcal{A}$ *and a function* $v : \mathcal{N}(\mathcal{A}) \to \Re$ *satisfying the following conditions:*

*1)* $v$ *is positive definite with respect to* $\mathcal{A}$ *on* $\mathcal{N}(\mathcal{A})$,

*2)* $\mathcal{D}^+v(X) < 0$ *for all* $X \in [\mathcal{N}(\mathcal{A}) - \mathcal{A}]$,

*and*

*3)* $\mathcal{D}^+v(X) = 0$ *for all* $X \in \mathcal{A}$. ∇∇∇

**Proof.** Let the system (4.1) have generalised motions $\underline{X}(t;X_0) \in C_0(\mathcal{I}_0 \times \mathcal{X}_0)$, $\mathcal{X}_0 \subseteq \mathcal{X} \subseteq \Re^n$ and $\mathcal{A}$, $\mathcal{A} \subset \mathcal{X}_0$, be compact.

*Necessity.* Let $\mathcal{A}$ be asymptotically stable, hence stable and attractive (Definition 1.6, Section 1.1.3). Let $\varepsilon \in \Re^+$ be arbitrary and $\delta = \delta(\varepsilon;\mathcal{A}) \in \Re^+$ obey Definition 1.3, Section 1.1.1. Let $\Delta > 0$ and $\tau(X_0, \zeta; \mathcal{A}) = \tau \in \Re_+$ obey Definition 1.5, Section 1.1.2. Now let $\xi = min\,(\delta, \Delta) \in \Re^+$, $\alpha \in ]1, +\infty[$ be arbitrary and fixed, $X \in [\mathcal{N}(\xi, \mathcal{A}) - \mathcal{A}]$ be arbitrary, $\zeta = \varphi^I\{\alpha^{-2}\varphi[\rho(X,\mathcal{A})]\} \in \Re^+$ for $\varphi \in \mathcal{K}_{[0,\varepsilon]}$.

$$v(X) = sup\,(\varphi\{\rho[\underline{X}(t+\sigma;t,X),\mathcal{A}]\}\frac{1+\alpha\sigma}{1+\sigma} : \sigma \in \Re_+).$$

The right hand side of this equation is independent of $t$ because $\underline{X}(t+\sigma;t,X) \equiv \underline{X}(\sigma;0,X) \equiv \underline{X}(\sigma;X)$ due to time-invariance of the system (4.1). This clarifies that $v$ does not depend on time $t$. Notice that

$$\psi(\sigma) = \frac{1+\alpha\sigma}{1+\sigma} \in C(\Re_+),\ sup\,[\psi(\sigma) : \sigma \in \Re_+] = \alpha.$$

These facts and $\underline{X}(t;X_0) \in C_t[\Re_+ \times \mathcal{N}(\xi,\mathcal{A})] \cap C_0[\Re_+ \times \mathcal{N}(\xi,\mathcal{A})]$ due to $\mathcal{N}(\xi,\mathcal{A}) \subseteq \mathcal{N}(\delta,\mathcal{A})$ and Definition 1.3, Section 1.1.1, prove

(i) $v(X) \in C[\mathcal{N}(\xi,\mathcal{A})]$.

Stability of $\mathcal{A}$ implies $\rho[\underline{X}(t;X_0),\mathcal{A}] = 0$ for all $t \in \Re_+$ and each $X_0 \in \mathcal{A}$ (Theorem 3.31, Section 3.6). Hence,

(ii) $v(X) = 0$ for all $X \in \mathcal{A}$.

The definition of $v$ implies directly

(iii) $v(X) \geq \varphi[\rho(X,\mathcal{A})]$ for all $X \in \mathcal{N}(\xi,\mathcal{A})$.

Let $\mathcal{N}(\mathcal{A}) = \mathcal{N}(\xi,\mathcal{A})$. Therefore, $\mathcal{N}(\mathcal{A})$ is a connected neighbourhood of $\mathcal{A}$. Altogether, (i)–(iii) show that $v$ is positively definite with respect to $\mathcal{A}$ on $\mathcal{N}(\mathcal{A})$, which proves necessity of 1). From the definitions of $\tau = \tau(X,\zeta;\mathcal{A})$ (Definition 1.5, Section 1.1.2) it follows

$$\rho[\underline{X}(t;X),\mathcal{A}] < \zeta \text{ for all } t \in ]\tau,+\infty[,$$

which together with the definition of $\zeta$ yield

$$\varphi\{\rho[\underline{X}(t;X),\mathcal{A}]\} < \varphi(\zeta) = \varphi(\varphi^I\{\alpha^{-2}\varphi[\rho(X,\mathcal{A})]\}) = \alpha^{-2}\varphi[\rho(X,\mathcal{A})]$$

for all $t \in ]\tau,+\infty[$. Hence,

$$\varphi\{\rho[\underline{X}(\sigma;X),\mathcal{A}]\}\frac{1+\alpha\sigma}{1+\sigma} < \{\alpha^{-2}\varphi[\rho(X,\mathcal{A})]\}\alpha < \varphi[\rho(X,\mathcal{A})]$$

for all $t \in ]\tau,+\infty[$. This result shows that there is $\hat{\sigma} \in [0,\tau]$ such that

$$v(X) = \varphi\{\rho[\underline{X}(t+\hat{\sigma};t,X),\mathcal{A}]\}\frac{1+\alpha\hat{\sigma}}{1+\hat{\sigma}}.$$

Let $\theta \in \Re^+$ be arbitrarily small and $X^* = \underline{X}(t+\theta;t,X) \equiv \underline{X}(\theta;0,X) \equiv \underline{X}(\theta;X)$. Then, by using Theorem 3.14 (Section 3.2.6), we derive

$$\begin{aligned} v(X^*) &= sup\left(\varphi\{\rho[\underline{X}(t+\theta+\sigma;t+\theta,X^*),\mathcal{A}]\}\frac{1+\alpha\sigma}{1+\sigma} : \sigma \in \Re_+\right) = \\ &= sup\left(\varphi\{\rho[\underline{X}(t+\theta+\sigma;t,X),\mathcal{A}]\}\frac{1+\alpha\sigma}{1+\sigma} : \sigma \in \Re_+\right) = \\ &= sup\left(\varphi\{\rho[\underline{X}(\theta+\sigma;X),\mathcal{A}]\}\frac{1+\alpha\sigma}{1+\sigma} : \sigma \in \Re_+\right) = \\ &= \varphi\{\rho[\underline{X}(\theta+\hat{\sigma};X),\mathcal{A}]\}\frac{1+\alpha\hat{\sigma}}{1+\hat{\sigma}}. \end{aligned}$$

Notice that (see Halanay [95]) for $\sigma = \theta + \hat{\sigma} \in [\theta, \tau+\theta] \subset \Re_+$

$$\frac{1+\alpha\hat{\sigma}}{1+\hat{\sigma}} = \frac{1+\alpha\sigma}{1+\sigma}\left\{1 - \frac{(\alpha-1)\theta}{(1+\hat{\sigma})(1+\alpha\sigma)}\right\} > 0.$$

Therefore, for $\sigma = \theta + \hat{\sigma}$,

$$\begin{aligned} v(X^*) &= v[\underline{X}(\theta;X)] = \\ &= \varphi\{\rho[\underline{X}(t+\sigma;t,X),\mathcal{A}]\}\frac{1+\alpha\sigma}{1+\sigma}\left\{1 - \frac{(\alpha-1)\theta}{(1+\hat{\sigma})(1+\alpha\sigma)}\right\} \leq \\ &\leq sup\left(\varphi\{\rho[\underline{X}(t+\sigma;t,X),\mathcal{A}]\}\frac{1+\alpha\sigma}{1+\sigma} : \sigma \in \Re_+\right) \cdot \\ &\qquad \cdot \left\{1 - \frac{(\alpha-1)\theta}{(1+\hat{\sigma})(1+\alpha\sigma)}\right\} = \\ &= v(X)\left\{1 - \frac{(\alpha-1)\theta}{(1+\hat{\sigma})(1+\alpha\sigma)}\right\}. \end{aligned}$$

Now, we set $v(X)$ on the left hand side of the preceding inequality, divide the new one by $\theta \in \Re^+$ and take $lim\,sup\,(\cdot)$ of both sides so as to get the following, by using $\hat{\sigma} = \sigma - \theta$ and $\sigma \in [\theta, \tau + \theta] \to [0^+, \tau]$ as $\theta \to 0^+$,

$$\begin{aligned}\mathcal{D}^+ v(X) &= lim\,sup\left\{\frac{v[\underline{X}(\theta; X)] - v(X)}{\theta} : \theta \to 0^+\right\} \leq \\ &\leq lim\,sup\left\{-\frac{\alpha - 1}{(1+\sigma-\theta)(1+\alpha\sigma)} v(X) : \theta \to 0^+\right\} \leq \\ &\leq -\frac{(\alpha-1)v(X)}{(1+\tau)(1+\alpha\tau)}, \quad \forall X \in \mathcal{N}(\mathcal{A}).\end{aligned}$$

Positive definiteness of $v$ with respect to $\mathcal{A}$ on $\mathcal{N}(\mathcal{A})$, $\alpha \in ]1, +\infty[$ and $\tau \in \Re_+$ linked with the preceding result prove $\mathcal{D}^+ v(X) < 0$ for all $X \in [\mathcal{N}(\mathcal{A}) - \mathcal{A}]$.

Hence, 2) is necessary. Let $X \in \mathcal{A}$. Then $\underline{X}(t; X) \in \mathcal{A}$ for all $t \in \Re_+$ that yields $v[\underline{X}(t; X)] = 0$ for all $t \in \Re_+$ and

$$\mathcal{D}^+ v(X) = lim\,sup\left\{\frac{v[\underline{X}(\theta; X)] - v(X)}{\theta} : \theta \to 0^+\right\} = lim\,sup\left(\frac{0-0}{\theta} : \theta \to 0^+\right) = 0$$

for all $X \in \mathcal{A}$, which verifies necessity of 3).

*Sufficiency.* Let the conditions of the theorem hold. Then all the conditions of Theorem 4.5 are satisfied implying stability of the set $\mathcal{A}$. Let $\sigma \in \Re^+$ and $\alpha \in \Re^+$ be defined as in Lemma 4.19, which is possible due to positive definiteness of $v$ with respect to $\mathcal{A}$ on $\mathcal{N}(\mathcal{A})$ (Lemma 4.5). Now all the conditions of Lemma 4.19 are satisfied so that $\mathcal{V}_\varsigma(\mathcal{A})$ is positively invariant and positively precompact relative to itself for any $\zeta \in ]0, \alpha[$. Let $\Delta \in \Re^+$ be such that $\mathcal{N}(\Delta, \mathcal{A}) \subseteq \mathcal{V}_\varsigma(\mathcal{A})$ for arbitrary $\zeta \in ]0, \alpha[$. Then, $X_0 \in \mathcal{N}(\Delta, \mathcal{A})$ implies $\underline{X}(t; X_0) \in \mathcal{V}_\varsigma(\mathcal{A})$ for all $t \in \Re_+$. From the conditions 2) and 3) follows $\mathcal{E}_0 = \{X : X \in \mathcal{V}_\varsigma(\mathcal{A}), \mathcal{D}^+ v(X) = 0\} = \mathcal{A}$ so that $\mathcal{M} \subseteq \mathcal{A}$ and $\mathcal{M}^0 = \overline{\mathcal{M} \cap \mathcal{V}_\varsigma(\mathcal{A})} \subseteq \mathcal{A}$. With these facts and $\mathcal{S} = \mathcal{V}_\varsigma(\mathcal{A})$ we conclude that the conditions 1)–6) of Lemma 4.17 are satisfied ($\mathcal{M}^0$ is compact because it is closed by the definition and subset of compact set $\mathcal{A}$). Hence, $\mathcal{M}^0$ is attractive and $\mathcal{V}_\varsigma(\mathcal{A}) \subseteq \mathcal{D}_a(\mathcal{M}^0)$, which proves attraction of $\mathcal{A}$ due to $\mathcal{M}^0 \subseteq \mathcal{A}$ and $\mathcal{V}_\varsigma(\mathcal{A}) \subseteq \mathcal{D}_a(\mathcal{A})$. $\mathcal{A}$ is stable and attractive. Therefore, it is asymptotically stable.

Notice that $\mathcal{M}^0 = \overline{\mathcal{M}^0} \subseteq \mathcal{A} = \overline{\mathcal{A}}$ and $v(X) = 0$ on $\mathcal{A}$ guarantee $v(X) = const.$ on $\partial\mathcal{M}^0$ so that all the conditions 1)–6) of Lemma 4.17 are fulfilled, which proves immediately (in another way) asymptotic stability of $\mathcal{A}$.

**Note 4.12** The classical form of Theorem 4.7, when it is not proved via the LaSalle principle differs from the form given here for the condition 2). The classical form of condition 2) reads as follows:

*2') there is a positive definite function $\psi$ with respect to $\mathcal{A}$ on $\mathcal{N}(\mathcal{A})$ such that*

$$\mathcal{D}^+ v(X) \leq -\psi(X) \quad \textit{for all} \quad X \in \mathcal{N}(\mathcal{A}).$$

Necessity of this condition is proved essentially in the proof of necessity of the conditions of Theorem 4.7 because $\psi$ defined by

$$\psi(X) = \frac{(\alpha - 1)v(X)}{(1+\tau)(1-\alpha\tau)}$$

is positive definite on $\mathcal{N}(\mathcal{A})$ and obeys the above condition 2′). For sufficiency we have just to conclude that the above condition 2′) guarantees validity of the condition 2) of Theorem 4.7. Altogether, Theorem 4.7 is valid; nevertheless we use the condition 2) or 2′).

### 4.6.2 Complete global asymptotic stability of sets

Barbashin and Krasovskii [6] extended the asymptotic stability conditions to those for global asymptotic stability by requiring radial unboundedness of the function $v$. This will be first stated in the framework of sets in an extended form related to complete global asymptotic stability.

**Theorem 4.8** *Let the system (4.1) have generalised motions* $\underline{X}(t; X_0) \in C_0(\mathcal{I}_0 \times \Re^n)$. *In order for a compact connected set* $\mathcal{A}$, $\mathcal{A} \subset \Re^n$, *to be completely globally asymptotically stable it is necessary and sufficient that there exists a function* $v : \Re^n \to \Re$ *satisfying the following conditions:*

*1)* $v$ *is globally positive definite with respect to* $\mathcal{A}$,

*2)* $v$ *is radially unbounded with respect to* $\mathcal{A}$,

*3)* $\mathcal{D}^+v(X) < 0$ *for all* $X \in [\Re^n - \mathcal{N}(\mathcal{A})]$,

*and*

*4)* $\mathcal{D}^+v(X) = 0$ *for all* $X \in \mathcal{A}$. ∇∇∇

**Proof.** Let the system (4.1) have generalised motions $\underline{X}(t; X_0) \in C_0(\mathcal{I}_0 \times \Re^n)$.

*Necessity.* Let a compact connected set $\mathcal{A}$ be completely globally asymptotically stable that implies both $\delta_M(\varepsilon, \mathcal{A}) \to +\infty$ as $\varepsilon \to +\infty$ (Definition 1.6, Section 1.1.3, and Definition 1.3, Section 1.1.1) and $\Delta = +\infty$ (Definition 1.6, Section 1.1.3, and Definition 1.5, Section 1.1.2). Hence, we may set $\xi = +\infty$ and $\varphi \in \mathcal{K}\Re$ in the proof of the necessity part of Theorem 4.7, which is then to be repeated so that necessity of the conditions 1), 3) and 4) immediately results. From (iii) of the proof of the necessity part of Theorem 4.7 follows now also necessity of the condition 2) due to $\varphi \in \mathcal{K}\Re$.

*Sufficiency.* Let all the conditions of Theorem 4.8 hold. Hence, all the conditions of Theorem 4.7 are valid for $\mathcal{N}(\mathcal{A}) = \Re^n$ with the additional property of $v$ – its radial unboundedness with respect to $\mathcal{A}$. Hence, $\mathcal{V}_\zeta(\mathcal{A})$ is positively precompact relative to itself and positively invariant for any $\zeta \in \Re^+$. Let $\varphi \in \mathcal{K}\Re$ obey $v(X) \geq \varphi[\rho(X, A)]$,

$\forall X \in \Re^n$. Let arbitrarily big $\varepsilon \in \Re^+$, $\varepsilon \to +\infty$ be permitted, and $\zeta = \varphi(\varepsilon) \in \Re^+$, $\zeta = \zeta(\varepsilon) \to +\infty$ as $\varepsilon \to +\infty$. Let $\delta(\varepsilon) = min\,[\rho(X, \mathcal{A}) : X \in \partial \breve{\mathcal{V}}_{\zeta(\varepsilon)}] \to +\infty$ as $\varepsilon \to +\infty$. Now, $X_0 \in \mathcal{N}[\delta(\varepsilon), \mathcal{A}]$ implies $\underline{X}(t; X_0) \in \mathcal{N}(\varepsilon, \mathcal{A})$ for all $t \in \Re_+$ because $\mathcal{N}[\delta(\varepsilon), \mathcal{A}] \subseteq \mathcal{V}_{\zeta(\varepsilon)}(\mathcal{A}) \subseteq \mathcal{N}(\varepsilon, \mathcal{A})$ and $\mathcal{V}_{\zeta(\varepsilon)}(\mathcal{A})$ is positively invariant. This proves that stability of $\mathcal{A}$ is global (Definition 1.2, Section 1.1.1). Global attraction of $\mathcal{A}$ results from $\Delta = +\infty$, which is implied by positive invariance of $\mathcal{V}_\zeta(\mathcal{A})$ for any $\zeta \in \Re^+$, the conditions 1), 3), and 4) due to Lemma 4.17.

**Note 4.13** As explained in Note 4.12, the condition 3) can be set in the classical form 3′),

*3′) there is a globally positive definite function $\psi$ with respect to $\mathcal{A}$ such that*

$$\mathcal{D}^+ v(X) \leq -\psi(X) \quad \textit{for all} \quad X \in \Re^n.$$

This note and Note 4.12 show that the conditions of Theorems 4.7 and 4.8 proved via LaSalle principle are more relaxed, less stringent than the classical conditions.

### 4.6.3 Asymptotic stability of $X = 0$

**Theorem 4.9** *Let the system (4.1) have generalised motions $\underline{X}(t; X_0) \in C_0(\mathcal{I}_0 \times \mathcal{X}_0)$, $\mathcal{O} \subset \mathring{\mathcal{X}}_0 \subseteq \mathcal{X}_0 \subseteq \mathcal{X} \subseteq \Re^n$. In order for $X = 0$ to be asymptotically stable it is necessary and sufficient that there exist its connected neighbourhood $\mathcal{N}$ and a function $v : \mathcal{N} \to \Re$ satisfying the following conditions:*

*1) $v$ is positive definite on $\mathcal{N}$,*

*2) $\mathcal{D}^+ v(X) < 0$ for all $(X \neq 0) \in \mathcal{N}$,*

*and*

*3) $\mathcal{D}^+ v(X) = 0$ for $X = 0$.* ∇∇∇

**Proof.** The set $\mathcal{O} = \{0\}$ is compact and connected and $\mathcal{O} \subset \mathring{\mathcal{X}}_0$. We may set $\mathcal{A} = \mathcal{O}$ so that Theorem 4.7 becomes Theorem 4.9.

Note 4.12 is applicable to Theorem 4.9.

### 4.6.4 Complete global asymptotic stability of $X = 0$

**Theorem 4.10** *Let the system (4.1) have generalised motions $\underline{X}(t; X_0) \in C_0(\mathcal{I}_0 \times \Re^n)$. In order for $X = 0$ to be completely globally asymptotically stable it is necessary and sufficient that there exists a function $v : \Re^n \to \Re$ satisfying the following conditions:*

*1) $v$ is globally positive definite,*

*2) $v$ is radially unbounded,*

*3)* $\mathcal{D}^+v(X) < 0$ *for all* $(X \neq 0) \in \Re^n$,

*and*

*4)* $\mathcal{D}^+v(X) = 0$ *for* $X = 0$. $\nabla\nabla\nabla$

**Proof.** The set $\mathcal{A} = \mathcal{O} = \{0\}$ is compact and connected. Theorem 4.10 directly results from Theorem 4.8.

Note 4.13 is valid also for Theorem 4.10.

### 4.6.5 Absolute stability of $X = 0$ of Lurie systems

#### Algebraic approach

Let the system (4.1) be in the Lurie form (4.3) (see Section 1.1.5),

$$\frac{dX}{dt} = AX + Bf(X), \tag{4.3a}$$

$$w = CX + Df(w). \tag{4.3b}$$

With $v^f = v(\cdot; f) : \Re^n \rightarrow \Re$ we denote a function, the functional form of which is governed by the functional form of the vector nonlinearity $f$. It will be a member of a functional family $\mathcal{F}_j$, $j \in \{0, 1\}$, defined by

$$\mathcal{F}_j = \left\{ v^f : v^f(X) \in C^{(j)}(\Re^n),\ f \in \mathcal{N}_j \right\},\ \ j \in \{0, 1\},$$

where

$$\mathcal{N}_j = \begin{cases} \mathcal{N}_0(\mathcal{L}) & ,\ j = 0, \\ \mathcal{N}_1(\mathcal{L}, \mathcal{M}) & ,\ j = 1. \end{cases}$$

Let $n_i(w) = n_i(w; f)$, $w = (w_1\ w_2 \ldots w_m)^T$,

$$n_i(w; f) = \begin{cases} 0 & ,\ w_i = 0 \\ w_i^{-1} f_i(w) & ,\ w_i \neq 0,\ \ i = 1, 2, \ldots, m, \end{cases}$$

and $N(w) = N(w; f)$,

$$N(w) = diag\,\{n_1(w)\ \ n_2(w)\ \ \ldots\ \ n_m(w)\},$$

so that

$$f(w) = N(w)w,$$

and

$$N(w) \in \mathcal{L}\ \text{ for all }\ w \in \Re^m.$$

**Assumption 4.1** $det(I - DL) \neq 0$ for every $L = diag\{l_1 \ l_2 \ \dots \ l_m\} \in (\mathcal{L} \cup \mathcal{M})$.

At first we establish the criteria for the system (4.3) to have the unique equilibrium state $X = 0$ for every $f \in \mathcal{N}_j$, $j \in \{0, 1\}$, which is necessary for absolute stability of $X = 0$ on $\mathcal{N}_j$. Uniqueness of the equilibrium state $X = 0$ of the system (4.3) is necessary for its absolute stability. Therefore, it should be tested before applying an absolute stability criterion.

**Theorem 4.11** *In order for $X = 0$ to be the unique equilibrium state of the Lurie system (4.3) on $\mathcal{N}_j$, $j \in \{0, 1\}$ in case $\mathcal{L} = [0, K)$ it is both necessary and sufficient that*

*1) det $A \neq 0$,*

*and*

*2) the matrix $[(D - CA^{-1}B)^T + (D - CA^{-1}B)]$ is negative definite.* ∇∇∇

**Proof.** From the definition of $\mathcal{N}_j$ (Section 1.1.5) it follows that the system (4.3) has the unique solution through $X = 0$. The condition (a) of the Theorem 3.22 (Section 3.3.4) is satisfied.

*Necessity.* Let $X = 0$ be the unique equilibrium state of the system (4.3) on $\mathcal{N}_j$ and $f \in \mathcal{N}_j$ be arbitrary and fixed. From the condition (b) of Theorem 3.22 (Section 3.3.4), $f(0) = 0$ due to $f \in \mathcal{N}_j$ and (4.3), we derive $AX = 0$ that has the unique solution $X = 0$ iff $det \ A \neq 0$. The condition 1) is necessary. For any $f \in \mathcal{N}_j$ the condition (b) of Theorem 3.22 (Section 3.3.4) applied to the system (4.3) reduces the latter to

$$0 = AX + Bf(w), \ \ w = CX + Df(w),$$

or, by solving for $X$, $X = -A^{-1}Bf(w)$,

$$w = (D - CA^{-1}B)f(w),$$

which multiplied by $f^T(w)$ on the left becomes

$$f^T(w)w = \frac{1}{2}f^T(w)\left[(D - CA^{-1}B)^T + (D - CA^{-1}B)\right]f(w).$$

From $f \in \mathcal{N}_j$ and $\mathcal{L} = [0, K)$ follows $f^T(w)w \geq 0$ for all $w \in \Re^m$. Therefore, for the last equation to have the unique solution $f(w) = 0$ it is necessary that $[(D - CA^{-1}B)^T + (D - CA^{-1}B)]$ is negative definite.

*Sufficiency.* Let $X^* \in \Re^n$ be an equilibrium state of the system (4.3) on $\mathcal{N}_j$. It obeys (Theorem 3.22, Section 3.3.4, and (4.3))

$$AX^* + Bf(w) = 0, \qquad w = CX^* + Df(w).$$

Let the conditions of Theorem 4.11 hold. Now, for any $f \in \mathcal{N}_j$,

$$X^* = -A^{-1}Bf(w) \tag{4.5a}$$

$$w = (D - CA^{-1}B)f(w). \tag{4.5b}$$

After multiplying (4.5b) by $f^T(w)$ on the left it takes the form of (4.6),

$$f^T(w)w = \frac{1}{2}f^T(w)\left[\left(D - CA^{-1}B\right)^T + \left(D - CA^{-1}B\right)\right]f(w). \tag{4.6}$$

The condition 2) and $f^T(w)w \geq 0$, $\forall w \in \Re^m$, $\forall f \in \mathcal{N}_j$, imply $f(w) = 0$ that yields $X^* = 0$ due to (4.5a).

By following [68], [69], [88, p. 46], we introduce the following notion.

**Definition 4.16** *A functional family* $\mathcal{F}_j = \left\{v^f : v^f(X) \in C^{(j)}(\Re^n), f \in \mathcal{N}_j\right\}$ *is a Lyapunov functional family of the Lurie system (4.3) on* $\mathcal{N}_j$, $j \in \{0,1\}$, *if and only if*

*1)* $v^f(0) = 0$ *for every* $v^f \in \mathcal{F}_j$,

*2) there is* $\varphi \in \mathcal{K}\Re$ *such that every* $v^f \in \mathcal{F}_j$ *obeys*

$$v^f(X) \geq \varphi(||X||) \quad \text{for all} \quad X \in \Re^n,$$

*and*

*3) for every* $f \in \mathcal{N}_j$ *the corresponding* $v^f \in \mathcal{F}_j$ *obeys a) and b),*

$$a) \quad \mathcal{D}^+v^f(X) < 0 \quad \text{for all} \quad (X \neq 0) \in \Re^n,$$

$$b) \quad \mathcal{D}^+v^f(X) = 0 \quad \text{for} \quad X = 0.$$

This definition and Theorem 4.10 imply the following absolute stability theorem in view of Definition 1.10 (Section 1.1.5), [68] and [88, p. 45, Theorem 14].

**Theorem 4.12** *For* $X = 0$ *of the Lurie system (4.3) to be absolutely stable on* $\mathcal{N}_j$, $j \in \{0,1\}$, *it is both necessary and sufficient that there exists a Lyapunov functional family for it.* ∇∇∇

Let

$$H = H^T \in \Re^{(n+m)\times(n+m)},$$

$$\Theta = diag\{\theta_1 \ \theta_2 \ \ldots \ \theta_m\} \in \Re^{m\times m}$$

$$E(L) = (I - DL)^{-1}C, \ E : \mathcal{L} \to \Re^{m\times n},$$

$$F(L) = A + BLE(L), \ F : \mathcal{L} \to \Re^{n\times n},$$

$$R(L) = \left[I \ \ E^T(L)\,L\right]H\left[I \ \ E^T(L)\,L\right]^T + E^T(L)\,L\Theta E(L), R : \mathcal{L} \to \Re^{n\times n},$$

$$Q(L,M) = (I - DM)^{-1} CF(L),\ Q : \mathcal{L} \times \mathcal{M} \rightarrow \Re^{m \times n},$$

$$T(L,M) = \left[F^T(L) \;\; Q^T(L,M)M\right] H \left[I \;\; E^T(L)L\right]^T + E^T(L)L\Theta Q(L,M) +$$
$$+ \left[I \;\; E^T(L)L\right] H \left[F^T(L) \;\; Q^T(L,M)M\right]^T + Q^T(L,M)\Theta LE(L).$$

**Theorem 4.13** *Let Assumption 4.1 hold. For the functional family $\mathcal{F}_1$, (4.7)*

$$\mathcal{F}_1 = \left\{v^f : v^f(X) = \left(X^T f^T\right) H \left(X^T f^T\right)^T + 2\int_0^w f^T(w)\Theta dw,\ \ f \in \mathcal{N}_1(\mathcal{L},\mathcal{M})\right\}, \tag{4.7}$$

*to be a Lyapunov functional family for the system (4.3) on $\mathcal{N}_1(\mathcal{L},\mathcal{M})$ it is both necessary and sufficient that*

*1) the matrix $R(L)$ is positive definite for every $L \in \mathcal{L}$,*

*and*

*2) the matrix $T(L,M)$ is negative definite for every $(L,M) \in \mathcal{L} \times \mathcal{M}$.* ∇∇∇

**Proof.** Let $f \in \mathcal{N}_1(\mathcal{L},\mathcal{M})$ be arbitrary. Notice that $f(w) = N(w)w$, $w = E(N)X$ for $N = N(w)$, $f^{(1)}(w) = f_w(w)w^{(1)}$, $w^{(1)} = CX^{(1)} + Df^{(1)} = CF(N)X + Df_w w^{(1)}$ so that $w^{(1)} = Q(N, f_w)X$. Furthermore, for every $X \in \Re^n$ there is $L \in \mathcal{L}$ such that $v^f(X) = X^T R(L)X$ and for any $X \in \Re^n$ there is $(L,M) \in \mathcal{L} \times \mathcal{M}$ such that $\dot{v}^f(X) = X^T T(L,M)X$. These equations prove the theorem due to Definition 4.16.

If $D = 0$ and $f \in \mathcal{N}_0(\mathcal{L})$ then we set $\mathcal{M} = \{0\}$ in Theorem 4.13 and obtain the following result in which $T(L)$ is given by (4.8)

$$T(L) = F^T(L)\left(H + C^T L\Theta C\right) + \left(H + C^T L\Theta C\right)F(L), H \in \Re^{n\times n}. \tag{4.8}$$

**Theorem 4.14** *For the functional family $\mathcal{F}_0$ (4.9),*

$$\mathcal{F}_0 = \left\{v^f : v^f(X) = X^T H X + 2\int_0^w f^T(w)\Theta dw,\ f \in \mathcal{N}_0(\mathcal{L})\right\}, \tag{4.9}$$

*to be a Lyapunov functional family on $\mathcal{N}_0(\mathcal{L})$ for the system (4.3) in case $D = 0$ it is both necessary and sufficient that*

*1) the matrix $(H + C^T L\Theta C)$ is positive definite for every $L \in \mathcal{L}$,*

*and*

*2) the matrix $T(L)$ (4.8) is negative definite for every $L \in \mathcal{L}$.* ∇∇∇

**Note 4.14** Theorems 4.13 and 4.14 present linear systems stability like conditions for absolute stability of Lurie nonlinear systems. Various other purely algebraic criteria for absolute stability of Lurie systems were established in [68], [69]. One of them will be presented. It does not require matrix definiteness tests for all $L \in \mathcal{L}$. However, the whole set of its conditions is only sufficient.

**Theorem 4.15** *In order for the functional family $\mathcal{F}_0$ (4.9) to be a Lyapunov functional family on $\mathcal{N}_0([0,K])$ for the system (4.3) in the case $D = 0$, it is necessary that the conditions 1), 2a) and 2b) hold and is sufficient that all the conditions 1) through 2d) are valid:*

1) *Real parts $Re\,\lambda_i(A)$ of all eigenvalues $\lambda_i(A)$ of the matrix $A$ are negative,*

$$Re\,\lambda_i(A) < 0 \quad \text{for every} \quad i = 1, 2, \ldots, n,$$

2) *The matrices $H$ and $\Theta$ are such that*

   a) *$H$ is positive definite,*

   b) *$\left(H + C^T K\Theta C\right)$ is positive definite,*

   c) *$\left(2K^{-1} - \Theta CB - B^T C^T \Theta\right)$ is positive definite,*

   d) *$\left[A^T H + HA + (A^T C^T \Theta + C^T + HB)(2K^{-1} - \Theta CB - B^T C^T \Theta)^{-1}(A^T C^T \Theta + C^T + HB)^T\right]$ is negative definite.*

∇∇∇

**Proof.** Let $D = 0$ in (4.3) and $\mathcal{L} = [0, K]$.

*Necessity.* $f(w) \equiv 0$ is in $\mathcal{N}_0(\mathcal{L})$, which reduces (4.3a) to $\dot{X} = AX$ implying the necessity of 1), and reduces $v^f$ to $v^f(X) = X^T HX$ that yields necessity of 2a) due to 2) of Definition 4.16 by accepting $\mathcal{F}_0$ (4.9) to be Lyapunov functional family on $\mathcal{N}_0(\mathcal{L})$ for the system (4.3). For the same reasons $f(w) = Kw \in \mathcal{N}_0(\mathcal{L})$ implies the condition 2b) by noting that then $v^f(X) = X^T(H + C^T K\Theta C)X$.

*Sufficiency.* Notice that $v^f$ obeys

$$\begin{aligned} min\,[X^T HX, X^T(H + C^T K\Theta C)X] \;&\leq\; v^f(X) \leq \\ &\leq\; max\,[X^T HX, X^T(H + C^T K\Theta C)X] \end{aligned}$$

for every $X \in \Re^n$ and each $f \in \mathcal{N}_0(\mathcal{L})$. This, 2a) and 2b) prove that $v^f$ is globally positive definite and radially unbounded for every $f \in \mathcal{N}_0(\mathcal{L})$. Let

$$\Gamma = 2K^{-1} - \Theta CB - B^T C^T \Theta, \; \Gamma \in \Re^{m\times m}, \tag{4.10a}$$

$$P = A^T C^T \Theta + C^T, \; P \in \Re^{n\times m}, \tag{4.10b}$$

$$Y^T Y = \Gamma, \; Y \in \Re^{m\times m}, \tag{4.10c}$$

$$XY = HB + P, X \in \Re^{n\times m}, \tag{4.10d}$$

$$N = A^T H + HA + XX^T, \; N \in \Re^{n\times n}. \tag{4.10e}$$

The condition 2c) implies $det\,\Gamma \neq 0$ so that $det\; Y \neq 0$ that enables solving the equation (4.10d) for $X$, $X = (HB{+}P)Y^{-1}$. The last equation and (4.10e) yield $N = A^T H{+}HA{+}(HB{+}P)(Y^TY)^{-1}(HB{+}P)^T = A^T H{+}HA{+}(HB{+}P)\Gamma^{-1}(HB{+}P)^T$ due to (4.10c). This result, (4.10a) and (4.10b) imply

$$N = A^T H + HA + \left(A^T C^T \Theta + C^T + HB\right)\left(2K^{-1} - \Theta CB - B^T C^T \Theta\right)^{-1} \cdot$$
$$\cdot \left(A^T C^T \Theta + C^T + HB\right)^T .$$

The condition 2d) shows that $N$ is negative definite. Now, we determine

$$\dot{v}^f(X) = \dot{X}^T HX + X^T H\dot{X} + f^T \Theta \dot{w} + \dot{w}^T \Theta f$$

along motions of the system (4.3) with $D = 0$, and add on and subtract from the right hand side the term $\Omega = 2f^T(w - K^{-1}f)$ by observing that $\Omega \geq 0$ for all $w \in \Re^m$ due to $f \in \mathcal{N}_0(\mathcal{L})$ and $\mathcal{L} = [0, K]$. After carrying out all calculations and using the equations (4.10) we get

$$\dot{v}^f(x) = X^T NX - ||X^T X - Yf(w)||^2 - \Omega \leq X^T NX, \;\; \forall X \in \Re^n, \; \forall f \in \mathcal{N}_0(\mathcal{L}).$$

This shows that $\dot{v}^f$ obeys conditions 3) of Definition 4.16 for every $f \in \mathcal{N}_0(\mathcal{L})$ because $N$ is negative definite.

### Matrix Popov criterion

The method by Popov [168] is based on the use of the frequency matrix function $F$ of the linear part of the Lurie system that will be considered for the case $D = 0$, that is,

$$\frac{dX}{dt} = Ax + Bf(w), \tag{4.11a}$$

$$w = CX. \tag{4.11b}$$

The matrix transfer function $G$ of the linear part of the system with input $f$ and output $(-w)$,

$$G(s) = C(A - sI)^{-1}B,$$

yields the matrix frequency function $F$ of the linear part of the system, $F(j\omega) = G(s)|_{s=j\omega}$, $\omega \in \Re$, $j = \sqrt{-1}$, that is

$$F(j\omega) = C(A - j\omega I)^{-1}B.$$

Its hermitian part $He\,F(j\omega)$ is now determined by

$$He\,F(j\omega) = \frac{1}{2}\left[F(j\omega) + F^T(-j\omega)\right].$$

The following lemma is known as the Yakubovich–Kalman lemma [168]:

**Lemma 4.20** *Let*

*1) the matrix A have all eigenvalues with negative real parts, that is*

$$Re\,\lambda_i(A) < 0 \quad for\ every \quad i = 1, 2, \ldots, n,$$

*2) the pair (A, B) be controllable, that is*

$$rank\left(B\ AB\ A^2B\ \ldots\ A^{n-1}B\right) = n,$$

*3) the pair (A, C) be observable, that is*

$$rank\left(C^T\ A^TC^T\ (A^T)^2C^T\ \ldots\ (A^T)^{n-1}C^T\right) = n,$$

*4) the matrices* $\Gamma$ *and P be defined by (4.10a) and (4.10b), respectively,*

*5)* $\Theta \in \Re_+^{m\times m}$ *be such that* $\Gamma$ *is positive definite,*

*and*

*6) N be negative definite.*

*For the existence of a unique positive definite matrix* $H = H^T$ *and matrices X and Y satisfying the Lurie matrix equations (4.10c)–(4.10e) it is necessary and sufficient that the matrix*

$$K^{-1} + He\left[(I + j2\omega\Theta)F(j\omega)\right]$$

*is positive definite for every* $\omega \in \Re_+$ *and* $\omega = +\infty$. ∇∇∇

The following theorem represents the matrix Popov criterion.

**Theorem 4.16** *Let the Lurie system (4.11) have the following properties:*

*1) the real parts of all eigenvalues of* $\mathcal{A}$ *are negative, that is*

$$Re\,\lambda_i(A) < 0 \quad for\ all \quad i = 1, 2, \ldots, n,$$

*2) the pair (A, B) is controllable, that is*

$$rank\left(B\ AB\ A^2B\ \ldots\ A^{n-1}B\right) = n,$$

*3) the pair (A, C) is observable, that is*

$$rank\left(C^T\ A^TC^T\ (A^T)^2C^T\ \ldots\ (A^T)^{n-1}C^T\right) = n,$$

*For the existence of a Lyapunov functional family on* $\mathcal{N}_0([0, K])$ *for the system (4.11) in the form of the family* $\mathcal{F}_0$*, (4.9), with* $\Theta \in \Re_+^{m\times m}$ *yielding positive definite* $\Gamma$*, (4.10a), it is necessary and sufficient that*

$$K^{-1} + He\left[(I + j2\omega\Theta)F(j\omega)\right]$$

*is positive definite for every* $\omega \in \Re_+$ *and for* $\omega = +\infty$. ∇∇∇

**Proof.** Let the system (4.11) have the properties 1)–3). Let $\Gamma$ and $P$ be defined by (4.10a) and (4.10b), respectively, $\Theta \in \Re_+^{m\times m}$ be such that $\Gamma$, (4.10a), is positive definite.

*Necessity.* Let there exist Lyapunov functional family on $\mathcal{N}_0([0,K])$ in the form $\mathcal{F}_0$, (4.9), for the system (4.11). Let $f(w) = 0$, $f \in \mathcal{N}_0([0,K])$. Hence, $H$ is positive definite. The equations (4.10) and

$$\Omega = 2f^T(w)\left[w - K^{-1}f(w)\right] \geq 0 \text{ for all } w \in \Re^n \text{ and any } f \in \mathcal{N}_0([0,K])$$

yield

$$\dot{v}^f(X) = X^T N X - \|X^T X - Y f(w)\|^2 - \Omega, \ \ \forall X \in \Re^m, \ \forall f \in \mathcal{N}_0([0,K]). \quad (4.12)$$

Since $\dot{v}^f(X) < 0$, $\forall (X \neq 0)$ and $f \in \mathcal{N}_0([0,K])$, (Definition 4.16), then $N$ is negative definite. Altogether, there are negative definite $N$, positive definite $H = H^T$, $X$ and $Y$ obeying (4.10c) through (4.10e). Lemma 4.20 implies now necessity of positive definiteness of $\{K^{-1} + He\left[(I + j2\omega\Theta)F(j\omega)\right]\}$ for every $\omega \in \Re_+$ and $\omega = +\infty$.

*Sufficiency.* Let all the conditions of Theorem 4.15 hold. Then, the positive definite $H = H^T$, $X$ and $Y$ obey (4.10c)–(4.10e) for negative definite $N$ due to Lemma 4.20. Hence, $\dot{v}^f$, (4.12), is negative for all $(X \neq 0) \in \Re^n$ and every $f \in \mathcal{N}_0([0,K])$. Since $\Theta \in \Re_+^{m\times m}$ and $f \in \mathcal{N}_0([0,K])$ then $2\int_0^w f^T(w)\Theta dw \geq 0$ for all $w \in \Re^m$ and each $f \in \mathcal{N}_0([0,K])$ so that $v^f(X) \geq X^T H X$. This shows that $v^f$ is globally positive definite and radially unbounded for every $f \in \mathcal{N}_0([0,K])$. All the conditions of Definition 4.16 are satisfied.

## 4.7 Exponential stability of $X = 0$

### 4.7.1 Krasovskii criterion

Krasovskii [116, pp. 70-74], established the criterion for exponential stability of $X = 0$ of the system (4.1) satisfying Assumption 4.2. The Krasovskii criterion will be presented as Theorem 4.17 without proof.

**Assumption 4.2** There are an open hyperball $\mathcal{B}_\Delta \subseteq \Re^n$ and $\gamma \in \Re^+$ such that the function $f$, (4.1), satisfies 1) and 2),

1) $f$ is continuously differentiable on $\mathcal{B}_\Delta$: $f(X) \in C^{(1)}(\mathcal{B}_\Delta)$,

2) the Jacobian matrix function $f_X = \left(\dfrac{\partial f_i}{\partial x_j}\right) : \Re^n \to \Re^{n\times n}$ of the function $f$ has bounded norm on $\mathcal{B}_\Delta$:

$$||f(X)|| < \gamma \text{ for all } X \in \mathcal{B}_\Delta.$$

**Theorem 4.17** *(a) Let Assumption 4.2 hold. For $X = 0$ of the system (4.1) to be exponentially stable it is necessary and sufficient that there exist an open hyperball $\mathcal{B}_\zeta \subseteq \Re^n$, a function $v : \mathcal{B}_\zeta \to \Re$ and $\eta_i \in \Re^+$, $i = 1, 2, 3, 4$, which satisfy 1)–4):*

*1) $v$ is continuously differentiable on $\mathcal{B}_\zeta$: $v(X) \in C^{(1)}(\mathcal{B}_\zeta)$,*

*2) $\eta_1||X||^2 \le v(X) \le \eta_2||X||^2$ for every $X \in \mathcal{B}_\zeta$,*

*3) $\dot{v}(X) \le -\eta_3||X||^2$ for every $X \in \mathcal{B}_\zeta$,*

*4) $||grad\ v(X)|| \le \eta_4||X||$ for every $X \in \mathcal{B}_\zeta$.*

*(b) Let Assumption 4.2 hold for $\Delta = +\infty$, that is $\mathcal{B}_\Delta = \Re^n$. For $X = 0$ of the system (4.1) to be globally exponentially stable it is necessary and sufficient that there exist a function $v : \Re^n \to \Re_+$ and $\eta_i \in \Re^+$, $i = 1, 2, 3, 4$, which satisfy 1)–4) of (a) for $\zeta = +\infty$, that is on $\mathcal{B}_\zeta = \Re^n$.*

$\nabla\nabla\nabla$

### 4.7.2 Yoshizawa criterion

Yoshizawa [217, Theorem 11.6, p. 47; Theorem 19.2, p. 93] generalised the Krasovskii criterion to the systems (4.1) satisfying

**Assumption 4.3** The function $f$, (4.1), has the following properties:

1) $f(0) = 0$,

2) there is an open hyperball $\mathcal{B}_\Delta \subseteq \Re^n$ such that $f$ is Lipschitzian on $\mathcal{B}_\Delta$ : $f(X) \in Lip\,(\mathcal{B}_\Delta)$.

**Theorem 4.18** *(a) Let Assumption 4.3 hold. For $X = 0$ of the system (4.1) to be exponentially stable it is necessary and sufficient that there exist an open hyperball $\mathcal{B}_\zeta \subseteq \Re^n$, a function $v : \mathcal{B}_\zeta \to \Re$ and $\eta_i \in \Re^+$, $i = 1, 2, 3$, obeying 1)–3),*

*1) $\eta_1||X|| \le v(X) \le \eta_2||X||$ for all $X \in \mathcal{B}_\zeta$,*

*2) $v$ is Lipschitzian on $\mathcal{B}_\zeta$: $v(X) \in Lip\,(\mathcal{B}_\zeta)$,*

*3) $\mathcal{D}^+v(X) \le -\eta_3||X||$ for all $X \in \mathcal{B}_\zeta$.*

*(b) Let Assumption 4.3 hold for $\Delta = +\infty$, that is $\mathcal{B}_\Delta = \Re^n$. For $X = 0$ of the system (4.1) to be globally exponentially stable it is necessary and sufficient that there exist a function $v : \Re^n \to \Re_+$ and $\eta_i \in \Re^+$, $i = 1, 2, 3$, which fulfil 1)–3) of (a) for $\zeta = +\infty$, that is on $\mathcal{B}_\zeta = \Re^n$.* $\nabla\nabla\nabla$

## 4.8 Stability domain estimates

Through this section we consider the system (4.1).

For various approaches to and results on stability domain estimates see [25], [30], [49], [51], [53], [54], [68], [70], [72], [89], [90], [98], [106], [107], [113], [115], [119], [120], [124], [128], [151], [158], [159], [180], [181], [184]–[187], [197], [204], [210]–[216].

### 4.8.1 Definitions of stability domain estimates

If the necessary and sufficient conditions for a set to be the stability domain of another set, or of $X = 0$, cannot be effectively applied, then we are looking for as large as possible estimates of stability domains in the sense of the following definition:

**Definition 4.17** *(a) A set $\mathcal{S}$, $\mathcal{S} \subseteq \Re^n$, is an estimate of the strict domain $\mathcal{D}_{sc}(\mathcal{A})$ of stability of a set $\mathcal{A} \subset \Re^n$, which is denoted by $\mathcal{E}_{sc}(\mathcal{A})$: $\mathcal{S} = \mathcal{E}_{sc}(\mathcal{A})$, if and only if*

*1) $\mathcal{S}$ is a connected neighbourhood of the set $\mathcal{A}$,*

*2) $\mathcal{S}$ is a subset of $\mathcal{D}_{sc}(\mathcal{A})$: $\mathcal{S} \subseteq \mathcal{D}_{sc}(\mathcal{A})$.*

*(b) $\mathcal{E}_{sc}(\mathcal{A})$ will be denoted by $\mathcal{E}_{sc}$: $\mathcal{E}_{sc}(\mathcal{A}) = \mathcal{E}_{sc}$ if and only if $\mathcal{A} = \mathcal{O} = \{0\}$.*

Notice that we allow any motion $\underline{X}$ starting from $\mathcal{E}_{sc}(\mathcal{A})$ at $t = 0$ to leave $\mathcal{E}_{sc}(\mathcal{A})$, $\underline{X}(t; X_0) \in \mathcal{E}_{sc}(\mathcal{A})$ at some $t \in \Re^+$ although $X_0 \in \mathcal{E}_{sc}(\mathcal{A})$, rather than to demand positive invariance of $\mathcal{E}_{sc}(\mathcal{A})$ by definition because it is not crucial. The crucial property is that $\mathcal{E}_{sc}(\mathcal{A}) \subseteq \mathcal{D}_{sc}(\mathcal{A})$, i.e. that for every $X_0 \in \mathcal{E}_{sc}(\mathcal{A})$ there is $\varepsilon \in \Re^+$ such that $\rho[\underline{X}(t; X_0), \mathcal{E}_{sc}(\mathcal{A})] < \varepsilon$ for all $t \in \Re_+$.

### 4.8.2 Estimates of the stability domain of a set

**Theorem 4.19** *Let the system (4.1) have generalised motions $\underline{X}(t; X_0) \in C_0(\mathcal{I}_0 \times \mathcal{X}_0)$, $\mathcal{X}_0 \subseteq \mathcal{X} \subseteq \Re^n$, and $\mathcal{A} \subset \Re^n$ be a compact connected set. In order for a set $\mathcal{S}$, $\overline{\mathcal{S}} \subseteq \mathcal{X}_0$, to be an estimate of the strict stability domain $\mathcal{D}_{sc}(\mathcal{A})$ of $\mathcal{A}$, that is $\mathcal{S} = \mathcal{E}_{sc}(\mathcal{A})$, it is sufficient that there are $\alpha > 0$, $\beta \in \Re^+$ and a function $v : \mathcal{N}(\alpha, \mathcal{A}) \to \Re_+$ satisfying 1)–5),*

*1) $\mathcal{S}$ is an open connected neighbourhood of $\mathcal{A}$ and subset of $\mathcal{N}(\alpha, \mathcal{A})$: $\mathcal{A} \subset \mathcal{S} \subseteq \mathcal{N}(\alpha, \mathcal{A})$,*

*2) $v$ is positive definite with respect to $\mathcal{A}$ on $\mathcal{N}(\alpha, \mathcal{A})$,*

*3) $\mathcal{V}_\beta(\mathcal{A}) = \mathcal{S}$,*

*4) $\mathcal{V}_\zeta(\mathcal{A})$ is bounded for every $\zeta \in ]0, \beta]$,*

*and*

*5)* $\mathcal{D}^+v(X) \leq 0$ *for all* $X \in \mathcal{N}(\alpha, \mathcal{A})$. $\nabla\nabla\nabla$

**Proof.** Let all the conditions of the theorem be valid. Then the set $\mathcal{A}$ is stable (Theorem 4.5). Let $\varepsilon \in \Re^+$ be arbitrary, $\xi \in \Re^+$ obey $\mathcal{V}_\xi(\mathcal{A}) \subseteq [\mathcal{N}(\varepsilon, \mathcal{A}) \cap \mathcal{S}]$. The conditions 2)–5) of the theorem guarantee positive invariance of $\mathcal{V}_\xi(\mathcal{A})$ (Lemma 4.18) so that for any $X_0 \in \mathcal{V}_\xi(\mathcal{A})$, $\underline{X}(t; X_0) \in \mathcal{V}_\xi(\mathcal{A})$ for all $t \in \Re_+$, hence $\underline{X}(t; X_0) \in \mathcal{N}(\varepsilon, \mathcal{A})$ for all $t \in \Re_+$. This implies $\mathcal{V}_\xi(\mathcal{A}) \subseteq \mathcal{D}_{sc}(\varepsilon; \mathcal{A})$ and $\mathcal{V}_\xi(\mathcal{A}) \subset \mathcal{D}_{sc}(\mathcal{A})$. Hence, $\mathcal{S} \cap \mathcal{D}_{sc}(\mathcal{A}) \neq \emptyset$. Let now $X_0 \in \mathcal{S}$ be arbitrary. Let $\gamma \in [v(X_0), \beta]$. The definition of $\mathcal{V}_\zeta(\mathcal{A})$ and the conditions 2)–4) guarantee $\mathcal{V}_{\zeta_1}(\mathcal{A}) \subset \mathcal{V}_{\zeta_2}(\mathcal{A})$ and $\partial\mathcal{V}_{\zeta_1}(\mathcal{A}) \cap \partial\mathcal{V}_{\zeta_2}(\mathcal{A}) = \emptyset$, $0 < \zeta_1 < \zeta_2 \leq \beta$ so that $X_0 \in \mathcal{V}_\gamma(\mathcal{A})$. Hence, $\mathcal{V}_\gamma(\mathcal{A})$ is also positively invariant. Let $\varepsilon^* \in \Re^+$ be such that $\mathcal{V}_\gamma(\mathcal{A}) \subseteq \mathcal{N}(\varepsilon^*, \mathcal{A})$. Such $\varepsilon^*$ exists because $\mathcal{V}_\gamma(\mathcal{A})$ is bounded (condition 4). Altogether, $X_0 \in \mathcal{V}_\gamma(\mathcal{A})$ implies $\underline{X}(t; X_0) \in \mathcal{V}_\gamma(\mathcal{A})$ for all $t \in \Re^+$, hence $\underline{X}(t; X_0) \in [\mathcal{S} \cap \mathcal{N}(\varepsilon^*, \mathcal{A})]$ for all $t \in \Re_+$. This proves the positive invariance of $\mathcal{S}$ and $\mathcal{V}_\gamma(\mathcal{A}) \subseteq \mathcal{D}_{sc}(\varepsilon^*, \mathcal{A}) \subseteq \mathcal{D}_{sc}(\mathcal{A})$, $\forall \gamma \in ]0, \beta]$, hence $\mathcal{S} \subseteq \mathcal{D}_{sc}(\mathcal{A})$.

If we cannot verify the existence of generalised motions of the system (4.1) then we can apply what follows.

**Theorem 4.20** *Let* $\mathcal{A} \subset \Re^n$ *be compact and connected. For a set* $\mathcal{S} \subseteq \Re^n$ *to be an estimate of the strict stability domain* $\mathcal{D}_{sc}(\mathcal{A})$ *of* $\mathcal{A}$, $\mathcal{S} = \mathcal{E}_{sc}(\mathcal{A})$, *it is sufficient that there are* $\beta \in \Re^+$ *and a function* $v : \Re^n \to \Re_+$ *such that*

*1)* $\mathcal{S}$ *is open connected neighbourhood of* $\mathcal{A}$: $\mathcal{A} \subset \mathring{\mathcal{S}} = \mathcal{S} \subseteq \Re^n$,

*2)* $v$ *is globally positive definite with respect to* $\mathcal{A}$ *and radially increasing with respect to* $\mathcal{A}$,

*3)* $\mathcal{V}_\beta(\mathcal{A}) = \mathcal{S}$,

*and*

*4)* $\mathcal{D}^+v(X) \leq 0$ *for all* $X \in \mathcal{S}$. $\nabla\nabla\nabla$

**Proof.** Let all the conditions of the theorem hold and $\varepsilon \in \Re^+$ be arbitrary. Let $\xi \in \Re^+$ obey $\mathcal{V}_\xi(\mathcal{A}) \subseteq [\mathcal{N}(\varepsilon, \mathcal{A}) \cap \mathcal{S}]$. The condition 2) guarantees boundedness of $\mathcal{V}_\zeta(\mathcal{A})$ for every $\zeta \in ]0, \psi[$, $\psi = sup\,[v(X) : X \in \Re^n]$, $\mathcal{V}_{\zeta_1}(\mathcal{A}) \subset \mathcal{V}_{\zeta_2}(\mathcal{A})$ and $\partial\mathcal{V}_{\zeta_1}(\mathcal{A}) \cap \partial\mathcal{V}_{\zeta_2}(\mathcal{A}) = \emptyset$, $\forall \zeta_i \in ]0, \psi[$, $\zeta_1 < \zeta_2$. This and the condition 4) prove positive invariance of $\mathcal{V}_\zeta(\mathcal{A})$ for every $\zeta \in ]0, \beta]$. Hence, $\mathcal{S}$ is also positively invariant due to the condition 3). Let $X_0 \in \mathcal{V}_\xi(\mathcal{A})$. Positive invariance of $\mathcal{V}_\xi(\mathcal{A})$ due to $\xi \in ]0, \beta]$ implied by $\mathcal{V}_\xi(\mathcal{A}) \subseteq \mathcal{S}$ and condition 3), and $\mathcal{V}_\xi(\mathcal{A}) \subseteq \mathcal{N}(\varepsilon, \mathcal{A})$ prove $\underline{X}(t; X_0) \in \mathcal{N}(\varepsilon, \mathcal{A})$, $\forall t \in \Re_+$. Therefore $\mathcal{V}_\xi(\mathcal{A}) \subseteq \mathcal{D}_s(\varepsilon, \mathcal{A})$ and the set $\mathcal{A}$ has the stability domain. Let $X_0 \in \mathcal{S}$ be arbitrary and $\gamma \in [v(X_0), \beta]$. Such $\gamma$ exists due to 1)–3). Obviously, $X_0 \in \mathcal{V}_\gamma(\mathcal{A})$. Positive invariance of $\mathcal{V}_\gamma(\mathcal{A})$ and its boundedness guarantee the existence of $\varepsilon^* \in \Re^+$ such that $\mathcal{V}_\gamma(\mathcal{A}) \subseteq \mathcal{N}(\varepsilon^*, \mathcal{A})$ and $\underline{X}(t; X_0) \in \mathcal{N}(\varepsilon^*, \mathcal{A})$, $\forall t \in \Re_+$. This, $\mathcal{V}_\xi(\mathcal{A}) \subseteq \mathcal{D}_{sc}(\varepsilon, \mathcal{A}) \subseteq \mathcal{D}_{sc}$, $\forall \xi \in ]0, \beta]$, condition 3) and positive invariance of $\mathcal{S}$ prove $\mathcal{S} \subseteq \mathcal{D}_{sc}(\mathcal{A})$.

**Theorem 4.21** *Let a set* $\mathcal{A} \subset \Re^n$ *be closed and uniquely bounded with a generating function* $u_1$ *obeying* $u_1(X) = \xi_1$ *on* $\partial\mathcal{A}$. *Let* $\mathcal{N}_u(\mathcal{A})$ *be its uniquely bounded* *neighbourhood with common centre $w$ of unique boundedness, $w \in \mathring{\mathcal{A}}$, and with a generating function $u_2$ satisfying $u_2(X) = \xi_2$ on $\partial\mathcal{N}_u(\mathcal{A})$, $\xi_i \in \Re^+$, $i = 1, 2$. Let $v : \Re^n \to \Re_+$ be defined by*

$$v(X) = \begin{cases} 0 & , X = w \\ \dfrac{u_1(X) - \xi_1}{u_1[\xi_2 u_2^{-1}(X)X] - \xi_1} \cdot \dfrac{u_1[\xi_2 u_2^{-1}(X)X]}{u_1(X)} \cdot \dfrac{\rho(X - w, \mathcal{A})}{\rho[\xi_2 u_2^{-1}(X)(X - w), \mathcal{A}]} & \\ & , (X \neq w) \in \Re^n. \end{cases}$$

*In order for the set $\mathcal{N}_u(\mathcal{A})$ to be an estimate of the strict domain $\mathcal{D}_{sc}(\mathcal{A})$ of stability of $\mathcal{A}$: $\mathcal{N}_u(\mathcal{A}) = \mathcal{E}_{sc}(\mathcal{A})$ it is sufficient that $\mathcal{D}^+v(X) \leq 0$ for all $X \in \overline{\mathcal{N}_u(\mathcal{A})}$.*

$\nabla\nabla\nabla$

**Proof.** Under the conditions of the theorem all the conditions of Lemma 4.13 are fulfilled. The conditions of Lemma 4.13, $u_2(X) = \xi_2$ on $\partial\mathcal{N}_u(\mathcal{A})$, $\overline{\mathcal{N}_u(\mathcal{A})} = \overline{\mathcal{V}}_{\xi_2}(\mathcal{A})$ and $\mathcal{D}^+v(X) \leq 0$ on $\overline{\mathcal{N}_u(\mathcal{A})}$ show that all the conditions of Theorem 4.19 are met for $\beta = \xi_2$ and $\mathcal{S} = \mathcal{N}_u(\mathcal{A})$.

In contrast to Theorems 4.19 and 4.20, Theorem 4.21 defines the function $v$. Notice that every estimate $\mathcal{E}_{sc}(\mathcal{A})$ of $\mathcal{D}_{sc}(\mathcal{A})$ is also an estimate of $\mathcal{D}_s(\mathcal{A})$.

### 4.8.3 Estimates of the stability domain of $X = 0$

**Theorem 4.22** *Let the system (4.1) have generalised motions $\underline{X}(t; X_0) \in C(\mathcal{I}_0 \times \mathcal{X}_0)$, $(X = 0) \in \mathring{\mathcal{X}}_0 \subseteq \mathcal{X} \subseteq \Re^n$. In order for a set $\mathcal{S}$, $\mathcal{S} \subseteq \mathcal{X}_0$, to be an estimate of the strict stability domain $\mathcal{D}_{sc}$ of $X = 0$, that is $\mathcal{S} = \mathcal{E}_{sc}$ it is sufficient that there are $\alpha > 0$, $\beta \in \Re^+$ and a function $v : \mathcal{N}(\alpha, 0) \to \Re_+$ satisfying 1)–5):*

*1) $\mathcal{S}$ is an open connected neighbourhood of $X = 0$ and subset of $\mathcal{N}(\alpha, 0)$: $\mathcal{O} \subset \mathcal{S} \subseteq \mathcal{N}(\alpha, 0)$,*

*2) $v$ is positive definite on $\mathcal{N}(\alpha, 0)$,*

*3) $\mathcal{V}_\beta = \mathcal{S}$,*

*4) $\mathcal{V}_\zeta$ is bounded for every $\zeta \in ]0, \beta]$,*

*and*

*5) $\mathcal{D}^+v(X) \leq 0$ for all $X \in \mathcal{N}(\alpha, \mathcal{A})$.* $\nabla\nabla\nabla$

**Proof.** Since $\mathcal{O} = \{0\}$ is connected we may set $\mathcal{A} = \mathcal{O}$ in Theorem 4.19 that then reduces to Theorem 4.22.

In case we wish to avoid testing the existence of the system's generalised motions, then we can use the following criterion.

**Theorem 4.23** *For a set $\mathcal{S} \subseteq \Re^n$ to be an estimate of the strict stability domain $\mathcal{D}_{sc}$ of $X = 0$, $\mathcal{S} = \mathcal{E}_s$, it is sufficient that there are $\beta \in \mathcal{R}^+$ and a function $v : \Re^n \to \Re_+$ such that:*

*1) $\mathcal{S}$ is a connected open neighbourhood of $\mathcal{O}$,*

*2) $v$ is globally positive definite and radially increasing,*

*3) $\mathcal{V}_\beta = \mathcal{S}$,*

*and*

*4) $\mathcal{D}^+ v(X) \leq 0$ for all $X \in \mathcal{S}$.* ∇∇∇

**Proof.** Let $\mathcal{A} = \mathcal{O}$. Now, Theorem 4.20 becomes Theorem 4.23.

Theorems 4.22 and 4.23 leave open the problem of finding functions $v$ obeying their conditions. This can be overcome under the following conditions.

**Theorem 4.24** *Let $\mathcal{N}_0$ be $\mathcal{O}$-uniquely bounded neighbourhood of $X = 0$ with a generating function $u$ satisfying $u(X) = \xi$ on $\partial\mathcal{N}_0$, $\xi \in \Re^+$. Let $v : \Re^n \to \Re_+$ be defined by $v(X) \equiv u(X)$. In order for the set $\mathcal{N}_0$ to be an estimate of the strict domain $\mathcal{D}_{sc}$ of stability of $X = 0$ it is sufficient that $\mathcal{D}^+ v(X) \leq 0$ for all $X \in \mathcal{N}_0$.* ∇∇∇

**Proof.** Lemmae 4.11 and 4.22 imply Theorem 4.24 for $\mathcal{A} = \mathcal{O}$.

## 4.9 Asymptotic stability domain and estimates

In what follows the system (4.1) is assumed.

### 4.9.1 Classical approach

The classical approach to determination of the asymptotic stability domain is based on the Lyapunov method used via the *existence* of a $v$ function with appropriate properties.

#### Zubov theorem

The Zubov theorem [218, Theorem 19, pp. 52-53], was most probably the first result of an exact determination of the asymptotic stability domain $\mathcal{D}_s$ of $X = 0$. It has had a great theoretical significance and influence on the study of the Lyapunov function construction problem [3], [38], [46], [101], [102], [131], [196]. For other approaches to this problem see [4], [5], [7], [26], [28], [31], [32], [36], [46], [47], [59], [97], [100]–[102], [104], [112], [117], [123], [126], [131], [132], [153]–[155], [160], [167], [170], [171], [174]–[176], [182], [183], [190], [199]–[203].

The Zubov theorem takes the following form in this framework:

**Theorem 4.25 (Zubov theorem)** *Let the function $f$ of the system (4.1) be continuously differentiable on a closed hyperball $\overline{\mathcal{B}}_\alpha$ for some $\alpha \in \Re^+$. In order for an open invariant set $\mathcal{S}$, $\mathcal{S} \subseteq \mathcal{B}_\alpha$, containing a certain neighbourhood of a closed invariant set $\mathcal{M}$, $\mathcal{M} \subset \mathcal{B}_\alpha$, to be the domain of asymptotic stability of the asymptotically stable set $\mathcal{M}$, it is necessary and sufficient that there exist two functions $v$ and $\psi$ having the following properties:*

*1) The function $v$ is defined and continuous on $\mathcal{S}$, and the function $\psi$ is defined and continuous on $\Re^n$.*

*2) $-1 < v(X) < 0$ for $X \in \mathcal{S}$; $\psi(X) > 0$ for $X \in \Re^n$, $\rho(X, \mathcal{M}) \neq 0$ and $\psi(X) = 0$ for $X \in \mathcal{M}$.*

*3) For any sufficiently small $\gamma_2 > 0$ it is possible to find quantities $\gamma_1$ and $\alpha_1$ such that*

$$v(X) < -\gamma_1 \text{ for } \rho(X, \mathcal{M}) \geq \gamma_2,$$

*and*

$$\psi(X) > \alpha_2 \text{ for } \rho(X, \mathcal{M}) \geq \gamma_2.$$

*4) The functions $v$ and $\psi$ tend to 0 as $\rho(X, \mathcal{M}) \to 0$.*

*5) If there exists a point $y \notin \mathcal{M}$, $y \in \partial\mathcal{S}$ then $lim\,[v(X) : \rho(X, y) \to 0] = -1$.*

*6)* $\dfrac{dv(X)}{dt} = \psi(X)\,[1 + v(X)].$ ∇∇∇

The Zubov theorem is valid for smooth systems described by continuously differentiable function $f$ in (4.1). The theorem assumes that the set $\mathcal{M}$ is asymptotically stable. Furthermore, the theorem is expressed in terms of the *existence* of the functions $v$ and $\psi$. It does not determine any guideline for a choice of either.

The Zubov theorem is directly applicable to the state $X = 0$. It suffices to set $\mathcal{M} = \{0\}$.

For the proof of the Zubov theorem see [218, p. 53].

### Knobloch–Kappel theorem

Knobloch and Kappel derived several interesting results in [114] for a special class of systems described by (4.1). They are those which obey the following assumption:

**Assumption 4.4** The function $f$, (4.1), has the following properties:

1) $f(0) = 0$,

2) $f$ is defined with continuous partial derivatives on $\Re^n$: $f(X) \in C^{(1)}(\Re^n)$,

3) the Jacobian matrix function $f_X$ of the function $f$ has at the origin all eigenvalues $\lambda_i[f_X(0)]$ with negative parts: $Re\,\lambda_i[f_X(0)] < 0$, $\forall i = 1, 2, \ldots, n$.

**Theorem 4.26 (Knobloch–Kappel theorem)** *Let Assumption 4.4 hold. Let* $\zeta : \Re^n \to \Re$ *be any scalar function obeying the following conditions:*

1) $\zeta$ *is defined and twice continuously partially differentiable on* $\Re^n$*:* $\zeta(X) \in C^{(2)}(\Re^n)$,

2) $\zeta(0) = 0$ *and* $\zeta(0) > 0$ *for all* $X \neq 0$,

3) *the function* $\zeta$ *has a positive lower limit on every subset of the set* $\{X : ||X|| \geq \varepsilon\}$, $\varepsilon > 0$.

*Then, there is a unique function* $v$ *that is continuously differentiable on* $\mathcal{D}$*:* $v(X) \in C^{(1)}(\mathcal{D})$ *and obeys*

$$[grad\, v(X)]^T f(X) = -\zeta(X) \quad for\ all \quad X \in \mathcal{D} \quad and \quad v(0) = 0.$$

*Besides,* $v$ *satisfies the following conditions:*

*(a)* $v(X) > 0$ *for* $X \neq 0$,

*and*

*(b)* $lim\,[v(X) : X \to \partial\mathcal{D}, X \in \mathcal{D}] = \infty$. ∇∇∇

The proof is given in [114, the proof of Theorem 10.3, p. 171].

Notice that the conditions of Assumption 4.4 assume knowledge that $X = 0$ is asymptotically stable.

Notice also that the choice of a function $\zeta$ obeying 1) through 3) of Theorem 4.26 is arbitrary, that it is not stated in terms of the existence of $v$ and assumes knowledge of $\mathcal{D}$.

### Vanelli–Vidyasagar results

Vanelli and Vidyasagar established a series of very interesting results on generation of a system Lyapunov function [198]. The main results will be presented without proofs that are given in [198].

The Vanelli–Vidyasagar approach and all their results in [198] are based on the following assumption.

**Assumption 4.5** (a) The system (4.1) has a unique solution $\underline{X}(t; X_0)$ through each initial state $X_0 \in \Re^n$.

(b) The solution $\underline{X}(t; X_0)$ depends continuously on $X_0$.

(c) The state $X = 0$ is asymptotically stable equilibrium state of the system.

The condition (c) of Assumption 4.5 shows that $X = 0$ has the domain $\mathcal{D}$ of asymptotic stability and the domain $\mathcal{D}_a$ of attraction so that (Theorem 3.35, Section 3.7) $\mathcal{D}_a = \mathcal{D} = \mathcal{D}_{ac} = \mathcal{D}_c$.

**Theorem 4.27** *Let Assumption 4.5 hold. If a set* $\mathcal{S} \subseteq \Re^n$ *containing* $\mathcal{O}$ *in its interior, a continuous function* $v : \mathcal{S} \to \Re_+$ *and a positive definite function* $\psi$ *on* $\mathcal{S}$ *are such that:*

(i) $v(0) = 0$, $v(X) > 0$ *for all* $X \in (\mathcal{S} - \mathcal{O})$: $v$ *is positive definite on* $\mathcal{S}$,

(ii) $\mathcal{D}_r v(X) = lim \left\{ \frac{v(\underline{X}(t;X)) - v(X)}{t} : t \to 0^+ \right\}$ *is well defined at all* $X \in \mathcal{S}$, *and satisfies* $\mathcal{D}_r v(X) = -\psi(X)$ *for all* $X \in \mathcal{S}$,

*and*

(iii) $v(X) \to +\infty$ *as* $X \to \partial\mathcal{S}$ *and/or as* $||X|| \to +\infty$,

*then* $\mathcal{S} = \mathcal{D}$. ∇∇∇

This theorem provides sufficient conditions for a set $\mathcal{S}$ to be the asymptotic stability domain $\mathcal{D}$. Under an additional requirement on the system, the conditions are also necessary:

**Theorem 4.28** *Let Assumption 4.5 hold and* $f$ *be Lipschitz continuous on* $\mathcal{D}$. *Then, in order for an open set* $\mathcal{S}$ *containing* $\mathcal{O}$ *to be the domain of asymptotic stability of* $X = 0$, *it is necessary and sufficient that there exist a continuous function* $v : \mathcal{S} \to \Re_+$ *and a positive definite function* $\psi$ *such that (i)–(iii) of Theorem 4.27 are satisfied.* ∇∇∇

Notice that this theorem requires knowledge of the Lipschitzian property of the function $f$ on $\mathcal{D}$, which essentially requests knowledge of asymptotic stability domain itself.

Both Theorem 4.27 and Theorem 4.28 state the conditions in the form of the *existence* of a function $v$ with the described properties. The same conclusion is valid for the next theorem that is based on the following important lemma:

**Lemma 4.21** *Let Assumption 4.5 hold. Suppose* $v$ *is a continuous function on some hyperball* $\mathcal{B}_\delta$ *such that* $v(0) = 0$, *and* $\mathcal{D}_r v$ *is negative definite. Then* $v$ *is positive definite.* ∇∇∇

**Theorem 4.29** *Let Assumption 4.5 hold. Suppose* $f$ *is Lipschitz continuous on* $\mathcal{D}$, *and suppose* $v$ *is a continuous function on some closed hyperball* $\overline{\mathcal{B}}_\delta$ *such that* $v(0) = 0$ *and* $\mathcal{D}_r v$ *is negative definite on* $\overline{\mathcal{B}}_\delta$. *Then there exists a (maximal Lyapunov) function* $v_m$ *that agrees with* $v$ *on* $\overline{\mathcal{B}}_\delta$ *and obeys (i)–(v):*

(i) $v_m : \Re^n \to (\Re_+ \cup \{\infty\})$,

(ii) $v_m(0) = 0$, $v_m(X) > 0$ *for all* $X \in (\mathcal{D} - \mathcal{O})$,

(iii) $v_m(X) < +\infty$ *iff* $X \in \mathcal{D}$,

(iv) $v_m(X) \to +\infty$ *as* $X \to \partial\mathcal{D}$ *and/or* $||X|| \to +\infty$,

(v) $\mathcal{D}_r v$ *is well defined and negative definite over* $\mathcal{D}$. ∇∇∇

Vanelli and Vidyasagar used [198] the preceding results by themselves to develop a numerical computer aided procedure to generate a system Lyapunov function. It is explained in Section 4.9.4.

### 4.9.2 Definition of asymptotic stability domain estimate

The classical results do not provide a complete solution to the problems of an exact generation of a system Lyapunov function $v$ and of an exact determination of the asymptotic stability domain. Therefore we are often satisfied with the best possible estimate.

**Definition 4.18** *(a) A set $\mathcal{S}$, $\mathcal{S} \subseteq \Re^n$, is an estimate of the strict domain $\mathcal{D}_{ac}(\mathcal{A})$ of attraction of the set $\mathcal{A}$, which is denoted by $\mathcal{E}_{ac}(\mathcal{A})$, $\mathcal{S} = \mathcal{E}_{ac}(\mathcal{A})$, if and only if*

*1) $\mathcal{S}$ is a connected neighbourhood of the set $\mathcal{A}$,*

*2) $\mathcal{S}$ is a subset of $\mathcal{D}_{ac}(\mathcal{A})$, $\mathcal{S} \subseteq \mathcal{D}_{ac}(\mathcal{A})$.*

*(b) $\mathcal{E}_{ac}(\mathcal{A})$ will be denoted by $\mathcal{E}_{ac}$, $\mathcal{E}_{ac}(\mathcal{A}) = \mathcal{E}_{ac}$ if and only if $\mathcal{A} = \mathcal{O}$.*

**Definition 4.19** *(a) A set $\mathcal{S}$, $\mathcal{S} \subseteq \Re^n$, is an estimate of the strict domain $\mathcal{D}_c(\mathcal{A})$ of asymptotic stability of a set $\mathcal{A}$, which is denoted by $\mathcal{E}_c(\mathcal{A})$, $\mathcal{S} = \mathcal{E}_c(\mathcal{A})$, if and only if*

*1) $\mathcal{S}$ is a connected neighbourhood of the set $\mathcal{A}$,*

*2) $\mathcal{S}$ is a subset of $\mathcal{D}_c(\mathcal{A})$, $\mathcal{S} \subseteq \mathcal{D}_c(\mathcal{A})$.*

*(b) $\mathcal{E}_c(\mathcal{A})$ will be denoted by $\mathcal{E}_c$, $\mathcal{E}_c(\mathcal{A}) = \mathcal{E}_c$ if and only if $\mathcal{A} = \mathcal{O}$.*

**Note 4.15** $\mathcal{E}_c(\mathcal{A}) = \mathcal{E}_{sc}(\mathcal{A}) \cap \mathcal{E}_{ac}(\mathcal{A})$ (Definitions 4.17–4.19).

### 4.9.3 Estimates of the asymptotic stability domain of a set

**Theorem 4.30** *Let the system (4.1) have generalised motions $\underline{X}(t; X_0) \in C_0(\mathcal{I}_0 \times \mathcal{X}_0)$, $\mathcal{X}_0 \subseteq \mathcal{X} \subseteq \Re^n$, and $\mathcal{A} \subset \Re^n$ be compact connected set. In order for a set $\mathcal{S}$, $\mathcal{S} \subseteq \mathcal{X}_0$, to be an estimate $\mathcal{E}_c(\mathcal{A})$ of the strict asymptotic stability domain $\mathcal{D}_c(\mathcal{A})$ of $\mathcal{A}$, that is $\mathcal{S} = \mathcal{E}_c(\mathcal{A})$, it is sufficient that there are $\alpha > 0$, $\beta \in \Re^+$ and a function $v : \mathcal{N}(\alpha, \mathcal{A}) \to \Re_+$ satisfying 1)–6),*

*1) $\mathcal{S}$ is an open connected neighbourhood of $\mathcal{A}$ and subset of $\mathcal{N}(\alpha, \mathcal{A})$: $\mathcal{A} \subset \mathcal{S} = \mathring{\mathcal{S}} \subseteq \mathcal{N}(\alpha, \mathcal{A})$,*

*2) $v$ is positive definite with respect to $\mathcal{A}$ on $\mathcal{N}(\alpha, \mathcal{A})$,*

*3) $\mathcal{V}_\beta(\mathcal{A}) = \mathcal{S}$,*

*4) $\mathcal{V}_\zeta(\mathcal{A})$ is bounded for every $\zeta \in ]0, \beta]$,*

*5) $\mathcal{D}^+v(X) \leq 0$ for all $X \in \mathcal{N}(\alpha, \mathcal{A})$,*

*and*

*6) $\mathcal{A}$ is the largest invariant set in $\mathcal{E}_0 = \{X : X \in \mathcal{N}(\alpha, \mathcal{A}), \mathcal{D}^+v(X) = 0\}$.*

$\nabla\nabla\nabla$

**Proof.** Let all the conditions 1)–6) hold. Then all the conditions of Theorem 4.18 are satisfied, which proves that $\mathcal{S}$ is an estimate of $\mathcal{D}_{sc}(\mathcal{A})$. From the proof of Theorem 4.19 it follows that $\mathcal{S}$ is positively invariant. The same was proved to hold for $\mathcal{V}_\zeta(\mathcal{A})$ for any $\zeta \in ]0, \beta]$ (Lemma 4.18). Hence, $\mathcal{S}$ is positively precompact relative to itself. This, the conditions 1), 2), 5), and 6), together with the existence of generalised motions prove $\underline{X}(t; X_0) \to \mathcal{A}$ as $t \to +\infty$ for every $X_0 \in \mathcal{S}$. Since $\mathcal{S}$ is positively invariant we may now conclude that $\mathcal{S} \subseteq \mathcal{D}_a(\mathcal{A})$. Altogether, $\mathcal{S} \subseteq \mathcal{D}_{ac}(\mathcal{A}) \cap \mathcal{D}_{sc}(\mathcal{A}) = \mathcal{D}_c(\mathcal{A})$ because $\mathcal{D}_a(\mathcal{A}) = \mathcal{D}_{ac}(\mathcal{A})$ (Theorem 4.35, Section 3.7).

Verification of the condition 6) can be difficult. To avoid it we state the following result.

**Theorem 4.31** *Let the system (4.1) have generalised motions $\underline{X}(t; X_0) \in C_0(\mathcal{I}_0 \times \mathcal{X}_0)$, $\mathcal{X}_0 \subseteq \mathcal{X} \subseteq \Re^n$, and $\mathcal{A} \subset \Re^n$ be a compact connected set. In order for a set $\mathcal{S}$, $\mathcal{S} \subseteq \mathcal{X}_0$, to be an estimate of the strict asymptotic stability domain $\mathcal{D}_c(\mathcal{A})$ of $\mathcal{A}$, that is $\mathcal{S} = \mathcal{E}_c(\mathcal{A})$, it is sufficient that there are $\alpha > 0$, $\beta \in \Re^+$ and a function $v : \mathcal{N}(\alpha, \mathcal{A}) \to \Re_+$ satisfying 1)–6),*

1) *$\mathcal{S}$ is an open connected neighbourhood of $\mathcal{A}$ and subset of $\mathcal{N}(\alpha, \mathcal{A})$: $\mathcal{A} \subset \mathcal{S} = \mathring{\mathcal{S}} \subseteq \mathcal{N}(\alpha, \mathcal{A})$,*

2) *$v$ is positive definite with respect to $\mathcal{A}$ on $\mathcal{N}(\alpha, \mathcal{A})$,*

3) *$\mathcal{V}_\beta(\mathcal{A}) = \mathcal{S}$,*

4) *$\mathcal{V}_\zeta(\mathcal{A})$ is bounded for every $\zeta \in ]0, \beta]$,*

5) *$\mathcal{D}^+v(X) = 0$ for all $X \in \mathcal{A}$,*

*and*

6) *$\mathcal{D}^+v(X) < 0$ for all $(X \notin \mathcal{A}) \in \mathcal{N}(\alpha, \mathcal{A})$.* $\nabla\nabla\nabla$

**Proof.** The conditions 5) and 6) show that $\mathcal{E}_0 = \{X : X \in \mathcal{N}(\alpha, \mathcal{A}), \mathcal{D}^+v(X) = 0\} = \mathcal{A}$. This shows that $\mathcal{A}$ can be the largest invariant set in $\mathcal{E}_0$, which together with conditions 1)–5) imply $\mathcal{S} \subseteq \mathcal{D}_c(\mathcal{A})$ (Theorem 4.30).

If we prefer to avoid verification of the existence of system generalised motions then we can use any of the following results.

**Theorem 4.32** *Let $\mathcal{A} \subset \Re^n$ be compact and connected. For a set $\mathcal{S} \subseteq \Re^n$ to be an estimate of the strict asymptotic stability domain $\mathcal{D}_c(\mathcal{A})$ of $\mathcal{A}$, $\mathcal{S} = \mathcal{E}_c(\mathcal{A})$, it is sufficient that there are $\beta \in \Re^+$ and a function $v : \Re^n \to \Re_+$ such that*

1) *$\mathcal{S}$ is an open connected neighbourhood of $\mathcal{A}$: $\mathcal{A} \subset \mathcal{S} = \mathring{\mathcal{S}} \subseteq \Re^n$,*

2) *$v$ is positive definite with respect to $\mathcal{A}$ and radially increasing with respect to $\mathcal{A}$,*

*3)* $\mathcal{V}_\beta(\mathcal{A}) = \mathcal{S}$,

*4)* $\mathcal{D}^+v(X) \leq 0$ *for all* $X \in \mathcal{S}$,

*and*

*5)* $\mathcal{A}$ *is the largest invariant set in* $\mathcal{E}_0 = \{X : X \in \mathcal{S}, \mathcal{D}^+v(X) = 0\}$. $\nabla\nabla\nabla$

**Proof.** Let all the conditions be valid. Then $\mathcal{S}$ is an estimate of the strict stability domain of $\mathcal{A}$, $\mathcal{S} = \mathcal{E}_{sc}(\mathcal{A})$, due to the conditions 1)–4) (Theorem 4.20). Besides, $\mathcal{S}$ is positively invariant and positively precompact relative to itself (the proof of Theorem 4.20), which together with the conditions 1), 2), 4), and 5) proves that $\underline{X}(t; X_0) \to \mathcal{A}$ as $t \to +\infty$ for every $X_0 \in \mathcal{S}$ (Lemma 4.18). Hence, $\mathcal{S}$ is an estimate of the strict attraction domain of $\mathcal{A}$, $\mathcal{S} = \mathcal{E}_{ac}(\mathcal{A})$. Altogether, $\mathcal{S} = \mathcal{E}_{sc}(\mathcal{A}) \cap \mathcal{E}_{ac}(\mathcal{A}) = \mathcal{E}_c(\mathcal{A})$.

The test of invariance of $\mathcal{A}$ can be omitted by sharpening the requirement on $\mathcal{D}^+v$.

**Theorem 4.33** *Let* $\mathcal{A} \subset \Re^n$ *be compact and connected. For a set* $\mathcal{S} \subseteq \Re^n$ *to be an estimate of the strict asymptotic stability domain* $\mathcal{D}_c(\mathcal{A})$ *of* $\mathcal{A}$, $\mathcal{S} = \mathcal{E}_c(\mathcal{A})$, *it is sufficient that there are* $\beta \in \Re^+$ *and a function* $v : \Re^n \to \Re_+$ *such that*

*1)* $\mathcal{S}$ *is an open connected neighbourhood of* $\mathcal{A}$: $\mathcal{A} \subset \mathcal{S} = \mathring{\mathcal{S}} \subseteq \Re^n$,

*2)* $v$ *is positive definite with respect to* $\mathcal{A}$ *and radially increasing with respect to* $\mathcal{A}$,

*3)* $\mathcal{V}_\beta(\mathcal{A}) = \mathcal{S}$,

*4)* $\mathcal{D}^+v(X) = 0$ *for all* $X \in \mathcal{A}$,

*and*

*5)* $\mathcal{D}^+v(X) < 0$ *for every* $(X \notin \mathcal{A}) \in \mathcal{S}$. $\nabla\nabla\nabla$

**Proof.** Let all the conditions hold. Hence, $\mathcal{S} = \mathcal{E}_{sc}(\mathcal{A})$ is an estimate of the strict stability domain of $\mathcal{A}$ (Theorem 4.20). It is also positively invariant and precompact relative to itself (the proof of Theorem 4.20), which together with the conditions 1), 2), 4), and 5) prove (Lemma 4.18) that $\mathcal{S} = \mathcal{E}_{ac}(\mathcal{A})$. Finally, $\mathcal{E}_c = \mathcal{E}_{sc}(\mathcal{A}) \cap \mathcal{E}_{ac}(\mathcal{A}) = \mathcal{E}_c(\mathcal{A})$.

The preceding results leave open the problem of finding an appropriate function $v$.

**Theorem 4.34** *Let a set* $\mathcal{A} \subset \Re^n$ *be closed and uniquely bounded with a generating function* $u_1$ *obeying* $u_1(X) = \xi_1$ *on* $\partial\mathcal{A}$. *Let* $\mathcal{N}_u(\mathcal{A})$ *be its uniquely bounded neighbourhood with the common centre* $w$ *of unique boundedness,* $w \in \mathring{\mathcal{A}}$, *and with a generating function* $u_2$ *satisfying* $u_2(X) = \xi_2$ *on* $\partial\mathcal{N}_u(\mathcal{A})$, $\xi_i \in \Re_+$, $i = 1, 2$. *Let* $v : \Re^n \to \Re_+$ *be defined by*

$$v(X) = \begin{cases} 0 & , X = w \\ \dfrac{u_1(X) - \xi_1}{u_1[\xi_2 u_2^{-1}(X)X] - \xi_1} \cdot \dfrac{u_1[\xi_2 u_2^{-1}(X)X]}{u_1(X)} \cdot \dfrac{\rho(X - w, \mathcal{A})}{\rho[\xi_2 u_2^{-1}(X)(X - w), \mathcal{A}]} \\ & , (X \neq w) \in \Re^n. \end{cases}$$

*In order for the neighbourhood $\mathcal{N}_u(\mathcal{A})$ to be an estimate of the strict domain $\mathcal{D}_c(\mathcal{A})$ of asymptotic stability of $\mathcal{A}$, $\mathcal{N}_u(\mathcal{A}) = \mathcal{E}_c(\mathcal{A})$, it is sufficient that*

1) $\mathcal{D}^+ v(X) < 0$ *for all* $X \in [\mathcal{N}_u(\mathcal{A}) - \mathcal{A}]$,

*and*

2) $\mathcal{D}^+ v(X) = 0$ *for all* $X \in \mathcal{A}$. ∇∇∇

**Proof.** Let all the conditions hold. Lemma 4.13 is therefore valid, which shows that all the conditions of Theorem 4.34 are satisfied for $\mathcal{S} = \mathcal{N}_u(\mathcal{A})$.

**Note 4.16** When the system (4.1) has generalised motions $\underline{X}(t; X_0) \in C_0(\mathcal{I}_0 \times \mathcal{X}_0)$ and $\mathcal{D}_c(\mathcal{A})$ then $\mathcal{D}(\mathcal{A}) = \mathcal{D}_c(\mathcal{A})$. Evidently $\mathcal{E}_c(\mathcal{A})$ is an estimate of both $\mathcal{D}_c(\mathcal{A})$ and $\mathcal{D}(\mathcal{A})$.

### 4.9.4 Estimates of the asymptotic stability domain of $X = 0$

#### General results

**Theorem 4.35** *Let the system (4.1) have generalised motions $\underline{X}(t; X_0) \in C_0(\mathcal{I}_0 \times \mathcal{X}_0)$, $\mathcal{X}_0 \subseteq \mathcal{X} \subseteq \Re^n$. In order for a set $\mathcal{S}$, $\mathcal{S} \subseteq \mathcal{X}_0$, to be an estimate of the strict asymptotic stability domain $\mathcal{D}_c$ of $X = 0$, that is $\mathcal{S} = \mathcal{E}_c$, it is sufficient that there are $\alpha > 0$ and $\beta \in \Re^+$ and a function $v : \mathcal{N}(\alpha, \mathcal{O}) \to \Re_+$ satisfying 1)–6),*

1) *$\mathcal{S}$ is an open connected neighbourhood of $\mathcal{O}$ and subset of $\mathcal{N}(\alpha, \mathcal{O})$, $\mathcal{O} \subset \mathcal{S} = \mathring{\mathcal{S}} \subseteq \mathcal{N}(\alpha, \mathcal{O})$,*

2) *$v$ is positive definite on $\mathcal{N}(\alpha, \mathcal{O})$,*

3) $\mathcal{V}_\beta = \mathcal{S}$,

4) *$\mathcal{V}_\zeta$ is bounded for every $\zeta \in ]0, \beta]$,*

5) $\mathcal{D}^+ v(X) \leq 0$ *for all* $X \in \mathcal{N}(\alpha, \mathcal{O})$,

*and*

6) *$\mathcal{O}$ is the largest invariant set in $\mathcal{E}_0 = \{X : X \in \mathcal{N}(\alpha, \mathcal{O}), \mathcal{D}^+ v(X) = 0\}$.*

∇∇∇

**Proof.** Let $\mathcal{A} = \mathcal{O}$. Theorem 4.30 becomes Theorem 4.35.

If we wish to avoid verification of the condition 6) then we can use the following result.

**Theorem 4.36** *Let the system (4.1) have generalised motions $\underline{X}(t; X_0) \in C_0(\mathcal{I}_0 \times \mathcal{X}_0)$, $\mathcal{X}_0 \subseteq \mathcal{X} \subseteq \Re^n$. In order for a set $\mathcal{S}$, $\mathcal{S} \subseteq \mathcal{X}_0$, to be an estimate of the strict asymptotic stability domain $\mathcal{D}_c$ of $X = 0$, that is $\mathcal{S} = \mathcal{E}_c$, it is sufficient that there are $\alpha > 0$, $\beta \in \Re^+$ and a function $v : \mathcal{N}(\alpha, \mathcal{O}) \to \Re_+$ satisfying 1)–6),*

*1) $\mathcal{S}$ is an open connected neighbourhood of $\mathcal{O}$ and subset of $\mathcal{N}(\alpha, \mathcal{O})$, $\mathcal{O} \subset \mathcal{S} = \mathring{\mathcal{S}} \subseteq \mathcal{N}(\alpha, \mathcal{O})$,*

*2) $v$ is positive definite on $\mathcal{N}(\alpha, \mathcal{O})$,*

*3) $\mathcal{V}_\beta = \mathcal{S}$,*

*4) $\mathcal{V}_\zeta$ is bounded for every $\zeta \in ]0, \beta]$,*

*5) $\mathcal{D}^+v(X) = 0$ for $X = 0$,*

*and*

*6) $\mathcal{D}^+v(X) < 0$ for all $(X \neq 0) \in \mathcal{N}(\alpha, \mathcal{O})$.* ∇∇∇

**Proof.** For $\mathcal{A} = \mathcal{O}$ Theorem 4.31 takes the form of Theorem 4.36.

The condition for the existence and continuity in $X_0 \in \mathcal{X}_0$ of system generalised motions can be omitted if $v$ is also radially increasing.

**Theorem 4.37** *For a set $\mathcal{S} \subseteq \Re^n$ to be an estimate of the strict asymptotic stability domain $\mathcal{D}_c$ of $X = 0$, $\mathcal{S} = \mathcal{E}_c$, it is sufficient that there are $\beta \in \Re^+$ and a function $v : \Re^n \to \Re_+$ such that*

*1) $\mathcal{S}$ is an open connected neighbourhood of $\mathcal{O}$: $\mathcal{O} \subset \mathcal{S} = \mathring{\mathcal{S}}$,*

*2) $v$ is globally positive definite and radially increasing,*

*3) $\mathcal{V}_\beta = \mathcal{S}$,*

*4) $\mathcal{D}^+v(X) \leq 0$ for all $X \in \mathcal{S}$,*

*and*

*5) $\mathcal{O}$ is the largest invariant set in $\mathcal{E}_0 = \{X : X \in \Re^n, \mathcal{D}^+v(X) = 0\}$.*

∇∇∇

**Proof.** Theorem 4.37 follows from Theorem 4.32 for $\mathcal{A} = \mathcal{O}$.

The test of condition 5) can be avoided as explained by the following result.

**Theorem 4.38** *For a set $\mathcal{S} \subseteq \Re^n$ to be an estimate of the strict asymptotic stability domain $\mathcal{D}_c$ of $X = 0$, $\mathcal{S} = \mathcal{E}_c$, it is sufficient that there are $\beta \in \Re^+$ and a function $v : \Re^n \to \Re_+$ such that*

*1) $\mathcal{S}$ is an open connected neighbourhood of $\mathcal{O}$: $\mathcal{O} \subset \mathcal{S} = \mathring{\mathcal{S}} \subseteq \Re^n$,*

*2) $v$ is globally positive definite and radially increasing,*

*3) $\mathcal{V}_\beta = \mathcal{S}$,*

*4) $\mathcal{D}^+ v(X) = 0$ for $X = 0$,*

*and*

*5) $\mathcal{D}^+ v(X) < 0$ for every $(X \neq 0) \in \mathcal{S}$.* ∇∇∇

**Proof.** Theorem 4.33 reduces to Theorem 4.38 in the case $\mathcal{A} = \mathcal{O}$.

The next Theorem relies on properties of $\mathcal{O}$-uniquely bounded sets, which determine the form of $v$.

**Theorem 4.39** *Let $\mathcal{N}_0$ be an $\mathcal{O}$-uniquely bounded neighbourhood of $X = 0$ with a generating function $u$ satisfying $u(X) = \xi$ on $\partial\mathcal{N}_0$, $\xi \in \Re^+$. Let $v : \Re^n \to \Re_+$ be defined by $v(X) \equiv u(X)$. In order for the neighbourhood $\mathcal{N}_0$ to be an estimate of the strict domain $\mathcal{D}_c$ of asymptotic stability of $X = 0$, $\mathcal{N}_0 = \mathcal{E}_c$, it is sufficient that*

*1) $\mathcal{D}^+ v(X) < 0$ for all $X \in (\mathcal{N}_0 - \mathcal{O})$,*

*and*

*2) $\mathcal{D}^+ v(X) = 0$ at $X = 0$.* ∇∇∇

**Proof.** Lemma 4.11 and Theorem 4.38 imply directly Theorem 4.39.

### Constructive computer Lyapunov function generation

In a series of papers [148], [149], [151] Michel *et al.* developed an algorithm for a computer construction of a system Lyapunov function and estimation of the asymptotic stability domain $\mathcal{D}$ for systems of the form (4.1) with $f(X) \in C^{(1)}(\Re^n)$. Besides, real parts of all eigenvalues $\lambda_i[f_X(0)]$ of $f_X(0)$ should be negative. For the sake of simplicity let

$$A = f_X(0)$$

and form the multiplicative semi-group $\mathcal{A}'$ of the infinite set $\mathcal{A} = \{M : M = \alpha(I+hA), h \in [0,\beta]\}$. $\alpha < 1$ and $\beta \in ]0,1[$ are to be chosen, where the multiplicative semigroup of the set $\mathcal{A} = \{M : M = \alpha(I + hA), h \in [0,\beta]\}$ is the set of all finite products of matrices in the set $\mathcal{A}$. Then, a constructive algorithm by Brayton and Tong [22], [23] is used by accepting an initial polyhedral neighbourhood $\mathcal{W}_0$ of the origin of $\Re^{n\times n}$ and by using a recursive formula given by

$$\mathcal{W}_{k+1} = \mathcal{H}\left[\cup\left(M_{k-1}^{j}\mathcal{W}_k\right) : j = 0, 1, \ldots, \infty\right],$$

where $\mathcal{H}[\mathcal{S}]$ is the convex hull of the set $\mathcal{S}$. A possible choice of $\mathcal{W}_0$ is such that the set $\mathcal{E}(\mathcal{W}_0)$ of all extreme points of the set $\mathcal{W}_0$ obeys

$$\mathcal{E}(\mathcal{W}_0) = \{w_i : w_i = (x_{i1}\, x_{i2}\, \ldots\, x_{in})^T \in \Re^n, x_{ii} = 1, x_{ij} = 0 \quad \text{for} \quad i \neq j;\; i, j = 1, 2, \ldots, n\},$$

and $M_{k-1}^{j}\mathcal{W}_k = \{X : X = M_{k-1}^{j}y, y \in \mathcal{W}_k\}$. The numbers $\alpha$ and $\beta$ are to be chosen so that all eigenvalues $\lambda_i(M)$ of $M = \alpha(I + hA)$ obey $|\lambda_i(M)| \leq K < 1$ for all $h \in [0, \beta]$. Let

$$\mathcal{W}^* = \cup\left[\mathcal{W}_k : k = 0, 1, \ldots, \infty\right] = \mathcal{H}\left[\cup\left(M\mathcal{W}_0\right) : M \in \mathcal{A}'\right].$$

The set $\mathcal{A}$ of matrices $M$ is called *stable* iff for every neighbourhood $\mathcal{U}, \mathcal{U} \subset \Re^n$, of the origin there exists another neighbourhood $\mathcal{Z}$, $\mathcal{Z} \subset \Re^n$, of the origin such that for every $M \in \mathcal{A}$ follows $M\mathcal{Z} \subseteq \mathcal{U}$, $M\mathcal{Z} = \{X : X = Mz, z \in \mathcal{Z}\}$. The set $\mathcal{A}$ is stable iff $\mathcal{W}^*$ is bounded [23]. In view of the choice of $\alpha$ and $\beta$ there is an integer $r$ such that [23]:

$$\mathcal{H}\left[\cup\left(M_{k-1}^{j}\mathcal{W}_k\right) : j = 0, 1, \ldots, \infty\right] = \mathcal{H}\left[\cup\left(M_{k-1}^{j}\mathcal{W}_k\right) : j = 0, 1, \ldots, r\right].$$

The number $r$ is the smallest natural number $\rho$ obeying

$$M_{k-1}\mathcal{H}\left[\cup\left(M_{k-1}^{j}\mathcal{W}_k\right) : j = 0, 1, \ldots, \rho\right] \subseteq \mathcal{H}\left[\cup\left(M_{k-1}^{j}\mathcal{W}_k\right) : j = 0, 1, \ldots, \rho\right].$$

The set $\mathcal{W}^*$ determines $||\cdot||_*$ as $||X||_* = inf\,\{\gamma : \gamma \geq 0, X \in \gamma\mathcal{W}^*\}$, which is accepted for a system Lyapunov function,

$$v(X) = ||X||_*.$$

Hence,

$$\partial\mathcal{W}^* = \{X : X \in \Re^n, ||X||_* = 1\}.$$

The set $\mathcal{W}^*$ is convex. Hence, for any $X \in \Re^n$ there exists an $y \in \partial\mathcal{W}^*$ such that $X = \gamma y$ for same $\gamma \geq 0$. If $\gamma < 1$ then $X$ is in the interior of $\mathcal{W}^*$, and if $\gamma > 1$ then $X$ is out of $\mathcal{W}^*$. Let $\zeta \in \Re_+$. Then, for arbitrarily large $\zeta$,

$$\mathcal{D}^+v(X) < 0 \;\text{ for all }\; (X \neq 0) \in \zeta\mathcal{W}^*$$

along motions of the linearised system $\dot{X} = AX$. We now determine the largest value $\zeta$ denoted by $\zeta_M$ such that

$$\mathcal{D}^+v(X) < 0 \;\text{ for all }\; (X \neq 0) \in \zeta_M\mathcal{W}^*$$

holds along motions of the system (4.1).

Finally we conclude that the function $v$, $v(X) = ||X||_*$, is positive definite on $\zeta_M \mathcal{W}^*$ and its Dini derivative $\mathcal{D}^+v(X)$ taken along motions of the system (4.1) obeys $\mathcal{D}^+v(X) < 0$ for all $(X \neq 0) \in \zeta_M \mathcal{W}^*$. We may conclude that the set $\zeta_M \mathcal{W}^*$ is an estimate of the asymptotic stability domain of the system (4.1) with $f(X) \in C^{(1)}(\Re^n)$ and $Re\, \lambda_i(A) = Re\, \lambda_i[f_X(0)] < 0$ for all $i = 1, 2, \ldots, n$.

### Vanelli–Vidyasagar approach

What follows further exposes the approach by Vanelli and Vidyasagar, which is based on their results presented in Section 4.9.1 [198]. It assumes validity of Assumption 4.5 as well as the following:

**Assumption 4.6** The function $f(4.1)$ is analytic and can be expressed in the form

$$f(X) = \sum_{i=1}^{\infty} f_i(X),$$

where $f_i$ is a homogenous function of degree $i$.

**Example 4.17** (a) $f_1(X) = AX$ represents a homogenous function of degree $i = 1$.

(b) $f_2(X) = XX^T\mathbf{1}$ represents a homogenous function of the second degree, $i = 2$.

(c) $f_3(X) = X^TXX$ represents a homogenous function of degree $i = 3$.

A tentative Lyapunov function $v$ for the system (4.1) is accepted in the form of a real rational function with $n(X)$ and $d(X)$ denoting its numerator and denominator polynomial, respectively,

$$v(X) = \frac{n(X)}{d(X)}, \tag{4.13}$$

where $n$ and $d$ are assumed in the next form that enables us to use a recursive procedure,

$$n(X) = \sum_{i=2}^{\infty} r_i(X), \quad \text{so that} \quad \dot{n}(X) = \left[\sum_{i=2}^{\infty} grad\, r_i(X)\right]^T \dot{X}, \tag{4.14a}$$

$$d(X) = 1 + \sum_{i=1}^{\infty} q_i(X), \quad \text{so that} \quad \dot{d}(X) = \left[\sum_{i=2}^{\infty} grad\, q_i(X)\right]^T \dot{X}. \tag{4.14b}$$

The functions $r_i$ and $q_i$ are homogenous of degree $i$.

From (4.1), (4.13) and (4.14) we get

$$\dot{v}(X) = \left\{\left[1 + \sum_{i=1}^{\infty} q_i(X)\right]\left[\sum_{i=2}^{\infty} grad\, r_i(X)\right]^T - \left[\sum_{i=2}^{\infty} r_i(X)\right]\left[\sum_{i=1}^{\infty} grad\, q_i(X)\right]^T\right\} f(X).$$

(4.15)

Let $G = G^T \in \Re^{n\times n}$ be any positive definite matrix. We wish to determine $v$ as a solution to (4.16),

$$\dot{v}(X) = -X^T GX, \tag{4.16a}$$

$$v(0) = 0. \tag{4.16b}$$

Equating the right-hand sides of (4.15) and (4.16a) and then equating the coefficients of the same degrees of the left and right hand sides we determine a general recursive relation (4.17b) that reduces to (4.17a) for degree 2, i.e. for $k = 2$,

$$(grad\, r_2)^T f_1 = -X^T GX, \quad k = 2, \tag{4.17a}$$

$$\sum_{i=2}^{k}(grad\, r_i)^T f_{k+1-i} + \sum_{i=1}^{k-2}\sum_{j=2}^{k-1}\left[q_i(grad\, r_j)^T - r_j(grad\, q_i)^T\right] f_{k+1-i-j} =$$

$$= -X^T GX\left(2q_{k-2} + \sum_{i=1}^{k-3} q_i q_{k-2-i}\right), \; k \geq 3. \tag{4.17b}$$

**Theorem 4.40 (Vanelli–Vidyasagar theorem)** *Let Assumptions 4.5 and 4.6 hold, and $X = 0$ of the linear system (4.18),*

$$\dot{X} = f_1(X) = AX \tag{4.18}$$

*be asymptotically stable. Let $G = G^T \in \Re^{n\times n}$ be any positive definite matrix and all $r_i$ and $q_i$ satisfy the recursive equations (4.17) and $[1+q_1(X)+\ldots+q_m(X)] \neq 0$ on some neighbourhood of $X = 0$* [1]. *Then,*

$$v_m(X) = \frac{r_2(X) + r_3(X) + \ldots + r_m(X)}{1 + q_1(X) + \ldots + q_m(X)} \tag{4.19}$$

*is a Lyapunov function of the system (4.1) for every $m \geq 2$.* ∇∇∇

**Proof [198].** Let

$$z(X) = \left[1 + \sum_{i=1}^{m-2} q_1(X)\right]^2 = [d(X)]^2. \tag{4.20}$$

Now, (4.19) and (4.20) yield along solutions of (4.1),

$$\dot{v}_m(X) = \frac{1}{z}\left[\left(1 + \sum_{i=1}^{m-2} q_i\right)\sum_{j=2}^{m}(grad\, r_j)^T - \left(\sum_{j=2}^{m} r_j\right)\sum_{i=1}^{m-2}(grad\, q_i)^T\right] f. \tag{4.21}$$

[1] The condition "$[1+q_1(X)+\ldots+q_m(X)] \neq 0$ on some neighbourhood of $X = 0$" is not stated in [198].

Expanding the right-hand side of (4.21) and using Assumption 4.6, we have

$$\dot{v}_m(X) = \frac{1}{z}\left\{(grad\, r_2)^T + \ldots + (grad\, r_m)^T + \right.$$

$$\left. + \sum_{i=1}^{m-2}\sum_{j=2}^{m}\left[q_i(grad\, r_j)^T - r_j(grad\, q_i)^T\right]\right\}\sum_{k=1}^{\infty} f_k =$$

$$= \sum_{k=1}^{\infty}\frac{1}{z}\left\{(grad\, r_2)^T + \ldots + (grad\, r_m)^T + \right.$$

$$\left. + \sum_{i=1}^{m-2}\sum_{j=2}^{m}\left[q_i(grad\, r_j)^T - r_j(grad\, q_i)^T\right]\right\} f_k =$$

$$= \frac{1}{z}\left\{(grad\, r_2)^T f_1 + \sum_{k=3}^{m}(grad\, r_2)^T f_{k-1} + \left[\sum_{j=3}^{k}\left((grad\, r_j)^T + \right.\right.\right.$$

$$\left.\left.\left. + \sum_{i=1}^{j-2}(q_i(grad\, r_{j-1})^T - r_{j-1}(grad\, q_i)^T)\right) f_{k-j+1}\right]\right\} +$$

$$+ \frac{1}{z}\left\{\sum_{k=m+1}^{\infty} f_{k-1}(grad\, r_2)^T + \left[\sum_{j=3}^{m}\left((grad\, r_j)^T + \right.\right.\right.$$

$$\left.\left.\left. + \sum_{i=1}^{j-2}(q_i(grad\, r_{j-1})^T - r_{j-1}(grad\, q_i)^T)\right)\right] f_{k-j+1}\right\}. \tag{4.22}$$

Substituting the right hand sides of equations (4.17) into (4.22) we obtain

$$\dot{v}_m(X) = \frac{1}{z}\left[-X^TGX - X^TGX\left(\sum_{k=3}^{m} 2q_{k-2} + \sum_{k=3}^{m}\sum_{i=1}^{k-3} q_i q_{k-2-i}\right)\right] +$$

$$+ \frac{1}{z}(\text{terms of degree} \geq n+1). \tag{4.23}$$

The numerator of the first term is

$$-X^TGX\left[1 + 2(q_1 + q_2 + \ldots + q_{m-2}) + q_1^2 + 2q_1q_2 + q_1q_{m-3} + \right.$$

$$\left. + q_2q_{m-4} + \ldots + q_{m-3}q_1\right]. \tag{4.24}$$

The terms in the parentheses are the terms and cross terms of $z$ of degree up to and including $m-2$. We add to and subtract from it the term $z$:

$$1 + 2\sum_{k=3}^{m} q_{k-2} + \sum_{k=3}^{m}\sum_{i=1}^{k-3} q_i q_{k-2-i} = z - \left(z - 1 - 2\sum_{k=3}^{m} q_{k-2} + \sum_{k=3}^{m}\sum_{i=1}^{k-3} q_i q_{k-2-i}\right).$$

(4.25)

The terms in the brackets on the right hand side of (4.25) are of a degree not less than $m-1$. We use (4.25) to rewrite (4.23) in view of (4.24) as follows:

$$\dot{v}_m(X) = -X^T GX + \frac{1}{z}\,(\text{terms of degree} \geq m-1). \tag{4.26}$$

From (4.26), we see that $\dot{v}_m$ is negative definite over a small neighbourhood of the origin. Lemma 4.21 now implies positive definiteness of $v_m$ due to the condition that $[1 + q_1(X) + \ldots + q_m(X)] \neq 0$ on some neighbourhood of $X = 0$.

### Vanelli–Vidyasagar recursive algorithm for generating estimates of $\mathcal{D}_c$ [198]

*Step 1.* Solve the Lyapunov matrix equation (4.27),

$$A^T H + HA = -G, \tag{4.27}$$

for arbitrarily chosen positive definite $G = G^T$. Then,

$$v_2(X) \equiv r_2(X) = X^T GX, \ (m = 2).$$

*Step 2.* Find the linear system representation form for $m \geq 3$ of (4.17b) by taking unknown coefficients as elements of an unknown vector $y$, that is, set (4.17b) in the equivalent form (4.28),

$$A_m y = b_m, \ m \geq 3. \tag{4.28}$$

*Step 3.* Let $e_m(y) = ||y||^2$ and solve the problem of minimizing $e_m(y)$ under the constraint (4.28). Let $y^*$ be the solution.

*Step 4.* Use $y^*$ to find coefficients of $r_m$ and $q_{m-2}$. If $e_m(y^*)$ is sufficiently small, go to the next step; otherwise, take $m+1$ as the new value for $m$ and repeat Steps 2–4.

*Step 5.* If $e_m(y^*) \leq \varepsilon$, go to Step 6; otherwise, solve the constrained minimization problem: $min\,[e_m(y) : y]$ under (4.28). Let $\zeta^*$ be the optimal value of $v_m$. Then,

$$\mathcal{S}_1 = \{X : v_m(X) < \zeta^*\}$$

is an approximation of $\mathcal{D}$, $\mathcal{S}_1 \subseteq \mathcal{D}$.

*Step 6.* The set

$$\mathcal{S}_2 = \left\{X : \sum_{i=1}^{m-2} q_i(X) > -1\right\}$$

is an approximation of $\mathcal{D}_c$, which may not be necessarily contained in $\mathcal{D}_c$ (that happens if there is $\hat{X} \in \mathcal{S}_2$ such that $v_m(\hat{X}) = 0$).

**Example 4.18 [198]** The following system is known as the state–space representation of the Van der Pol equation,

$$\begin{aligned}\dot{x}_1 &= x_2,\\ \dot{x}_2 &= x_1 - x_2 + x_1^2 x_2.\end{aligned}$$

Then, for $m = 4$,

$$e_4 = 0.12295,$$

$$v_4(X) = \frac{r_2(X) + r_3(X) + r_4(X)}{1 + q_1(X) + q_2(X)},$$

$$r_2(X) = \frac{3}{2}x_1^2 - x_1 x_2 + x_2^2,$$

$$r_3 = 0,$$

$$r_4 = -0.3186x_1^4 + 0.7124x_1^3 x_2 - 0.1459x_1^2 x_2^2 + 0.1409x_1 x_2^3 - 0.03769x_2^4,$$

$$q_1 = 0,$$

$$q_2 = -0.2362x_1^2 + 0.31747x_1 x_2 - 0.1091'x_2^2,$$

$$\zeta^* = 5.4413,$$

$$\mathcal{S}_1 = \{X : v_4(X) < \zeta^*\} \subset \mathcal{D}_c.$$

The strict asymptotic stability domain $\mathcal{D}_c$ and its approximation $\mathcal{S}_1$ are shown in Fig. 4.1.

### Estimate of the domain of asymptotic stability with finite attraction time

**Definition 4.20** *(a) The state $X = 0$ of the system (4.1) has the domain of asymptotic stability with finite attraction time $\tau_a = \tau_a(X_0)$, which is denoted by $\mathcal{D}_\tau$, if and only if*

*1) it has the asymptotic stability domain $\mathcal{D}$,*

*2) there is finite time $\tau_a = \tau_a(X_0)$ such that $\underline{X}(t; X_0) = 0$ for all $t \in [\tau_a, \infty[$ provided only that $X_0 \in \mathcal{D}_\tau$,*

*3) $\mathcal{D}_\tau$ is a neighbourhood of $X = 0$.*

*(b) A set $\mathcal{S}$, $\mathcal{S} \subseteq \Re^n$, is an estimate $\mathcal{E}_\tau$ of $\mathcal{D}_\tau$, $\mathcal{S} = \mathcal{E}_\tau$, if and only if*

*1) $\mathcal{S}$ is a neighbourhood of $X = 0$,*

*2) $\mathcal{S}$ is a subset of $\mathcal{D}_\tau$.*

**Note 4.17** From Definitions 2.1, 2.3 and 2.5 of Section 2.2 and Definition 4.20 follows $\mathcal{D}_\tau \subseteq \mathcal{D}$.

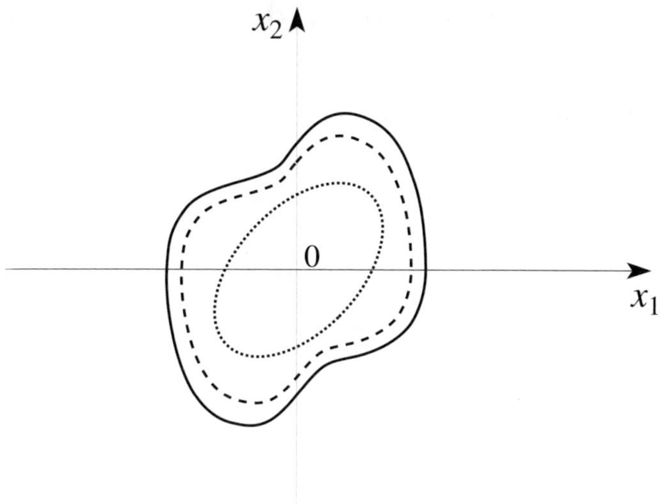


Figure 4.1: The strict asymptotic stability domain $\mathcal{D}_c$ (—) and its approximation $\mathcal{S}_1$ (- - -) for the Van der Pol system (Example 4.18) (reprinted from [198] with kind permission from Elsevier Science Ltd, UK).

**Theorem 4.41** *For a set $\mathcal{S} \subseteq \Re^n$ to be an estimate of the domain $\mathcal{D}_\tau$ of asymptotic stability of $X = 0$ with finite attraction time $\tau_a$, $\mathcal{S} = \mathcal{E}_\tau$, it is sufficient that there are $\beta \in \Re^+$, a real number $\rho \in ]0,1[$ and a function $v : \Re^n \to \Re^+$ such that*

*1) $\mathcal{S}$ is open connected neighbourhood of $\mathcal{O}$: $\mathcal{O} \subseteq \mathring{\mathcal{S}} = \mathcal{S}$,*

*2) $v$ is positive definite and radially increasing,*

*3) $\mathcal{V}_\beta = \mathcal{S}$,*

*4) $\mathcal{D}^+ v(X) \leq -[v(X)]^\rho$ for all $X \in \mathcal{S}$.*

∇∇∇

**Proof.** Let all the conditions hold. Hence, Theorem 4.38 is satisfied, which shows $\mathcal{S} = \mathcal{E}$ – an estimate of $\mathcal{D}$. Integrating $\mathcal{D}^+ v[\underline{X}(t;X_0)] \leq -\{v[\underline{X}(t;X_0)]\}^\rho$ for any $X_0 \in \mathcal{S}$ we get the maximal generalised solution $v[\underline{X}(t;X_0)]$ as

$$v[\underline{X}(t;X_0)] = [v(X_0)]^{1-\rho} - (1-\rho)t$$

for $t \in [0, \tau_a(X_0)]$, and $v[\underline{X}(t;X_0)] = 0$ for $t \in [\tau_a(X_0), +\infty[$, where

$$\tau_a(X_0) = \left[\frac{v(X_0)}{1-\rho}\right] \in [0, +\infty[$$

for every $X_0 \in \mathcal{S}$. The function $v$ is globally positive definite (hence, positive definite on $\mathcal{S}$) because it is both positive definite and radially increasing. Positive definiteness of $v$ on $\mathcal{S}$ implies now

$$\underline{X}(t;X_0) = 0 \ \text{ for all } \ t \in [\tau_a(X_0), +\infty[ \ \text{ and every } \ X_0 \in \mathcal{S},$$

which proves $\mathcal{S} = \mathcal{E}_\tau$.

# 4.10 Exponential stability domain estimate

## 4.10.1 Definition of exponential stability domain estimate

**Definition 4.21** *(a) A set* $\mathcal{S}$, $\mathcal{S} \subseteq \Re^n$, *is an estimate of the [strict] domain* $\mathcal{D}_e(\mathcal{A};\beta,\gamma)$ $[\mathcal{D}_{ec}(\mathcal{A};\beta,\gamma)]$ *of exponential stability of a set* $\mathcal{A}$, *which is denoted by* $\mathcal{E}_e(\mathcal{A};\beta,\gamma)$ $[\mathcal{E}_{ec}(\mathcal{A};\beta,\gamma)]$, $\mathcal{S} = \mathcal{E}_e(\mathcal{A};\beta,\gamma)$ $[\mathcal{S} = \mathcal{E}_{ec}(\mathcal{A};\beta,\gamma)]$, *if and only if*

*1)* $\mathcal{S}$ *is [connected] neighbourhood of the set* $\mathcal{A}$,

*2)* $\mathcal{S}$ *is subset of* $\mathcal{D}_e(\mathcal{A};\beta,\gamma)$ *[of* $\mathcal{D}_{ec}(\mathcal{A};\beta,\gamma)$*], respectively.*

*(b)* $\mathcal{E}_e(\mathcal{A};\beta,\gamma)$ *will be denoted by* $\mathcal{E}_e(\mathcal{A})$ *if and only if both* $\beta$ *and* $\gamma$ *are prespecified and fixed. The same holds for* $\mathcal{E}_{ec}(\mathcal{A};\beta,\gamma)$.

*(c)* $\mathcal{E}_e(\mathcal{A};\beta,\gamma)$ *will be denoted by* $\mathcal{E}_e(\beta,\gamma)$ *if and only if* $\mathcal{A} = \mathcal{O}$. *The same holds for* $\mathcal{E}_{ec}(\mathcal{A};\beta,\gamma)$.

*(d)* $\mathcal{E}_e(\beta,\gamma)$ *will be denoted by* $\mathcal{E}_e$ *if and only if both* $\beta$ *and* $\gamma$ *are prespecified and fixed. The same is valid for* $\mathcal{E}_{ec}(\beta,\gamma)$.

## 4.10.2 Estimates of the exponential stability domain of a set

**Theorem 4.42** *Let the system (4.1) have generalised motions* $\underline{X}(t;X_0) \in C_0(\mathcal{I}_0 \times \mathcal{X}_0)$, $\mathcal{X}_0 \subseteq \mathcal{X} \subseteq \Re^n$, *and* $\mathcal{A} \subset \Re^n$ *be compact connected set. In order for a set* $\mathcal{S}$, $\mathcal{S} \subseteq \mathcal{X}_0$, *to be an estimate of the exponential stability domain* $\mathcal{D}_e(\mathcal{A};\beta,\gamma)$ *of* $\mathcal{A}$, *that is* $\mathcal{S} = \mathcal{E}_e(\mathcal{A};\beta,\gamma)$, *it is sufficient that there are* $\zeta > 0$, $\xi \in \Re^+$, $\eta \in \Re^+$ *and a function* $v : \mathcal{N}(\zeta,\mathcal{A}) \to \Re_+$ *satisfying 1)–5),*

*1)* $\mathcal{S}$ *is an open connected neighbourhood of* $\mathcal{A}$ *and subset of* $\mathcal{N}(\zeta,\mathcal{A})$: $\mathcal{A} \subset \mathcal{S} \subseteq \mathring{\mathcal{S}} \subseteq \mathcal{N}(\zeta,\mathcal{A})$,

*2)* $\eta\rho(X,\mathcal{A}) \leq v(X) \leq \beta\eta\rho(X,\mathcal{A})$ *for all* $X \in \mathcal{N}(\zeta,\mathcal{A})$,

*3)* $\mathcal{V}_\xi(\mathcal{A}) = \mathcal{S}$,

*4)* $\mathcal{V}_\mu(\mathcal{A})$ *is bounded for every* $\mu \in ]0,\xi]$,

*5)* $\mathcal{D}^+v(X) \leq -\gamma v(X)$ *for all* $X \in \mathcal{N}(\zeta,\mathcal{A})$. ∇∇∇

**Proof.** Let all the conditions hold. All the conditions of Theorem 4.32 are satisfied. The proof of Theorem 4.32 shows that $\mathcal{S}$ is positively invariant. Let $X_0 \in \mathcal{S}$ so that $\underline{X}(t;X_0) \in \mathcal{S}$ for all $t \in \Re_+$. Integrating $\mathcal{D}^+v(X) \leq -\gamma v(X)$ for $X = \underline{X}(t;X_0)$ we get $v[\underline{X}(t;X_0)] \leq v(X_0)\,exp\,(-\gamma t)$, $\forall t \in \Re_+$. This result combined with the condition 2) yield $\rho[\underline{X}(t;X_0),\mathcal{A}] \leq \beta\rho(X_0,\mathcal{A})\,exp\,(-\gamma t)$ for all $t \in \Re_+$ and arbitrary $X_0 \in \mathcal{S}$. Hence, $\mathcal{S} = \mathcal{E}_e(\mathcal{A};\beta,\gamma)$.

If we wish to avoid a test of the existence and $C_0$-continuity of system generalised motions, we may use the following result.

**Theorem 4.43** *Let $\mathcal{A} \subset \Re^n$ be compact and connected. For a set $\mathcal{S} \subseteq \Re^n$ to be an estimate of the exponential stability domain $\mathcal{D}_e(\mathcal{A};\beta,\gamma)$ of $\mathcal{A}$, $\mathcal{S} = \mathcal{E}_e(\mathcal{A};\beta,\gamma)$, it is sufficient that there are $\xi \in \Re^+$, $\eta \in \Re^+$ and a function $v : \Re^n \to \Re_+$ such that*

1) *$\mathcal{S}$ is an open connected neighbourhood of $\mathcal{A}$: $\mathcal{A} \subset \mathcal{S} = \mathring{\mathcal{S}} \subseteq \Re^n$,*

2) *$v$ is radially increasing with respect to $\mathcal{A}$ and $\eta\rho(X,\mathcal{A}) \le v(X) \le \beta\eta\rho(X,\mathcal{A})$ for all $X \in \Re^n$,*

3) *$\mathcal{V}_\xi(\mathcal{A}) = \mathcal{S}$,*

4) *$\mathcal{D}^+v(X) \le -\gamma v(X)$ for all $X \in \mathcal{S}$.* ∇∇∇

**Proof.** Let all the conditions hold. Theorem 4.20 is satisfied. The proof of Theorem 4.20 verifies the positive invariance of $\mathcal{S}$ so that $\underline{X}(t;X_0) \in \mathcal{S}$ for all $t \in \Re_+$ and every $X_0 \in \mathcal{S}$. Integrating $\mathcal{D}^+v(X) \le -\gamma v(X)$ and then using $\eta\rho(X,\mathcal{A}) \le v(X) \le \beta\eta\rho(X,\mathcal{A})$ for any $X \in \Re^n$, with $X = \underline{X}(t;X_0)$, we prove $\rho[\underline{X}(t;X_0),\mathcal{A}] \le \beta\rho(X_0,\mathcal{A})\,exp\,(-\gamma t)$ for all $t \in \Re_+$ and every $X_0 \in \mathcal{S}$.

### 4.10.3 Estimates of the exponential stability domain of $X = 0$

**Theorem 4.44** *Let the system (4.1) have generalised motions $\underline{X}(t;X_0) \in C_0(\mathcal{I}_0 \times \mathcal{X}_0)$, $\mathcal{X}_0 \subseteq \mathcal{X} \subseteq \Re^n$. In order for a set $\mathcal{S}$, $\mathcal{S} \subseteq \mathcal{X}_0$, to be an estimate of the exponential stability domain $\mathcal{D}_e(\beta,\gamma)$ of $X = 0$, $\mathcal{S} = \mathcal{E}_e(\beta,\gamma)$, it is sufficient that there are $\zeta > 0$, $\xi \in \Re^+$, $\eta \in \Re^+$ and a function $v : \mathcal{N}(\zeta,\mathcal{O}) \to \Re_+$ satisfying 1)–5),*

1) *$\mathcal{S}$ is an open connected neighbourhood of $\mathcal{O}$ and subset of $\mathcal{N}(\zeta,\mathcal{O})$: $\mathcal{O} \subset \mathcal{S} = \mathring{\mathcal{S}} \subseteq \mathcal{N}(\zeta,\mathcal{O})$,*

2) *$\eta||X|| \le v(X) \le \beta\eta||X||$ for all $X \in \mathcal{N}(\zeta,\mathcal{A})$,*

3) *$\mathcal{V}_\xi = \mathcal{S}$,*

4) *$\mathcal{V}_\mu$ is bounded for every $\mu \in ]0,\xi]$,*

5) *$\mathcal{D}^+v(X) \le -\gamma v(X)$ for all $X \in \mathcal{N}(\zeta,\mathcal{O})$.* ∇∇∇

**Proof.** Let $\mathcal{A} = \mathcal{O}$. Theorem 4.42 takes the form of Theorem 4.44.

If we do not require $C_0$-continuity of system generalised motions and/or wish to avoid a test of their existence, then the following theorem is applicable.

**Theorem 4.45** *For a set $\mathcal{S} \subseteq \Re^n$ to be an estimate of the exponential stability domain $\mathcal{D}_e(\beta,\gamma)$ of $X = 0$, $\mathcal{S} = \mathcal{E}_e(\beta,\gamma)$, it is sufficient that there are $\xi \in \Re^+$, $\eta \in \Re^+$ and a function $v : \Re^n \to \Re_+$ such that*

1) *$\mathcal{S}$ is an open connected neighbourhood of $X = 0$: $\mathcal{O} \subset \mathcal{S} = \mathring{\mathcal{S}} \subseteq \Re^n$,*

2) *$v$ is radially increasing and $\eta||X|| \le v(X) \le \beta\eta||X||$ for all $X \in \Re^n$,*

3) *$\mathcal{V}_\xi = \mathcal{S}$,*

4) *$\mathcal{D}^+v(X) \le -\gamma v(X)$ for all $X \in \mathcal{S}$.* ∇∇∇

**Proof.** Theorem 4.43 becomes Theorem 4.45 as soon as $\mathcal{A} = \mathcal{O}$.

## 4.11 Estimates of the asymptotic stability domains on $\mathcal{N}_{(\cdot)}(\cdot\cdot)$

In this section we will consider Lurie systems described by (4.29) (see Section 1.1.5),

$$\frac{dX}{dt} = AX + Bf(w), \tag{4.29a}$$

$$w = CX + Df(w), \tag{4.29b}$$

where $f \in \mathcal{N}_i(\mathcal{L}, \mathcal{M}; \mathcal{S})$, $i \in \{0, 1\}$, and $\mathcal{S}$ will be determined by $\underline{w}_k < 0$ and $\overline{w}_k > 0$, $\forall k = 1, 2, \ldots, m$, via $\underline{W} = (\underline{w}_1\, \underline{w}_2 \ldots \underline{w}_m)^T$, $\overline{W} = (\overline{w}_1\, \overline{w}_2 \ldots \overline{w}_m)^T$ and $\mathcal{S} = \{X : \underline{W} \leq [(I - DL)^{-1}CX] \leq \overline{W}, det(I - DL) \neq 0, \forall L \in \mathcal{L}\}$. If and only if both $\underline{W} = -\infty\mathbf{1}$ and $\overline{W} = +\infty\mathbf{1}$ then $\mathcal{S} = \Re^n$. Otherwise, $\mathcal{S} \subset \Re^n$.

### 4.11.1 Definition of estimate of the asymptotic stability domain on $\mathcal{N}_{(\cdot)}(\cdot\cdot)$

**Definition 4.22** *(a) A set $\mathcal{S}^*$, $\mathcal{S}^* \subseteq \Re^n$, is an estimate of the domain $\mathcal{D}_i(\mathcal{L}, \mathcal{M}; \mathcal{A}; \mathcal{S})$ of asymptotic stability on $\mathcal{N}_i(\mathcal{L}, \mathcal{M}; \mathcal{S})$ of a set $\mathcal{A}$, $\mathcal{A} \subset \Re^n$, which is denoted by $\mathcal{E}_i(\mathcal{L}, \mathcal{M}; \mathcal{A}; \mathcal{S})$, $\mathcal{S}^* = \mathcal{E}_i(\mathcal{L}, \mathcal{M}; \mathcal{A}; \mathcal{S})$, if and only if*

*1) $\mathcal{S}^*$ is a neighbourhood of the set $\mathcal{A}$,*

*2) $\mathcal{S}^*$ is a subset of $\mathcal{D}_i(\mathcal{L}, \mathcal{M}; \mathcal{A}; \mathcal{S})$, $i \in \{0, 1, 2\}$.*

*(b) $\mathcal{E}_i(\mathcal{L}, \mathcal{M}; \mathcal{A}; \mathcal{S})$ will be denoted by $\mathcal{E}_i(\mathcal{A})$ if and only if $\mathcal{L}$,$\mathcal{M}$ and $\mathcal{S}$ are prespecified and fixed.*

*(c) $\mathcal{E}_i(\mathcal{L}, \mathcal{M}; \mathcal{A}; \mathcal{S})$ will be denoted by $\mathcal{E}_i(\mathcal{L}, \mathcal{M}; \mathcal{S})$ if and only if $\mathcal{A} = \mathcal{O}$.*

*(d) $\mathcal{E}_i(\mathcal{L}, \mathcal{M}; \mathcal{S})$ will be denoted by $\mathcal{E}_i$ if and only if $\mathcal{L}$, $\mathcal{M}$ and $\mathcal{S}$ are prespecified and fixed.*

### 4.11.2 Algebraic approach

Two classes of Lurie systems (4.29), will be differed:

a) the general class characterized by $D \neq 0$,

and

b) the special class specified by $D = 0$.

In both cases $\Theta$, $E(L)$, $F(L)$, $R(L)$ and $Q(L,M)$ will be used as defined in Section 4.6.5. However, $T(L,M)$ will be defined differently according to the case considered.

In both cases $K_1 = 0 \in \Re^{m\times m}$, $K_2 \in \Re_+^{m\times m}$, $\mathcal{L} = [0, K_2]$, $M_1 \in \Re_-^{m\times m}$, $M_2 \in \Re_+^{m\times m}$ and $\mathcal{M} = [M_1, M_2]$.

**The general case**

In this case $H = (H_{ij}) = H^T \in \Re^{(n+m)\times(n+m)}$, $i, j = 1, 2$, and

$$T(L,M) = \left[F^T(L)\, Q^T(L,M)M\right] H \left[I\; E^T(L)L\right]^T + E^T(L)L\Theta Q(L,M) + \\ + \left[I\; E^T(L)L\right] H \left[F^T(L)\, Q^T(L,M)M\right]^T + Q^T(L,M)\Theta L E(L).$$

Notice that $C = (c_1 c_2 \ldots c_m)^T$ with $c_k \in \Re^n$ for every $k = 1, 2, \ldots, m$, $H_{11} = H_{11}^T \in \Re^{n\times n}$, $H_{12} = H_{21}^T \in \Re^{n\times m}$, $H_{22} = H_{22}^T \in \Re^{m\times m}$; $\Theta = 0$, $H_{12} = 0$ and/or $H_{22} = 0$ are permitted.

**Theorem 4.46** *Let Assumption 4.1 hold,*

*1) A be nonsingular with negative real parts of all its eigenvalues,*

*2)* $\left[\left(D - CA^{-1}B\right)^T + \left(D - CA^{-1}B\right)\right]$ *be negative definite,*

*3) $R(L)$ be positive definite for every $L \in \mathcal{L}$,*

*4) $T(L,M)$ be negative definite for every $(L,M) \in \mathcal{L}\times\mathcal{M}$,*

*5)* $\zeta = min\left\{\left[min\left(\underline{w}_i^2, \overline{w}_i^2\right)\right]\left[c_i^T[R(L)]^{-1}c_i\right]^{-1} : L \in \mathcal{L}, i \in \{1, 2, \ldots, m\}\right\}$,

*6) set $\mathcal{E}_1$ (4.30),*

$$\mathcal{E}_1 = \cap\left[\{X : X^T R(L) X < \zeta\} : L \in \mathcal{L}\right], \tag{4.30}$$

*be a neighbourhood of $X = 0$.*

*Then the set $\mathcal{E}_1$ is estimate $\mathcal{E}_1(\mathcal{L},\mathcal{M};\mathcal{S})$, $\mathcal{E}_1 = \mathcal{E}_1(\mathcal{L},\mathcal{M};\mathcal{S})$, of the domain $\mathcal{D}(\mathcal{L},\mathcal{M};\mathcal{S})$ of asymptotic stability of $X = 0$ of the system (4.29) on $\mathcal{N}_1(\mathcal{L},\mathcal{M};\mathcal{S})$.*
$\nabla\nabla\nabla$

**Proof.** Let $f \in \mathcal{N}_1(\mathcal{L},\mathcal{M};\mathcal{S})$ and all the conditions hold. Then $X = 0$ is the unique equilibrium state of (4.29) in $\mathcal{S}$ on $\mathcal{N}_1(\mathcal{L},\mathcal{M};\mathcal{S})$ (Theorem 4.11). Let $\mathcal{F}_1$ be defined by (4.7) and $v^f \in \mathcal{F}_1$. For every $X \in \Re^n$ there is $(L,M) \in \mathcal{L}\times\mathcal{M}$ such that $v^f(X) = X^T R(L) X$ and $\dot{v}^f(X) = X^T T(L,M) X$ (proof of Theorem 4.13). The conditions 3) and 4) now prove positive definiteness of $v^f$ and negative definiteness of $\dot{v}^f$ on $\mathcal{S}$. From condition 5) and (4.30) it follows that

$$\mathcal{E}_1 \subseteq \mathcal{S}. \tag{4.31}$$

Besides,

$$\mathcal{V}_\zeta^f = \left\{X : v^f(X) = (X^T f^T)H(X^T f^T)^T + 2\int_0^w f^T(w)\Theta dw < \zeta\right\} \subseteq \mathcal{S} \quad (4.32)$$

due to the definition of $\zeta$. Let $X_0 \in \mathcal{E}_1$ be arbitrary. Then $X_0 \in \mathcal{V}_\zeta^f$ because $\mathcal{E}_1 \subseteq \mathcal{V}_\zeta^f$ in view of (4.30) and the definitions of $R(L)$ and $v^f$. Positive definiteness of $v^f$ and negative definiteness of $\dot{v}^f$ on $\mathcal{S}$ and $\mathcal{V}_\zeta^f \subseteq \mathcal{S}$ (4.32) imply positive invariance of $\mathcal{V}_\zeta^f$, which together with $X_0 \in \mathcal{V}_\zeta^f$ imply both the existence of $\varepsilon \in \Re_+$ such that $\|\underline{X}(t; X_0; f)\| < \varepsilon$ for all $t \in \Re_+$ and $lim[\|\underline{X}(t; X_0; f)\| : t \to +\infty] = 0$. Hence, $\mathcal{V}_\zeta^f \subseteq \mathcal{D}(\mathcal{L}, \mathcal{M}; \mathcal{S})$ that proves $\mathcal{E}_1 \subseteq \mathcal{D}(\mathcal{L}, \mathcal{M}; \mathcal{S})$ due to $\mathcal{E}_1 \subseteq \mathcal{V}_\zeta^f$. This and condition 6) show that $\mathcal{E}_1$ is an estimate of $\mathcal{D}(\mathcal{L}, \mathcal{M}; \mathcal{S})$.

Let $\mathcal{L}^v$ denote the set of all vertices of $\mathcal{L} = [0, K]$, $K = K_2 = diag\{k_{21}\, k_{22} \ldots k_{2m}\} = diag\{k_1\, k_2 \ldots k_m\}$ is given, $\mathcal{L}^v = \{L : L = diag\{\alpha_1\, \alpha_2 \ldots \alpha_m\}, \alpha_i \in \{0, k_i\}, i = 1, 2, \ldots, m\}$.

By following [69] we deduce from Theorem 4.46 the following criterion based on $\mathcal{F}_0 = \{v^f : v^f(X) = v(X) = X^T H_{11} X, H_{11} = H_{11} \in \Re^{n \times n}\}$. This means $\Theta = 0$, $H_{12} = H_{21}^T = 0$, $H_{22} = 0$.

**Theorem 4.47** *Let*

*1) $det\ A \neq 0$,*

*2) $\left[(D - CA^{-1}B)^T + (D - CA^{-1}B)\right]$ be negative definite,*

*3) $H_{11} = H_{11}^T$ be positive definite,*

*4) $T(L) = [F^T(L)H_{11} + H_{11}F(L)]$ be negative definite for every $L \in \mathcal{L}^v$ for $F(L) = A + BL$,*

*5) $\zeta = min\left\{\left[min\left(\underline{w}_i^2, \overline{w}_i^2\right)\right]\left(c_i^T H_{11}^{-1} c_i\right)^{-1} : i \in \{1, 2, \ldots, m\}\right\}$,*

*Then the set $\mathcal{E}_0 = \{X : X^T H_{11} X < \zeta\}$ is estimate $\mathcal{E}_0(\mathcal{L}; \mathcal{S})$, $\mathcal{E}_0 = \mathcal{E}_0(\mathcal{L}; \mathcal{S})$, of the domain $\mathcal{D}(\mathcal{L}; \mathcal{S})$ of asymptotic stability of $X = 0$ of the system (4.29) on $\mathcal{N}_0(\mathcal{L}; \mathcal{S})$.*
$\nabla\nabla\nabla$

**Proof.** Let all the conditions hold. Since $\mathcal{L}$ is compact and $T(L) = [F^T(L)H_{11} + H_{11}F(L)]$ is linear in $L$ due to linearity of $F(L) = (A + BL)$ in $L$ then the condition 4) guarantees negative definiteness of $T(L)$ for every $L \in \mathcal{L}$. This fact shows that Theorem 4.46 reduces to Theorem 4.47 for $\Theta = 0$, $H_{12} = H_{21}^T = 0$, $H_{22} = 0$, which are accepted in the statement of Theorem 4.47 in view of $\mathcal{E}_0 = \{X : X^T H_{11} X < \zeta\}$.

**The special class of Lurie systems**

Let $D = 0$ in (4.29b) and $f \in \mathcal{N}_0(\mathcal{L};\mathcal{S})$. Since $f$ may not be differentiable then we use a Lyapunov functional family in the form (4.9) that yields $T(\mathcal{L})$ in the form (4.8). Notice that $H = H^T \in \Re^{n\times n}$ now represents $H_{11}$ of $H$ used in the preceding section. Since $D = 0$ then $R(L) = H + L\Theta$.

**Theorem 4.48** *Let $D = 0$,*

*1) det $A \neq 0$,*

*2) $\left[(CA^{-1}B)^T + (CA^{-1}B)\right]$ be positive definite,*

*3) $R(L)$ be positive definite for every $L \in \mathcal{L}$,*

*4) $T(L)$ (4.8) be negative definite for every $L \in \mathcal{L}$,*

*5) $\zeta = min\left\{\left[min\left(\underline{w}_i^2, \overline{w}_i^2\right)\right]\left[c_i^T[R(L)]^{-1}c_i\right]^{-1} : L \in \mathcal{L}, i \in \{1,2,\ldots,m\}\right\}$,*

*6) set $\mathcal{E}_0$ (4.33),*

$$\mathcal{E}_0 = \cap\left[\{X : X^T R(L)X < \zeta\} : L \in \mathcal{L}\right], \tag{4.33}$$

*be a neighbourhood of $X = 0$.*

*Then the set $\mathcal{E}_0$ is an estimate $\mathcal{E}_0(\mathcal{L};\mathcal{S})$, $\mathcal{E}_0 = \mathcal{E}_0(\mathcal{L};\mathcal{S})$, of the domain $\mathcal{D}(\mathcal{L};\mathcal{S})$ of asymptotic stability of $X = 0$ of the system (4.29) on $\mathcal{N}_0(\mathcal{L};\mathcal{S})$.* ∇∇∇

**Proof.** Let $f \in \mathcal{N}_0(\mathcal{L};\mathcal{S})$ and all the conditions hold. Then $X = 0$ is the unique equilibrium state of (4.29) in $\mathcal{S}$ on $\mathcal{N}_0(\mathcal{L};\mathcal{S})$ (Theorem 4.11). Let $\mathcal{F}_0$ be defined by (4.9) and $v^f \in \mathcal{F}_0$. For every $X \in \Re^n$ there is $L \in \mathcal{L}$ such that $v^f(X) = X^T R(L)X$ and $\dot{v}^f(X) = X^T T(L)X$ (proof of Theorem 4.14). The conditions 3) and 4) together with $f \in \mathcal{N}_0(\mathcal{L};\mathcal{S})$ prove the positive definiteness of $v^f$ and negative definiteness of $\dot{v}^f$ on $\mathcal{S}$. From the conditions 5) and 6) follows $\mathcal{E}_0 \subseteq \mathcal{S}$. Besides,

$$\mathcal{V}_\zeta^f = \left\{X : v^f(X) = X^T HX + 2\int_0^w f^T(w)\Theta dw < \zeta\right\} \supseteq \mathcal{E}_0,$$

due to the condition 5) and definitions of $v^f$, $R(L)$ and $\mathcal{E}_0$. Hence, $v^f[\underline{X}(t;X_0;f)] \leq v^f(X_0)$ for all $t \in \Re_+$ and $v^f[\underline{X}(t;X_0)] \to 0$ as $t \to +\infty$ in view of $v^f > 0$ and $\dot{v}^f < 0$ on $\mathcal{N}_0(\mathcal{L};\mathcal{S})$, which prove the positive invariance of $\mathcal{V}_\zeta^f$ and $\mathcal{V}_\zeta^f \subseteq \mathcal{D}(\mathcal{L};\mathcal{S})$. This, $\mathcal{E}_0 \subseteq \mathcal{V}_\zeta^f$ and the condition 6) prove that $\mathcal{E}_0$ is an estimate of $\mathcal{D}(\mathcal{L};\mathcal{S})$.

We can avoid definiteness tests for every $L \in \mathcal{L}$ if we use Theorem 4.49.

**Theorem 4.49** *Let $D = 0$, $\Theta$ be either non-positive or non-negative diagonal,*

*1) real parts of all the eigenvalues of $A$ be negative,*

*2) $H$, $(H + C^T K\Theta C)$ and $(2K^{-1} - \Theta CB - B^T C^T\Theta)$ be positive definite,*

3) $\left[A^T H + HA + (A^T C^T \Theta + C^T + HB)\left(2K^{-1} - \Theta CB - B^T C^T \Theta\right)^{-1}\left(A^T C^T \Theta + C^T + HB\right)^T\right]$ be negative definite,

4) $\zeta = \left\{min\left[min\left(\underline{w}_i^2, \overline{w}_i^2\right)\right]\left(min\left[\left(c_i^T H^{-1} c_i\right)^{-1}, c_i^T\left(H + C^T K\Theta C\right)^{-1} c_i\right]\right) : i = 1, 2, \ldots, m\right\}$.

*Then the set* $\mathcal{E}_0 = \{X : X^T HX < \zeta\}$ *iff* $\Theta \leq 0$ *and* $\mathcal{E}_0 = \{X : X^T(H + C^T K\Theta C)X < \zeta\}$ *iff* $\Theta \geq 0$ *is estimate* $\mathcal{E}_0(\mathcal{L}; \mathcal{S})$*, of the domain* $\mathcal{D}(\mathcal{L}; \mathcal{S})$ *of asymptotic stability of* $X = 0$ *of the system (4.29) on* $\mathcal{N}_0(\mathcal{L}; \mathcal{S})$. ∇∇∇

**Proof.** Let all the conditions be valid. Hence, $\mathcal{F}_0$ (4.9) is a Lyapunov functional family on $\mathcal{N}_0(\mathcal{L}; \mathcal{S})$ for the system (4.29) with $D = 0$ (proof of the sufficiency part of Theorem 4.15). Let $v^f \in \mathcal{F}_0$. The condition 4) guarantees

$$\mathcal{V}_\zeta^f = \left\{X : v^f(X) = X^T HX + 2\int_0^w f^T(w)\Theta dw < \zeta\right\} \subseteq \mathcal{S}.$$

Besides, condition 4) and the definition of $\mathcal{E}_0$ yield

$$\mathcal{E}_0 \subseteq \mathcal{V}_\zeta^f.$$

Since, $v^f > 0$ and $\dot{v}^f < 0$ on $\mathcal{S}$ and $\mathcal{V}_\zeta^f \subseteq \mathcal{S}$ then $\mathcal{V}_\zeta^f$ is positively invariantt, $v^f[\underline{X}(t; X_0; f)] \leq v^f(X_0)$ for all $t \in \Re_+$, $v^f[\underline{X}(t; X_0; f)] \to 0$ as $t \to +\infty$, which imply $\mathcal{V}_\zeta^f \subseteq \mathcal{D}(\mathcal{L}; \mathcal{S})$. This, $\mathcal{E}_0 \subseteq \mathcal{V}_\zeta^f$ and the fact that $\mathcal{E}_0$ is a neighbourhood of $X = 0$ due to positive definiteness of both $H$ and $(H + C^T K\Theta C)$ prove that $\mathcal{E}_0$ is an estimate of $\mathcal{D}(\mathcal{L}; \mathcal{S})$.

### 4.11.3 Frequency domain approach

What follows generalises results by Popov [168], Šiljak [185], Šiljak and Weissenberger [186], [187], Walker and McClamroch [204] and Weissenberger [210]–[215].

We allow vector nonlinearity $f$ in what follows, which is based on the matrix Popov criterion (Theorem 4.16).

**Theorem 4.50** *Let* $D = 0$,

1) *the real parts of all eigenvalues of the matrix* $A$ *be negative,*

2) *the pair* $(A, B)$ *be controllable,*

3) *the pair* $(A, C)$ *be observable,*

4) $\Theta \in \Re_+^{m \times m}$ *be such that*

$$K^{-1} + He\left[(I + j2\omega\Theta)F(j\omega)\right]$$

*is positive definite for all* $\omega \geq 0$,

5) $\zeta = min\left\{\left[min\left(\underline{w}_i^2, \overline{w}_i^2\right)\right]\left(c_i^T H^{-1} c_i\right)^{-1} : i = 1, 2, \ldots, m\right\}.$

*Then the set* $\mathcal{E}_0 = \{X : X^T(H + C^T K\Theta C)X < \zeta\}$ *is the estimate* $\mathcal{E}_0(\mathcal{L}; \mathcal{S})$, $\mathcal{E}_0 = \mathcal{E}_0(\mathcal{L}; \mathcal{S})$, *of the domain* $\mathcal{D}(\mathcal{L}; \mathcal{S})$ *of asymptotic stability of* $X = 0$ *of the system (4.29) on* $\mathcal{N}_0(\mathcal{L}; \mathcal{S})$. ∇∇∇

**Proof.** Let all the conditions hold. Then $\mathcal{F}_0$ (4.9) is Lyapunov functional family for the system (4.29) on $\mathcal{N}_0(\mathcal{L}; \mathcal{S})$ (proof of Theorem 4.16). From this point on we repeat the rest of the proof of Theorem 4.49.

This theorem broadens essentially the class of systems to which the Popov frequency approach can be effectively used. For example, nonlinearities can be sinusoidal and/or sigmoidal. The latter show a direct application of Theorem 4.50 to neural networks.

# Chapter 5

# Novel development of the Lyapunov method

## 5.1 Introductory comment

The fundamental results of the Lyapunov (second, direct) method are expressed in terms of the *existence* of a function $v$ with appropriate properties. They do not clarify how to construct the function $v$ for a given system (5.1),

$$\frac{dX}{dt} = f(X). \tag{5.1}$$

They do not provide necessary and sufficient conditions for

1) an exact direct generation of the function $v$,

and for

2) an exact direct determination of a stability domain

when the system (5.1) is prespecified. In order to avoid solving the problems raised from 1) and 2) we usually *a priori* accept a form of the $v$-function and then look for conditions on the system under which the so chosen $v$ will satisfy the corresponding fundamental theorem of the Lyapunov method. Unfortunately, the available forms of $v$ used effectively so far are few. Their use can be inadequate for a given system, which results in too conservative, restrictive conditions on the system.

It is intuitively clear that the form of the function $v$ depends on the form of the function $f$ defining the system (5.1). The problems of exact direct (from $f$) construction of the function $v$ and of exact direct determination of a stability domain have been until recently ([71], [72], [74], [75]–[86].) unsolved fundamental problems of the Lyapunov stability theory. This chapter aims to present the original results on these problems, which open a new direction for the use and development of the Lyapunov method.

This chapter is based on [71], [72], [74]–[86].

## 5.2 Systems with differentiable motions

### 5.2.1 Smoothness property

**Definition 5.1 Weak smoothness property:**

*(i)* *There is an open neighbourhood* $\mathcal{S}$ *of* $X = 0$, $\mathcal{S} \in \Re^n$, *such that for every* $X_0 \in \mathcal{S}$

*(a)* *the system (5.1) has the unique solution* $\underline{X}(t; X_0)$ *through* $X_0$ *at* $t = 0$ *over time interval* $I_0$,

*and*

*(b)* *the motion* $\underline{X}(t; X_0)$ *is defined, continuous and differentiable in* $(t, X_0) \in I_0 \times \mathcal{S}$.

*(ii)* *for every* $X_0 \in (\Re^n - \mathcal{S})$ *every motion* $\underline{X}(t; X_0)$ *of the system (5.1) is continuous in* $t \in I_0$.

*Weak smoothness property is global iff* $\mathcal{S} = \Re^n$.

**Strong smoothness property:**

*(i)* *The system (5.1) has weak smoothness property.*

*(ii)* *If the boundary* $\partial\mathcal{S}$ *of* $\mathcal{S}$ *is non-empty then every motion of the system (5.1) passing through* $X_0 \in \partial\mathcal{S}$ *at* $t = 0$ *obeys* $inf\,[||\underline{X}(t; X_0)|| : t \in I_0] > 0$ *for every* $X_0 \in \partial\mathcal{S}$.

**Specific smoothness property:**

*There is an open neighbourhood* $\mathcal{S}$ *of* $X = 0$ *such that both*

*(i)* *the function* $f$ *is differentiable on* $\mathcal{S}$: $f(X) \in C^{(1)}(\mathcal{S})$,

*and*

*(ii)* *there is* $\gamma \in \Re^+$ *such that* $sup\,[||f_X(X)|| : X \in \mathcal{S}] < \gamma$, *where* $f_X$ *is the Jacobian of* $f$.

### 5.2.2 Two-stage approach

**Distinction between problems**

The two-stage approach means the following:

1) At first a system Lyapunov function $v$ is generated exactly, at one shot, directly via ordinary differential equations (5.1) and (5.2),

$$\frac{dv(X)}{dt} = -q(X), \tag{5.2a}$$

$$v(0) = 0, \tag{5.2b}$$

or equivalently via the partial differential equation (5.3a)

$$[grad\, v(X)]^T f(X) = -q(X), \tag{5.3a}$$

so that it obeys the boundary condition (5.3b) [i.e. (5.2b)],

$$v(0) = 0. \tag{5.3b}$$

Such a function $v$ need not explicitly provide information about the asymptotic stability domain $\mathcal{D}$ of $X = 0$. Properties of $v$ are governed by properties of $q$ for a given system (5.1), i.e. for a given $f$.

2) At the second stage we use the same function $q$ in a specific way to be clarified later on and resolve the equations (5.1) and (5.2), or (5.3), for a new form of $v$ that will yield exact determination of the asymptotic stability domain $\mathcal{D}$.

This approach was developed for the systems (5.1) with the global weak smoothness property or with the weak smoothness property or with the strong smoothness property in [71], [72], [74]–[76], [78], [79].

The properties and the form of a function solution $v$ to (5.1), (5.2), or (5.3) are governed by properties of both $f$ and $q$. Once the system (5.1) is given, the function $f$ is fixed. We meet a question how to choose a suitable function $q$. A selection of $q$ should meet two goals:

a) it should guarantee the existence of a solution $v$ to the equations (5.1) and (5.2), or equivalently to (5.3),

and

b) it should give such properties to $v$ that we can draw conclusions about asymptotic stability and, if aimed, the asymptotic stability domain.

**a)** The problem of the existence of a function solution $v$ to the equations (5.1), (5.2), or (5.3), is purely mathematical problem that is independent of stability properties of the system (5.1). This is illustrated as follows:

**Example 5.1** Let a first order system (5.1) be defined by

$$\frac{dx}{dt} = \alpha x^{5/3}, \alpha \in \{-1, +1\}.$$

The zero state $x = 0$ is globally asymptotically stable for $\alpha = -1$ and unstable for $\alpha = +1$.

We shall make two different choices of the function $q$ and look for a solution $v$ to (5.2):

1) $q(x) = x^{2/3}$.
The equation (5.3a) now takes the following form:

$$\frac{dv}{dx}\alpha x^{5/3} = -x^{2/3}.$$

The solution to this equation is

$$v(x) = \frac{1}{\alpha}\, ln\, x.$$

It is not defined at $x \leq 0$ nevertheless what is $\alpha \in \{-1, +1\}$. Hence, the solution to (5.3a) does not obey (5.3b), $v(0) = -\frac{\infty}{\alpha} \neq 0$, in this case. Notice that both $f$ and $q$ are continuous everywhere, $f(x) = \alpha x^{5/3}$ and $q(x) = x^{2/3}$. Besides, $f$ is differentiable on $\Re^1$ and $q$ on $(\Re^1 - \mathcal{O})$. Notice that $q$ is not Lipschitzian at $x = 0$.

2) $q(x) = |x|$.
The equation (5.2a) (and (5.3a)) becomes now

$$\frac{dv}{dx}\alpha x^{5/3} = -|x|.$$

The solution to this equation is found as

$$v(x) = -\frac{3}{\alpha}|x|^{1/3}.$$

This solution obeys (5.2b) (and (5.3b)),

$$v(0) = 0.$$

Certainly, $f$ is differentiable on $\Re^1$ (it is unchanged), and $q$ is continuous on $\Re^1$ and differentiable on $(\Re^1 - \mathcal{O})$. In this case $q$ is obviously Lipschitzian at $x = 0$.

This example illustrates the challenging complexity of the following purely mathematical problem: what are the (necessary and) sufficient conditions for the function $q$ to guarantee the existence of a solution function $v$ to the equations (5.1), (5.2), or equivalently to (5.3).

A study of this problem is out of the scope of this book.

The preceding analysis emphasizes the importance of determining the family of all functions $q$ for which the equations (5.3) have a solution.

**Note 5.1** The function $q$ need not possess a (semi-)definiteness property in order to guarantee existence of a solution function $v$ to the equations (5.3). To illustrate this we can accept $q(x) = x$ in Example 5.1, which yields a solution $v$ to (5.3) as

$$v(x) = -\frac{3}{\alpha}x^{1/3} \in [C(\Re^1) \cap C^{(1)}(\Re^1 - \mathcal{O})].$$

**b)** Let $Q^*$ be the family of all functions $q$ guaranteeing existence of a solution function $v$ to the equations (5.1), (5.2), or (5.3). The goal of what follows is to determine a subfamily $Q$ of $Q^*$, $Q \subset Q^*$, and necessary and sufficient properties of a solution function $v$ to (5.3) for arbitrary $q \in Q$ in order for $X = 0$ to be an asymptotically stable state of the system (5.1), or for both $X = 0$ to be asymptotically stable and a set $\mathcal{S}$ to be its domain of asymptotic stability.

**Family $Q(\rho; f)$**

**Definition 5.2** *(a) A function $q : \Re^n \to \Re$ belongs to the family $Q(\rho; f)$ for $\rho \in \Re^+$ if and only if*

1) *$q$ is differentiable on $\overline{\mathcal{B}_\rho}$:$q(X) \in C^{(1)}(\overline{\mathcal{B}_\rho})$ if $\overline{\mathcal{B}_\rho} \subset \Re^n$, or on $\Re^n$ : $q(x) \in C^{(1)}(\Re^n)$ if $\mathcal{B}_\rho = \Re^n$*

*and*

2) *the equations (5.1), (5.2), equivalently (5.3), have a solution $v$ that is well defined in $\Re$ and continuous on $\overline{\mathcal{B}_\mu}$ for some $\mu \in ]0, +\infty[$, $\mu = \mu(q, f)$; or equivalently, the equations (5.4),*

$$[grad\, \nu(X)]^T f(X) = -q(X)\,[1 - \nu(X)], \tag{5.4a}$$

$$\nu(0) = 0, \tag{5.4b}$$

*have a solution $\nu$ that is well defined in $\Re$ and continuous on $\overline{\mathcal{B}_\mu}$.*

*(b) A function $q : \Re^n \to \Re$ belongs to the family $Q^1(\rho; f)$ if and only if $q \in Q(\rho; f)$ and a solution $v$ to (5.3), or equivalently a solution $\nu$ to (5.4), is also differentiable on $\mathcal{B}_\mu$.*

*(c) $Q(\rho; f)$ and $Q^1(\rho; f)$ will be denoted by $Q(f)$ and $Q^1(f)$, respectively if and only if $\mathcal{B}_\rho = \Re^n$.*

*(d) The argument $f$ will be omitted if and only if $f$ is given and fixed.*

**Note 5.2** Let $v$ and $\nu$ be solutions to (5.3) and (5.4), respectively. Then they obey (Vanelli and Vidyasagar [198])

$$\nu(X) = 1 - exp\,[-v(X)].$$

and

$$v(X) = -\,ln\,[1 - \nu(X)].$$

To verify this let $\nu$ be replaced by $1 - exp\,[-v(X)]$ in (5.4) so that

$$\frac{d\nu}{dt} = -(-1)\,exp\,[-v(X)]\frac{dv}{dt} = -q(X)\{1 - 1 + exp\,[-v(X)]\},$$

or,

$$\frac{dv(X)}{dt} = [grad\, v(X)]^T f(X) = -q(X) \tag{5.2a}$$

and

$$0 = \nu(0) = 1 - exp\,[-v(0)].$$

Hence,

$$v(0) = 0. \tag{5.2b}$$

We have proved that the equations (5.4) become (5.3) when $\nu$ is replaced by $1 - exp\,(-v)$ in them. Analogously, the equations (5.3) take the form of the equations (5.4) when $v$ is replaced by $-ln\,(1-\nu)$ in them. Altogether, the equations (5.3) and (5.4) are equivalent. The equation (5.3a) is linear in $v$ and $q$, which is simpler than the equation (5.4a) that is bilinear in $\nu$ and $q$. However, if for some $X^*$, $v(X^*) = +\infty$ then $\nu(X^*) = 1$. The latter is more suitable for computation than the former.

### Results

**Theorem 5.1** *In order for the state $X = 0$ of the system (5.1) with the strong smoothness property to have the domain $\mathcal{D}$ of asymptotic stability and for a set $\mathcal{N}$, $\mathcal{N} \subseteq \Re$, to be the domain of its asymptotic stability, $\mathcal{N} = \mathcal{D}$, it is both necessary and sufficient that*

*1) the set $\mathcal{N}$ is an open connected neighbourhood of $X = 0$ and $\mathcal{N} \subseteq \mathcal{S}$,*

*2) $f(X) = 0$ for $X \in \mathcal{N}$ iff $X = 0$,*

*and*

*3) (a) for arbitrarily chosen $\rho \in \Re^+$ satisfying $\overline{\mathcal{B}}_\rho \subset \mathcal{N}$, for arbitrary positive definite $p \in Q(\rho; f)$ and for $\eta = min\,(\mu, \rho)$ the equations (5.5),*

$$\frac{dv(X)}{dt} = [grad\, v(X)]^T f(X) = -w(X), \tag{5.5a}$$

$$w(X) = \begin{cases} p(X) & if\ \ X \in \overline{\mathcal{B}}_\eta, \\ \dfrac{||X||}{\eta}\, p\left(\dfrac{\eta X}{||X||}\right) & if\ \ X \in (\mathcal{S} - \overline{\mathcal{B}}_\eta), \end{cases} \tag{5.5b}$$

$$v(0) = 0, \tag{5.5c}$$

*have the unique solution function $v$ with the following features:*

*(i) $v$ is positive definite on $\mathcal{N}$,*

*and*

*(ii) if the boundary $\partial\mathcal{N}$ of $\mathcal{N}$ is non-empty then $v(X) \to +\infty$ as $X \to \partial\mathcal{N}$, $X \in \mathcal{N}$,*

*or,*

*(b) for arbitrarily chosen* $\rho \in \Re^+$ *satisfying* $\overline{\mathcal{B}}_\rho \subset \mathcal{N}$*, for arbitrary positive definite* $p \in Q^1(\rho; f)$ *and for* $\eta = min\,(\mu, \rho)$ *the equations (5.5) have the unique solution function* $v$ *with the following features:*

*(i)* $v$ *is positive definite on* $\mathcal{N}$ *and differentiable on* $(\mathcal{N} - \partial\mathcal{B}_\eta)$,

*and*

*(ii) if the boundary* $\partial\mathcal{N}$ *on* $\mathcal{N}$ *is non-empty then* $v(X) \to +\infty$ *as* $X \to \partial\mathcal{N}$, $X \in \mathcal{N}$. ∇∇∇

**Proof.** *Necessity.* Let $X = 0$ of the system (5.1) with the strong smoothness property have the asymptotic stability domain $\mathcal{D}$. Then, $\mathcal{D}_a = \mathcal{D}_{ac} = \mathcal{D} = \mathcal{D}_c \subseteq \mathcal{D}_{sc} \subseteq \mathcal{D}_s$ (Theorem 3.35, Section 3.7) and $\mathcal{D}_a$ is a neighbourhood of $X = 0$ (Definition 2.4, Section 2.2.3). $\mathcal{S}$ is also a neighbourhood of $X = 0$ in view of the strong smoothness property. Hence, $\mathcal{D}_a \cap \mathcal{S} \neq \emptyset$. Let us prove $\mathcal{D}_a \subseteq \mathcal{S}$. If $\partial\mathcal{S} = \emptyset$ then $\mathcal{S} = \Re^n$ and $\mathcal{D}_a \subseteq \mathcal{S}$ due to $\mathcal{D}_a \subseteq \Re^n$. If $\partial\mathcal{S} \neq \emptyset$ then we shall consider $X_0 \in \partial\mathcal{S}$ and $X_0^* \in (\Re^n - \overline{\mathcal{S}})$. If $X_0 \in \partial\mathcal{S}$ then $X_0 \notin \mathcal{D}_a$ due to (ii) of the strong smoothness property. Therefore, $\partial\mathcal{S} \cap \mathcal{D}_a = \emptyset$. If $X_0^* \in (\Re^n - \overline{\mathcal{S}})$ then for $\underline{X}(t; X_0^*) \to 0$ as $t \to +\infty$ it is necessary that there is $t^* \in \Re_+$ such that $\underline{X}(t^*; X_0^*) \in \partial\mathcal{S}$ because $\mathcal{D}_a$ and $\mathcal{S}$ are neighbourhoods of $X = 0$, $X_0^* \notin \overline{\mathcal{S}}$ and the motion $\underline{X}(t; X_0^*)$ is continuous in $t \in \Re_+$ due to (ii) of the weak smoothness property ensured by (i) of the strong smoothness property. However, $\underline{X}(t^*; X_0^*) \in \partial\mathcal{S}$ implies that $\underline{X}(t; X_0^*)$ does not converge to $X = 0$ because of (ii) of the strong smoothness property. This yields $X_0^* \notin \mathcal{D}_a$ and $(\Re^n - \overline{\mathcal{S}}) \cap \mathcal{D}_a = \emptyset$. By connecting all the results, that is that $\mathcal{D}_a \cap \mathcal{S} \neq \emptyset$, $\mathcal{D}_a \cap \partial\mathcal{S} = \emptyset$ and $\mathcal{D}_a \cap (\Re^n - \overline{\mathcal{S}}) = \emptyset$ we conclude that $\mathcal{D}_a \subseteq \mathcal{S}$. Therefore, $\mathcal{D} \subseteq \mathcal{S}$ due to $\mathcal{D} = \mathcal{D}_a$. Let $\mathcal{N} = \mathcal{D} = \mathcal{D}_{ac}$ so that $\mathcal{N} \subseteq \mathcal{S}$ and $\mathcal{N}$ is an open connected neighbourhood of $X = 0$ due to (i-b) of the weak smoothness property, $\mathcal{N} = \mathcal{D} = \mathcal{D}_{ac}$ and invariance of $\mathcal{D}_a$ with respect to system motions (Theorem 3.34, Section 3.7). This proves the necessity of the condition 1). From $\mathcal{N} = \mathcal{D} = \mathcal{D}_a \subseteq \mathcal{D}_s$ it results (Theorem 3.28, Section 3.4, and Theorem 3.35, Section 3.7) that $X = 0$ is the unique equilibrium state of the system (5.1) in $\mathcal{N}$, which implies $f(X) = 0$ for $X \in \mathcal{N} = \mathcal{D}$ iff $X = 0$ (Theorem 3.22, Section 3.3.4) and proves the necessity of the condition 2).

Since $\mathcal{N} = \mathcal{D}$ then $I_0 = \Re_+$ for every $X_0 \in \mathcal{N}$, $I_0 = I_0(X_0)$, due to definitions of $\mathcal{D}$ and $\mathcal{D}_a$.

Let $\rho \in \Re^+$ satisfying $\overline{\mathcal{B}}_\rho \subset \mathcal{N}$ be arbitrarily chosen. Let positive definite $p \in Q(\rho; f)$ be arbitrarily selected. Then, the equation (5.5a) has a solution $v$ defined on $\overline{\mathcal{B}}_\eta$ for $\eta$ given by

$$\eta = min\,(\mu, \rho) > 0,$$

because (5.5a) reduces to (5.3a) on $\overline{\mathcal{B}}_\eta$ for so defined $\eta$. Since $\eta < +\infty$ due to its definition and $\rho < +\infty$ then $\overline{\mathcal{B}}_\eta$ is compact. This and (a-2) of Definition 5.2 imply

$$|v(X)| < +\infty \ \text{ for every } \ X \in \overline{\mathcal{B}}_\eta. \tag{5.6}$$

Let $X_0 \in \mathcal{N}$ and $\tau \in \Re_+$, $\tau = \tau(X_0, \eta)$, obey (5.7),

$$\underline{X}(t; X_0) \in \overline{\mathcal{B}}_\eta \ \text{ for all } \ t \in [\tau, +\infty[. \tag{5.7}$$

Such $\tau$ exists for $X_0 \in \mathcal{N}$, $\mathcal{N} = \mathcal{D} = \mathcal{D}_a$, and the definition of $\mathcal{D}_a$. Let (5.5a) be integrated from $t \in \Re_+$ to $+\infty$ so that

$$v[\underline{X}(\infty; X_0)] - v[\underline{X}(t; X_0)] = -\int_t^{+\infty} w[\underline{X}(\sigma; X_0)]d\sigma \quad \text{for every} \quad (t, X_0) \in \Re_+ \times \mathcal{N}. \tag{5.8}$$

$\mathcal{N} = \mathcal{D}$ and $X_0 \in \mathcal{N}$ imply $\underline{X}(\infty; X_0) = 0$ (the definitions of $\mathcal{D}_a$ and $\mathcal{D}$). Let, in view of (5.5c),

$$v(0) = 0, \tag{5.9}$$

so that (5.8) and (5.9) yield (5.10),

$$v[\underline{X}(t; X_0)] = \int_t^{\tau} w[\underline{X}(\sigma; X_0)]d\sigma + \int_{\tau}^{+\infty} w[\underline{X}(\sigma; X_0)]d\sigma \quad \text{for every} \quad (t, X_0) \in [0, \tau] \times \mathcal{N}. \tag{5.10}$$

Positive definiteness of $w$ on $\mathcal{N}$, invariance of $\mathcal{D}_a$ with respect to system motions, $\mathcal{N} = \mathcal{D} = \mathcal{D}_a$, (i) of the weak smoothness property and compactness of $[t, \tau]$ for every $t \in [0, \tau]$ and $\tau \in \Re_+$ yield

$$\left| \int_t^{\tau} w[\underline{X}(\sigma; X_0)]d\sigma \right| < +\infty \quad \text{for every} \quad (t, X_0) \in [0, \tau] \times \overline{\mathcal{B}}_\eta. \tag{5.11}$$

Integrating (5.5a) now from $\tau$ to $+\infty$, using $\underline{X}(\infty; X^*) = 0$ for $X^* \in \overline{\mathcal{B}}_\eta$ due to $\overline{\mathcal{B}}_\eta \subset \mathcal{D}$, and utilizing (5.9) we get

$$|v(X^*)| = \left| \int_{\tau}^{+\infty} w[\underline{X}(\sigma; X^*)]d\sigma \right| < +\infty. \tag{5.12}$$

Let $X^* = \underline{X}(\tau; X_0) \in \overline{\mathcal{B}}_\eta$ for $X_0 \in \mathcal{N}$. This and (5.10)–(5.12) prove (5.13),

$$|v[\underline{X}(t; X_0)]| < +\infty \quad \text{for every} \quad (t, X_0) \in [0, \tau] \times \mathcal{N}. \tag{5.13}$$

Let $t = 0$ and $X_0 = X$ be set in (5.13) that becomes (5.14),

$$|v(X)| < +\infty \quad \text{for every} \quad X \in \mathcal{N}. \tag{5.14}$$

Invariance of $\mathcal{D}$, $\mathcal{D} = \mathcal{D}_a$, the strong smoothness property, $\overline{\mathcal{B}}_\eta \subset \mathcal{N}$, differentiability of $w$ on $\overline{\mathcal{B}}_\eta$ due to differentiability of $p$ on $\overline{\mathcal{B}}_\eta$ and (5.5b), and (5.14) prove continuity of the function $v$ on $\overline{\mathcal{B}}_\eta$, and its differentiability on $(\mathcal{N} - \partial\mathcal{B}_\eta)$ in the case $p \in Q^1(\rho; f)$:

$$v(X) \in \begin{cases} C(\overline{\mathcal{B}}_\eta) & \text{if } p \in Q(\rho; f) \\ C^{(1)}(\mathcal{B}_\eta) & \text{if } p \in Q^1(\rho; f). \end{cases} \tag{5.15}$$

Let $\hat{X}_0 \in (\mathcal{N} - \mathcal{B}_\eta)$ and let $\psi = \psi(X_0) \in \Re_+$ denote the last moment when $||\underline{X}(t; \hat{X}_0)|| = \eta$. Such $\psi$ exists because $\hat{X}_0 \in \mathcal{N}$, $\mathcal{N} = \mathcal{D}$, and (i-b) of the weak smoothness property. Hence, continuity of both system motions on $\Re_+ \times \mathcal{N}$ and the function $w$ on $\mathcal{N}$, $\mathcal{N} = \mathcal{D}_a$, invariance of $\mathcal{D}_a$ with respect to system motions and compactness of $[t, \psi]$ and $[\psi, \tau]$ for every $t \in [0, \psi]$ imply

$$\left| \int_t^{\psi} w[\underline{X}(\sigma; \hat{X}_0)]d\sigma \right| < +\infty \text{ for every } (t, \hat{X}_0) \in [0, \psi] \times (\mathcal{N} - \mathcal{B}_\eta),$$

and

$$\left| \int_{\psi}^{\tau} w[\underline{X}(\sigma; \hat{X}_0)]d\sigma \right| < +\infty \text{ for every } \hat{X}_0 \in (\mathcal{N} - \mathcal{B}_\eta).$$

From the definition of $\eta$, (5.6) and (5.7) follows

$$\left| \int_{\tau}^{+\infty} w[\underline{X}(\sigma; \hat{X}_0)]d\sigma \right| < +\infty \text{ for every } \hat{X}_0 \in (\mathcal{N} - \mathcal{B}_\eta).$$

These inequalities, $\mathcal{B}_\eta \subseteq \mathcal{B}_\rho$, (5.10), continuity of the function $w$ on $\mathcal{N}$ due to $p \in Q(\rho; f)$ and (5.4b), and continuity of motions $\underline{X}$ on $\Re_+ \times \mathcal{N}$ prove continuity of the function $v$ on $(\mathcal{N} - \overline{\mathcal{B}}_\eta)$. Its differentiability on $(\mathcal{N} - \overline{\mathcal{B}}_\eta)$ in case $p \in Q^1(\rho; f)$ is implied by $\mathcal{B}_\eta \subseteq \mathcal{B}_\rho$, by differentiability of the function $w$ on $\mathcal{N}$, which is ensured by $p \in Q^1(\rho; f)$ and (5.5b), and by differentiability of motions on $\Re_+ \times \mathcal{N}$ (due to (i-b) of the weak smoothness property),

$$v(X) \in \begin{cases} C(\mathcal{N} - \mathcal{B}_\eta),\ p \in Q(\rho; f) \\ C(\mathcal{N} - \mathcal{B}_\eta) \cap C^{(1)}(\mathcal{N} - \overline{\mathcal{B}}_\eta),\ p \in Q^1(\rho; f). \end{cases}$$

This result and (5.14) prove both

$$|v(X)| < +\infty \text{ for every } X \in \mathcal{N} \tag{5.16}$$

and

$$v(X) \in \begin{cases} C(\mathcal{N}),\ p \in Q(\rho; f) \\ C(\mathcal{N}) \cap C^{(1)}(\mathcal{N} - \partial\mathcal{B}_\eta),\ p \in Q^1(\rho; f). \end{cases} \tag{5.17}$$

(5.10), (5.16) and (5.17) prove the existence of a solution function $v$ to the equations (5.5) for arbitrarily chosen $p \in [Q(\rho; f) \cup Q^1(\rho; f)]$, where $\rho$ was arbitrarily selected provided $\overline{\mathcal{B}}_\rho \subset \mathcal{N}$. Let $v_1$ and $v_2$ be two such solutions to (5.5). Uniqueness of the system motion $\underline{X}(t; X_0)$ for every $X_0 \in \mathcal{N}$ due to (i-a) of the weak smoothness property, invariance of $\mathcal{N} = \mathcal{D}_a$ with respect to system motions and uniqueness of $w$ on $\mathcal{N}$ due to its positive definiteness on $\mathcal{N}$ yield

$$v_i(X_0) = \int_0^{+\infty} w[\underline{X}(\sigma; X_0)]d\sigma \text{ for every } i = 1, 2 \text{ and every } X_0 \in \mathcal{N},$$

$$v_1(X_0) = v_2(X_0) \text{ for every } X_0 \in \mathcal{N}.$$

This proves the uniqueness of the solution $v$ to (5.5). Invariance of $\mathcal{D}_a$ with respect to system motions, $\mathcal{N} = \mathcal{D} = \mathcal{D}_a$, positive definiteness of $w$ on $\mathcal{N}$ and (5.18) obtained from (5.10) for $t = 0$ and $X = X_0$,

$$v(X) = \int_0^{\infty} w[\underline{X}(\sigma; X)]d\sigma \text{ for every } X \in \mathcal{N}, \tag{5.18}$$

imply (5.19):

$$v(X) > 0 \text{ for every } (X \neq 0) \in \mathcal{N}. \tag{5.19}$$

(5.9), (5.16), (5.17) and (5.19) prove the necessity of the conditions (3a-i) and (3b-i) for $p \in Q(\rho; f)$ and $p \in Q^1(\rho; f)$, respectively. Let $X_1, X_2, \ldots, X_k, \ldots$ be a sequence converging to $\hat{X}$, $X_k \to \hat{X}$ as $k \to +\infty$, where $X_k \in \mathcal{N}$ and $\hat{X} \in \partial\mathcal{N}$ in case $\partial\mathcal{N}$ is non-empty. Let $\tau_k = \tau(X_k, \eta) \in \Re_+$ be the first instant obeying $\underline{X}(t; X_k) \in \overline{\mathcal{B}}_\eta$ for all $t \in [\tau_k, +\infty[$. Existence of such $\tau_k$ is guaranteed by $X_k \in \mathcal{N}$ and $\mathcal{N} = \mathcal{D} = \mathcal{D}_a$ (due to definitions of $\mathcal{D}_a$ and $\mathcal{D}$). Continuity of motions $\underline{X}$ in $(t, X_0) \in \Re_+ \times \mathcal{N}$ due to (i-b) of the weak smoothness property and $\mathring{\mathcal{N}} = \mathcal{N} = \mathcal{D} = \mathcal{D}_a$ imply (Theorem 3.34, Section 3.7) $\tau(X_k, \eta) \to +\infty$ as $k \to +\infty$. Let $m$ be such a natural number that $X_k \in (\mathcal{N} - \overline{\mathcal{B}}_\eta)$ for all $k = m, m+1, \ldots$ Such $m$ exists because $\mathcal{N}$ is open, $\overline{\mathcal{B}}_\eta \subset \mathcal{N}$ and $X_k \to \partial\mathcal{N}$ as $k \to +\infty$. Let $\alpha$ be defined by (5.20):

$$\alpha = min\,[w(X) : X \in (\mathcal{N} - \mathcal{B}_\eta)]. \tag{5.20}$$

Positive definiteness of $p$ on $\overline{\mathcal{B}}_\eta$ and (5.5b) imply $\alpha \in \Re^+$. Combining (5.10) and (5.20) we get

$$v(X_k) \geq \int_0^{\tau_k} \alpha d\sigma + \int_{\tau_k}^{+\infty} w[\underline{X}(\sigma; X_k)]d\sigma.$$

This inequality, positive definiteness of $w$ on $\overline{\mathcal{B}}_\eta$ and $\underline{X}(t; X_k) \in \overline{\mathcal{B}}_\eta$ for all $t \in [\tau_k, +\infty[$ imply

$$v(X_k) > \alpha\tau_k,\ \tau_k = \tau(X_k, \eta),\ X_k \in \mathcal{N}, k = m, m+1, \ldots \tag{5.21}$$

Since $\tau_k \to +\infty$ as $k \to +\infty$ then (5.21) shows that

$$v(X_k) \to +\infty \text{ as } k \to +\infty,$$

or, in view of the definition of $X_k$,

$$v(X) \to +\infty \text{ as } X \to \partial\mathcal{N}, X \in \mathcal{N},$$

which proves the necessity of the conditions (3a-ii) and (3b-ii) for $p \in Q(\rho; f)$ and $p \in Q^1(\rho; f)$, respectively.

*Sufficiency.* (a) Let all the conditions hold for $p \in Q(\rho; f)$. Then, $X = 0$ is asymptotically stable (Theorem 4.9, Section 4.6.3). Hence it has the domains $\mathcal{D}_s$, $\mathcal{D}_a$ and $\mathcal{D}$, which are related by $\mathcal{D}_a = \mathcal{D}_{ac} = \mathcal{D} = \mathcal{D}_c \subseteq \mathcal{D}_{sc} \subseteq \mathcal{D}_s$. This follows from Theorem 3.35, Section 3.7. Besides, $\mathcal{D}$ is a neighbourhood of $X = 0$ by the definition. Hence, $\mathcal{N} \cap \mathcal{D} \neq \emptyset$. If both $\partial\mathcal{N}$ and $\partial\mathcal{D}$ are empty then $\mathcal{D} = \mathcal{N} = \Re^n$. If $\partial\mathcal{D} \neq \emptyset$ then we can assume $\partial\mathcal{D} \cap \mathcal{N} \neq \emptyset$. Let $\hat{X} \in \partial\mathcal{D} \cap \mathcal{N}$. Hence $\hat{X} \in \mathcal{N}$. From the proof of the necessity part it follows that $v(X) \to +\infty$ as $X \to \hat{X}$, $X \in \mathcal{D}$, which contradicts continuity of $v$ on $\mathcal{N}$. Hence, $\partial\mathcal{D} \cap \mathcal{N} = \emptyset$ so that $\mathcal{N} \subseteq \mathcal{D}$. Let it be assumed that $\partial\mathcal{N} \cap \mathcal{D} \neq \emptyset$. From the condition (3-a-ii) it follows that for any $X^* \in \partial\mathcal{N}$, $v(X) \to +\infty$ as $X \to X^*$, $X \in \mathcal{N}$, which contradicts continuity of $v$ obtained via (5.5) as shown in the proof of the necessity part. Hence, $\partial\mathcal{N} \cap \mathcal{D} = \emptyset$. This, $\partial\mathcal{D} \cap \mathcal{N} = \emptyset$ and the fact that both $\mathcal{D}$ and $\mathcal{N}$ are open connected neighbourhoods of $X = 0$ prove $\mathcal{D} = \mathcal{N}$ in case $\partial\mathcal{D} \neq \emptyset$ and $\partial\mathcal{N} \neq \emptyset$. If $\partial\mathcal{D} = \emptyset$ then $\partial\mathcal{N}$ must be empty, and if $\partial\mathcal{N} = \emptyset$ then $\partial\mathcal{D}$ must be empty as shown above. Altogether, $\mathcal{D} = \mathcal{N}$.

In what follows $dv(x)/dt$ will be written for short because

$$\frac{dv(x)}{dt} = [grad\, v(x)]^T f(x).$$

The preceding theorem, like that for time invariant linear systems (Lyapunov [127]), permits free, arbitrary selection of both $\rho \in \Re^+$ satisfying $\overline{\mathcal{B}}_\rho \subset \mathcal{N}$ and positive definite function $p \in Q(\rho; f)$. Then the function $v$ is determined via (5.5). If and only if it has the properties specified under 3), then $\mathcal{N}$ is the domain $\mathcal{D}$ of asymptotic stability of $X = 0$ of the system (5.1). However, if it does not obey 3) then $X = 0$ does not have the domain of asymptotic stability. This means that in such a case $X = 0$ is not asymptotically stable as illustrated by the following result.

**Theorem 5.2** *For the state $X = 0$ of the system (5.1) possessing the weak smoothness property to be asymptotically stable it is both necessary and sufficient that, for any $\zeta \in \Re_+$ and any function $p \in Q(\zeta; f)$, there exists the unique solution function $v$ to (5.22) with (5.22a) determined along system motions,*

$$\frac{d}{dt}v(x) = -p(x), \tag{5.22a}$$

$$v(0) = 0, \tag{5.22b}$$

*which is also positive definite, and differentiable if $p \in Q^1(\zeta; f)$.* ∇∇∇

**Proof.** *Necessity.* Let the system (5.1) possess weak smoothness property. Let $X = 0$ be asymptotically stable. Then it has $\mathcal{D}_a$, $\mathcal{D}_s$ and $\mathcal{D}$, and $\mathcal{D}_a \cap \mathcal{S} \neq \emptyset$, $\mathcal{D}_s \cap \mathcal{S} \neq \emptyset$ and $\mathcal{D} \cap \mathcal{S} \neq \emptyset$ because $\mathcal{D}_a$, $\mathcal{D}_s$, $\mathcal{D}$ and $\mathcal{S}$ are neighbourhoods of $X = 0$. Let any $\zeta \in \Re^+$ be chosen and a positive definite $p \in Q(\zeta; f)$ be arbitrarily selected, that is that, on an open connected neighbourhood $\mathcal{A}$ of $X = 0$, $\mathcal{A} \subseteq \mathcal{B}_\zeta$, it obeys

$$1.\ p(X) \in C^{(1)}(\mathcal{A}), \quad 2.\ p(0) = 0, \quad \text{and} \quad 3.\ p(X) > 0 \ \text{ for } \ (X \neq 0) \in \mathcal{A}.$$

The set $\mathcal{L} = \mathcal{A} \cap \mathcal{D} \subseteq \mathcal{D}$ is also open connected neighbourhood of $X = 0$ (see the proof of Theorem 5.1 for such property of $\mathcal{D}$). Let $\varepsilon$ satisfying $\overline{\mathcal{B}}_\varepsilon \subset \mathcal{L}$ be arbitrarily chosen. Then $\overline{\mathcal{B}}_\varepsilon \subset \mathcal{D}$. Let $\rho \in ]0, min\,(\varepsilon, \zeta)[$ fulfilling $\overline{\mathcal{B}}_\rho \subset \mathcal{D}_s(\varepsilon)$ be also arbitrarily accepted. From the proofs of (5.11) and (5.15) we prove that $v$ defined by (5.22) has the following properties since $\overline{\mathcal{B}}_\rho \subset \mathcal{D}_s(\varepsilon) \subseteq \mathcal{B}_\varepsilon \subset \mathcal{L}$,

$$|v(X)| < +\infty \text{ for every } X \in \overline{\mathcal{B}}_\rho, \tag{5.23a}$$

$$v(X) \in \begin{cases} C(\overline{\mathcal{B}}_\rho), p \in Q(\zeta; f) \\ C(\overline{\mathcal{B}}_\rho) \cup C^{(1)}(\mathcal{B}_\rho), p \in Q^1(\zeta; f). \end{cases} \tag{5.23b}$$

Definitions 5.2 and 5.3, $\mathcal{B}_\varepsilon \subset \mathcal{A} \cap \mathcal{D}$, positive definiteness of $p$ on $\mathcal{A}$, $\underline{X}(t; X_0) \in \mathcal{B}_\varepsilon$ for all $(t, X_0) \in \Re_+ \times \overline{\mathcal{B}}_\rho$ due to $\overline{\mathcal{B}}_\rho \subset \mathcal{D}_s(\varepsilon)$, (5.22a) used after integration from $t = 0$ to $t = +\infty$ together with (5.22b) and $\underline{X}(\infty; X_0) = 0$ imply

$$v(X_0) > 0 \text{ for every } (X_0 \neq 0) \in \overline{\mathcal{B}}_\rho. \tag{5.24}$$

Now, (5.22b)–(5.24) prove positive definiteness of the function solution $v$ to (5.22) on any $\overline{\mathcal{B}}_\rho \subset \mathcal{D}_s(\varepsilon)$ and its differentiability on $\mathcal{B}_\rho$. Its uniqueness is proved in the same way as in the proof of Theorem 5.1. This completes the proof of the necessity part.

*Sufficiency.* Sufficiency of the conditions of Theorem 5.2 for asymptotic stability of $X = 0$ of the system (5.1) is due to Lyapunov [127] (Theorem 3.9, Section 3.7). This completes the proof of Theorem 5.2.

From the computational point of view the condition that $v(X) \to +\infty$ as $X \to \partial\mathcal{N}$, $X \in \mathcal{N}$, is not suitable. It can be overcome as follows.

**Theorem 5.3** *For the state $X = 0$ of the system (5.1) with the strong smoothness property to have the domain of asymptotic stability $\mathcal{D}$, and for a set $\mathcal{N}$, $\mathcal{N} \subseteq \Re^n$, to be the domain of asymptotic stability: $\mathcal{N} = \mathcal{D}$, it is both necessary and sufficient that*

1) *the set $\mathcal{N}$ is an open connected neighbourhood of $X = 0$ and $\mathcal{N} \subseteq \mathcal{S}$,*

2) *$f(X) = 0$ for $X \in \mathcal{N}$ iff $X = 0$, and*

3) *(a) for arbitrary $\rho \in \Re^+$ obeying $\overline{\mathcal{B}}_\rho \subset \mathcal{N}$ and for arbitrarily chosen $p \in Q(\rho; f)$ the equations (5.25) with (5.25a) determined along system motions and with $\eta = min\,(\mu, \rho)$,*

$$\frac{d}{dt}\nu(X) = -w(X), \tag{5.25a}$$

$$w(X) = \begin{cases} p(X)[1 - \nu(x)], & X \in \overline{\mathcal{B}}_\eta \\ \dfrac{\|X\|}{\eta}\, p\left(\dfrac{\eta X}{\|X\|}\right)[1 - \nu(X)], & X \in (\mathcal{S} - \mathcal{B}_\eta), \end{cases} \tag{5.25b}$$

$$\nu(0) = 0, \tag{5.25c}$$

*have the unique solution function $\nu$ with the following features:*

*(i) $\nu$ is positive definite on $\mathcal{N}$,*

*(ii) if the boundary $\partial\mathcal{N}$ of $\mathcal{N}$ is non-empty then $\nu(X) \to 1$ as $X \to \partial\mathcal{N}$, $X \in \mathcal{N}$,*

*or,*

*(b) for arbitrary $\rho \in \Re^+$ obeying $\overline{\mathcal{B}}_\rho \subset \mathcal{N}$ and for arbitrarily chosen $q \in Q^1(\rho; f)$ the equations (5.25) with (5.25a) determined along system motions and with $\eta = \min(\mu, \rho)$ have the unique solution function $\nu$ with the following features:*

*(i) $\nu$ is positive definite on $\mathcal{N}$ and differentiable on $(\mathcal{N} - \partial\mathcal{B}_\eta)$,*

*and*

*(ii) if the boundary $\partial\mathcal{N}$ of $\mathcal{N}$ is non-empty then $\nu(X) \to 1$ as $X \to \partial\mathcal{N}$, $X \in \mathcal{N}$.* ∇∇∇

**Proof.** Since the equations (5.4) and (5.25) are equivalent (Note 5.2) then Theorem 5.3 results directly from Theorem 5.1.

Differently from Theorem 5.1, Theorem 5.2 permits the system (5.1) to possess the weak smoothness property only. For such systems the necessary and sufficient conditions for determination of both their Lyapunov functions and asymptotic stability domains are discovered as follows.

**Theorem 5.4** *For the state $X = 0$ of the system (5.1) possessing the weak smoothness property to have the domain of asymptotic stability $\mathcal{D}$ and for a subset $\mathcal{N}$ of $\mathcal{S}$, $\mathcal{N} \subseteq \mathcal{S}$, to be equal to $\mathcal{D}$: $\mathcal{N} = \mathcal{D}$, it is both necessary and sufficient that*

*1) the set $\mathcal{N}$ is an open connected neighbourhood of $X = 0$,*

*2) $f(X) = 0$ for $X \in \mathcal{N}$ iff $X = 0$, and*

*3) (a) for arbitrary $\rho \in \Re^+$ satisfying $\overline{\mathcal{B}}_\rho \subset \mathcal{N}$ and for arbitrary positive definite function $p \in Q(\rho; f)$ the equations (5.26) with (5.26a) determined along system motions and for $\eta = \min(\nu, \rho)$,*

$$\frac{d}{dt}v(X) = -w(X), \tag{5.26a}$$

$$w(X) = \begin{cases} p(X) & \text{if } X \in \overline{\mathcal{B}}_\eta \\ \dfrac{\|X\|}{\eta} p\left(\dfrac{\eta X}{\|X\|}\right) & \text{if } X \in (\Re^n - \mathcal{B}_\eta), \end{cases} \tag{5.26b}$$

$$v(0) = 0, \tag{5.26c}$$

*have the unique solution $v$ with the following features:*

*(i) $v$ is positive definite on $\mathcal{N}$,*

*and*

*(ii) if the boundary $\partial\mathcal{N}$ of $\mathcal{N}$ is non-empty then $v(X) \to +\infty$ as $X \to \partial\mathcal{N}$, $X \in \mathcal{N}$,*

*or,*

*(b) for arbitrary* $\rho \in \Re^+$ *satisfying* $\overline{\mathcal{B}}_\rho \subset \mathcal{N}$ *and for arbitrary positive definite function* $p \in Q^1(\rho; f)$ *the equations (5.26) with (5.26a) determined along system motions and for* $\eta = min(\mu, \rho)$ *have the unique solution function* $v$ *with the following features:*

*(i)* $v$ *is positive definite on* $\mathcal{N}$ *and differentiable on* $(\mathcal{N} - \partial\mathcal{B}_\eta)$,

*and*

*(ii) if the boundary* $\partial\mathcal{N}$ *on* $\mathcal{N}$ *is non-empty then* $v(X) \to +\infty$ *as* $X \to \partial\mathcal{N}$, $X \in \mathcal{N}$. ∇∇∇

**Proof.** *Necessity.* Let the system (5.1) possess the weak smoothness property. Let $X = 0$ have the asymptotic stability domain $\mathcal{D}$ and let $\mathcal{N} \subseteq \mathcal{S}$ be equal to $\mathcal{D}$: $\mathcal{N} = \mathcal{D}$. From this point on we have to repeat the proof of Theorem 5.1 to show that the conditions 1)–3) of Theorem 5.4 hold. In such a way we complete the proof of the necessity part.

*Sufficiency.* Sufficiency of the conditions for $\mathcal{N} = \mathcal{D}$ results from Theorem 5.1.

If we wish to avoid computational difficulties due to $v(x) \to +\infty$ as $X \to \partial\mathcal{N}$, $X \in \mathcal{N}$, then we can use the following theorem that follows from Theorem 5.4 and Note 5.2.

**Theorem 5.5** *For the state* $X = 0$ *of the system (5.1) possessing the weak smoothness property to have the domain of asymptotic stability* $\mathcal{D}$ *and for a subset* $\mathcal{N}$ *of* $\mathcal{S}$, $\mathcal{N} \subseteq \mathcal{S}$, *to be equal to* $\mathcal{D}$: $\mathcal{N} = \mathcal{D}$ *it is both necessary and sufficient that*

*1) the set* $\mathcal{N}$ *is an open connected neighbourhood of* $X = 0$,

*2)* $f(X) = 0$ *for* $X \in \mathcal{N}$ *iff* $X = 0$, *and*

*3) (a) for arbitrary* $\rho \in \Re^+$ *satisfying* $\overline{\mathcal{B}}_\rho \subset \mathcal{N}$ *and for arbitrary positive definite* $p \in Q(\rho; f)$ *the equations (5.27) with (5.27a) determined along system motions and for*

$$\eta = min(\mu, \rho)$$

$$\frac{d}{dt}\nu(X) = -w(X), \tag{5.27a}$$

$$w(X) = \begin{cases} p(X)\,[1 - \nu(x)], & X \in \overline{\mathcal{B}}_\nu \\ \dfrac{||X||}{\eta}\, p\left(\dfrac{\eta X}{||X||}\right)[1 - \nu(X)], & X \in (\Re^n - \mathcal{B}_\eta), \end{cases} \tag{5.27b}$$

$$\nu(0) = 0, \tag{5.27c}$$

*have the unique solution function* $\nu$ *with the following features:*

*(i)* $\nu$ *is positive definite on* $\mathcal{N}$,

*(ii) if the boundary $\partial\mathcal{N}$ of $\mathcal{N}$ is non-empty then $\nu(X) \to +1$ as $X \to \partial\mathcal{N}$, $X \in \mathcal{N}$,*

*or*

*(b) for arbitrary $\rho \in \Re^+$ satisfying $\overline{\mathcal{B}}_\rho \subset \mathcal{N}$ and for arbitrary positive definite $p \in Q^1(\rho; f)$, the equations (5.27) with (5.27a) determined along system motions and for $\eta = \min(\mu, \rho)$ have the unique solution function $\nu$ with the following features:*

*(i) $\nu$ is positive definite on $\mathcal{N}$ and differentiable on $(\mathcal{N} - \partial\mathcal{B})$,*

*(ii) if the boundary $\partial\mathcal{N}$ on $\mathcal{N}$ is non-empty then $\nu(X) \to +1$ as $X \to \partial\mathcal{N}$, $X \in \mathcal{N}$.* ∇∇∇

## Examples

The following examples are taken from [78].

**Example 5.2** Let a first ($n = 1$) order system (5.1) be described by

$$\frac{dx}{dt} = (x^2 - 1)x, \quad f(x) = (x^2 - 1)x \in C^{(1)}(\Re^1). \tag{5.28}$$

It possesses the strong smoothness property on $\Re^1$. Now, from $f(x) = (x^2 - 1)x = 0$ it results that the system (5.24) has three equilibrium states:

$$x_{1e} = -1, \quad x_{2e} = 0, \quad x_{3e} = +1.$$

They suggest to select $\rho \in \,]0, +1[$ and $\mathcal{S} = \,]-1, +1[$ because $\partial\mathcal{S} = \{-1, +1\} = \{x_{1e}, x_{3e}\}$ which shows that (ii) of the strong smoothness property holds on $\mathcal{S} = \,]-1, +1[$. Hence, the system (5.28) has strong smoothness property also on $\mathcal{S} = \,]-1, +1[$.

Two different forms of the function $p$ will be considered.

a) By referring to the form of the function $f$ (5.28) let

$$p(x) = (1 - x^2)x^2. \tag{5.29}$$

The function $p$ is differentiable and positive definite on $\mathcal{S} = \,]-1, +1[$. Now, for any $\rho \in \,]0, 1[$, the function $w$ is determined by (5.4) and (5.29)

$$w(x) = \begin{cases} (1 - x^2)x^2, & |x| \le p \\ \dfrac{|x|}{\rho}\left[\left(1 - \dfrac{\rho^2 x^2}{|x|^2}\right)\dfrac{\rho^2 x^2}{|x|^2}\right] = \rho(1 - \rho^2)|x|, & |x| \in [\rho, 1[. \end{cases} \tag{5.30}$$

It is also positive definite on $\mathcal{S} = \,]-1, +1[$ and differentiable on the set $\mathcal{B} = \{x : x \in \,]-1, 1[, \, |x| \ne \rho < 1\}$. Now, from (5.4), (5.28) and (5.30) it results that

$$v(x) = \begin{cases} \dfrac{1}{2}x^2, & \text{for } |x| \in [0, \rho] \\ \left\{\dfrac{\rho(1 - \rho^2)}{2}\, ln\left[\dfrac{(1 - \rho)(1 + |x|)}{(1 + \rho)(1 - |x|)}\right] + \dfrac{1}{2}\rho^2\right\} & \text{for } |x| \in [\rho, +1[. \end{cases} \tag{5.31}$$

The function $v$ determined by (5.31) is positive definite on $\mathcal{S} = ]-1,+1[$, differentiable on the set $\mathcal{B}$ and $v(x) \to +\infty$ as $x \to \partial\mathcal{S}$, $x \in \mathcal{S}$. We conclude that all the conditions of Theorem 5.1 are satisfied for $\mathcal{N} = \mathcal{S} = ]-1,+1[$. Hence, the set $\mathcal{N}$ is the asymptotic stability domain of $x = 0$ of the system (5.28) that is $\mathcal{N} = \mathcal{D} = ]-1,+1[$.

b) Let the function $p$ be selected in another form,

$$p(x) = x^2. \tag{5.32}$$

Obviously, the function $p$ is differentiable and positive definite on $\mathcal{S} = ]-1,+1[$. The function $w$ is now determined by arbitrarily chosen $\rho \in ]0,1[$, (5.4b) and (5.32):

$$w(x) = \begin{cases} x^2, & \text{for } |x| \in [0,\rho] \\ \dfrac{|x|}{\rho}\left(\dfrac{\rho|x|}{|x|}\right)^2 = \rho|x| & \text{for } |x| \in [\rho,+1[. \end{cases} \tag{5.33}$$

Now, (5.27a), (5.32) and (5.33) yield

$$v(x) = \begin{cases} \dfrac{1}{2} ln \dfrac{1}{1-x^2} & \text{for } |x| \in [0,\rho] \\ \left\{\dfrac{\rho}{2} ln \left[\dfrac{(1+|x|)(1-\rho)}{(1-|x|)(1+\rho)}\right] + \dfrac{1}{2} ln \dfrac{1}{1-\rho^2}\right\} & \text{for } |x| \in [\rho,+1[. \end{cases} \tag{5.34}$$

The function $v$ defined by (5.34) obeys evidently all the conditions of the Theorem 5.1 on $\mathcal{N} = \mathcal{S} = ]-1,+1[$. Hence, the domain $\mathcal{D}$ of asymptotic stability of $x = 0$ of the system (5.1) equals $\mathcal{S} = ]-1,+1[$:

$$\mathcal{D} = ]-1,+1[.$$

**Example 5.3** Let

$$\frac{dx}{dt} = -(x^2-1)x. \tag{5.35}$$

In this case $f(x) = -(x^2-1)x \in C^{(1)}(\Re^1)$. The strong smoothness property is valid on $\Re^1$. Equilibrium states are again $x_{1e} = -1$, $x_{2e} = 0$ and $x_{3e} = +1$ so that $\mathcal{S} = ]-1,+1[$ is selected. The system (5.35) has strong smoothness property also on $\mathcal{S}$. Let the function $p$ be defined by (5.29). Repeating the procedure explained in Example 5.2 under a) we get

$$v(x) = -\frac{1}{2}x^2 \text{ for } |x| \in [0,\rho[, \ \rho \in ]0,1[.$$

Hence, the function $v$ is not positive definite. Theorem 5.2 is not satisfied. Since its conditions are necessary for $x = 0$ to be asymptotically stable then the result is that $x = 0$ is not asymptotically stable. It does not have the asymptotic stability domain. This result and Theorem 3.35 (Section 3.7) show that $x = 0$ does not have the attraction domain, i.e. it is not attractive. In this example it is also unstable.

**Example 5.4** Let

$$\begin{pmatrix} \dot{x}_1 \\ \dot{x}_2 \end{pmatrix} = \begin{pmatrix} -x_1 - x_2\, exp\,(||x||) \\ -x_2 - x_1\, exp\,(||x||) \end{pmatrix}. \tag{5.36}$$

The system (5.36) possesses the weak smoothness property on $\Re^2$. Let $p(x) = 8||x||^2$, so that (5.22) yields $v(x) = 4||x||^2$. Hence, $x = 0$ of the system (5.36) is asymptotically stable (Theorem 5.2).

**Example 5.5** Let

$$\begin{pmatrix} \dot{x}_1 \\ \dot{x}_2 \end{pmatrix} = \begin{pmatrix} -(x_1 + x_2\, sin^2 x_2)(1 + x_1^4) \\ -(1 - 2x_1 x_2)\, sin\, x_2 \end{pmatrix}. \tag{5.37}$$

The system (5.37) has the weak smoothness property. Let $p(x) = 2x_1^2 + sin^2 x_2$. Now, (5.22) yields $v(x) = arctan\, x_1^2 + 1 - cos\, x_2$. All the conditions of Theorem 5.2 are satisfied. Hence, $x = 0$ of the system (5.37) is asymptotically stable.

### 5.2.3 Approach via $\mathcal{O}$-uniquely bounded sets

This section presents results and examples from [84].

#### Introductory comment

The two-stage approach requires the partial differential equation $[grad\, v(X)]^T f(X) = -w(X)$ to be solved twice in order to determine the asymptotic stability domain $\mathcal{D}$. It should be first solved for $w(X) = p(X)$ on $\overline{\mathcal{B}}_\eta$. Afterwards it should be solved for $w(X) = \dfrac{||X||}{\eta}\, p\left(\dfrac{\eta X}{||X||}\right)$. The same procedure holds if $w(X) = p(X)\,[1 - \nu(X)]$ is first accepted to find $\nu$ from $[grad \nu(X)]^T f(X) = -w(X)$ on $\overline{\mathcal{B}}_\eta$. Once it has been done we should solve the same equation for $w(X) = \dfrac{||X||}{\eta}\, p\left(\dfrac{\eta X}{||X||}\right)[1 - \nu(X)]$. In order to avoid such double solving of the partial differential equation we need another approach.

#### Family $Q(\mathcal{S}; f)$

If we replace $\mathcal{B}_\rho$ by $\mathcal{S}$ in Definition 5.2 we get the following:

**Definition 5.3** *(a) A function $q : \Re^n \to \Re$ belongs to the family $Q(\mathcal{S}; f)$ if and only if*

*1) $q$ is differentiable on $\mathcal{S}$: $q(X) \in C^{(1)}(\mathcal{S})$,*

*and*

2) *the equations (5.1), (5.2), or equivalently (5.3), have a solution* $v$ *that is well defined in* $\Re$ *and continuous for every* $X \in \overline{\mathcal{B}}_\mu$ *for some* $\mu \in ]0, +\infty[$, $\mu = \mu(q, f)$*; or equivalently, the equations (5.4) have a solution function* $\nu$ *that is well defined in* $\Re$ *and continuous on* $\overline{\mathcal{B}}_\mu$.

(b) *A function* $q : \Re^n \to \Re$ *belongs to the family* $Q^1(\mathcal{S}; f)$ *if and only if* $q \in Q(\mathcal{S}; f)$ *and a solution* $v$ *to (5.1) and (5.2), or equivalently to (5.3), or equivalently a solution* $\nu$ *to (5.4), is also differentiable on* $\overline{\mathcal{B}}_\mu$.

(c) $Q(\mathcal{S}; f)$ *and* $Q^1(\mathcal{S}; f)$ *will be denoted by* $Q(f)$ *and* $Q^1(f)$ *if and only if* $\mathcal{S} = \Re^n$.

(d) *The argument* $f$ *will be omitted if and only if* $f$ *is given and fixed.*

### Results

**Theorem 5.6** *In order for the state* $X = 0$ *of the system (5.1) with the strong smoothness property to have the domain of asymptotic stability* $\mathcal{D}$ *and for a set* $\mathcal{N}$, $\mathcal{N} \subseteq \Re^n$, *to be the domain of its asymptotic stability,* $\mathcal{N} = \mathcal{D}$, *it is both necessary and sufficient that*

1) *the set* $\mathcal{N}$ *is an open connected neighbourhood of* $X = 0$ *and* $\mathcal{N} \subseteq \mathcal{S}$,

2) $f(X) = 0$ *for* $X \in \mathcal{N}$ *iff* $X = 0$,

*and*

3) *for arbitrarily selected* $\mathcal{O}$*-uniquely bounded set* $\mathcal{U}$, $\overline{\mathcal{U}} \subset \mathcal{S}$, *with the generating function* $u$ *on* $\mathcal{S}$ *obeying* $u \in Q(\mathcal{S}; f)$, *the equations (5.3) for* $q = u$ *have the unique solution function* $v$ *on* $\mathcal{N}$ *with the following properties:*

   (i) $v$ *is positive definite on* $\mathcal{N}$,

   *and*

   (ii) *if the boundary* $\partial\mathcal{N}$ *of* $\mathcal{N}$ *is non-empty then* $v(X) \to +\infty$ *as* $X \to \partial\mathcal{N}$, $X \in \mathcal{N}$,

*or*

4) *for arbitrarily selected* $\mathcal{O}$*-uniquely bounded set* $\mathcal{U}$, $\overline{\mathcal{U}} \subset \mathcal{S}$, *with the generating function* $u$ *on* $\mathcal{S}$ *obeying* $u \in Q^1(\mathcal{S}; f)$, *the equations (5.3) for* $q = u$ *have the unique solution function* $v$ *on* $\mathcal{N}$ *with the following properties:*

   (i) $v$ *is differentiable and positive definite on* $\mathcal{N}$,

   *and*

   (ii) *if the boundary* $\partial\mathcal{N}$ *on* $\mathcal{N}$ *is non-empty then* $v(X) \to +\infty$ *as* $X \to \partial\mathcal{N}$, $X \in \mathcal{N}$.

$\nabla\nabla\nabla$

**Proof.** *Necessity.* Let $X = 0$ of the system (5.1) with the strong smoothness property have the asymptotic stability domain $\mathcal{D}$. Definitions of $\mathcal{D}_a$ and $\mathcal{D}$ show that it has also the attraction domain $\mathcal{D}_a$, $\mathcal{D} \subseteq \mathcal{D}_a$. It is a neighbourhood of $X = 0$ due to definition of $\mathcal{D}_a$, and $\mathcal{S}$ is a neighbourhood of $X = 0$ in view of the smoothness property. Hence, $\mathcal{D}_a \cap \mathcal{S} \neq \emptyset$. Let us prove $\mathcal{D}_a \subseteq \mathcal{S}$. If $\partial\mathcal{S} = \emptyset$ then $\mathcal{S} = \Re^n$ and $\mathcal{D}_a \subseteq \mathcal{S}$ due to $\mathcal{D}_a \subseteq \Re^n$. If $\partial\mathcal{S} \neq \emptyset$ then we shall consider both $X_0 \in \partial\mathcal{S}$ and $X_0^* \in (\Re^n - \overline{\mathcal{S}})$. If $X_0 \in \partial\mathcal{S}$ then $X_0 \notin \mathcal{D}_a$ due to (ii) of the strong smoothness property. Therefore, $\partial\mathcal{S} \cap \mathcal{D} = \emptyset$. If $X_0^* \in (\Re^n - \overline{\mathcal{S}})$ then for $\underline{X}(t; X_0^*) \to 0$ as $t \to +\infty$ it is necessary that there is $t^* \in \Re_+$ such that $X(t^*; X_0^*) \in \partial\mathcal{S}$ because $\mathcal{D}$ and $\mathcal{S}$ are neighbourhoods of $X = 0$, $X_0^* \notin \overline{\mathcal{S}}$, and the motion $\underline{X}(t; X_0)$ is continuous in $t \in \Re_+$ due to (ii) of weak smoothness property ensured by (i) of the strong smoothness property. However, $\underline{X}(t^*; X_0^*) \in \partial\mathcal{S}$ implies that $X(t; X_0)$ does not converge to $X = 0$ because of (ii) of the strong smoothness property. This yields $X_0^* \notin \mathcal{D}$ and $(\Re^n - \overline{\mathcal{S}}) \cap \mathcal{D} = \emptyset$. By connecting all the results, that is that $\mathcal{D}_a \cap \mathcal{S} \neq \emptyset$, $\mathcal{D}_a \cap \partial\mathcal{S} = \emptyset$, $\mathcal{D}_a \cap (\Re^n - \overline{\mathcal{S}}) = \emptyset$ we conclude that $\mathcal{D}_a \subseteq \mathcal{S}$. Therefore, $\mathcal{D} = \mathcal{D}_a$ (Theorem 3.35, Section 3.7) and $\mathcal{D}_a = \mathcal{S}$. Let $\mathcal{N} = \mathcal{D}$ so that $\mathcal{N} \subseteq \mathcal{S}$. Hence, $\mathcal{N}$ is an open connected neighbourhood of $X = 0$ due to (i-b) of the weak smoothness property, $\mathcal{N} = \mathcal{D} = \mathcal{D}_a$, and invariance of $\mathcal{D}_a$ with respect to system motions (Theorem 3.34, Section 3.7). This proves the necessity of the condition 1). From $\mathcal{N} = \mathcal{D} = \mathcal{D}_a \subseteq \mathcal{D}_s$ the result (Theorem 3.28, Section 3.4; Theorem 3.35, Section 3.7) is that $X = 0$ is the unique equilibrium state of the system (5.1) in $\mathcal{N}$, which implies $f(X) = 0$ for $X \in \mathcal{N} = \mathcal{D}$ iff $X = 0$ (Theorem 3.22, Section 3.34) and proves the necessity of the condition 2).

From $\mathcal{N} = \mathcal{D}$ it follows that the interval $I_0$ of existence of $\underline{X}(t; X_0)$ equals $\Re_+$, $I_0 = \Re_+$, for every $X_0 \in \mathcal{N}$ due to definitions of $\mathcal{D}_a$, $\mathcal{D}_s$ and $\mathcal{D}$. Let $\mathcal{U}$ be arbitrarily selected open $\mathcal{O}$-uniquely bounded set obeying both $\overline{\mathcal{U}} \subset \mathcal{N}$ and that its generating function $u$ on $\mathcal{S}$ obeys $u \in Q(\mathcal{S}; f)$. Such a set $\mathcal{U}$ exists because $\mathcal{S}$ is an open neighbourhood of $X = 0$ (Lemma 4.10, Section 4.3.1). Property UB (Section 4.3.2), the same lemma and Definition 4.13 (Section 4.3.2) show that the function $u$ is also positive definite on $\mathcal{S}$. Since $\mathcal{N} = \mathcal{D} \subseteq \mathcal{S}$, then the function $u$ is the differentiable positive definite generating function on $\mathcal{N}$, too. The property of $u \in Q(\mathcal{S}; f)$ guarantees existence of $\mu > 0$ such that there exists a solution function $v$ to the equations (5.3) for $q = u$, which is well defined in $\Re$ and continuous for every $X \in \overline{\mathcal{B}}_\mu$, that is that

$$|v(X)| < +\infty \quad \text{for every} \quad X \in \overline{\mathcal{B}}_\mu. \tag{5.38}$$

Let $\zeta \in ]0, +\infty[$ be such that

$$\overline{\mathcal{U}}_\zeta \subseteq \overline{\mathcal{B}}_\mu. \tag{5.39}$$

The existence of such $\zeta$ is assured by Lemma 4.11 (Section 4.3.3). Let $\tau \in ]0, +\infty[$, $\tau = \tau(X_0; f; u; \zeta)$, be such that for any $X_0 \in \mathcal{N}$ the condition (5.40) holds:

$$\underline{X}(t; X_0) \in \overline{\mathcal{U}}_\zeta, \quad \text{for every} \quad t \in [\tau, +\infty[. \tag{5.40}$$

Such $\tau$ exists in view of definitions of $\mathcal{D}_a$ and $\mathcal{D}$, $\mathcal{D}_a = \mathcal{D}$, $\mathcal{N} = \mathcal{D}$ and $X_0 \in \mathcal{N}$. Notice that $X_0 \in \mathcal{N}$ implies also

$$\underline{X}(+\infty; X_0) = 0. \tag{5.41}$$

After integrating (5.3a) for $q = u$ from $t \in \Re_+$ to $+\infty$ we derive (5.42):

$$v[\underline{X}(+\infty; X_0)] - v[\underline{X}(t; X_0)] = -\int_t^{+\infty} u[\underline{X}(\sigma; X_0)]d\sigma \text{ for every } (t; X_0) \in \Re_+ \times \mathcal{N}. \tag{5.42}$$

Since $u \in Q(\mathcal{S}; f)$ then (5.43) holds:

$$v(0) = 0. \tag{5.43}$$

Now, (5.41)–(5.43) yield (5.44):

$$v[\underline{X}(t; X_0)] = \int_t^{+\infty} u[\underline{X}(\sigma; X_0)]d\sigma \text{ for every } (t; X_0) \in \Re_+ \times \mathcal{N}. \tag{5.44}$$

This can be rewritten in the following form:

$$v[\underline{X}(t; X_0)] = \int_t^{\tau} u[\underline{X}(\sigma; X_0)]d\sigma + \int_{\tau}^{+\infty} u[\underline{X}(\sigma; X_0)]d\sigma \text{ for every } (t; X_0) \in \Re_+ \times \mathcal{N}. \tag{5.45}$$

Positive invariance of $\mathcal{D}$ with respect to system motions, $\mathcal{N} = \mathcal{D}$, continuity of the motions $\underline{X}$ due to the smoothness property, continuity of $u$ on $\mathcal{N}$ in view of its differentiability on $\mathcal{N}$, the definition of $\tau$ (5.40), (5.38), and compactness of $[\tau - t]$ for any $t \in \Re_+$ prove (5.46):

$$\left| \int_t^{\tau} u[\underline{X}(\sigma; X_0)]d\sigma \right| < +\infty \text{ for every } (t; X_0) \in \Re_+ \times \mathcal{N}. \tag{5.46}$$

From (5.38)–(5.40) it follows that (5.47)

$$\left| \int_t^{+\infty} u[\underline{X}(\sigma; X_0)]d\sigma \right| < +\infty \text{ for every } X_0 \in \mathcal{N}. \tag{5.47}$$

Altogether (5.45)–(5.47) prove boundedness of $v[\underline{X}(t; X_0)]$ expressed by (5.48):

$$|v[\underline{X}(t; X_0)]| < +\infty \text{ for every } (t; X_0) \in \Re_+ \times \mathcal{N}. \tag{5.48}$$

Hence, by setting $t = 0$ and $X_0 = X$ in (5.48) we derive (5.49):

$$|v(X)| < +\infty \text{ for every } X \in \mathcal{N}. \tag{5.49}$$

Differentiability of the motion $\underline{X}$ in $X_0 \in \mathcal{N}$, differentiability of $u$ in $X \in \mathcal{N}$, positive invariance of $\mathcal{N} = \mathcal{D}$ with respect to system motions, (5.44) and (5.48) prove continuity of $v$ in $X \in \mathcal{N}$ (5.50):

$$v(X) \in C(\mathcal{N}). \tag{5.50}$$

Let now $u \in Q^1(\mathcal{S}; f)$ be arbitrarily chosen. Let $v_X = (\frac{\partial v}{\partial x_1} \frac{\partial v}{\partial x_2} \cdots \frac{\partial v}{\partial x_n})^T$ and $u_X$ be analogously defined. From (5.45) it follows that $v(X) = \int_0^{\tau} u[\underline{X}(\sigma; X_0)]d\sigma +$

$v(\hat{X})$, $\hat{X} = \underline{X}(\tau; X_0)$. Since $\hat{X} \in \overline{\mathcal{B}}_\mu$ then $u \in Q^1(\mathcal{S}; f)$ implies $v(\hat{X}) \in C^{(1)}(\overline{\mathcal{B}}_\mu)$, which together with b) of the weak smoothness property implies $v[\underline{X}(\tau; X)] \in C^{(1)}$, so that there exists $\alpha(X) \in ]0, +\infty[$, $\forall X \in \mathcal{N}$, such that $\|v_X[\underline{X}(\tau; X_0)]\| < \alpha(X)$, $\forall X \in \mathcal{N}$. Now, a-1) of Definition 5.3, b) of the weak smoothness property and compactness of $[0, \tau]$ imply existence of $\beta(X) \in ]0, +\infty[$, $\forall X \in \mathcal{N}$, such that $\left\| \int_0^\tau u_X[\underline{X}(\sigma; X)] d\sigma \right\| < \beta(X)$, $\forall X \in \mathcal{N}$. Altogether, $\|v_X(X)\| < \alpha(X) + \beta(X) < +\infty$, $\forall X \in \mathcal{N}$. Therefore,

$$v(X) \in C^{(1)}(\mathcal{N}). \tag{5.51}$$

Positive invariance of $\mathcal{N}$ with respect to system motions, positive definiteness of $u$ on $\mathcal{N}$ and (5.44) imply (5.52):

$$v(X) > 0 \text{ for all } (X \neq 0) \in \mathcal{N}. \tag{5.52}$$

Now, (5.43), (5.50) and (5.52) prove the necessity of the condition 3i), and (5.43), (5.51) and (5.52) prove the necessity of the condition 4i).

Let $v_1$ and $v_2$ be two solutions to (5.3). Hence,

$$v_1(X_0) - v_2(X_0) = \int_0^\infty \{u[\underline{X}_1(\sigma; X_0)] - u[\underline{X}_2(\sigma; X_0)]\} d\sigma \text{ for every } X_0 \in \mathcal{N}.$$

Since $u(X)$ is uniquely determined by $X \in \mathcal{N}$ due to (a) of Property UB and Definition 4.13 (Section 4.3.2), and the motion of the system is unique through $X_0$, then $\underline{X}_1(\sigma; X_0) \equiv \underline{X}_2(\sigma; X_0)$ and $u[\underline{X}_1(\sigma; X_0)] \equiv u[\underline{X}_2(\sigma; X_0)]$ so that

$$v_1(X_0) - v_2(X_0) = 0 \text{ for every } X_0 \in \mathcal{N}.$$

This proves the uniqueness of the solution $v$ to (5.3) and completes the proof of 3i) and 4i).

Let $X_1, X_2, \ldots, X_k, \ldots$ be a sequence converging to $\tilde{X}$, $X_k \to \tilde{X}$ as $k \to +\infty$, where $X_k \in \mathcal{N}$ for all $k = 1, 2, \ldots$; $\partial\mathcal{N}$ be non-empty and $\tilde{X} \in \partial\mathcal{N}$. Let $\zeta \in ]0, +\infty[$ be arbitrarily chosen so that $\mathcal{U}_\zeta = \{X : u(X) < \zeta\} \subseteq \mathcal{U}$. Such $\zeta$ exists because the set $\mathcal{U}$ is $\mathcal{O}$-uniquely bounded and the function $u$ is its generating function on $\mathcal{N}$ (Definition 4.12, Property UB, Lemma 4.10, Definition 4.11 and Definition 4.13 (Sections 4.2 and 4.3)). The set $\mathcal{U}_\zeta$ is connected open neighbourhood of $X = 0$ (Lemma 4.11, Section 4.3.3). Let $T_k$, $T_k = T(X_k, \zeta) \in [0, +\infty[$ be the first instant obeying (5.53):

$$\underline{X}(t; X_k) \in \overline{\mathcal{U}}_\zeta \text{ for all } t \in [T_k, +\infty[. \tag{5.53}$$

The existence of such $T_k$ is guaranteed by $X_k \in \mathcal{N}$ and $\mathcal{N} = \mathcal{D}$ (definition of $\mathcal{D}_a$ and definition of $\mathcal{D}$). Continuity of the motions $\underline{X}$ in $(t, X_0) \in \Re_+ \times \mathcal{N}$ due to the strong smoothness property and $\mathring{\mathcal{N}} = \mathcal{N} = \mathcal{D}$ (Theorem 3.35, Section 3.7) and $\mathcal{D} \subseteq \mathcal{S}$ imply $T_k \to +\infty$ as $k \to +\infty$ (Theorem 5.34, Section 3.7). Let $m$ be such a natural number that $X_k \in (\mathcal{N} - \overline{\mathcal{U}}_\zeta)$ for all $k = m, m+1, \ldots$. Such $m$ exists because $\mathcal{N}$ is open, $\overline{\mathcal{U}}_\zeta \subset \mathcal{N}$ and $X_k \to \partial\mathcal{N}$ as $k \to +\infty$. Let $\hat{\alpha}$ be defined by (5.54):

$$\hat{\alpha} = min\,[u(X) : X \in (\mathcal{N} - \mathcal{U}_\zeta)]. \tag{5.54}$$

The $\mathcal{O}$-unique boundedness of the set $\mathcal{U}_\zeta$ and the fact that the function $u$ is its generating function on $\mathcal{N}$ (Lemma 4.11, Section 4.3.3) guarantee [4 and 5 of Property UB and Lemma 4.10 (Section 4.3.2)] that $\hat{\alpha}$ defined by (5.54) satisfies (5.55):

$$\hat{\alpha} = \zeta \in ]0, +\infty[. \tag{5.55}$$

From (5.45) we get (5.56) after replacing $\tau$ by $T_k$:

$$v[\underline{X}(t; X_k)] = \int_t^{T_k} u[\underline{X}(\sigma; X_k)]d\sigma + \int_{T_k}^{+\infty} u[\underline{X}(\sigma; X_k)]d\sigma \text{ for } (t; X_k) \in \Re_+ \times \mathcal{N},$$

$$\text{and for } k = m, m+1, \ldots \tag{5.56}$$

Setting $t = 0$ in (5.56) and using (5.54) and (5.55) we derive

$$v(X_k) \geq \int_0^{T_k} \zeta d\sigma + \int_{T_k}^{+\infty} u[\underline{X}(\sigma; X_k)]d\sigma \text{ for } X_k \in \mathcal{N} \text{ and all } k = m, m+1, \ldots \tag{5.57}$$

Positive invariance of $\mathcal{N} = \mathcal{D}$ with respect to system motions, positive definiteness of $u$ on $\mathcal{N}$ and (5.57) imply

$$v(X_k) \geq \zeta T_k, \text{ for } X_k \in \mathcal{N} \text{ and all } k = m, m+1, \ldots$$

Since $T_k \to +\infty$ as $k \to +\infty$ then

$$v(X_k) \to +\infty \text{ as } X_k \to \partial\mathcal{N} \text{ due to } k \to +\infty,\ X_k \in \mathcal{N},$$

which proves necessity of the condition (3-ii).

*Sufficiency.* Let the conditions of the theorem to be proved hold. The generating function $u$ of the selected $\mathcal{O}$-uniquely bounded set $\mathcal{U}$ is positive definite on $\mathcal{S}$ (Lemma 4.11, Section 4.3.3). Since the function $v$ is positive definite solution to (5.3) on $\mathcal{N}$, and $\mathcal{N}$ is an open connected neighbourhood of $X = 0$, it follows that $X = 0$ is asymptotically stable (Theorem 4.9, Section 4.6.3). Hence, $X = 0$ has the asymptotic stability domain. Since the function $v$ is the solution to (5.3) obtained for positive definite $u \in Q(\mathcal{S}; f)$ then $v(X) \in C(\mathcal{D})$ and $v(X) \to +\infty$ as $X \to \partial\mathcal{D}$ in case $\partial\mathcal{D} \neq \emptyset$ (see the proof of the necessity part). These facts, positive definiteness of $v$ on $\mathcal{N}$, $v(X) \to +\infty$ as $X \to \partial\mathcal{N}$ if $\partial\mathcal{N} \neq \emptyset$, and the fact that both $\mathcal{D}$ and $\mathcal{N}$ are connected neighbourhoods of $X = 0$, prove $\mathcal{D} = \mathcal{N}$ (for details see the proof of sufficiency part of Theorem 5.6).

In view of equivalence of equations (5.3) and (5.4) (Note 5.2) Theorem 5.6 can be set in the following form:

**Theorem 5.7** *In order for the state $X = 0$ of the system (5.1) with the strong smoothness property to have the domain of asymptotic stability $\mathcal{D}$ and for a set $\mathcal{N}$, $\mathcal{N} \subseteq \Re^n$, to be the domain of its asymptotic stability, $\mathcal{N} = \mathcal{D}$, it is both necessary and sufficient that*

1) *the set $\mathcal{N}$ is an open connected neighbourhood of $X = 0$ and $\mathcal{N} \subseteq \mathcal{S}$,*

2) *$f(X) = 0$ for $X \in \mathcal{N}$ iff $X = 0$,*

*and*

3) *for arbitrarily selected $\mathcal{O}$-uniquely bounded set $\mathcal{U}$, $\overline{\mathcal{U}} \subset \mathcal{S}$, with the generating function $u$ on $\mathcal{S}$ obeying $u \in Q(\mathcal{S}; f)$, the equations (5.4) for $q = u$ have the unique solution function $\nu$ on $\mathcal{N}$ with the following properties:*

   (i) *$\nu$ is positive definite on $\mathcal{N}$,*

   (ii) *if the boundary $\partial\mathcal{N}$ of $\mathcal{N}$ is non-empty then $\nu(X) \to 1$ as $X \to \partial\mathcal{N}$, $X \in \mathcal{N}$,*

*or*

4) *for arbitrarily selected $\mathcal{O}$-uniquely bounded set $\mathcal{U}$, $\overline{\mathcal{U}} \subset \mathcal{S}$, with the differentiable generating function $u$ on $\mathcal{S}$ obeying $u \in Q^1(\mathcal{S}; f)$, the equations (5.4) with $q = u$ have the unique solution function $\nu$ on $\mathcal{N}$ with the following properties:*

   (i) *$\nu$ is differentiable and positive definite on $\mathcal{N}$,*

   *and*

   (ii) *if the boundary $\partial\mathcal{N}$ of $\mathcal{N}$ is non-empty then $\nu(X) \to 1$ as $X \to \partial\mathcal{N}$, $X \in \mathcal{N}$.* ∇∇∇

However, the conditions slightly change if the system (5.1) possesses the weak smoothness property rather than the strong smoothness property.

**Theorem 5.8** *For the state $X = 0$ of the system (5.1) possessing the weak smoothness property to have the domain of asymptotic stability $\mathcal{D}$, and for a subset $\mathcal{N}$ of $\mathcal{S}$, $\mathcal{N} \subseteq \mathcal{S}$ to be equal to $\mathcal{D}$: $\mathcal{N} = \mathcal{D}$, it is both necessary and sufficient that*

1) *the set $\mathcal{N}$ is an open connected neighbourhood of $X = 0$,*

2) *$f(X) = 0$ for $X \in \mathcal{N}$ iff $X = 0$,*

*and*

3) *for arbitrarily selected $\mathcal{O}$-uniquely bounded set $\mathcal{U}$, $\overline{\mathcal{U}} \subset \mathcal{S}$, with the generating function $u$ obeying $u \in Q(\Re^n; f)$, the equations (5.3) have the unique solution function $v$ on $\mathcal{N}$ with the following properties:*

   (i) *$v$ is positive definite on $\mathcal{N}$,*

   *and*

   (ii) *if the boundary $\partial\mathcal{N}$ of $\mathcal{N}$ is non-empty then $v(X) \to +\infty$ as $X \to \partial\mathcal{N}$, $X \in \mathcal{N}$,*

*or*

*4) for arbitrarily selected $\mathcal{O}$-uniquely bounded set $\mathcal{U}$, $\overline{\mathcal{U}} \subset \mathcal{S}$, with the generating function $u$ obeying $u \in Q^1(\Re^n; f)$, the equations (5.3) have the unique solution function $v$ on $\mathcal{N}$ with the following properties:*

*(i) $v$ is differentiable and positive definite on $\mathcal{N}$,*

*and*

*(ii) if the boundary $\partial\mathcal{N}$ of $\mathcal{N}$ is non-empty then $v(X) \to +\infty$ as $X \to \partial\mathcal{N}$, $X \in \mathcal{N}$.* $\nabla\nabla\nabla$

**Proof.** *Necessity.* Let the system (5.1) possess the weak smoothness property. Let $X = 0$ have the asymptotic stability domain $\mathcal{D}$ and let $\mathcal{N} \subseteq \mathcal{S}$ be equal to $\mathcal{D}$: $\mathcal{N} = \mathcal{D}$. Let an $\mathcal{O}$-uniquely bounded set $\mathcal{U}$, $\overline{\mathcal{U}} \subset \mathcal{S}$, with generating function $u$ obeying $u \in Q(\Re^n; f)$, $[u \in Q^1(\Re^n; f)]$, be arbitrarily selected. From this point on we have to repeat the proof of Theorem 5.6 in order to show that conditions 1)–3), [1), 2), 4)], respectively, of Theorem 5.8 hold. In such a way we complete easily the proof of the necessity part.

*Sufficiency.* Let the system (5.1) possess the weak smoothness property and the conditions 1)–3), or 1), 2), 4), hold. Then, $X = 0$ of the system (5.1) is asymptotically stable (Theorem 4.9, Section 4.6.3). Therefore, $X = 0$ has the domain $\mathcal{D}$ of asymptotic stability. Let $X_0 \in (\Re^n - \mathcal{N})$. Continuity of $\underline{X}(t; X_0)$ in $t \in I_0$ due to the weak smoothness property, positive definiteness of $u$ on $\Re^n$ (Lemma 4.11, Section 4.3.3), negativeness of $\dot{v}(X)$ on $(\Re^n - \mathcal{N})$ in view of positive definiteness of $u$ on $\Re^n$ and (5.3a), and the condition 3(ii) imply $\underline{X}(t; X_0) \in (\Re^n - \mathcal{N})$ for all $t \in I_0$. Therefore, $\mathcal{D} \subseteq \mathcal{N}$. Furthermore, (5.3a) and positive definiteness of $u$ on $\Re^n$ imply (see the proof of the necessity part of Theorem 5.6) $v(X) \to +\infty$ as $X \to \partial\mathcal{D}$, $X \in \mathcal{D}$, which together with the condition 3i) proves $\partial\mathcal{D} \cap \mathcal{N} = \emptyset$. This result, $\mathcal{D} \subseteq \mathcal{N}$ and the fact that $\mathcal{D}$ is non-empty neighbourhood of $X = 0$ imply $\mathcal{D} = \mathcal{N}$ and complete the proof.

In Theorem 5.8 it is prespecified that the set $\mathcal{N}$ is a subset of the set $\mathcal{S}$ because the weak smoothness property only is required rather than the strong smoothness property.

If we prefer to solve (5.4) rather than (5.3) then we use the following criterion that follows from Theorem 5.8 due to equivalency between (5.3) and (5.4).

**Theorem 5.9** *For the state $X = 0$ of the system (5.1) possessing the weak smoothness property to have the domain of asymptotic stability $\mathcal{D}$ and for a subset $\mathcal{N}$ of $\mathcal{S}$, $\mathcal{N} \subseteq \mathcal{S}$, to be equal to $\mathcal{D}$: $\mathcal{N} = \mathcal{D}$, it is both necessary and sufficient that*

*1) the set $\mathcal{N}$ is an open connected neighbourhood of $X = 0$,*

*2) $f(X) = 0$ for $X \in \mathcal{N}$ iff $X = 0$,*

*and*

*3) for arbitrarily selected $\mathcal{O}$-uniquely bounded set $\mathcal{U}$, $\overline{\mathcal{U}} \subset \mathcal{S}$, with the generating function $u$ obeying $u \in Q(\Re^n; f)$, the equations (5.4) have the unique solution function $\nu$ on $\mathcal{N}$ with the following properties:*

*(i) $\nu$ is positive definite on $\mathcal{N}$,*

*and*

*(ii) if the boundary $\partial\mathcal{N}$ of $\mathcal{N}$ is non-empty then $\nu(X) \rightarrow 1$ as $X \rightarrow \partial\mathcal{N}$, $X \in \mathcal{N}$,*

*or*

*4) for arbitrarily selected $\mathcal{O}$-uniquely bounded set $\mathcal{U}$, $\overline{\mathcal{U}} \subset \mathcal{S}$, with the generating function $u$ on $\Re^n$ obeying $u \in Q^1(\Re^n; f)$, the equations (5.4) have the unique solution function $\nu$ on $\mathcal{N}$ with the following properties:*

*(i) $\nu$ is differentiable and positive definite on $\mathcal{N}$,*

*and*

*(ii) if the boundary $\partial\mathcal{N}$ of $\mathcal{N}$ is non-empty then $\nu(X) \rightarrow 1$ as $X \rightarrow \partial\mathcal{N}$, $X \in \mathcal{N}$.* $\nabla\nabla\nabla$

## Examples

**Example 5.6** Let $n = 1$ and $f(X) = -sin\, X$ so that (5.1) becomes (5.58),

$$\frac{dX}{dt} = -sin\, X. \tag{5.58}$$

The function $f$ is differentiable on $\Re^1$. Hence, the system (5.58) possesses the strong smoothness property on $\Re^1$.

The equilibrium states are determined by $X_{ek} = k\pi$, $k = 0, \pm 1, \pm 2, \ldots$ This suggests $\mathcal{S} = ]-\pi, \pi[$ and $\mathcal{U} = \left\{X : sin^2\left(\frac{1}{2}X\right) < \alpha\right\}$ for $\alpha \in ]0, 1[$. The system has the strong smoothness property also on $\mathcal{S}$. The function $u$ defined by $u(X) = sin^2\left(\frac{1}{2}X\right)$ is the differentiable generating function on $\mathcal{S}$ of the $\mathcal{O}$-uniquely bounded set $\mathcal{U}$. Besides, the function $u$ is positive definite on $\mathcal{S}$. The definition of the function $u$, (5.3a) and (5.58) yield

$$\frac{dv}{dX} = \frac{1}{2} tan\left(\frac{1}{2}X\right),$$

which has the solution obeying $v(0) = 0$ (5.3b) determined by

$$v(X) = -\,ln\, cos\left(\frac{1}{2}X\right).$$

Evidently, $u \in Q(]-\pi, \pi[; -sin(\cdot))$. The function $v$ is differentiable and positive definite on $\mathcal{S}$ and $v(X) \rightarrow +\infty$ as $X \rightarrow \partial\mathcal{S} = \{-\pi, \pi\}$, $X \in \mathcal{S}$. All the conditions of Theorem 5.6 are satisfied for $\mathcal{N} = \mathcal{S} = ]-\pi, \pi[$. Hence, the set $\mathcal{D} = \mathcal{N} = \mathcal{S} = ]-\pi, \pi[$ is the domain of asymptotic stability of $X = 0$ of the system (5.58). Notice that

$f^2(X)$ is differentiable and positive definite on $\mathcal{N} = \mathcal{S} = ]-\pi, \pi[$ but $u(X) \neq f^2(X)$ in this example because $f^2(X) = sin^2 X$ cannot be the generating function on $\mathcal{S}$ of any $\mathcal{O}$-uniquely bounded set. It is not strictly increasing in $X$ for $|X| \in \left[\frac{\pi}{2}, \pi\right[$. Notice also that $X_e = -\pi$ and $X_e = \pi$ are unstable and not attractive due to the domain of asymptotic stability of $X = 0$: $\mathcal{D} = ]-\pi, \pi[$.

If we use Theorem 5.7 then the equations (5.4) take the following form:

$$\frac{d\nu}{1-\nu} = \frac{1}{2} tan\left(\frac{1}{2}X\right) dx$$

and

$$\nu(0) = 0.$$

The solution $\nu(X) = 1 - cos\left(\frac{1}{2}X\right)$ obeys all the conditions of Theorem 5.7.

Motions of system (5.58) for $|X_0| < \pi$ are defined by

$$\underline{X}(t; X_0) = 2\, arctan\left[exp\,(-t)\, tan\left(\frac{1}{2}X_0\right)\right],\ |X_0| < \pi.$$

It is now simple to verify the above result.

**Example 5.7** Let $f(X) = sin\, X$ and

$$\frac{dX}{dt} = sin\, X. \tag{5.59}$$

This system possesses the strong smoothness property on $\mathcal{S} = \Re^1$. Its equilibrium states are $X_{ek} = k\pi$, $k = 0, \pm1, \pm2, \ldots$ Hence, $\mathcal{N} = ]-\pi, \pi[$ is accepted and $\mathcal{U} = \{X : sin^2\left(\frac{1}{2}X\right) < \alpha\}$, $\alpha \in ]0, 1[$. By repeating the procedure explained in Example 5.6 we derive

$$v(X) = ln\, cos\left(\frac{1}{2}X\right).$$

The function $v$ is negative definite on $\mathcal{N}$. Thus, it does not satisfy either Theorem 5.6 or Theorem 5.7, which shows that $X = 0$ of the system (5.59) does not have the asymptotic stability domain $\mathcal{D} \subseteq \Re^1$, or equivalently that it is not asymptotically stable. In fact, it is unstable and not attractive.

**Example 5.8** Let

$$\frac{dX}{dt} = -tan\, X. \tag{5.60}$$

The system possesses the strong smoothness property on $\mathcal{S} = \left]-\frac{\pi}{2}, \frac{\pi}{2}\right[$ but not on $\Re^1$. The equilibrium state $X = 0$ is unique in $\mathcal{S}$. This and the form of $|f(X)| = |tan\, X|$ suggest $\mathcal{U} = \{X : tan^2 X < 1\} \subset \mathcal{S}$. The generating function on $\mathcal{S}$ of the

$\mathcal{O}$-uniquely bounded set $\mathcal{U}$ is $u(X) = tan^2 X$. It is differentiable on $\mathcal{S}$. Now, (5.3a), (5.60) and the definition of $u$ yield

$$\frac{dv}{dX} = tan\, X.$$

Its solution satisfying $v(0) = 0$ (5.3b) is found as

$$v(X) = -\, ln\, cos\, X. \tag{5.61}$$

This shows that $u \in Q\left(\left]-\frac{\pi}{2}, \frac{\pi}{2}\right[; -tan(\cdot)\right)$. The function $v$ (5.61) is differentiable and positive definite on $\mathcal{S}$ and

$$v(X) \to +\infty \text{ as } X \to \partial\mathcal{S},\ X \in \mathcal{S},\ \partial\mathcal{S} = \left\{-\frac{\pi}{2}, \frac{\pi}{2}\right\}.$$

All the conditions of Theorem 5.6 are satisfied for $\mathcal{N} = \mathcal{S}$. Hence, $\mathcal{D} = \left]-\frac{\pi}{2}, \frac{\pi}{2}\right[$ is the domain of asymptotic stability of $X = 0$ of the system (5.60). In this example $u(X) = f^2(X)$.

Motions of the system are defined by:

$$\underline{X}(t; X_0) = \begin{cases} X_0 \text{ for } X_0 = k\pi, \\ arcsin\,(exp\,(-t)\, sin\, X_0),\ X_0 \in \left]\frac{2k-1}{2}\pi, \frac{2k+1}{2}\pi\right[, k = 0, \pm 1, \ldots \end{cases}$$

Using this it is easy to verify $\mathcal{D} = \left]-\frac{\pi}{2}, \frac{\pi}{2}\right[$.

**Example 5.9** The function $f$ defined by $f(X) = (X^2 - 4)X$ is differentiable on $\Re^1$. Therefore, the system (5.62),

$$\frac{dX}{dt} = (X^2 - 4)X, \tag{5.62}$$

possesses the strong smoothness property on $\Re^1$. The equilibrium states are: $X_{e1} = -2$, $X_{e2} = 0$ and $X_{e3} = 2$. They suggest $\mathcal{S} = ]-2, 2[$. Hence, the system (5.62) has the strong smoothness property also on $\mathcal{S}$. The form of the function $f$ suggest $\mathcal{U} = \{X : X^2 < \alpha\}$ for $\alpha \in ]0, 2[$ so that the generating function on $\mathcal{S}$ of the $\mathcal{O}$-uniquely bounded set $\mathcal{U}$ is $u(X) = X^2$. It is differentiable and positive definite on $\mathcal{S}$. The definition of the function $u$, (5.3a) and (5.62) result in

$$\frac{dv}{dX} = \frac{X}{4 - X^2}. \tag{5.63}$$

The solution of (5.3) obeying $v(0) = 0$ (5.3b) is derived as

$$v(X) = -\frac{1}{2}\, ln\left[\frac{4 - X^2}{4}\right]. \tag{5.64}$$

Hence, $u \in Q(]-2, 2[; (X^2 - 4)X)$. The function $v$ (5.64) is positive definite and differentiable on $\mathcal{S} = ]-2, 2[$ and $v(X) \to +\infty$ as $X \to \partial\mathcal{S} = \{-2, 2\}$, $X \in \mathcal{S}$. All the conditions of Theorem 5.6 are satisfied for $\mathcal{N} = \mathcal{S} = ]-2, 2[$. The set $\mathcal{N} = ]-2, 2[$ is the domain of asymptotic stability of $X = 0$ of the system (5.63) (Theorem 5.6).

**Example 5.10** Let

$$\frac{dX}{dt} = -X(1 + X^4). \tag{5.65}$$

Since the function $f$, $f(X) = -X(1 + X^4)$, is differentiable on $\Re^1$ then the system (5.65) has the strong smoothness property on $\mathcal{S} = \Re^1$. It has the unique equilibrium state $X_e = 0$. Let $\mathcal{U} = \{X : X^2 < 1\}$. The function $u$, $u = X^2$, is the generating function on $(\mathcal{S} = \Re^1)$ of the $\mathcal{O}$-uniquely bounded set $\mathcal{U}$. The solution of the equations (5.3) in this case is

$$v(X) = \frac{1}{2} \, arctan \, X^2. \tag{5.66}$$

Hence, $u \in Q(\Re^1; f)$. The function $v$ obeys all the conditions of Theorem 5.6 on $\mathcal{N} = \mathcal{S} = \Re^1$. Hence, the asymptotic stability domain $\mathcal{D}$ of $X = 0$ of the system (5.65) is the whole state space,

$$\mathcal{D} = \Re^1,$$

which shows that $X = 0$ is globally asymptotically stable. In this connection it is emphasized that the system Lyapunov function $v$ (5.66) is not radially unbounded because

$$v(X) \to \frac{\pi}{4} \quad \text{as} \quad |X| \to +\infty.$$

This result does not contradict the well-known Barbashin–Krasovskii criterion for the global asymptotic stability [6], which requires existence of radially unbounded globally positive definite function $v$ with globally negative definite derivative $\dot{v}$. However, it does not require that every function $v$ proving the global asymptotic stability is radially unbounded. In order to show that Theorem 5.6 leads to a radially unbounded Lyapunov function under an appropriate choice of $\mathcal{U}$, let $\mathcal{U} = \{X : 4X^4 < 1\}$. The generating function of $\mathcal{O}$-uniquely bounded set $\mathcal{U}$ is $u_1(X) = 4X^4$, which with the equations (5.3) implies

$$v_1(X) = ln\left(1 + X^4\right).$$

Hence, $u_1 \in Q(\Re^1; f)$. This function $v_1$ obeys also all the conditions of Theorem 5.6 on $\mathcal{N} = \mathcal{S} = \Re^1$. In addition it is radially unbounded, that is

$$v_1(X) \to +\infty \quad \text{as} \quad |X| \to +\infty.$$

Radial unboundedness of $v$ is not a necessary condition for global asymptotic stability in case $v$ is constructed via Theorem 5.6 – Theorem 5.9.

**Example 5.11** A second order system described by (5.67),

$$\begin{aligned} \frac{dX_1}{dt} &= -\frac{X_1(1 - X_1^2)^2(1 - X_2^2)}{1 + X_2^2}, \\ \frac{dX_2}{dt} &= -\frac{X_2(1 - X_1^2)(1 - X_2^2)^2}{1 + X_1^2}, \end{aligned} \tag{5.67}$$

has the strong smoothness property on $\Re^2$ because $f : \Re^2 \to \Re^2$,

$$f(X) = \begin{bmatrix} -\dfrac{X_1\left(1-X_1^2\right)^2\left(1-X_2^2\right)}{1+X_2^2} \\ -\dfrac{X_2\left(1-X_1^2\right)\left(1-X_2^2\right)^2}{1+X_1^2} \end{bmatrix} \in C^{(1)}(\Re^2). \tag{5.68}$$

Obviously the set $\mathcal{S}_e$ of the equilibrium states is determined by

$$\mathcal{S}_e = \{X : (X = 0) \ \text{ or } \ (X_1 = \pm 1, X_2 \in \Re) \ \text{ or } \ (X_1 \in \Re, X_2 = \pm 1)\}.$$

Let $\mathcal{S} = \{X : |X_1| < 1, |X_2| < 1\}$. Then, $X = 0$ is the unique equilibrium state of the system (5.67) in $\mathcal{S}$, and the system possesses the strong smoothness property also on $\mathcal{S}$. By referring to (5.67) and the set $\mathcal{S}$ we accept $\mathcal{U} = \mathcal{B}_1 = \{X : ||X||^2 < 1\} \subset \mathcal{S}$. The set $\mathcal{U}$ is $\mathcal{O}$-uniquely bounded with the differentiable generating function $u$ on $\mathcal{S} : u(X) = ||X||^2$. Such a choice of the function $u$ and (5.3) leads to

$$v(X) = \frac{1}{2}\frac{X_1^2 + X_2^2}{(1-X_1^2)(1-X_2^2)}. \tag{5.69}$$

Hence, $u \in Q(\mathcal{S}; f)$. The function $v$ is differentiable and positive definite on $\mathcal{S}$ and

$$v(X) \to +\infty \ \text{ as } \ X \to \partial\mathcal{S}, \ \ \partial\mathcal{S} = \big\{X : (|X_1| = 1, \ |X_2| \le 1) \tag{5.70}$$
$$\text{or} \ \ (|X_1| \le 1, \ |X_2| = 1)\big\}.$$

The set $\mathcal{N} = \mathcal{S}$ and the function $v$ (5.69) obey all the conditions of Theorem 5.6, which shows that

$$\mathcal{D} = \{X : |X_1| < 1, \ |X_2| < 1\}.$$

**Example 5.12** Let

$$\dot{X}_1 = -X_1(4 - X_1^2)(1 + X_2^2), \qquad \dot{X}_2 = -X_2(4 - X_2^2)(1 + X_1^2). \tag{5.71}$$

The system (5.70) satisfies the requirements for the strong smoothness property on $\Re^2$. The set $\mathcal{S}_e$ of the equilibrium states is found in the following form

$$\mathcal{S}_e = \{X : X_1 \in \{-2, 0, 2\} \ \text{ and } \ X_2 \in \{-2, 0, 2\}\}.$$

Let $\mathcal{S} = \{X : |X_1| < 2, |X_2| < 2\}$. The system $(\mathcal{S})$ possesses the strong smoothness property also on $\mathcal{S}$. Let $\mathcal{U} = \{X : 2[X_1^2(1 + X_2^2) + X_2^2(1 + X_1^2)] < 2\} \subset \mathcal{S}$ be selected. It is $\mathcal{O}$-uniquely bounded set with the differentiable generating function $u$ on $\mathcal{S} : u(X) = 2[X_1^2(1 + X_2^2) + X_2^2(1 + X_1^2)]$. Such a form of the function $u$ and (5.3) imply

$$v(X) = -\,ln\left[\frac{(4 - X_1^2)(4 - X_2^2)}{16}\right]. \tag{5.72}$$

Hence, $u \in Q(\mathcal{S}; f)$. The function $v$ (5.71) is differentiable and positive definite on $\mathcal{S}$. Besides,

$$v(X) \to +\infty \text{ as } X \to \partial\mathcal{S}, \;\; \partial\mathcal{S} = \{X : (|X_1| = 2, \; |X_2| \leq 2)$$
$$\text{or} \;\; (|X_1| \leq 2, |X_2| = 2)\}.$$

The function $v$ (5.71) and the set $\mathcal{N} = \mathcal{S}$ obey all the conditions of Theorem 5.6 which verifies that

$$\mathcal{D} = \{X : |X_1| < 2 \text{ and } |X_2| < 2\}.$$

**Example 5.13** The system (5.72),

$$\frac{dX_1}{dt} = -sin\, X_1\, cos^3 X_1\, cos^2 X_2, \qquad \frac{dX_2}{dt} = -sin\, X_2\, cos^2 X_1\, cos^3 X_2, \tag{5.73}$$

obeys all the requirements for the strong smoothness property on $\Re^2$. The set $\mathcal{S}_e$ of all the equilibrium states is found in the form:

$$\mathcal{S}_e = \{X : X_i \in \{k\pi : k = 0, \pm1, \ldots\} \cup \left\{\frac{2k+1}{2}\pi : k = 0, \pm1, \ldots\right\}, i = 1, 2\}.$$

Let $\mathcal{S} = \{X : |X_1| < \frac{\pi}{2}, |X_2| < \frac{\pi}{2}\}$. The system (5.72) has the strong smoothness property on $\mathcal{S}$. Let $\mathcal{U} = \{X : 2[sin^2 X_1(1 + sin^2 X_2) + sin^2 X_2(1 + sin^2 X_1)] < 2\}$. It is $\mathcal{O}$-uniquely bounded set with the differentiable generating function $u$ on $\mathcal{S}$ : $u(X) = 2[sin^2 X_1(1 + sin^2 X_2) + sin^2 X_2(1 + sin^2 X_1)]$. Hence, the function $v$ (5.3) is found in the following form:

$$v(X) = \frac{sin^2 X_1 + sin^2 X_2}{cos^2 X_1\, cos^2 X_2}. \tag{5.74}$$

Therefore, $u \in Q(\mathcal{S}; f)$. The function $v$ (5.73) is differentiable and positive definite on $\mathcal{S}$. Furthermore,

$$v(X) \to +\infty \text{ as } X \to \partial\mathcal{S}, \;\; \partial\mathcal{S} = \left\{X : \left(|X_1| = \frac{\pi}{2}, \; |X_2| \leq \frac{\pi}{2}\right)\right.$$
$$\left.\text{or} \;\; \left(|X_1| \leq \frac{\pi}{2}, \; |X_2| = \frac{\pi}{2}\right)\right\}.$$

The function $v$ and the set $\mathcal{N} = \mathcal{S}$ satisfy all the conditions of Theorem 5.6. The domain $\mathcal{D}$ of asymptotic stability of $X = 0$ of the system (5.72) is the set $\mathcal{S}$:

$$\mathcal{D} = \left\{X : |X_1| < \frac{\pi}{2}, \; |X_2| < \frac{\pi}{2}\right\}.$$

**Comment 5.1** *The classical forms of a Lyapunov function are*

- *quadratic form*

$$v(X) = X^T H X,$$

- *Lurie form (Section 4.11.2, (5.32))*

$$v^f(X) = (X^T f^T)H(X^T f^T)^T + \int_0^w f^T(w)\Theta dw,$$

*and*

- *the natural form (Section 4.4.1, Example 4.16)*

$$v(X) = \mathsf{b}^T|X|.$$

*However, Lyapunov functions constructed directly very often do not have a classical form as shown by (5.34), by*

$$v(X) = -\ln\cos\left(\frac{1}{2}X\right) \quad \text{and} \quad \nu(X) = 1 - \cos\left(\frac{1}{2}X\right),$$

*of Example 5.6, by (5.64), by (5.66), by (5.69), by (5.71) and by (5.73).*

### 5.2.4 General one-shot approach

What follows is deduced from [81] by accepting $\mathcal{I} = \{0\}$, and from [83].

#### Introductory comments

Like the $\mathcal{O}$-uniquely bounded set approach, the general one-shot approach is aimed to provide the necessary and sufficient conditions for a system Lyapunov function construction with a simultaneous exact determination of the asymptotic stability domain by solving $\dot{v} = -p$, or equivalently $\dot{\nu} = -(1-\nu)p$, only once. Unlike the former the latter does not require use of an $\mathcal{O}$-uniquely bounded set that gives the following property to the generating function $u$:

$$\min\left[u(X) : X \in (\mathcal{S} - \mathcal{B}_\alpha)\right] \geq \beta > 0,$$

for every $\alpha \in \Re^+$ for which $\mathcal{S}$ is neighbourhood of $\mathcal{B}_\alpha$. This property of $u$ is used to prove that $v(x) \to +\infty$ (or equivalently $\nu(X) \to 1$) as $X \to \partial\mathcal{N}$, $X \in \mathcal{N}$, in case $\partial\mathcal{N}$ is nonempty. This property of $u$ will be used in what follows, too.

#### Family $Q(\mathcal{S}; f)$

In the one-shot approach the family $Q(\mathcal{S}, f)$ is determined differently than in the preceding approaches.

**Definition 5.4** *(a) $Q(\mathcal{S}; f)$ is the family of all the functions $q : \Re^n \to \Re$ with the following properties:*

*1) $q$ is continuous on $\mathcal{S}$,*

*2) for any $\alpha > 0$ such that $\mathcal{S}$ is a neighbourhood of $\mathcal{B}_\alpha = \{X : ||X|| < \alpha\}$, there is $\beta > 0$, $\beta = \beta(\alpha, q, \mathcal{S})$, obeying $inf\ [q(X) : X \in (\mathcal{S} - \mathcal{B}_\alpha] = \beta$,*

*3) there is $\mu \in\ ]0, +\infty[$, $\mu = \mu(q; f)$, such that there exists a solution $v : \Re^n \to \Re$ to the equations (5.1) and (5.2), or (5.3), with (5.2a), or (5.3a), determined along motions of the system (5.1), which is defined and bounded on $\overline{\mathcal{B}}_\mu$.*

*(b) $Q^1(\mathcal{S}; f)$ is the family of all the functions $q \in Q(\mathcal{S}; f)$ for which the solution $v$ to (5.1) and (5.2), or (5.3), is continuously differentiable on the closure $\overline{\mathcal{B}}_\mu$ of $\mathcal{B}_\mu$.*

**Results**

**Theorem 5.10** *In order for the state $X = 0$ of the system (5.1) with the strong smoothness property to have the domain of asymptotic stability $\mathcal{D}$ and for a set $\mathcal{N}$ to be the domain of its asymptotic stability, $\mathcal{N} = \mathcal{D}$ it is both necessary and sufficient that*

*1) the set $\mathcal{N}$ is an open connected neighbourhood of $X = 0$ and $\mathcal{N} \subseteq \mathcal{S}$,*

*2) $f(X) = 0$ for $X \in \mathcal{N}$ iff $X = 0$,*

*and*

*3) for arbitrary positive definite $q \in Q(\mathcal{S}; f)$, the equations (5.3) with (5.3a) determined along system motions have the unique solution function $v$ with the following features:*

*(i) $v$ is positive definite on $\mathcal{N}$,*

*and*

*(ii) if the boundary $\partial\mathcal{N}$ of $\mathcal{N}$ is non-empty then $v(X) \to +\infty$ as $X \to \partial\mathcal{N}$, $X \in \mathcal{N}$,*

*or*

*4) for arbitrary positive definite $q \in Q^1(\mathcal{S}; f)$, the equations (5.3) with (5.3a) determined along system motions have the unique solution function $v$ with the following features:*

*(i) $v$ is positive definite and differentiable on $\mathcal{N}$,*

*and*

*(ii) if the boundary $\partial\mathcal{N}$ of $\mathcal{N}$ is non-empty then $v(X) \to +\infty$ as $X \to \partial\mathcal{N}$, $X \in \mathcal{N}$.* $\nabla\nabla\nabla$

**Proof.** *Necessity.* Let $X = 0$ of the system (5.1) with the strong smoothness property have the asymptotic stability domain $\mathcal{D}$. Hence, it has also the attraction domain $\mathcal{D}_a$, $\mathcal{D} \subseteq \mathcal{D}_a$. It is a neighbourhood of $X = 0$ in view of the strong smoothness property. Hence, $\mathcal{D}_a \cap \mathcal{S} \neq \emptyset$. Let us prove $\mathcal{D}_a \subseteq \mathcal{S}$. If $\partial\mathcal{S} = \emptyset$ then $\mathcal{S} = \Re^n$ and $\mathcal{D}_a \subseteq \mathcal{S}$ due to $\mathcal{D}_a \subseteq \Re^n$. If $\partial\mathcal{S} \neq \emptyset$ then we shall consider $X_0 \in \partial\mathcal{S}$

and $X_0^* \in (\Re^n - \overline{\mathcal{S}})$. If $X_0 \in \partial\mathcal{S}$ then $X_0 \notin \mathcal{D}_a$ due to (ii) of the strong smoothness property. Therefore, $\partial\mathcal{S} \cap \mathcal{D} = \emptyset$. If $X_0^* \in (\Re^n - \overline{\mathcal{S}})$ then for $\underline{X}(t; X_0^*) \to 0$ as $t \to +\infty$ it is necessary that there is $t^* \in \Re_+$ such that $\underline{X}(t^*; X_0^*) \in \partial\mathcal{S}$ because $\mathcal{D}$ and $\mathcal{S}$ are neighbourhoods of $X = 0$, $X_0^* \notin \overline{\mathcal{S}}$ and the motion $\underline{X}(t; X_0^*)$ is continuous in $t \in \Re_+$ due to (ii) of the weak smoothness property implied by (i) of the strong smoothness property. However, $\underline{X}(t^*; X_0^*) \in \partial\mathcal{S}$ shows that $\underline{X}(t; X_0^*)$ does not converge to $X = 0$ because of (ii) of the strong smoothness property. This yields $X_0^* \notin \mathcal{D}$ and $(\Re^n - \overline{\mathcal{S}}) \cap \mathcal{D} = \emptyset$. By connecting all the results, that is that $\mathcal{D}_a \cap \mathcal{S} \neq \emptyset$, $\mathcal{D}_a \cap \partial\mathcal{S} = \emptyset$ and $\mathcal{D}_a \cap (\Re^n - \overline{\mathcal{S}}) = \emptyset$ we conclude that $\mathcal{D}_a \subseteq \mathcal{S}$. Therefore, $\mathcal{D} = \mathcal{D}_a$ and $\mathcal{D} \subseteq \mathcal{S}$. Let $\mathcal{N} = \mathcal{D}$ so that $\mathcal{N} \subseteq \mathcal{S}$. Hence, $\mathcal{N}$ is open connected neighbourhood of $X = 0$ due to (i-b) of the weak smoothness property, $\mathcal{N} = \mathcal{D} = \mathcal{D}_a$, and invariance of $\mathcal{D}_a$ with respect to system motions. This proves necessity of the condition 1). From $\mathcal{N} = \mathcal{D} = \mathcal{D}_a \subseteq \mathcal{D}_s$ it follows that $X = 0$ is the unique equilibrium state of the system (5.1) in $\mathcal{N}$, which implies $f(X) = 0$ for $X \in \mathcal{N} = \mathcal{D}$ iff $X = 0$ and proves the necessity of the condition 2).

Since $\mathcal{N} = \mathcal{D}$ then $I_0 = \Re_+$, for every $X_0 \in \mathcal{N}$. Let $q \in Q(\mathcal{S}; f)$ be arbitrarily chosen. Then, the equation (5.3a) has a solution $v$ defined on $\overline{\mathcal{B}}_\mu$ in view of (a-3) of Definition 5.4. Since $\mu < +\infty$ due to Definition 5.4 then $\overline{\mathcal{B}}_\mu$ is compact. This and (a-1) of Definition 5.4 imply (5.74),

$$|v(x)| < +\infty \quad \text{for every} \quad X \in \overline{\mathcal{B}}_\mu. \tag{5.75}$$

Let $X_0 \in \mathcal{N}$, $\eta \in \,]0, \mu]$, $\mathcal{B}_\eta \subseteq \mathcal{N}$, and $\tau \in [0, +\infty[$, $\tau = \tau(X_0, \eta)$, obey (5.75)

$$\underline{X}(t; X_0) \in \overline{\mathcal{B}}_\eta \quad \text{for all} \quad t \in [\tau, +\infty[. \tag{5.76}$$

Such $\tau$ exists because $X_0 \in \mathcal{N}$ and $\mathcal{N} = \mathcal{D} = \mathcal{D}_a$. Let (5.3a) be integrated from $t \in \Re_+$ to $+\infty$ so that

$$v[\underline{X}(\infty; X_0)] - v[\underline{X}(t; X_0)] = -\int_t^{+\infty} q[\underline{X}(\sigma; X_0)]d\sigma \quad \text{for every} \quad (t; X_0) \in \Re_+ \times \mathcal{N}. \tag{5.77}$$

$\mathcal{N} = \mathcal{D}$ and $X_0 \in \mathcal{N}$ imply $X(\infty; X_0) = 0$. Let, in view of (5.2b),

$$v(0) = 0, \tag{5.78}$$

so that (5.76) and (5.77) yield (5.78),

$$v[\underline{X}(t; X_0)] = \int_t^{\tau} q[\underline{X}(\sigma; X_0)]d\sigma + \int_\tau^{+\infty} q[\underline{X}(\sigma; X_0)]d\sigma \;\text{for every}\; (t; X_0) \in [0, \tau] \times \mathcal{N}. \tag{5.79}$$

Positive definiteness of $q$ on $\mathcal{N}$, invariance of $\mathcal{D}_a$ with respect to system motions, $\mathcal{N} = \mathcal{D} = \mathcal{D}_a$, $\mathcal{B}_\eta \subseteq \mathcal{N} \cap \mathcal{B}_\mu$, (i) of the weak smoothness property, compactness of $[t, \tau]$ for every $t \in [0, \tau]$ and $\tau \in \Re_+$ yield

$$\left| \int_t^{\tau} q[\underline{X}(\sigma; X_0)]d\sigma \right| < +\infty \quad \text{for every} \quad (t, X_0) \in [0, \tau] \times \mathcal{N},$$

which together with $\eta > 0$ (5.74), (5.75) and (5.78) proves

$$|v[\underline{X}(t;X_0)]| < +\infty \text{ for every } (t, X_0) \in \Re_+ \times \mathcal{N}. \tag{5.80}$$

Hence, for $(t = 0) \in \Re_+$,

$$|v(X_0)| < +\infty \text{ for every } X_0 \in \mathcal{N}. \tag{5.81}$$

The invariance of $\mathcal{N} = \mathcal{D} = \mathcal{D}_a$, the weak smoothness property, $\mathcal{N} \subseteq \mathcal{S}$, continuity of $q$ on $\mathcal{S}$ due to $q \in Q(\mathcal{S}; f)$ and (5.80) prove continuity of the function $v$ on $\mathcal{N}$:

$$v(X) \in C(\mathcal{N}). \tag{5.82}$$

Now, (5.76), (5.77) and (5.80) complete the proof of the existence of a solution function $v$ to (5.2) obeying (5.81) for every $q \in Q(\mathcal{S}; f)$. Let now $q \in Q^{(1)}(\mathcal{S}; f)$ be arbitrarily chosen. Let $v_X = (\partial v/\partial X_1 \ldots \partial v/\partial X_2)^T$ and $q_X$ be analogously defined. From (5.78) it follows that $v(X) = \int_0^\tau q[\underline{X}(\sigma; X)]d\sigma + v(\hat{X})$, $\hat{X} = \underline{X}(\tau; X_0)$. Since $\hat{X} \in \overline{\mathcal{B}}_\mu$ then $q \in Q^{(1)}(\mathcal{S}; f)$ implies $v(X) \in C^{(1)}(\overline{\mathcal{B}}_\mu)$ which together with b) of the weak smoothness property implies $v[\underline{X}(\tau; X_0)] \in C^{(1)}(\mathcal{N})$ so that there is $\alpha(X) \in ]0, +\infty[$, $\forall X \in \mathcal{N}$, such that $\|v_X[\underline{X}(\tau; X_0)]\| < \alpha(X)$, $\forall X \in \mathcal{N}$. Now, 1) of Definition 5.4, b) of the weak smoothness property and compactness of $[0, \tau]$ imply existence of $\beta(X) \in ]0, +\infty[$, $\forall X \in \mathcal{N}$, such that $\| \int_0^\tau q_X[\underline{X}(\sigma; X_0)]d\sigma\| < \beta(X)$, $\forall X \in \mathcal{N}$. Altogether, $\|v_X(X)\| < \alpha(X) + \beta(X) < +\infty$, $\forall X \in \mathcal{N}$. Hence,

$$v(X) \in C^{(1)}(\mathcal{N}). \tag{5.83}$$

Now, (5.76), (5.77) and (5.80) prove the existence of a solution function $v$ to (5.3) satisfying (5.82) for every $q \in Q^{(1)}(\mathcal{S}; f)$.

Let $v_1$ and $v_2$ be two solutions to (5.3). Uniqueness of the system motion $\underline{X}(t; X_0)$ for every $X_0 \in \mathcal{N}$, invariance of $\mathcal{N} = \mathcal{D}_a$, $\mathcal{N} \subseteq \mathcal{S}$, and uniqueness of $q$ on $\mathcal{N}$ due to its positive definiteness on $\mathcal{N}$ yield

$$v_i(X_0) = \int_0^{+\infty} w[\underline{X}(\sigma; X_0)]d\sigma \text{ for every } i = 1, 2 \text{ and every } X_0 \in \mathcal{N}, \tag{5.84}$$

which implies

$$v_1(X_0) = v_2(X_0) \text{ for every } X_0 \in \mathcal{N}. \tag{5.85}$$

This proves the uniqueness of the solution $v$ to (5.2). The invariance of $\mathcal{D}_a$, $\mathcal{N} = \mathcal{D}_a$, and positive definiteness of $q$ on $\mathcal{N}$ and (5.78) imply

$$v(X) > 0 \text{ for all } (X \neq 0) \in \mathcal{N}. \tag{5.86}$$

(5.77), (5.81) and (5.85) prove the necessity of the condition 3 (i); and (5.77), (5.82) and (5.85) prove the necessity of the condition 4 (i). Let $\partial\mathcal{N}$ be non-empty. Let $X_1, X_2, \ldots, X_k, \ldots$ be a sequence converging to $\tilde{X}$, $X_k \to \tilde{X}$ as $k \to +\infty$, where $X_k \in \mathcal{N}$ and $\tilde{X} \in \partial\mathcal{N}$. Let $\tau_k = \tau(X_k, \eta) \in [0, +\infty[$ be the first instant obeying $\underline{X}(t; X_k) \in \overline{\mathcal{B}}_\eta$ for all $t \in [\tau(X_k, \eta), +\infty[$ and for $\eta$ as defined above. The existence of such $\tau(X_k, \eta)$ is guaranteed by $X_k \in \mathcal{N}$ and $\mathcal{N} = \mathcal{D}$. Continuity of motions $\underline{X}$

in $(t, X_0) \in \Re_+ \times \mathcal{N}$ due to (i-b) of the weak smoothness property, $\mathcal{N} \subseteq \mathcal{S}$, and $\overset{\circ}{\mathcal{N}} = \mathcal{N} = \mathcal{D}$ imply $\tau(X_k, \eta) \to +\infty$ as $k \to +\infty$. Let $m$ be such a natural number that $X_k \in (\mathcal{N} - \overline{\mathcal{B}}_\eta)$ for all $k = m, m+1, \ldots$ Such $m$ exists because $\mathcal{N}$ is open, $\overline{\mathcal{B}}_\eta \subset \mathcal{N}$ and $X_k \to \partial\mathcal{N}$ as $k \to +\infty$. Let $\gamma$ be defined by (5.86), and let $\alpha = \eta$ in Definition 5.4:

$$\gamma = min\,[q(X) : X \in (\mathcal{N} - \mathcal{B}_\eta]. \tag{5.87}$$

Positive definiteness of $q$ on $\mathcal{S}$ and $\mathcal{N} \subseteq \mathcal{S}$ imply $\gamma \in ]0, +\infty[$ due to 2) of Definition 5.4. Combining (5.78) and (5.86) we prove

$$v(X_k) \geq \int_0^{\tau_k} \gamma dt + \int_{\tau_k}^{+\infty} q[\underline{X}(t; X_k)]dt. \tag{5.88}$$

This inequality, positive definiteness of $q$ on $\mathcal{S}$ and $\underline{X}(t; X_k) \in \overline{\mathcal{B}}_\eta$ for all $t \in [\tau_k, +\infty[$ imply

$$v(X_k) \geq \gamma\tau_k, \tau_k = \tau(X_k, \eta), k = m, m+1, \ldots \tag{5.89}$$

Since $\tau_k \to +\infty$ as $k \to +\infty$ then (5.88) shows that

$$v(X_k) \to +\infty \;\text{ as }\; k \to +\infty. \tag{5.90}$$

Now, (5.89) and the definition of $X_k$, yield

$$v(X) \to +\infty \;\text{ as }\; X \to \partial\mathcal{N}, X \in \mathcal{N},$$

which proves necessity of the condition 3 (ii) and of the condition 4 (ii).

*Sufficiency.* Let all the conditions 1)–3) of the theorem hold. Hence, all the conditions of Theorem 4.26 (Section 4.9) due to Theorem 4.9 (Section 4.6.3) are satisfied, which proves that the set $\mathcal{N}$ is the domain of attraction of $X = 0$, $\mathcal{N} = \mathcal{D}_a$. Since $\mathcal{D}_a = \mathcal{D}$ then $\mathcal{N} = \mathcal{D}$, i.e. the set $\mathcal{N}$ is the domain of the asymptotic stability of $X = 0$ of the system (5.1).

Equivalency of (5.3) and (5.4), and the preceding theorem yield

**Theorem 5.11** *In order for the state $X = 0$ of the system (5.1) with the strong smoothness property to have the domain of asymptotic stability $\mathcal{D}$ and for a set $\mathcal{N}$ to be the domain of its asymptotic stability, $\mathcal{N} = \mathcal{D}$, it is both necessary and sufficient that*

*1) the set $\mathcal{N}$ is an open connected neighbourhood of $X = 0$ and $\mathcal{N} \subseteq \mathcal{S}$,*

*2) $f(X) = 0$ for $X \in \mathcal{N}$ iff $X = 0$,*

*and*

*3) for arbitrary positive definite $q \in Q(\mathcal{S}; f)$, the equations (5.4) have the unique solution function $\nu$ with the following features:*

*(i)* $\nu$ *is positive definite on* $\mathcal{N}$,

*and*

*(ii) if the boundary* $\partial\mathcal{N}$ *of* $\mathcal{N}$ *is non-empty then* $\nu(X) \to 1$ *as* $X \to \partial\mathcal{N}$, $X \in \mathcal{N}$,

*or*

*4) for arbitrary positive definite* $q \in Q^1(\mathcal{S}; f)$, *the equations (5.4) have the unique solution function* $\nu$ *with the following features:*

*(i)* $\nu$ *is positive definite and differentiable on* $\mathcal{N}$,

*and*

*(ii) if the boundary* $\partial\mathcal{N}$ *of* $\mathcal{N}$ *is non-empty then* $\nu(X) \to 1$ *as* $X \to \partial\mathcal{N}$, $X \in \mathcal{N}$. ▽▽▽

If the system (5.1) possesses only the weak smoothness property then the conditions slightly change.

**Theorem 5.12** *For the state* $X = 0$ *of the system (5.1) possessing the weak smoothness property to have the domain of asymptotic stability* $\mathcal{D}$ *and for a subset* $\mathcal{N}$ *of* $\mathcal{S}$ *to be equal to* $\mathcal{D}$: $\mathcal{N} = \mathcal{D}$, *it is both necessary and sufficient that*

*1) the set* $\mathcal{N}$ *is an open connected neighbourhood of* $X = 0$,

*2)* $f(X) = 0$ *for* $X \in \mathcal{N}$ *iff* $X = 0$,

*and*

*3) for arbitrary positive definite function* $q \in Q(\Re^n; f)$, *the equations (5.3) with (5.3a) determined along system motions have the unique solution function* $v$ *with the following features:*

*(i)* $v$ *is positive definite on* $\mathcal{N}$,

*and*

*(ii) if the boundary* $\partial\mathcal{N}$ *of* $\mathcal{N}$ *is non-empty then* $v(X) \to +\infty$ *as* $X \to \partial\mathcal{N}$, $X \in \mathcal{N}$,

*or*

*4) for arbitrary positive definite function* $q \in Q^1(\Re^n; f)$, *the equations (5.3) with (5.3a) determined along system motions have the unique solution function* $v$ *with the following features:*

*(i)* $v$ *is positive definite and differentiable on* $\mathcal{N}$,

*and*

*(ii) if the boundary* $\partial\mathcal{N}$ *of* $\mathcal{N}$ *is non-empty then* $v(X) \to +\infty$ *as* $X \to \partial\mathcal{N}$, $X \in \mathcal{N}$. ▽▽▽

**Proof.** *Necessity.* Let the system (5.1) possess the weak smoothness property. Let $X = 0$ have the asymptotic stability domain $\mathcal{D}$ and let $\mathcal{N} \subseteq \mathcal{S}$ be equal to $\mathcal{D}$: $\mathcal{N} = \mathcal{D}$. From this point on we have to repeat the proof of Theorem 5.10 to show that the conditions 1)–3) or 1), 2) and 4) of Theorem 5.11 hold. In such a way we complete the proof of the necessity part.

*Sufficiency.* Let the system (5.1) possess the weak smoothness property and the conditions 1)–3) or 1), 2) and 4) be valid. Then, $X = 0$ of the system (5.1) is asymptotically stable. Therefore, $X = 0$ has the domain $\mathcal{D}$ of asymptotic stability. Let $X_0 \in (\Re^n - \mathcal{N})$. Continuity of $\underline{X}(t; X_0)$ in $t \in I_0$ due to the weak smoothness property, positive definiteness of $q$ on $\Re^n$, hence, negativeness of $\dot{v}(X)$ on $(\Re^n - \{0\})$ and 3 (ii) or 4 (ii) imply $\underline{X}(t; X_0) \in (\Re^n - \mathcal{N})$ for all $t \in I_0$. Therefore, $\mathcal{D} \subseteq \mathcal{N}$. Furthermore, (5.3a) and positive definiteness of $q$ on $\Re^n$ imply (see the proof of the necessity part of Theorem 5.10) $v(X) \to +\infty$ as $X \to \partial\mathcal{D}$, $X \in \mathcal{D}$, which together with the condition 3 (i) or 4 (i) proves $\partial\mathcal{D} \cap \mathcal{N} = \emptyset$. This result, $\mathcal{D} \subseteq \mathcal{N}$ and the fact that $\mathcal{D}$ is non-empty neighbourhood of $X = 0$ imply $\mathcal{D} = \mathcal{N}$ and complete the proof.

In order to get bounded Lyapunov function we can solve (5.4) instead of (5.3). Then Theorem 5.12 becomes Theorem 5.13.

**Theorem 5.13** *For the state $X = 0$ of the system (5.1) possessing the weak smoothness property to have the domain of asymptotic stability $\mathcal{D}$ and for a subset $\mathcal{N}$ of $\mathcal{S}$ to be equal to $\mathcal{D}$: $\mathcal{N} = \mathcal{D}$, it is both necessary and sufficient that*

1) *the set $\mathcal{N}$ is an open connected neighbourhood of $X = 0$,*

2) *$f(X) = 0$ for $X \in \mathcal{N}$ iff $X = 0$,*

*and*

3) *for arbitrary positive definite function $q \in Q(\Re^n; f)$, the equations (5.4) have the unique solution function $\nu$ with the following features:*

   *(i) $\nu$ is positive definite on $\mathcal{N}$,*

   *and*

   *(ii) if the boundary $\partial\mathcal{N}$ of $\mathcal{N}$ is non-empty then $\nu(X) \to 1$ as $X \to \partial\mathcal{N}$, $X \in \mathcal{N}$,*

*or*

4) *for arbitrary positive definite function $q \in Q^1(\Re^n; f)$, the equations (5.4) have the unique solution function $\nu$ with the following features:*

   *(i) $\nu$ is positive definite and differentiable on $\mathcal{N}$,*

   *and*

   *(ii) if the boundary $\partial\mathcal{N}$ of $\mathcal{N}$ is non-empty then $\nu(X) \to 1$ as $X \to \partial\mathcal{N}$, $X \in \mathcal{N}$.* ▽▽▽

**Examples**

**Example 5.14** Let $f$ be defined by

$$f(X) = \begin{pmatrix} -X_1(1 - X_1^2 - X_2^2)^2 \\ -X_2(1 - X_1^2 - X_2^2)^2 \end{pmatrix} \in C^{(1)}(\Re^2), \tag{5.91}$$

so that the system (5.1), (5.90) has the unique solution $\underline{X}(t; X_0)$ through every $X_0 \in \Re^2$, which is defined, continuous and differentiable in $(t, X_0) \in I_0 \times \Re^2$. Hence, the system possesses the weak smoothness property (on $\Re^2$). The set $\mathcal{S}_e$ of all the equilibrium states is easily found as

$$\mathcal{S}_e = \{X : (X = 0) \ \text{ or } \ (X_1^2 + X_2^2 = 1)\}. \tag{5.92}$$

Hence,

$$||\underline{X}(t; X_e)|| = ||X_e|| \quad \text{for all} \quad t \in \Re_+ \quad \text{and every} \quad X_e \in \mathcal{S}_e.$$

This suggests to accept the set $\mathcal{S}$ as

$$\mathcal{S} = \{X : X_1^2 + X_2^2 < 1\} = \mathcal{B}_1 \subset \Re^2. \tag{5.93}$$

Now, (5.91) and (5.92) imply

$$inf\,[||\underline{X}(t; X_0)|| : t \in \Re_+] = ||X_0|| = 1 > 0 \ \text{ for every } \ X_0 \in \partial\mathcal{S}.$$

This, (5.92) and the weak smoothness property (on $\Re^2$) show that the system possesses also the strong smoothness property. Let the function $q$ be determined by

$$q(X) = (X_1^2 + X_2^2) \in C^{(1)}(\Re^2). \tag{5.94}$$

Hence, it is differentiable and positive definite on $\mathcal{S}$. For any $\alpha \in \,]0, 1[$ there is $\beta > 0$ obeying

$$\beta = inf\,[q(X) : X \in (\mathcal{B}_1 - \mathcal{B}_\alpha)] =$$

$$= inf\,[(X_1^2 + X_2^2) : X \in \{X : 0 < \alpha \leq (X_1^2 + X_2^2) < 1\}] = \alpha > 0.$$

In this case $\beta = \alpha$. Hence, 1) and 2) of Definition 5.4 are satisfied. The equations (5.2) with $q$ defined by (5.93) have a solution

$$v(X) = \frac{1}{2} \frac{X_1^2 + X_2^2}{1 - X_1^2 - X_2^2}, \tag{5.95}$$

which is defined and continuously differentiable on $\overline{\mathcal{B}}_\mu$ for any $\mu \in \,]0, 1[$. Hence, the condition 4) of Definition 5.4 is also satisfied, which completely verifies $q \in Q^1(\mathcal{S}; f)$. Let us verify now the conditions of the Theorem 5.10.

1) the set $\mathcal{S} = \mathcal{B}_1$ (5.92) is open connected neighbourhood of $X = 0$,

2) $f(X) = 0$ for $X \in \mathcal{S}$ iff $X = 0$,

3) for the function $q$ (5.93) we concluded that it belongs to $Q^1(\mathcal{S}; f)$ and leads to the solution $v$ of (5.2) in the form (5.94). The function $v$ (5.94) obeys (i) and (ii),

(i) $v$ is positive definite and differentiable on $\mathcal{S}$,

(ii) $v(X) \to +\infty$ as $X \to \partial\mathcal{S}$ [i.e. as $(X_1^2 + X_2^2) \to 1$].

The conditions 1), 2) and 4) of Theorem 5.1 are satisfied for

$$\mathcal{N} = \mathcal{S} = \mathcal{B}_1.$$

Hence, the set $\mathcal{N} = \mathcal{B}_1$ is the domain $\mathcal{D}$ of asymptotic stability of $X = 0$ of the system (5.1), (5.90):

$$\mathcal{D} = \mathcal{B}_1 = \{X : X_1^2 + X_2^2 < 1\}.$$

**Example 5.15** The second order system (5.95),

$$\frac{dX}{dt} = (10 - X_1^2 + 2X_1X_2 - 4X_2^2)\begin{bmatrix} -2X_1 + 4X_2 \\ X_1 - 4X_2 \end{bmatrix} = f(X), \tag{5.96}$$

has the unique solution $\underline{X}(t; X_0)$ through every $X_0 \in \Re^2$, which is defined, continuous and differentiable in $(t, X_0) \in I_0 \times \Re^2$. Therefore, the system possesses the weak smoothness property on $\Re^2$. It has the unique equilibrium state $X = 0$ in the set $\mathcal{S} = \{X : X_1^2 - 2X_1X_2 + 4X_2^2 < 10\}$. Other equilibrium states compose the boundary $\partial\mathcal{S} = \{X : X_1^2 - 2X_1X_2 + 4X_2^2 = 10\}$ of the set $\mathcal{S}$. Therefore, $\mathcal{N} \subseteq \mathcal{S}$. Notice that the system has also the strong smoothness property on $\mathcal{S}$. We may apply any of the theorems. Let the function $q$ be accepted in the form (5.96):

$$q(X) = X^T G X, \quad G = \begin{bmatrix} 6 & -14 \\ -14 & 40 \end{bmatrix} = G^T > 0. \tag{5.97}$$

The function $q$ (5.96) obeys 1) and 2) of Definition 5.4. The solution $v$ of (5.2) for such a choice of the function $q$ is

$$v(X) = ln\frac{10}{10 - X^T H X}, \quad H = \begin{bmatrix} 1 & -1 \\ -1 & 4 \end{bmatrix} = H^T > 0. \tag{5.98}$$

The function $v$ is continuously differentiable on $\mathcal{S}$, which shows that the function $q$ obeys 3) and 4) of Definition 5.1. Moreover, the function $v$ is positive definite on

$\mathcal{S}$ and $v(X) \to +\infty$ as $X \to \partial\mathcal{S}$, $X \in \mathcal{S}$. All conditions of Theorem 5.10 through Theorem 5.13 corresponding to $q \in Q^1(\mathcal{S}, f)$ have been verified. Any of them shows that $X = 0$ of the system (5.95) is asymptotically stable. They clarify that the set $\mathcal{N} = \mathcal{S}$ is the asymptotic stability domain of $X = 0$ of the system,

$$\mathcal{D} = \{X : X_1^2 - 2X_1X_2 + 4X_2^2 < 10\}.$$

### 5.2.5 Exponential stability

What follows is based on [85].

#### Introductory comment

In this section the following problem is addressed and solved.

**Problem.** *What are the necessary and sufficient conditions for an exact algorithmic one-shot construction of a Lyapunov function $v$ for the system (5.1) with the specific smoothness property such that it and its total time derivative $\dot{v}$ along system motions prove the following estimate:*

$$||x(t;x_0)|| \leq \alpha||x_0||\, exp\,(-\beta t), \quad for\ every \quad (t, x_0) \in \Re_+ \times \mathcal{B}_\delta, \tag{5.99}$$

*for some numbers $\alpha \in [1, +\infty[$ and $\beta \in ]0, +\infty[$, and for $\delta$ obeying (5.99),*

$$0 < \delta \leq \alpha^{-1}\Delta, \tag{5.100}$$

*where $\Delta > 0$ satisfies $\mathcal{B}_\Delta \subseteq \mathcal{S}$.*

#### Problem solution

In this framework the system will be assumed to possess the specific smoothness property (Definition 5.1).

The complete solution to the problem reads as follows:

**Theorem 5.14** *Let the system (5.1) possess the specific smoothness property and $\Delta > 0$ obey $\mathcal{B}_\Delta \subseteq \mathcal{S}$. In order for system solutions to obey (5.98) for some numbers $\alpha \in \Re^+$, $\alpha \geq 1$, $\beta \in \Re^+$ and $\delta$ given by (5.99) it is necessary and sufficient that for arbitrarily chosen positive definite differentiable function $p$ obeying (5.100),*

$$\mu_1||X||^k \leq p(X) \leq \mu_2||X||^k, \;\; \forall X \in \mathcal{B}_\Delta, \tag{5.101}$$

*for some numbers $\mu_i \in \Re^+$, $i = 1, 2$, and some natural number $k$ there exists a unique solution function $v$ to (5.101) with (5.101a) determined along system motions,*

$$\frac{dv(X)}{dt} = -p(X), \quad \forall X \in \mathcal{B}_\Delta, \tag{5.102a}$$

$$v(0) = 0, \tag{5.101b}$$

*which has the following properties:*

*(i) $v$ is differentiable on $\mathcal{B}_\delta$: $f(X) \in C^{(1)}(\mathcal{B}_\delta)$,*

*and*

*(ii) it satisfies*

$$\eta_1 ||X||^k \leq v(X) \leq \eta_2 ||X||^k, \quad \forall X \in \mathcal{B}_\delta, \tag{5.102}$$

*for some numbers $\eta_i \in \Re^+$, $i = 1, 2$.* ∇∇∇

**Proof.** Let the system (5.1) possess the specific smoothness property and $\Delta > 0$ obey $\mathcal{B}_\Delta \subseteq \mathcal{S}$.

*Necessity.* The proof of the necessity part is a modification of the proof of Theorem 11.1 by Krasovskii [116]. Let the estimate (5.98) hold for some numbers $\alpha \in \Re^+$, $\alpha \geq 1$, $\beta \in \Re^+$, for $\delta$ obeying (5.99). Let us choose arbitrary differentiable positive definite function $p$ on $\mathcal{B}_\Delta$ obeying (5.100) for some positive numbers $\mu_i$, $i = 1, 2$, and some natural number $k$. Let a function $v$ be determined by (5.101). Then,

$$v[\underline{X}(t; X_0)] - v(X_0) = -\int_0^t p[\underline{X}(\tau; X_0)]d\tau, \quad \forall (t, X_0) \in \Re_+ \times \mathcal{B}_\delta.$$

This, (5.98) and (5.101b) imply (5.103):

$$v(X_0) = \int_0^{+\infty} p[\underline{X}(\tau; X_0)]d\tau, \quad \forall X_0 \in \mathcal{B}_\delta. \tag{5.103}$$

The specific smoothness property (Definition 5.1) guarantees the existence and uniqueness of motions of the system (5.1) on $\Re_+$ for every $X_0 \in \mathcal{B}_\delta$. Besides, it ensures validity of (5.104) (inequality (4.11) by Krasovskii [116]),

$$||\underline{X}(t; X_0)|| \geq ||X_0|| \, exp\,(-n^2 \gamma t), \quad \forall (t, X_0) \in \Re_+ \times \mathcal{B}_\delta. \tag{5.104}$$

Let $\eta_1 = (k\gamma n^2)^{-1}\mu_1 \in \Re^+$ and $\eta_2 = \alpha^k (k\beta)^{-1}\mu_2 \in \Re^+$. Now, (5.98), (5.99), (5.103) and (5.104) imply (5.105):

$$\eta_1 ||X_0||^k \leq v(X_0) \leq \eta_2 ||X_0||^k, \quad \forall X_0 \in \mathcal{B}_\delta. \tag{5.105}$$

By substituting $X_0$ with $X$, $X \in \mathcal{B}_\delta$, in (5.105) we prove (5.102). Besides, (5.103) and (5.105) for $X_0 = X$ prove that $v$ is defined on $\mathcal{B}_\delta$. Continuity and differentiability of $\underline{X}(t; X_0)$ on $\Re_+ \times \mathcal{B}_\delta$ due to the specific smoothness property, together with

(5.101a) and (5.103) prove continuity and continuous differentiability of the function $v$ on $\mathcal{B}_\delta$ (for all the details see the proof of Theorem 11.1 by Krasovskii [116]). Furthermore, uniqueness of the motions $\underline{X}$ of the system (5.1), and (5.103) prove the uniqueness of the solution function $v$ to (5.101), which completes the proof of the necessity part.

*Sufficiency.* Let all the conditions of the theorem to be proved hold. From (5.100), (5.101a) and (5.102) we derive:

$$\dot{v}[\underline{X}(t;X_0)] \leq -\mu_1\eta_2^{-1}v[\underline{X}(t;X_0)],\ \ \forall[t,\underline{X}(t;X_0)] \in \Re_+ \times \mathcal{B}_\Delta.$$

A solution to this inequality satisfies the following:

$$v[\underline{X}(t;X_0)] \leq v(X_0)\,exp\,(-\mu_1\eta_2^{-1}t),\ \ \forall[t,\underline{X}(t;X_0)] \in \Re_+ \times \mathcal{B}_\Delta.$$

Combining this with (5.102) we derive

$$||\underline{X}(t;X_0)|| \leq \alpha||X_0||\,exp\,(-\beta t),\ \ \forall(t,\alpha X_0) \in \Re_+ \times \mathcal{B}_\Delta, \tag{5.106a}$$

and

$$\alpha = (\eta_1^{-1}\eta_2)^{1/k} \in [1,+\infty[,\quad \beta = \mu_1(k\eta_2)^{-1} \in \Re^+. \tag{5.106b}$$

Let now $\delta$ be chosen as

$$0 < \delta \leq \alpha^{-1}\Delta, \tag{5.106c}$$

so that $||X_0|| < \delta$ implies $\alpha||X_0|| < \Delta$ that together with (5.106a) proves (5.98), and (5.106c) proves (5.99).

**Example 5.16** Let a third order system be described by

$$\frac{dX}{dt} = \begin{pmatrix} -x_1 - 2x_1^3 + 6x_1x_2^2 - 10x_1x_3^2 \\ -10x_1^2x_2 - x_2 - 2x_2^3 + 6x_2x_3^2 \\ 6x_1^2x_3 - 10x_2^2x_3 - x_3 - 2x_3^3 \end{pmatrix} = f(X).$$

The function $f$ is differentiable on $\Re^3$: $f(X) \in C^{(1)}(\Re^3)$. Its Jacobian $f_X$ has bounded norm on $\mathcal{B}_\Delta$ for any $\Delta \in \Re^+$, that is for any $\Delta \in \Re^+$ there is $\gamma \in \Re^+$, $\gamma = \gamma(\Delta)$ such that

$$sup\,[||f_X(X)|| : X \in \mathcal{B}_\Delta] < \gamma(\Delta) < +\infty.$$

Hence, the system possesses the specific smoothness property (Definition 5.1) with $\mathcal{S} = \mathcal{B}_\Delta$. Let at first

$$p(X) = 2||X||^2(1 + 2||X||^2).$$

The function $p$ obeys (5.100) for

$$k = 2,\quad \mu_1 = 2,\quad \mu_2 = 2(1 + 2\Delta^2).$$

For such a function $p$, the solution $v$ to (5.101) is

$$v(X) = ||X||^2.$$

Hence, the function $v$ obeys Theorem 5.14 for

$$\eta_1 = \eta_2 = 1,$$

so that (5.98) is satisfied for

$$\alpha = 1, \quad \beta = \frac{1}{4(1+2\Delta^2)},$$

and for $\delta \in ]0, \Delta]$.

Let the second choice of $p$ be

$$p(X) = 2||X||^2(1+2||X||^2)^2.$$

so that (5.100) holds for

$$k = 2, \quad \mu_1 = 2, \quad \mu_2 = 2(1+2\Delta^2)^2.$$

The solution $v$ to (5.101) is found as

$$v(X) = ||X||^2(1+||X||^2).$$

It obeys (5.102) for

$$\eta_1 = 1, \quad \eta_2 = 1+\Delta^2.$$

In this case (5.98) is valid for

$$\alpha = \sqrt{1+\Delta^2} \in [1, +\infty[, \quad \beta = \frac{1}{1+\Delta^2},$$

and for $\delta \in \left]0, \frac{\Delta}{\sqrt{1+\Delta^2}}\right]$.

If we select

$$p(X) = \frac{2||X||^2}{1+2||X||^2},$$

which yields

$$k = \mu_1 = \mu_2 = 2,$$

then the solution $v$ to (5.101) is

$$v(X) = \frac{||X||^2}{1 + 2||X||^2}.$$

It satisfies (5.102) for

$$\eta_1 = \frac{1}{1 + 2\Delta^2}, \quad \eta_2 = 1,$$

so that (5.98) holds for

$$\alpha = (1 + 2\Delta^2) \in [1, +\infty[, \quad \beta = 1, \quad \delta = \frac{\Delta}{1 + 2\Delta^2}.$$

This example shows dependence of $v$, $\alpha$, $\beta$ and $\delta$ on the choice of the function $p$.

## 5.3 Systems with continuous motions (generalised motions)

### 5.3.1 Smoothness property

**Definition 5.5 Weak smoothness property:**

*(i) There is an open neighbourhood $\mathcal{S}$ of $X = 0$, $\mathcal{S} \subseteq \Re^n$, such that for every $X_0 \in \mathcal{S}$*

*(a) the system (5.1) has the unique solution $\underline{X}(t; X_0)$ through $X_0$ at $t = 0$ over a time interval $I_0$, and*

*(b) the motion $\underline{X}(t; X_0)$ is defined and continuous in $(t, X_0) \in I_0 \times \mathcal{S}$.*

*(ii) For every $X_0 \in (\Re^n - \mathcal{S})$ every motion $\underline{X}(t; X_0)$ of the system (5.1) is continuous in $t \in I_0$.*

**Strong smoothness property:**

*(i) The system (5.1) has the weak smoothness property.*

*(ii) If the boundary $\partial\mathcal{S}$ of $\mathcal{S}$ is non-empty then every motion of the system (5.1) passing through $X_0 \in \partial\mathcal{S}$ at $t = 0$ obeys $inf\ [||X(t; X_0)|| : t \in I_0] > 0$ for every $X_0 \in \partial\mathcal{S}$.*

Throughout this section (5.3) a smoothness property will be used in the sense of Definition 5.5 (rather than in the sense of Definition 5.1).

### 5.3.2 Approach via $\mathcal{O}$-uniquely bounded sets

This section exposes results and examples from [80]. In what follows we shall make use of the following definition.

**Definition 5.6** *A function $u : \Re^n \to \Re$ belongs to the family $\mathcal{E}(\mathcal{S}; f)$ if and only if*

*1) $u$ is continuous on $\mathcal{S}$: $u(X) \in C(\mathcal{S})$,*

*and*

*2) the equations (5.107) with (5.107a) taken along motions of the system (5.1),*

$$\mathcal{D}^+ v(X) = -u(X), \tag{5.107a}$$

$$v(0) = 0, \tag{5.107b}$$

*have a solution $v$ that is well defined in $\Re$ and continuous for every $X \in \overline{\mathcal{B}}_\mu$ for some $\mu \in ]0, +\infty[$, $\mu = \mu(u, f)$.*

**Theorem 5.15** *In order for the state $X = 0$ of the system (5.1) with the strong smoothness property to have the domain of asymptotic stability $\mathcal{D}$ and for a set $\mathcal{N}$, $\mathcal{N} \subseteq \Re^n$, to be the domain of its asymptotic stability, $\mathcal{N} = \mathcal{D}$, it is both necessary and sufficient that*

*1) the set $\mathcal{N}$ is an open connected neighbourhood of $X = 0$ and $\mathcal{N} \subseteq \mathcal{S}$,*

*2) $f(X) = 0$ for $X \in \mathcal{N}$ iff $X = 0$,*

*and either*

*3) for arbitrarily selected $\mathcal{O}$-uniquely bounded set $\mathcal{U}$, $\overline{\mathcal{U}} \subseteq \overline{\mathcal{S}}$, with the generating function $u$ on $\mathcal{S}$ obeying $u \in \mathcal{E}(\mathcal{S}; f)$, the equations (5.107) have the unique solution function $v$ on $\mathcal{N}$ with the following properties:*

*(i) $v$ is positive definite on $\mathcal{N}$,*

*and*

*(ii) if the boundary $\partial\mathcal{N}$ of $\mathcal{N}$ is non-empty then $v(X) \to +\infty$ as $X \to \partial\mathcal{N}$, $X \in \mathcal{N}$,*

*or*

*4) for arbitrarily selected $\mathcal{O}$-uniquely bounded set $\mathcal{U}$, $\overline{\mathcal{U}} \subseteq \overline{\mathcal{S}}$, with the generating function $u$ on $\mathcal{S}$ obeying $u \in \mathcal{E}(\mathcal{S}; f)$, the equations (5.108) with (5.108a) taken along system motions,*

$$\mathcal{D}^+ \nu(X) = -u(X)\,[1 - \nu(X)], \tag{5.108a}$$

$$\nu(0) = 0, \tag{5.108b}$$

*have the unique solution function $\nu$ on $\mathcal{N}$ with the following properties:*

*(i) $\nu$ is positive definite on $\mathcal{N}$,*

*and*

*(ii) if the boundary $\partial\mathcal{N}$ of $\mathcal{N}$ is non-empty then $\nu(X) \rightarrow 1$ as $X \rightarrow \partial\mathcal{N}$, $X \in \mathcal{N}$.* $\nabla\nabla\nabla$

**Proof.** 1)–3) of the statement are proved in the same way as 1)–3) of Theorem 5.6. Equivalency of (5.107) and (5.108) then proves 4).

The conditions slightly change if the system (5.1) possesses the weak smoothness property rather than the strong smoothness property.

**Theorem 5.16** *For the state $X = 0$ of the system (5.1) possessing the weak smoothness property to have the domain of asymptotic stability $\mathcal{D}$ and for a subset $\mathcal{N}$ of $\mathcal{S}$, $\mathcal{N} \subseteq \mathcal{S}$, to be equal to $\mathcal{D}$: $\mathcal{N} = \mathcal{D}$, it is both necessary and sufficient that*

*1) the set $\mathcal{N}$ is an open connected neighbourhood of $X = 0$,*

*2) $f(X) = 0$ for $X \in \mathcal{N}$ iff $X = 0$,*

*and either*

*3) for arbitrarily selected $\mathcal{O}$-uniquely bounded set $\mathcal{U}$, $\overline{\mathcal{U}} \subseteq \overline{\mathcal{S}}$, with the generating function $u$ on $\Re^n$ obeying $u \in \mathcal{E}(\mathcal{S}; f)$, the equations (5.107) have the unique solution function $v$ on $\mathcal{N}$ with the following properties:*

*(i) $v$ is positive definite on $\mathcal{N}$,*

*and*

*(ii) if the boundary $\partial\mathcal{N}$ of $\mathcal{N}$ is non-empty then $v(X) \rightarrow +\infty$ as $X \rightarrow \partial\mathcal{N}$, $X \in \mathcal{N}$,*

*or*

*4) for arbitrarily selected $\mathcal{O}$-uniquely bounded set $\mathcal{U}$, $\overline{\mathcal{U}} \subseteq \overline{\mathcal{S}}$, with the generating function $u$ on $\Re^n$ obeying $u \in \mathcal{E}(\mathcal{S}; f)$, the equations (5.108) have the unique solution function $\nu$ on $\mathcal{N}$ with the following properties:*

*(i) $\nu$ is positive definite on $\mathcal{N}$,*

*and*

*(ii) if the boundary $\partial\mathcal{N}$ of $\mathcal{N}$ is non-empty then $X \rightarrow \partial\mathcal{N}$, $X \in \mathcal{N}$ implies $\nu(X) \rightarrow 1$.* $\nabla\nabla\nabla$

**Proof.** This theorem is proved in the same way as Theorem 5.8 and Theorem 5.9.

**Example 5.17** Let $n = 1$,

$$\frac{dx}{dt} = -x + h(x), \quad h(x) = \begin{cases} x|x| & \text{for } |x| \in [0, 1], \\ x\sqrt{x} & \text{for } |x| \in [1, +\infty[. \end{cases}$$

The system possesses the strong smoothness property because $f(x) = -x + h(x)$ is Lipschitzian on $\Re^1$. The equilibrium states are $x_{e1} = -1$, $x_{e2} = 0$ and $x_{e3} = +1$. They suggest $\mathcal{S} = ]-1, +1[$ and $\mathcal{U} = \{x : |x| < \alpha\} = ]-\alpha, +\alpha[$ for $\alpha \in ]0, 1[$. The generating function $u$ on $\mathcal{N}$, $u(x) = |x|$, of the $\mathcal{O}$-uniquely bounded set $\mathcal{U}$ and (5.2) yield

$$\mathcal{D}^+ v(x) = -|x|, \quad x \in \mathcal{S}.$$

Its solution $v$ is

$$v(x) = -\ln(1 - |x|), \quad x \in \mathcal{S}.$$

The function $v(x)$ and the set $\mathcal{N} = \mathcal{S} = ]-1, +1[$ obey all the requirements of Theorem 5.15, that is that

1) $\mathcal{N} = ]-1, +1[$ is an open connected neighbourhood of $X = 0$ and $\mathcal{N} = \mathcal{S}$,

2) $f(x) = -x + h(x) = 0$ for $x \in \mathcal{N}$ iff $x = 0$,

3) (i) $v(x) = 0$ for $x \in \mathcal{N}$ iff $x = 0$,

   $v(x) \in C(\mathcal{N})$,

   $v(x) > 0$ for every $(x \neq 0) \in \mathcal{N}$,

   which prove positive definiteness of $v$ on $\mathcal{N}$,

   (ii) $v(x) \to +\infty$ as $x \to \partial\mathcal{N} = \{-1, +1\}$, $x \in \mathcal{N}$.

Hence, $\mathcal{N} = ]-1, +1[$ is the domain $\mathcal{D}$ of asymptotic stability of $x = 0$,

$$\mathcal{D} = ]-1, +1[.$$

Notice that $|f(x)| = |x|(1 - |x|)$, $x \in \mathcal{N}$, is not a generating function on $\mathcal{N}$ of any $\mathcal{O}$-uniquely bounded set because it is not radially increasing on $\mathcal{N}$.

**Example 5.18** Let the function $h$ be defined as in Example 5.17 and

$$\frac{dx}{dt} = x - h(x).$$

It is clear that the system possesses the strong smoothness property on $\Re^1$ and has the equilibrium states $x_{e1} = -1$, $x_{e2} = 0$ and $x_{e3} = +1$ (see Example 5.17). Let again $\mathcal{U} = \{x : |x| < \alpha\} = ]-\alpha, +\alpha[$ for $\alpha \in ]0, +\infty[$ so that $u(x) = |x|$. From (5.107a) and $u(x) = |x|$ we get

$$\mathcal{D}^+ v(x) = -|x|, \quad x \in \mathcal{N}.$$

Integrating this equation along motions of the system we derive

$$v(x) = \ln(1 - |x|), \quad x \in \mathcal{N},$$

which is negative definite on $\mathcal{N}$ and, thus, does not satisfy the necessary and sufficient conditions for asymptotic stability of $x = 0$ of the system. Hence, $x = 0$ of the system is not asymptotically stable and does not have the asymptotic stability domain.

### 5.3.3 General one-shot approach

In what follows we shall extend the general one-shot approach to systems with continuous (generalised) rather than with differentiable motions [77], [82].

In this framework we shall use the family $\mathcal{E}(\mathcal{S}; f)$ defined as follows:

**Definition 5.7** *$\mathcal{E}(\mathcal{S}; f)$ is the family of all the functions $e$, $e : \Re^n \to \Re$, with the following properties:*

*1) $e$ is continuous on $\mathcal{S}$: $e(X) \in C(\mathcal{S})$,*

*2) for any $\alpha > 0$ for which the set $\mathcal{S}$ is neighbourhood of $\mathcal{B}_\alpha$ there is $\beta > 0$, $\beta = \beta(\alpha; e; \mathcal{S})$, which obeys*

$$inf\,[e(X) : X \in (\mathcal{S} - \mathcal{B}_\alpha)] = \beta,$$

*3) there is $\mu \in \Re^+$, $\mu = \mu(e; f)$, such that there exists a solution $v$ to the equations (5.109) with (5.109a) taken along motions of the system (5.1),*

$$\mathcal{D}^+ v(X) = -e(X), \tag{5.109a}$$

$$v(0) = 0, \tag{5.109b}$$

*which is defined in $\Re$ and continuous on $\overline{\mathcal{B}}_\mu$, or equivalently, there exists a solution $\nu$ to the equations (5.110), with (5.110a) taken along motions of the system (5.1),*

$$\mathcal{D}^+ \nu(X) = -e(X)\,[1 - \nu(X)], \tag{5.110a}$$

$$\nu(0) = 0, \tag{5.110b}$$

*which is defined in $\Re$ and continuous on $\overline{\mathcal{B}}_\mu$.*

**Theorem 5.17** *In order for the state $X = 0$ of the system (5.1) with the strong smoothness property to have the domain of asymptotic stability $\mathcal{D}$ and for a set $\mathcal{N}$, $\mathcal{N} \subseteq \Re^n$, to be the domain of its asymptotic stability, $\mathcal{N} = \mathcal{D}$, it is both necessary and sufficient that*

*1) the set $\mathcal{N}$ is an open connected neighbourhood of $X = 0$ and $\mathcal{N} \subseteq \mathcal{S}$,*

*2) $f(X) = 0$ for $X \in \mathcal{N}$ iff $X = 0$,*

*and*

*3) for arbitrarily selected positive definite function $e$ obeying $e \in \mathcal{E}(\mathcal{S}; f)$, the equations (5.109) [equivalently (5.110)] have the unique solution function $v$ [equivalently $\nu$] on $\mathcal{N}$ with the following properties:*

*(i) $v$ [equivalently $\nu$] is positive definite on $\mathcal{N}$,*

*and*

*(ii) if the boundary $\partial\mathcal{N}$ of $\mathcal{N}$ is non-empty then $v(X) \to +\infty$ [equivalently $\nu(X) \to 1$], respectively, as $X \to \partial\mathcal{N}$, $X \in \mathcal{N}$.* ∇∇∇

**Proof.** This theorem is proved in the same way [77], [82] as Theorem 5.10.

**Theorem 5.18** *For the state $X = 0$ of the system (5.1) with the weak smoothness property to have the domain of asymptotic stability $\mathcal{D}$ and for a subset $\mathcal{N}$ of $\mathcal{S}$, $\mathcal{N} \subseteq \mathcal{S}$, to be equal to $\mathcal{D}$: $\mathcal{N} = \mathcal{D}$, it is both necessary and sufficient that*

1) *the set $\mathcal{N}$ is an open connected neighbourhood of $X = 0$,*

2) *$f(X) = 0$ for $X \in \mathcal{N}$ iff $X = 0$,*

*and*

3) *for arbitrarily selected function $e$ obeying $e \in \mathcal{E}(\mathcal{S}; f)$ the equations (5.109) [equivalently (5.110)] have the unique solution function $v$ [equivalently $\nu$] on $\mathcal{N}$ with the following properties:*

   (i) *$v$ [equivalently $\nu$] is positive definite on $\mathcal{N}$,*

   *and*

   (ii) *if the boundary $\partial\mathcal{N}$ of $\mathcal{N}$ is non-empty then $v(X) \to +\infty$ [equivalently $\nu(X) \to 1$], respectively, as $X \to \partial\mathcal{N}$, $X \in \mathcal{N}$.* ∇∇∇

**Proof.** The proof is carried out in the same way as that of Theorem 5.12 [77], [82].

**Example 5.19** Let the second order system be given by

$$\frac{dx_1}{dt} = -\frac{(10-|x_1|)^2(6-|x_2|)}{30+2|x_2|}x_1 = f_1(X),$$

$$\frac{dx_2}{dt} = -\frac{(10-|x_1|)(6-|x_2|)^2}{12+3|x_1|}x_2 = f_2(X),$$

$$f = (f_1\ f_2)^T.$$

The system has a unique (generalised) motion $\underline{X}(t; X_0)$ for every $X_0 \in \Re^2$, which is continuous in $t \in \Re_+$. The set $\mathcal{S}_e$ of the equilibrium states is found as

$$\mathcal{S}_e = \{X : X \in \Re^2,\ (X = 0) \text{ or } (|x_1| = 10 \text{ and } x_2 \in \Re) \text{ or } (x_1 \in \Re \text{ and } |x_2| = 6)\}.$$

The set $\mathcal{S}_e$ suggest the set $\mathcal{S}$ as

$$\mathcal{S} = \{X : X \in \Re^2,\ |x_1| < 10 \text{ and } |x_2| < 6\}.$$

Since the set $\mathcal{S}_e$ is positive invariant and $\partial\mathcal{S} \subset \mathcal{S}_e$ then it is easy to verify that the system possesses the strong smoothness property with $\mathcal{S}$ determined above. Let

$$e(X) = |x_1| + |x_2|,$$

which yields the solution to (5.109) as

$$v(X) = \frac{3|x_1| + 2|x_2|}{(10-|x_1|)(6-|x_2|)}.$$

The function $v$ is well defined on $\overline{\mathcal{B}}_\mu$ for $\mu \in ]0, 5.99[$. Since

$$inf\,[e(X) : X \in (\mathcal{S} - \mathcal{B}_\alpha)] = \alpha > 0, \quad \text{for any} \quad \alpha \in ]0, 5.99[,$$

then we conclude that the positive definite function $e$ obeys $e \in \mathcal{E}(\mathcal{S}; f)$. The function $v$ satisfies all the conditions of Theorem 5.17 on $\mathcal{S}$. Since $X = 0$ is a unique equilibrium state of the system in $\mathcal{S}$, then we may conclude that the set $\mathcal{S}$ is the domain $\mathcal{D}$ of asymptotic stability of $X = 0$,

$$\mathcal{N} = \mathcal{S} = \mathcal{D} = \{X : X \in \Re^2, \ \ |x_1| < 10 \ \text{ and } \ |x_2| < 6\}.$$

**Example 5.20** Let the system (5.1) take the following specific form:

$$\frac{dx_1}{dt} = -x_1(1 + x_2^2)\,cos^2|x_1| = f_1(X),$$

$$\frac{dx_2}{dt} = -x_2(1 + x_2^2)(1 + |x_1|) = f_2(X),$$

$$f = (f_1 \ f_2)^T.$$

The system possesses the weak smoothness property on $\Re^2$: $\mathcal{S} = \Re^2$. Let

$$e(X) = |x_1|(1 + x_2^2) + |x_2|(1 + |x_1|).$$

It satisfies 1) and 2) of Definition 5.7 on $\Re^2$. Furthermore, the solution function $v$ to (5.109) is obtained in the form of

$$v(X) = tan\,|x_1| + arctan\,|x_2|,$$

which completes the proof of $e \in \mathcal{E}(\Re^2; f)$. The function $v$ is positive definite on $\mathcal{N}$,

$$\mathcal{N} = \left\{X : X \in \Re^2, \ \ |x_1| < \frac{\pi}{2} \ \text{ and } \ x_2 \in \Re\right\}.$$

The functions $e$ and $v$, and the set $\mathcal{N}$ fulfil all the conditions of Theorem 5.18. Hence, $X = 0$ of the system has the domain $\mathcal{D}$ of asymptotic stability given by $\mathcal{D} = \mathcal{N}$, that is

$$\mathcal{D} = \left\{X : X \in \Re^2, \ \ |x_1| < \frac{\pi}{2} \ \text{ and } \ x_2 \in \Re\right\}.$$

**Example 5.21** Let

$$\frac{dx_1}{dt} = -x_1(1 + x_1^2)(2 + sin\,x_2) = f_1(X),$$

$$\frac{dx_2}{dt} = -x_2^3\,exp\,|x_2|\,(4 + cos\,x_1) = f_2(X),$$

$$f = (f_1 \ f_2)^T.$$

This system has the weak smoothness property on $\Re^2$. The function $e$,

$$e(X) = |x_1|(2 + \sin x_2) + |x_2|^3(4 + \cos x_1),$$

obeys $e \in \mathcal{E}(\Re^2; f)$ because it yields the solution $v$ to (5.109) in the following form:

$$v(X) = \arctan |x_1| + 1 - \exp(-|x_2|).$$

The functions $v$ and $e$, and the set $\mathcal{N} = \Re^2$ satisfy all the conditions of Theorem 5.18. Therefore, $X = 0$ of the system has the domain of asymptotic stability,

$$\mathcal{D} = \Re^2.$$

This means that $x = 0$ is globally asymptotically stable, which is proved by using non-radially unbounded function $v$. This is possible because $v$ is solution to (5.109) for $e \in \mathcal{E}(\Re^2; f)$ and $X = 0$ is the unique equilibrium state in $\Re^2$.

The obtained result does not contradict the Barbashin–Krasovskii theorem [6] on asymptotic stability in the whole, which requires only the *existence* of a globally positive definite radially unbounded function with globally negative definite derivative along system motions. In order to show that such a function $v$ can be generated by using Theorem 5.18 and (5.109) let the function $e$ have the form either under (i) or (ii),

(i) $e(X) = 2x_1^2(1 + x_1^2)(2 + \sin x_2) + 2x_2^4 \exp |x_2| (4 + \cos x_1)$, which yields the solution $v$ to (5.109) as

$$v(X) = ||X||^2,$$

(ii) $e(X) = |x_1|(1 + x_1^2)(2 + \sin x_2) + |x_2|^3 \exp |x_2| (4 + \cos x_1)$, which leads via (5.109) to

$$v(X) = |x_1| + |x_2|.$$

In both cases the function $v$ is globally positive definite and radially unbounded.

## 5.4 Conclusion

The new methodology based on the Lyapunov (direct) method [127] presented in this chapter for nonlinear systems described by (5.1) and with a corresponding smoothness property, is characterized by the following in the case where $X = 0$ is unique equilibrium state in its neighbourhood $\mathcal{N}$:

1. A function $q$ can be *arbitrarily* chosen from an appropriate functional family so that $q$ is positive definite.

2. Either properties of a solution $v$ to $\dot{v} = -q$ with $v(0) = 0$, or properties of a solution $\nu$ to $\dot{\nu} = -q(1 - \nu)$ with $\nu(0) = 0$ are to be tested. If and only if, for a chosen positive definite $q$ from the functional family determined in the chapter for different cases, the solution $v$ (or $\nu$) is positive definite then $X = 0$ is asymptotically stable. Hence, if $v$ (or $\nu$) is not positive definite then $X = 0$ is not asymptotically stable. The procedure ends in either case after testing properties of $v$ (or $\nu$) only once.

3. If we wish to verify whether an open connected neighbourhood $\mathcal{N}$ of $X = 0$ is the domain of asymptotic stability of $X = 0$ then, in addition to positive definiteness of $v$ (or $\nu$) on $\mathcal{N}$, it is both necessary and sufficient that $v(X) \to \infty$ [or $\nu(X) \to 1$] as $X \to \partial\mathcal{N}$, $X \in \mathcal{N}$.

In this way we achieve both

a) an exact construction of a system Lyapunov function,

and

b) an exact determination of the asymptotic stability domain of $X = 0$.

The same holds in the case where we test asymptotic stability of a set [81].

The methodology presented herein is inverse to Lyapunov's methodology for nonlinear systems [127] that starts with a choice of $v$ (or $\nu$) and verification of properties of $\dot{v}$ (or $\dot{\nu}$). If $\dot{v}$ (or $\dot{\nu}$) does not possess the requested properties, the $v$ (or $\nu$) should be reselected without any guideline how to reselect them. Therefore, Lyapunov's original methodology allows the procedure to be endless.

It should be noticed that Lyapunov's original methodology for nonlinear systems is inverse to his own methodology for time-invariant linear systems [127].

However, the methodology presented in this chapter for nonlinear systems is the same as Lyapunov's methodology for time-invariant linear systems. This ensures consistency of the methodology based on the Lyapunov method and makes it complete.

It should be also noticed that the methodology presented herein separates the stability problem reduced to the problem of testing the properties of $v$ (or $\nu$) from the purely mathematical problems of the existence of the solution $v$ to $\dot{v} = -q$ with $v(0) = 0$ [or $\nu$ to $\dot{\nu} = -q(1-\nu)$ with $\nu(0) = 0$] and of solving these equations. These mathematical problems are not considered in this chapter but in the preceding chapter (Section 4.9.4) and deserve further mathematical research.

# Chapter 6

# Foundations of practical stability domains

## 6.1 Introductory comment

A practical stability analysis is based on the Lyapunov method in a broad sense. Qualitative properties of the system (generalised) motions are studied via behaviour of a continuous function $v$ along the motions by using its total time derivative rather than to determine the motions themselves. This is the basis of the Lyapunov method in a broad sense. Its closer sense requires the function $v$ to be positive definite, and generally its total time derivative to possess a (semi-)definiteness property. Such demands are not imposed on the function $v$ in the framework of practical stability. We use instead the extremal (minimal or maximal) values of the function $v$ on appropriate sets or on their boundaries. Then, by using the total time derivative of $v$, we are able to compare instantaneous values of $v$ along motions with its extremal values of importance over an accepted (finite or infinite) time interval.

What follows is further development of the methodology established by Weiss and Infante [208], Michel [144]–[146] and Grujić [60]–[62], and opening of new practical stability study compared with that of [11], [29], [43], [44], [52], [60]–[63], [91], [93], [99], [105], [109]–[111], [121], [133], [144]–[146], [150], [152], [156], [179], [194], [205], [207]–[209]. The novelty is related to conditions for a set to be an estimate of the domain of a requested practical stability property.

## 6.2 System aggregation function and sets

### 6.2.1 System description and sets

We follow Sections 1.2 and 2.3 so that systems to be studied are generally in forced regimes and described by (6.1),

$$\frac{dX}{dt} = f(X, \mathtt{i}), \ \ f : \Re^n \times \Re^l \to \Re^n. \tag{6.1}$$

Our interest is in system (generalised) motions $\underline{X}(t; X_0; \mathbf{i})$ over a (finite or infinite) time interval $\Re_\tau = [0, \tau[\subseteq \Re_+$ with respect to some of, or all of the following sets possessing a non-empty interior:

$\mathcal{X}_A$ is the set of admitted system states over $\Re_\tau$,

$\mathcal{X}_F$ is the set of admitted system states over $\mathcal{T}_{\tau s} = [\tau_s, \tau[\subset \Re_\tau$, where $\tau_s \in \overline{\Re}_\tau$ so that $\tau_s = 0$ implies $\mathcal{X}_F = \mathcal{X}_A$ and $\tau_s = \tau$ means $\mathcal{X}_F = \Re^n$, otherwise $\mathcal{X}_F \subset \mathcal{X}_A$,

$\mathcal{I}$ is the set of permitted inputs (vector functions) $\mathbf{i}$ over $\Re_\tau$.

The sets $\mathcal{X}_A$ and $\mathcal{X}_F$ are (usually) prespecified. The set $\mathcal{I}$ and times $\tau$ and $\tau_s$ can be given or to be determined in general, but in what follows they will be assumed given.

The sets $\mathcal{X}_A$, $\mathcal{X}_F$ and $\mathcal{I}$ will be assumed connected throughout this chapter.

### 6.2.2 Definition of estimates of practical stability domains of systems

**Definition 6.1** *(a) A set $\mathcal{S}$ is an estimate of the domain $\mathcal{D}_{ps}(\tau, \mathcal{X}_A, \mathcal{I})$ of practical stability of the system (6.1) with respect to $\{\tau, \mathcal{X}_A, \mathcal{I}\}$, which is denoted by $\mathcal{E}_{ps}(\tau, \mathcal{X}_A, \mathcal{I})$, $\mathcal{S} = \mathcal{E}_{ps}(\tau, \mathcal{X}_A, \mathcal{I})$, if and only if*

*(i) the interior $\mathring{\mathcal{S}}$ of $\mathcal{S}$ is non-empty and $\mathcal{S}$ is connected,*

*(ii) $\mathcal{S}$ is subset of $\mathcal{D}_{ps}(\tau, \mathcal{X}_A, \mathcal{I})$: $\mathcal{S} \subseteq \mathcal{D}_{ps}(\tau, \mathcal{X}_A, \mathcal{I})$.*

*(b) A set $\mathcal{S}$ is an estimate of the domain $\mathcal{D}_{pc}(\tau, \tau_s, \mathcal{X}_F, \mathcal{I})$ of practical contraction with the settling time $\tau_s$ of the system (6.1) with respect to $\{\tau, \mathcal{X}_F, \mathcal{I}\}$, which is denoted by $\mathcal{E}_{pc}(\tau, \tau_s, \mathcal{X}_F, \mathcal{I})$, $\mathcal{S} = \mathcal{E}_{pc}(\tau, \tau_s, \mathcal{X}_F, \mathcal{I})$, if and only if*

*(i) the interior $\mathring{\mathcal{S}}$ of $\mathcal{S}$ is non-empty and $\mathcal{S}$ is connected,*

*(ii) $\mathcal{S}$ is subset of $\mathcal{D}_{pc}(\tau, \tau_s, \mathcal{X}_F, \mathcal{I})$: $\mathcal{S} \subseteq \mathcal{D}_{pc}(\tau, \tau_s, \mathcal{X}_F, \mathcal{I})$.*

*(c) A set $\mathcal{S}$ is an estimate of the domain $\mathcal{D}_p(\tau, \tau_s, \mathcal{X}_A, \mathcal{X}_F, \mathcal{I})$ of practical stability with the settling time $\tau_s$ of the system (6.1) with respect to $\{\tau, \mathcal{X}_A, \mathcal{X}_F, \mathcal{I}\}$, which is denoted by $\mathcal{E}_p(\tau, \tau_s, \mathcal{X}_A, \mathcal{X}_F, \mathcal{I})$, $\mathcal{S} = \mathcal{E}_p(\tau, \tau_s, \mathcal{X}_A, \mathcal{X}_F, \mathcal{I})$, if and only if*

*(i) the interior $\mathring{\mathcal{S}}$ of $\mathcal{S}$ is non-empty and $\mathcal{S}$ is connected,*

*(ii) $\mathcal{S}$ is subset of $\mathcal{D}_p(\tau, \tau_s, \mathcal{X}_A, \mathcal{X}_F, \mathcal{I})$.*

*(d) The arguments $\tau, \tau_s, \mathcal{X}_A, \mathcal{X}_F$ and $\mathcal{I}$ may be omitted if and only if they are prespecified and fixed.*

**Note 6.1** If $\mathcal{E}_{ps}$ and $\mathcal{E}_{pc}$ are estimates of $\mathcal{D}_{ps}$ and $\mathcal{D}_{pc}$, respectively, then for $\mathcal{S} = \mathcal{E}_{ps} \cap \mathcal{E}_{pc}$ to be an estimate $\mathcal{E}_p$ of $\mathcal{D}_p$ it is both necessary and sufficient that $\mathcal{S}$ is connected with non-empty interior $\mathring{\mathcal{S}}$.

### 6.2.3 System aggregation function extrema and sets

Let $\mathcal{Z} \subseteq \mathring{\mathcal{X}}_A$ denote a finite set or $\mathcal{Z} = \emptyset$.

**Definition 6.2** *A function $v : \Re \times \Re^n \to \Re$ is an aggregation function of the system (6.1) if and only if*

1) $v(t, X) \in C[\Re_+ \times (\Re^n - \mathcal{Z})]$,

2) $v(t, X) \in [\Re \cup \{-\infty\}]$ *for all* $t \in \Re_+$ *and every* $X \in \mathcal{Z}$,

3) *properties of $v$ and $\mathcal{D}^+ v$ yield to a clear conclusion about a qualitative dynamical property of the system.*

**Note 6.2** A Lyapunov function of a system is its aggregation function, but vice versa need not be true.

**Note 6.3** Since $v$ can depend on time $t$ we present the definition of $\mathcal{D}^+ v(t, X)$ as

$$\mathcal{D}^+ v(t, X) = lim\ sup \left\{ \frac{v[t+\theta, \underline{X}(t+\theta; t, X; \mathtt{i})] - v(t, X)}{\theta} : \theta \to 0^+ \right\}.$$

If $v(t, X) \in C^{(1,1)}(\Re_+ \times \Re^n)$ then

$$\mathcal{D}^+ v(t, X) = \frac{dv(t, X)}{dt} = \dot{v}(t, X) = \frac{\partial}{\partial t} v(t, X) + [grad\, v(t, X)]^T f(X, \mathtt{i}).$$

The following notation is usual for extreme values of $v$ on sets $\mathcal{X}_{(.)}$ in the framework of practical stability:

$$\begin{aligned}
v_{M(\cdot)}(t) &= sup\left[v(t, X) : x \in \mathcal{X}_{(\cdot)}\right], \ v_{\overline{M}(\cdot)}(t) = sup\left[v(t, X) : x \in \overline{\mathcal{X}}_{(\cdot)}\right], \\
v_{m(\cdot)}(t) &= inf\left[v(t, X) : x \in \mathcal{X}_{(\cdot)}\right], \ v_{\overline{m}(\cdot)}(t) = inf\left[v(t, X) : x \in \overline{\mathcal{X}}_{(\cdot)}\right], \\
v^{\partial}_{m(\cdot)}(t) &= inf\left[v(t, X) : x \in \partial\mathcal{X}_{(\cdot)}\right], \\
\dot{v}_{\overline{M}(\cdot)\tau} &= sup\left[\mathcal{D}^+ v(t, X, \mathtt{i}) : (t, X, \mathtt{i}) \in \Re_\tau \times \overline{\mathcal{X}}_{(\cdot)} \times \mathcal{I}\right].
\end{aligned}$$

**Example 6.1** Let $\mathcal{X}_A = \{X : ||X|| < e\} = \mathcal{B}_e$, $\mathcal{X}_F = \{X : ||X|| < 1\} = \mathcal{B}_1$ and $v(t, X) = 2\, ln\,[(1+t)||X||] \in C[\Re_+ \times (\Re^n - \mathcal{O})]$. In this case $\mathcal{Z} = \mathcal{O}$. Now we calculate the extrema of $v$ on $\mathcal{X}_A$, $\overline{\mathcal{X}}_A$, $\partial\mathcal{X}_A$, $\mathcal{X}_F$, $\overline{\mathcal{X}}_F$ and $\partial\mathcal{X}_F$ as follows:

$$\begin{aligned}
v_{MA}(t) &= sup\,[v(t, X) : x \in \mathcal{X}_A] = sup\,\{2\, ln\,[(1+t)||X||] : X \in \mathcal{B}_e\} = 2\, ln\,[e(1+t)], \\
v_{\overline{M}A}(t) &= sup\,[v(t, X) : x \in \overline{\mathcal{X}}_A] = sup\,\{2\, ln\,[(1+t)||X||] : X \in \overline{\mathcal{B}}_e\} = 2\, ln\,[e(1+t)], \\
v_{mA}(t) &= inf\,[v(t, X) : x \in \mathcal{X}_A] = inf\,\{2\, ln\,[(1+t)||X||] : X \in \mathcal{B}_e\} = -\infty, \\
v_{\overline{m}A}(t) &= inf\,[v(t, X) : x \in \overline{\mathcal{X}}_A] = inf\,\{2\, ln\,[(1+t)||X||] : X \in \overline{\mathcal{B}}_e\} = -\infty, \\
v^{\partial}_{mA}(t) &= inf\,[v(t, X) : x \in \partial\mathcal{X}_A] = inf\,\{2\, ln\,[(1+t)||X||] : X \in \partial\mathcal{B}_e\} = 2\, ln\,[e(1+t)],
\end{aligned}$$

and

$$v_{MF}(t) = sup\,[v(t,X) : x \in \mathcal{X}_F] = sup\,\{2\,ln\,[(1+t)||X||] : X \in \mathcal{B}_1\} = 2\,ln\,(1+t),$$

$$v_{\overline{M}f}(t) = sup\,[v(t,X) : x \in \overline{\mathcal{X}}_f] = sup\,\{2\,ln\,[(1+t)||X||] : X \in \overline{\mathcal{B}}_1\} = 2\,ln\,(1+t),$$

$$v_{mF}(t) = inf\,[v(t,X) : x \in \mathcal{X}_F] = inf\,\{2\,ln\,[(1+t)||X||] : X \in \mathcal{B}_1\} = -\infty,$$

$$v_{\overline{m}F}(t) = inf\,[v(t,X) : x \in \overline{\mathcal{X}}_F] = inf\,\{2\,ln\,[(1+t)||X||] : X \in \overline{\mathcal{B}}_1\} = -\infty,$$

$$v^{\partial}_{mF}(t) = inf\,[v(t,X) : x \in \partial\mathcal{X}_F] = inf\,\{2\,ln\,[(1+t)||X||] : X \in \partial\mathcal{B}_1\} = 2\,ln\,(1+t).$$

## 6.3 Estimate of the system practical stability domain

A set $\mathcal{Z}$, $\mathcal{Z} \subset \Re^n$, will be called "at most finite" iff it is either finite (the number of its elements is finite) or empty.

**Theorem 6.1** *Let the system (6.1) have generalised motions* $\underline{X}(t;X_0;\mathtt{i}) \in C_t(\Re_\tau \times \overline{\mathcal{X}}_A \times \mathcal{I})$. *In order for a compact set* $\mathcal{S}$, $\mathcal{S} \subseteq \mathcal{X}_A$, *to be the estimate* $\mathcal{E}_{ps}(\tau,\mathcal{X}_A,\mathcal{I})$ *of the practical stability domain* $\mathcal{D}_{ps}(\tau,\mathcal{X}_A,\mathcal{I})$ *of the system (6.1) with respect to* $\{\tau,\mathcal{X}_A,\mathcal{I}\}$: $\mathcal{S} = \mathcal{E}_{ps}(\tau,\mathcal{X}_A,\mathcal{I})$, *it is sufficient that*

1) *$\mathcal{S}$ is connected and with non-empty interior:* $\mathring{\mathcal{S}} \neq \emptyset$,

2) *there are at most a finite subset* $\mathcal{Z}$ *of* $\mathcal{S}$ *and a function* $v : \Re_\tau \times \overline{\mathcal{X}}_A \to (\Re \cup \{-\infty\})$ *such that*

   a) $v(t,X) \in C[\Re_\tau \times (\overline{\mathcal{X}}_A - \mathcal{Z})]$; $v^{\partial}_{mA}(t) < v(t,X)$, $\forall (t,X) \in \Re_\tau \times (\Re^n \times \overline{\mathcal{X}}_A)$,

   b) *there exists a function* $\psi : \Re_\tau \to \Re$, *which is integrable over* $\Re_\tau$ *and obeys both*

   (i) $\mathcal{D}^+ v(t,X,\mathtt{i}) < \psi(t)$ *for all* $(t,X) \in \Re_\tau \times (\overline{\mathcal{X}}_A - \mathcal{Z})$ *and every* $i \in \mathcal{I}$, *and*

   (ii) $\int_0^t \psi(\sigma)d\sigma \leq v^{\partial}_{mA}(t) - v_{\overline{M}\mathcal{S}}(t)$ *for all* $t \in \Re_\tau$. ∇∇∇

**Proof.** Let all the conditions hold. From the conditions under 2) we easily derive

$$v[t,\underline{X}(t;X_0;\mathtt{i})] < v(X_0) + v^{\partial}_{mA}(t) - v_{\overline{M}\mathcal{S}}(t),\ \ \forall (t,X_0) \in \Re_\tau \times \mathcal{S}, v(X_0) = v(0,X_0).$$

By the definition of $v$-extrema:

$$v(X_0) \leq v_{\overline{M}\mathcal{S}},\ \ \forall X_0 \in \mathcal{S}, v_{\overline{M}\mathcal{S}} = v_{\overline{M}\mathcal{S}}(0).$$

The preceding inequalities yield

$$v[t,\underline{X}(t;X_0;\mathtt{i})] < v^{\partial}_{mA}(t),\ \ \forall (t,X_0) \in \Re_\tau \times \mathcal{S}.$$

From $\underline{X}(t;X_0;\mathtt{i}) \in C(\Re_\tau \times \mathcal{X}_A \times \mathcal{I})$, connectedness of $\mathcal{X}_A$, $\mathcal{S} \subseteq \mathcal{X}_A$ and $X_0 \in \mathcal{S}$ it follows that $\underline{X}(t;X_0;\mathtt{i})$ can leave $\overline{\mathcal{X}}_A$ only by passing through its boundary $\partial\mathcal{X}_A$. Let be assumed that there is $(X_0,\mathtt{i},\zeta) \in \mathcal{S} \times \mathcal{I} \times \Re_\tau$ such that $\underline{X}(\zeta;X_0;\mathtt{i}) \in \partial\mathcal{X}_A$. From the last inequality we deduce

$$v[\zeta,\underline{X}(\zeta;X_0;\mathtt{i})] < v^{\partial}_{mA}(\zeta)$$

that implies $\underline{X}(\zeta;X_0;\mathtt{i}) \notin \partial\mathcal{X}_A$. This contradicts the definition of $(X_0,\mathtt{i},\zeta) \in \mathcal{S} \times \mathcal{I} \times \Re_\tau$. Hence, such $(X_0,\mathtt{i},\zeta) \in \mathcal{S} \times \mathcal{I} \times \Re_\tau$ does not exist, which proves

$$\underline{X}(t;X_0;\mathtt{i}) \in \mathcal{X}_A \text{ for all } t \in \Re_\tau \text{ and every } (X_0,\mathtt{i}) \in \mathcal{S} \times \mathcal{I}.$$

This, connectedness of $\mathcal{S}$, $\mathring{\mathcal{S}} \neq \emptyset$ and $\mathcal{S} \subseteq \mathcal{X}_A$ prove

$$\mathcal{S} = \mathcal{E}_{ps}(\tau,\mathcal{X}_A,\mathcal{I}).$$

Let

$$\dot{v}_{\overline{M}AS} = sup\left[\mathcal{D}^+v(t,X,\mathtt{i}) : (t,X,\mathtt{i}) \in \Re_\tau \times (\overline{\mathcal{X}}_A - \mathring{\mathcal{S}}) \times \mathcal{I}\right].$$

**Theorem 6.2** *Let the system (6.1) have generalised motions* $\underline{X}(t;X_0;\mathtt{i}) \in C_t(\Re_\tau \times \overline{\mathcal{X}}_A \times \mathcal{I})$. *In order for a compact set* $\mathcal{S}$, $\mathcal{S} \subseteq \mathcal{X}_A$, *to be the estimate* $\mathcal{E}_{ps}(\tau,\mathcal{X}_A,\mathcal{I})$ *of the domain* $\mathcal{D}_{ps}(\tau,\mathcal{X}_A,\mathcal{I})$ *of practical stability of the system (6.1) with respect to* $\{\tau,\mathcal{X}_A,\mathcal{I}\}$: $\mathcal{S} = \mathcal{E}_{ps}(\tau,\mathcal{X}_A,\mathcal{I})$, $\mathcal{X}_A = \overline{\mathcal{X}}_A$ *it is sufficient that*

*1)* $\mathcal{S}$ *is connected and with non-empty interior:* $\mathring{\mathcal{S}} \neq \emptyset$,

*and*

*2) there are at most finite subset* $\mathcal{Z}$ *of* $\mathcal{S}$ *and a function* $v : \Re_\tau \times \overline{\mathcal{X}}_A \to (\Re \cup \{-\infty\})$ *such that*

*a)* $v(t,X) \in C[\Re_\tau \times (\overline{\mathcal{X}}_A - \mathcal{Z})]$; $v^{\partial}_{mA}(t) < v(t,X)$, $\forall (t,X) \in \Re_\tau \times (\Re^n \times \overline{\mathcal{X}}_A)$,

*b)* $v_{\overline{M}S}(0) \leq v^{\partial}_{mA}(t) - t\dot{v}_{\overline{M}AS}$, $\forall t \in \Re_\tau$. ∇∇∇

**Proof.** Let all the conditions hold. Let $(\mathtt{i},X_0) \in \mathcal{I} \times \mathcal{S}$ be such that there are $t_1 \in ]0,\tau[$ and $\sigma \in ]t_1,\tau]$, $t_1 = t_1(X_0;\mathcal{I},\mathcal{S})$, $\sigma = \sigma(X_0;\mathcal{I},\mathcal{S};t_1)$, which obey $\underline{X}(t;X_0;\mathtt{i}) \in (\mathcal{X}_A - \mathring{\mathcal{S}})$, $\forall t \in [t_1,\sigma[$, and $\sigma$ determines the largest interval $[t_1,\sigma[$.

From the definition of $\dot{v}_{\overline{M}AS}$ follows

$$\mathcal{D}^+v(t,X,\mathtt{i}) \leq \dot{v}_{\overline{M}AS}, \quad \forall (t,X,\mathtt{i}) \in ]t_1,\tau[ \times (\overline{\mathcal{X}}_A - \mathring{\mathcal{S}}) \times \mathcal{I}.$$

Integrating this inequality we get

$$v[\underline{X}(t;X_0;\mathtt{i})] - v(X_0) \leq t\dot{v}_{\overline{M}AS}, \quad \forall t \in ]t_1,\sigma[,$$

which with the conditions under 2) of the theorem and the definition of $t_1$ yields

$$v[\underline{X}(t;X_0;\mathtt{i})] \leq v^{\partial}_{mA}(t) + v(X_0) - v_{\overline{M}S}(0), \quad \forall (t,X_0,\mathtt{i}) \in ]t_1,\sigma[ \times \mathcal{S} \times \mathcal{I}.$$

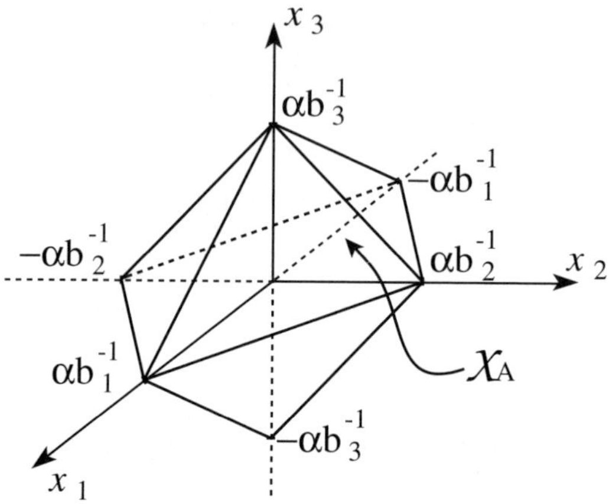


Figure 6.1: The set $\mathcal{X}_A = \{X : \mathtt{b}^T|X| \leq \alpha\}$, (6.2), in the case $n = 3$.

By the definition

$$v_{\overline{M}\mathcal{S}}(0) \geq v(X_0), \;\; \forall X_0 \in \mathcal{S}.$$

The preceding two inequalities combined imply

$$v[\underline{X}(t; X_0; \mathtt{i})] \leq v^{\partial}_{mA}(t), \;\; \forall (t, X_0, \mathtt{i}) \in ]t_1, \sigma[\times \mathcal{S} \times \mathcal{I}.$$

This, $\underline{X}(t; X_0; \mathtt{i}) \in C_t(\Re_\tau \times \mathcal{X}_A \times \mathcal{I})$, 2a) and the definitions of $t_1$ and $\sigma$ prove

$$\underline{X}(t; X_0; \mathtt{i}) \in \mathcal{X}_A, \;\; \forall (t, X_0, \mathtt{i}) \in \Re_\tau \times \mathcal{S} \times \mathcal{I},$$

which together with the condition 1) proves that $\mathcal{S} = \mathcal{E}_{ps}(\tau, \mathcal{X}_A, \mathcal{I})$ (Definition 6.1a).

Let (Fig. 6.1)

$$\mathcal{X}_A = \{X : \mathtt{b}^T|X| = \sum_{i=1}^{n} b_i|x_i| \leq \alpha\}, \; \alpha \in \Re^+, \; \mathtt{b} > 0, \tag{6.2}$$

where $\mathtt{b} = (b_1 \; b_2 \; \ldots \; b_n)^T$. If

$$f(0) = 0 \tag{6.3}$$

then the following transformation by Michel *et al.* [148] is valid:

$$f(X, \mathtt{i}) = H(X, \mathtt{i})X, \tag{6.4a}$$

$$H(\cdot) : \Re^n \times \Re^l \to \Re^{n \times n}, \quad H(\cdot) = [h_{ij}(\cdot)],$$

$$h_{ij}(X, \mathtt{i}) = \left\{ \begin{array}{cc} \dfrac{f_i(X, \mathtt{i})x_j}{||X||^2} & , \; ||X|| \neq 0, \\ 0 & , \; ||X|| = 0 \end{array} \right\}, \tag{6.4b}$$

so that the system (6.1), (6.3) can be set in the form (6.5),

$$\frac{dX}{dt} = H(X, \mathtt{i})X. \tag{6.5}$$

Let

$$A = (a_{ij}) \in \Re^{n\times n}, \tag{6.6a}$$

$$a_{ij} = \delta_{ij}\, sup\,[h_{ii}(X,\mathtt{i}) : (X,\mathtt{i}) \in (\mathcal{X}_A - \mathring{\mathcal{S}}) \times \mathcal{I}] +$$

$$+ \ (1-\delta_{ij})\, max\,\{0, sup\,[h_{ij}(X,\mathtt{i})\, sign\, x_j : (X,\mathtt{i}) \in (\mathcal{X}_A - \mathring{\mathcal{S}}) \times \mathcal{I}]\}. \tag{6.6b}$$

**Theorem 6.3** *Let the system (6.5) have generalised motions* $\underline{X}(t; X_0; \mathtt{i}) \in C_t(\Re_\tau \times \mathcal{X}_A \times \mathcal{I})$. *In order for the set* $\mathcal{S} = \{X : \mathtt{b}^T|X| \le \zeta\alpha\}$, $\zeta \in\, ]0,1]$, *to be estimate* $\mathcal{E}_{ps}(\tau, \mathcal{X}_A, \mathcal{I})$ *of the domain* $\mathcal{D}_{ps}(\tau, \mathcal{X}_A, \mathcal{I})$ *of practical stability of the system (6.5) with respect to* $\{\tau, \mathcal{X}_A\ (6.2), \mathcal{I}\}$: $\mathcal{S} = \mathcal{E}_{ps}(\tau, \mathcal{X}_A, \mathcal{I})$, *it is sufficient that*

$$\gamma t \le (1-\zeta)\alpha, \quad \forall t \in \Re_\tau,$$

*where*

$$\gamma = \sum_{k=0}^{n} max\left(0, \alpha\frac{c_k}{b_k}\right),$$

*and* $\mathtt{c} = (c_1\ c_2\ \ldots c_n)^T$ *is determined by*

$$\mathtt{c} = A^T\mathtt{b}.$$

∇∇∇

**Proof.** Under the conditions of the theorem statement it follows that 2a) of Theorem 6.2 is valid, $v_{\overline{M}S} = \beta = \zeta\alpha$, $v^{\partial}_{MA} = \alpha$ for $v(X) = \mathtt{b}^T|X|$, and

$$\mathcal{D}^+ v(X,\mathtt{i}) = \mathtt{b}^T\mathcal{D}^+|X| = \mathtt{b}^T\Omega(X,\dot{X})H(X,\mathtt{i})X \le \mathtt{b}^T A|X| = \mathtt{c}^T|X| \le \gamma,$$

$$\forall (X,\mathtt{i}) \in \overline{\mathcal{X}}_A \times \mathcal{I},$$

so that

$$\dot{v}_{\overline{M}AS} = \gamma.$$

Hence, the condition $\gamma t \le (1-\zeta)\alpha$, $\forall t \in \Re_\tau$, can be set in the following form:

$$v_{\overline{M}S} \le v^{\partial}_{MA} - t\dot{v}_{\overline{M}AS}, \quad \forall t \in \Re_\tau,$$

which shows that the condition 2b) of Theorem 6.2 holds. The set $\mathcal{S} = \{X : \mathtt{b}^T|X| \le \zeta\alpha\}$ obviously is connected with non-empty interior because $\zeta\alpha \in \Re^+$ and $\mathtt{b} > 0$ so that the condition 1) of the Theorem 6.2 is also satisfied. All conditions of the Theorem 6.2 are valid, which proves the statement of Theorem 6.3.

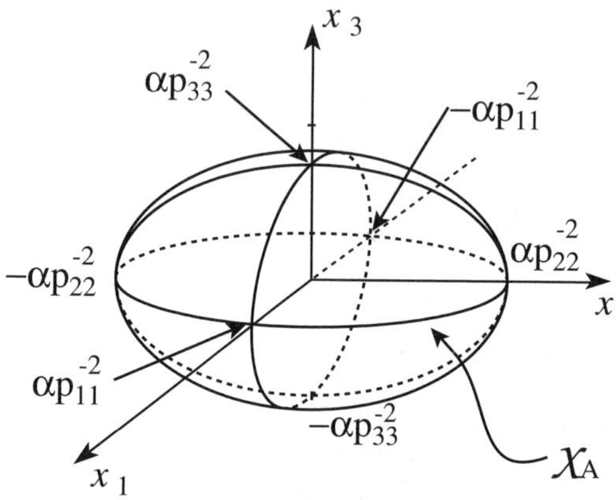


Figure 6.2: The set $\mathcal{X}_A = \{X : X^T PX \leq \alpha\}$, (6.7), in the case $n = 3$.

**Comment 6.1** *Stability of the matrix A is NOT required in Theorem 6.3. This is due to possible finite $\tau$: $\tau < \infty$, and represents an essential difference from Lyapunov's conditions.*

Let (Fig. 6.2)

$$\mathcal{X}_A = \{X : X^T PX \leq \alpha\},\ \alpha \in \Re^+,\ P = P^T = (p_{ij}) > 0 \tag{6.7}$$

and

$$K = (k_{ij}) \in \Re^{n \times n}, \tag{6.8a}$$

$$k_{ij} = \delta_{ij}\, sup\,[w_{ii}(X, \mathtt{i}) : (X, \mathtt{i}) \in (\mathcal{X}_A - \mathring{\mathcal{S}}) \times \mathcal{I}] +$$

$$+\ (1 - \delta_{ij})\, max\{0, sup\,[w_{ij}(X, \mathtt{i})\, sign(x_i x_j) : (X, \mathtt{i}) \in (\mathcal{X}_A - \mathring{\mathcal{S}}) \times \mathcal{I}]\}, \tag{6.8b}$$

$$W(\cdot) = [w_{ij}(\cdot)] = H^T(\cdot)P + PH(\cdot). \tag{6.8c}$$

**Theorem 6.4** *Let the system (6.5) have generalised motions $\underline{X}(t; X_0; \mathtt{i}) \in C_t(\Re_\tau \times \mathcal{X}_A \times \mathcal{I})$. In order for the set $\mathcal{S} = \{X : X^T PX \leq \zeta\alpha\}$, $\zeta \in\, ]0, 1]$ to be estimate $\mathcal{E}_{ps}(\tau, \mathcal{X}_A, \mathcal{I})$ of the domain $\mathcal{D}_{ps}(\tau, \mathcal{X}_A, \mathcal{I})$ of practical stability of the system (6.5) with respect to $\{\tau, \mathcal{X}_A$ (6.7)$, \mathcal{I}\}$: $\mathcal{S} = \mathcal{E}_{ps}(\tau, \mathcal{X}_A, \mathcal{I})$, it is sufficient that*

$$\Theta t \leq (1 - \zeta)\alpha,\ \ \forall t \in \Re_\tau,$$

*where*

$$\Theta = \begin{cases} \alpha \dfrac{\Lambda(K)}{\lambda(P)} & if\ \Lambda(K) \geq 0, \\ 0 & if\ \Lambda(K) \leq 0. \end{cases}$$

∇∇∇

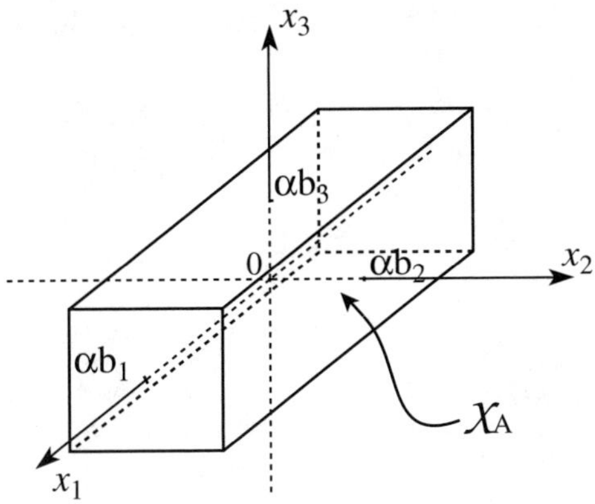


Figure 6.3: The set $\mathcal{X}_A = \left\{ X : max\left(\frac{|x_i|}{b_i} : i = 1, 2, \ldots, n\right) \leq \alpha \right\}$, (6.9), in case $n = 3$.

**Proof.** The set $\mathcal{X}_A$ (6.7) obeys the condition 1) of Theorem 6.2 due to $\alpha \in \Re^+$ and $P = P^T > 0$. Let $v(X) = X^T P X$ so that 2a) of Theorem 6.2 holds,

$$v_{\overline{M}S} = \zeta\alpha = \beta, \; v^{\partial}_{MA} = \alpha$$

and

$$\begin{aligned} \dot{v}(X, \mathtt{i}) &= X^T \left[ PH(X, \mathtt{i}) + H^T(X, \mathtt{i})P \right] X = \\ &= X^T W(X, \mathtt{i}) X \leq |X|^T K |X|, \;\; \forall (X, \mathtt{i}) \in \mathcal{X}_A \times \mathcal{I}. \end{aligned}$$

This result, the fact that $||X||^2 \leq \alpha/\lambda(P)$ for all $X \in \mathcal{X}_A$ and the definition of $\Theta$, yield

$$\dot{v}(X, \mathtt{i}) \leq \Theta, \;\; \forall (X, \mathtt{i}) \in \mathcal{X}_A \times \mathcal{I}, \dot{v}(X, \mathtt{i}) \leq \zeta\Theta, \;\; \forall (X, \mathtt{i}) \in \mathcal{S} \times \mathcal{I}.$$

Hence,

$$\dot{v}_{\overline{M}AS} = \Theta.$$

This result, $v_{\overline{M}S} = \beta$ and $v^{\partial}_{MA} = \alpha$ together with the condition $\Theta t \leq (1 - \zeta)\alpha$, $\forall t \in \Re_\tau$, yield

$$v_{\overline{M}S} \leq v^{\partial}_{MA} - t\dot{v}_{\overline{M}AS}, \;\; \forall t \in \Re_\tau.$$

All the conditions of Theorem 6.2 hold, which completes the proof of Theorem 6.4.

Let (Fig. 6.3)

$$\mathcal{X}_A = \left\{ X : max\left(\frac{|x_i|}{b_i} : i = 1, 2, \ldots, n\right) \leq \alpha \right\}, \;\; \alpha \in \Re^+, \;\; \mathtt{b} > 0. \tag{6.9}$$

We will look for an estimate of the practical stability domain in the same form. Therefore,

$$\mathcal{S} = \left\{ X : max \left( \frac{|x_i|}{\zeta_i b_i} : i = 1, 2, \ldots, n \right) \leq \alpha \right\}, \ \zeta_i \in ]0, 1], \ \ i = 1, 2, \ldots, n, \quad (6.10)$$

is accepted for such a tentative estimate. Let

$$\mathsf{c} = A\mathsf{b}, \ \mathsf{c} = (c_1 \ c_2 \ \ldots \ c_n)^T. \quad (6.11)$$

**Theorem 6.5** *Let the system (6.5) have generalised motions* $\underline{X}(t; X_0; \mathtt{i}) \in C_t(\Re_\tau \times \mathcal{X}_a \times \mathcal{I})$. *In order for the set* $\mathcal{S}$, *(6.10), to be estimate* $\mathcal{E}_{ps}(\tau, \mathcal{X}_A, \mathcal{I})$ *of the domain* $\mathcal{D}_{ps}(\tau, \mathcal{X}_A, \mathcal{I})$ *of practical stability of the system (6.5) with respect to* $\{\tau, \mathcal{X}_A(6.9), \mathcal{I}\}$: *it is sufficient that*

$$\tau c_i \leq (1 - \zeta_i) b_i, \ \ i = 1, 2, \ldots, n.$$

$\nabla\nabla\nabla$

**Proof.** Let the condition of the theorem hold. The set $\mathcal{S}$ (6.10) is compact and connected. This and $\underline{X}(t; X_0; \mathtt{i}) \in C_t(\Re_\tau \times \mathcal{X}_a \times \mathcal{I})$ show that any motion starting from the set $\mathcal{S}$ (6.10), which leaves the set $\mathcal{S}$ after some time $t \in ]0, \tau[$ must be at the boundary $\partial\mathcal{S}$ of the set $\mathcal{S}$ at some time $t_1 \in \Re_\tau$. Let $t_1 \in [0, \sigma[$ and $\sigma \in ]0, \tau]$ obey

$$\underline{X}(t; X_0; \mathtt{i}) \in (\mathcal{X}_A - \mathring{\mathcal{S}}), \ \ \forall t \in [t_1, \sigma[,$$

where $\sigma$ determines the largest interval $[t_1, \sigma[$. Now we estimate the upper bound of $\mathcal{D}^+|x_i|$ along motions of the system (6.5) by using (6.6),

$$\mathcal{D}^+|x_i| \leq \sum_{j=1}^{n} a_{ij}|x_j|, \ \ \forall X \in (\mathcal{X}_A - \mathcal{S}), \ \ i = 1, 2, \ldots, n.$$

and make use of

$$\sum_{j=1}^{n} a_{ij}|x_j| \leq \alpha \sum_{j=1}^{n} a_{ij} b_j, \ \ \forall X \in (\mathcal{X}_A - \mathcal{S}), \ \ i = 1, 2, \ldots, n,$$

by combining them,

$$\mathcal{D}^+|x_i| \leq \alpha \sum_{j=1}^{n} a_{ij} b_j, \ \ \forall X \in (\mathcal{X}_A - \mathcal{S}), \ \ i = 1, 2, \ldots, n.$$

After integrating this inequality from $t_1$ to $t \in ]t_1, \sigma[$ we derive

$$|\underline{x}_i(t; X_0; \mathtt{i})| - |\underline{x}_i(t_1; X_0; \mathtt{i})| \leq \alpha(t - t_1)c_i,$$
$$\forall (t, X_0, \mathtt{i}) \in [t_1, \sigma[ \times \mathcal{S} \times \mathcal{I}, \ \ i = 1, 2, \ldots, n, \quad (6.12)$$

where $c_i$ is the i'th element of $\mathsf{c} = A\mathsf{b}$.

If $c_i \leq 0$ then (6.12) yield

$$|\underline{x}_i(t; X_0; \mathtt{i})| \leq |\underline{x}_i(t_1; X_0; \mathtt{i})|, \ \forall (t, X_0, \mathtt{i}) \in [t_1, \sigma[\times \mathcal{S} \times \mathcal{I}, \ \ c_i \leq 0. \tag{6.13a}$$

Hence, the assumed $t_1$ does not exist.

If $c_i \geq 0$ then $\alpha(t - t_i)c_i < \tau$ so that from (6.12) follows:

$$|\underline{x}_i(t; X_0; \mathtt{i})| \leq |\underline{x}_i(t_1; X_0; \mathtt{i})| + \alpha\tau c_i = \alpha(\zeta_i b_i + \tau c_i), \ \ c_i \geq 0, \tag{6.13b}$$

where the equality is due to $|\underline{X}(t_1; X_0; \mathtt{i})| \in \partial\mathcal{S}$ that yields $|\underline{x}_i(t; X_0; \mathtt{i})| = \alpha\zeta_i b_i$. If $t_1$ as defined above does not exist then $X_0 \in \mathcal{S}$, $\underline{X}(t; X_0; \mathtt{i}) \in C_t(\Re_\tau \times \mathcal{X}_a \times \mathcal{I})$, connectedness and compactness of $\mathcal{S}$ imply $\underline{X}(t; X_0; \mathtt{i}) \in \mathcal{S}$, $\forall t \in \Re_\tau$. In that case $\underline{X}(t; X_0; \mathtt{i}) \in \mathcal{X}_A$, $\forall t \in \Re_\tau$, due to $\mathcal{S} \subseteq \mathcal{X}_A$. Hence, we continue considering (6.13b). Since $\alpha(\zeta_i b_i + \tau c_i) \leq \alpha b_i, \forall i = 1, 2, \ldots, n$, due to the condition $\tau c_i \leq (1 - \zeta_i) b_i$ and $\alpha \in \Re^+$, then (6.13b) yields

$$\frac{|\underline{x}_i(t; X_0; \mathtt{i})|}{b_i} \leq \alpha, c_i \geq 0.$$

This together with connectedness and compactness of $\mathcal{X}_A$ and $\underline{X}(t; X_0; \mathtt{i}) \in C_t(\Re_\tau \times \mathcal{X}_a \times \mathcal{I})$, and (6.14) ensure that $\underline{X}(t; X_0; \mathcal{I}) \in \mathcal{X}_A$ also in the case when there exist $t_1$ and $\sigma$ as defined above. Altogether,

$$\underline{X}(t; X_0; \mathtt{i}) \in \mathcal{X}_A, \ \ \forall (t, X_0, \mathtt{i}) \in \Re_\tau \times \mathcal{S} \times \mathcal{I},$$

which proves the theorem in view of Definition 6.1(a).

It is interesting to note that the set $\mathcal{X}_A$ is $\mathcal{O}$-uniquely bounded in all three cases with the generating function $u(\cdot)$ determined by

a) $u(X) = \mathtt{b}^T |X|$ in case $\mathcal{X}_A$ is defined by (6.2),

b) $u(X) = X^T P X$ in case $\mathcal{X}_A$ is defined by (6.7),

c) $u(X) = max\left(\frac{|x_i|}{b_i} : i = 1, 2, \ldots, n\right)$ in case $\mathcal{X}_A$ is defined by (6.9).

We terminate this section with a result corresponding to the general case of a uniquely bounded set $\mathcal{X}_A$ with a generating function $u(\cdot)$ (Definition 4.13, section 4.3.2):

$$u(X) \begin{Bmatrix} < \alpha, & \forall X \in \mathring{\mathcal{X}}_A \\ = \alpha, & \forall X \in \partial\mathcal{X}_A \\ > \alpha, & \forall X \in (\Re^n - \mathcal{X}_A), \ \mathcal{X}_A = \overline{\mathcal{X}}_a, \ \ \alpha \in \Re^+, \end{Bmatrix} \tag{6.14}$$

and with an example. Let

$$\mathcal{S} = \{X : u(X) \leq \zeta\alpha\}, \zeta \in ]0, 1], \tag{6.15}$$

and

$$\dot{u}_{MAS} = max\,[\mathcal{D}^+ u(X) : X \in (\mathcal{X}_A - \mathring{\mathcal{S}})]. \tag{6.16}$$

**Theorem 6.6** *Let the system (6.1) have generalised motions* $\underline{X}(t; X_0; \mathrm{i}) \in C_t(\Re_\tau \times \mathcal{X}_a \times \mathcal{I})$ *where* $\mathcal{X}_A$ *is a closed uniquely bounded set with the generating function* $u$*, (6.14). In order for the set* $\mathcal{S}$*, (6.15), to be estimate* $\mathcal{E}_{ps}(\tau, \mathcal{X}_A, \mathcal{I})$ *of the domain* $\mathcal{D}_{ps}(\tau, \mathcal{X}_A, \mathcal{I})$ *of practical stability of the system (6.1) with respect to* $\{\tau, \mathcal{X}_A(6.14), \mathcal{I}\}$ *it is sufficient that*

$$t\dot{u}_{MAS} = (1-\zeta)\alpha, \ \ \forall t \in \Re_\tau.$$

∇∇∇

**Proof.** Let all the conditions hold. Hence, $\mathcal{S}$ is connected with a non-empty interior because $\zeta \in ]0,1]$ and due to Lemma 4.11 (section 4.3.3). The definitions (6.14) and (6.15) of $\mathcal{X}_A$ and $\mathcal{S}$ yield $v(\cdot) = u(\cdot)$ (in Theorem 6.2):

$$v_{\overline{M}S} = \zeta\alpha, v_{mA}^{\partial} = \alpha \ \text{ and } \ \dot{v}_{\overline{M}AS} = \dot{u}_{MAS}.$$

These equations show that the condition $t\dot{u}_{MAS} < (1-\zeta)\alpha$ becomes

$$v_{\overline{M}S} < v_{mA}^{\partial} - t\dot{v}_{\overline{M}AS}.$$

Hence, all the conditions of Theorem 6.2 are satisfied, which proves Theorem 6.6.

Notice that the condition of the theorem is satisfied for $\zeta = 1$, i.e. for $\mathcal{S} = \mathcal{X}_A$, as soon as $\dot{u}_{MAS} \leq 0$.

**Example 6.2** In order to illustrate the meaning of "at most finite subset $\mathcal{Z}$ of $\mathcal{S}$" in the Theorem 6.1 and Theorem 6.2 and its usefulness, we consider the problem of finding the greatest $\tau$ and the largest set $\mathcal{S}$ to be an estimate of $\mathcal{D}_{ps}(\tau, \mathcal{X}_A, \mathcal{I})$ of simple linear second order system ($X \in \Re^2$):

$$\frac{dX}{dt} = \begin{bmatrix} -3.1 & 2 \\ -2 & -3.1 \end{bmatrix} X,$$

with

$$\mathcal{X}_A = \{X : ||X|| \leq e\}.$$

Evidently, $\mathcal{I} = \emptyset$ in this example. The function $v$ is selected as

$$v(X) = ln\,||X||^2,$$

so that

$$v(X) \in C(\Re^n - \{0\}).$$

Hence, $\mathcal{Z} = \{0\}$ and $v(0) = -\infty$. In this case, we can apply Theorem 6.2:

$$v_{MA}^{\partial} = ln\,||X||^2|_{||X||^2=e^2} = ln\ e^2 = 2,$$

and, for $X \neq 0$,

$$\dot{v}(X) = \frac{1}{||X||^2} 2X^T \dot{X} = \frac{1}{||X||^2} 2X^T \begin{bmatrix} -3.1 & 2 \\ -2 & -3.1 \end{bmatrix} X =$$

$$= \frac{2}{||X||^2}[-3.1(x_1^2 + x_2^2)] = -6.2.$$

Hence,

$$\dot{v}_{\overline{M}AS} = -6.2.$$

The function $v$ obeys the condition 2a) of Theorem 6.2 for any neighbourhood $\mathcal{S}$ of $X = 0$ because then $\mathcal{Z} = \{0\} \subset \mathcal{S}$. The largest neighbourhood $\mathcal{S}$ and the greatest $\tau$ obeying the conditions 1) and 2b) of the same theorem are to obey $v_{\overline{M}S} \leq v_{mA}^{\partial} - t\dot{v}_{\overline{M}AS}$ for all $t \in \Re_\tau$, which in this case means: $v_{\overline{M}S} \leq 2 - t(-6.2)$ or $v_{\overline{M}S} \leq 2 + 6.2t$ for all $t \in \Re_\tau$. Since $min\,[2 + 6.2t : t \in \Re_\tau] = (2 + 6.2t)|_{t=0}$ and $(t = 0) \in \Re_\tau$ then $\overline{v}_{\overline{M}S}$ should obey $\overline{v}_{\overline{M}S} \leq 2$. This is satisfied by the set $\mathcal{X}_A$ itself. Hence, $\tau = \infty$ and

$$\mathcal{S} = \mathcal{D}_{ps}(\infty, \mathcal{X}_A, \{0\}) = \mathcal{X}_A.$$

**Example 6.3 [61]** Let function $f(\cdot)$ in (6.1) have the next specific form:

$$f(X, \mathtt{i}) = \begin{bmatrix} -0.2 & & \\ 0 & 0 & \\ 0 & -1.0 & 2.0 \\ 0 & 0 & -15 \end{bmatrix} X +$$

$$+ \begin{bmatrix} -0.2 & 0 & 0 & 0 \\ 0 & -2.0 & 0 & 0 \\ 0 & 0 & 2 & -100 \end{bmatrix} \begin{bmatrix} N_1[\sigma_1(X)] \\ N_2[\sigma_2(X)] \\ N_3[\sigma_3(X), \dot{\sigma}_3(X, \dot{X})] \\ N_4[\sigma_4(X, \mathtt{i})] \end{bmatrix},$$

where

$$\begin{bmatrix} \sigma_1(X) \\ \sigma_2(X) \\ \sigma_3(X) \\ \sigma_4(X, \mathtt{i}) \end{bmatrix} = \begin{bmatrix} 0.5 & 0.5 & 0 \\ 0 & 1 & 0 \\ 0 & 1 & 0 \\ \mathtt{i} & \frac{1}{7}|x_3| & 10^{-3}|x_1| \end{bmatrix} X,$$

$$N_1(\sigma_1) = \left\{ \begin{array}{ll} 6.9 & , \sigma_1 \geq 3.75 \\ \frac{13.8}{7.5}\sigma_1 & , |\sigma_1| \leq 3.75 \\ -6.9 & , \sigma_1 \leq -3.75 \end{array} \right\},$$

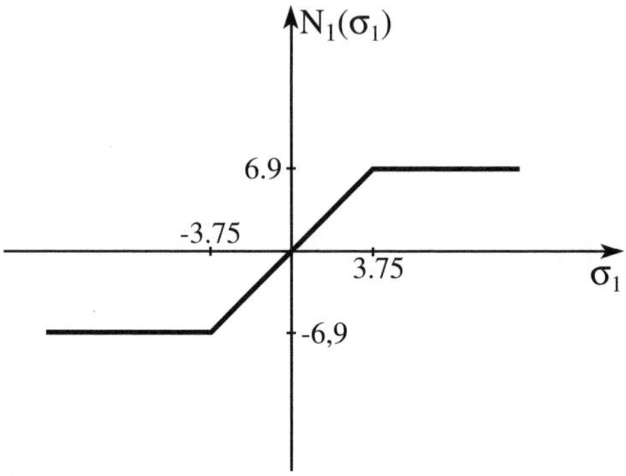


Figure 6.4: Nonlinearity $N_1(\cdot)$ graph.

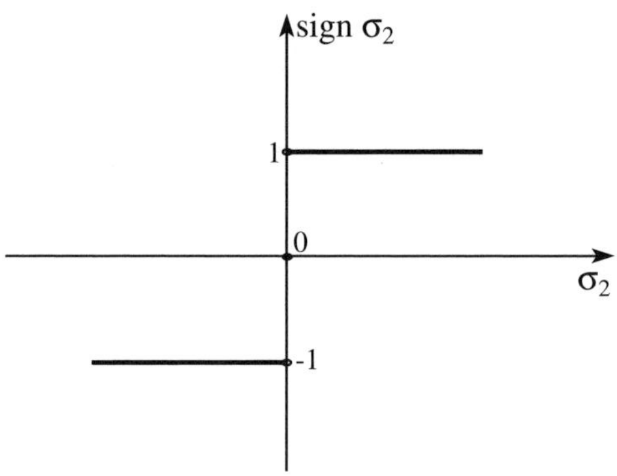


Figure 6.5: Nonlinearity $sgn(\cdot)$ graph.

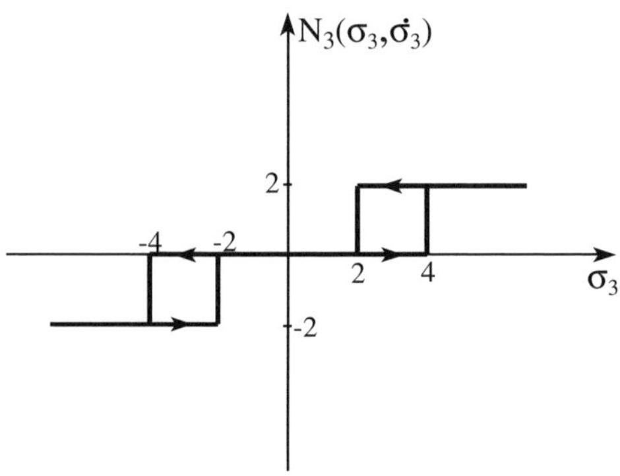


Figure 6.6: Nonlinearity $N_3(\cdot)$ graph.

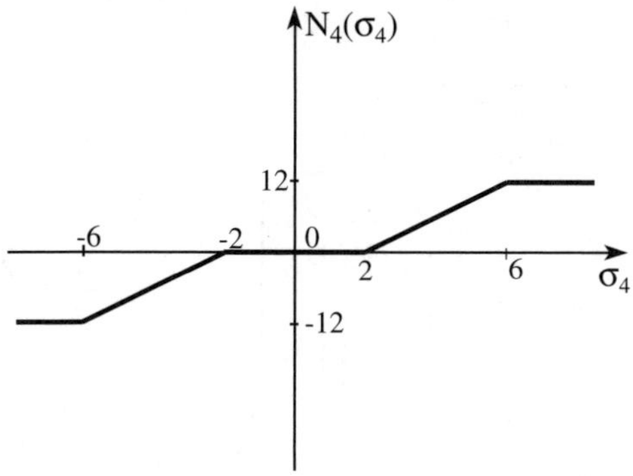


Figure 6.7: Nonlinearity $N_4(\cdot)$ graph.

$$N_2(\sigma_2) = 60\ sign\ \sigma_2,$$

$$N_3(\sigma_3, \dot{\sigma}_3) = \left\{ \begin{array}{ll} 2\ sgn\ \sigma_3 & , \left\{ \begin{array}{ll} |\sigma_3| > 4 & , \text{ or} \\ |\sigma_3| > 2 & , \sigma_3\dot{\sigma}_3 < 0 \end{array} \right\}, \\ 0 & , \left\{ \begin{array}{ll} |\sigma_3| < 4 & , \sigma_3\dot{\sigma}_3 > 0 \text{ or} \\ |\sigma_3| < 2 & \end{array} \right\} \end{array} \right\},$$

$$N_4(\sigma_4) = \left\{ \begin{array}{cl} 12\ sgn\ \sigma_4 & , |\sigma_4| \leq 6, \\ 3(|\sigma_4| - 2) sgn\ \sigma_4 & , 2 \leq |\sigma_4| \leq 6, \\ 0 & , |\sigma_4| \leq 2 \end{array} \right\},$$

$$\mathtt{i}(t) = (2.5\ sin^2 3t).$$

In this example

$$\mathcal{X}_A = \left\{ X : X \in \Re^3, \begin{bmatrix} |x_1| \\ |x_2| \\ |x_3| \end{bmatrix} \leq \begin{bmatrix} 16 \\ 30 \\ 70 \end{bmatrix} \right\}.$$

The set $\mathcal{X}_A$ can be expressed in the form (6.9) for which we set

$$\alpha = 2,\ b_1 = 8,\ b_2 = 15,\ b_3 = 35,$$

so that

$$\mathcal{X}_A = \left\{ X : X \in \Re^3,\ max\left( \frac{|x_i|}{b_i} : i = 1, 2, 3 \right) \leq \alpha \right\}.$$

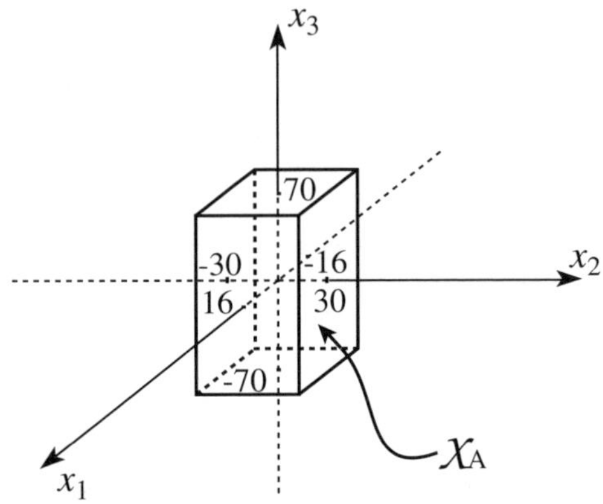


Figure 6.8: Set $\mathcal{X}_A$.

The function $H(\cdot)$ is found as

$$H(X,\mathtt{i}) = \begin{bmatrix} -0.2-0.1\frac{N_1(\sigma_1)}{\sigma_1} & 0 & -250\frac{N_4(\sigma_4)}{\sigma_4}sin^2 3t \\ -0.1\frac{N_1(\sigma_1)}{\sigma_1} & -1-20\frac{N_2(\sigma_2)}{\sigma_2} & 2\frac{N_3(\sigma_3,\dot{\sigma}_3)}{\sigma_3} - \\ & & -\frac{1}{7}\times 100\frac{N_4(\sigma_4)}{\sigma_4}|x_3| \\ 0 & 2 & -15-0.1\frac{N_4(\sigma_4)}{\sigma_4}|x_1| \end{bmatrix}^T,$$

so that the matrix $A$, (5.6), is determined by

$$A = \begin{pmatrix} -0.5 & 0.2 & 0 \\ 0 & -41.0 & 2 \\ 0 & 30.0 & -15 \end{pmatrix}.$$

In this case,

$$\mathsf{c} = A\mathsf{b} = \begin{pmatrix} -0.5 & 0.2 & 0 \\ 0 & -41.0 & 2 \\ 0 & 30.0 & -15 \end{pmatrix}\begin{pmatrix} 8 \\ 15 \\ 35 \end{pmatrix} = \begin{pmatrix} -1 \\ -1365 \\ -75 \end{pmatrix} < 0.$$

Hence, $\tau c_i < 0$ for any $\tau > 0$. The condition $\tau c_i \leq (1-\zeta_i)b_i$, $i = 1,2,3$, of Theorem 6.5 is satisfied for any $\zeta \in\, ]0,1]$. The result is that the set $\mathcal{S} = \mathcal{X}_A$ is (itself) the (maximal) estimate of $\mathcal{D}_{ps}(\tau, \mathcal{X}_A; \mathtt{i})$, i.e.

$$\mathcal{D}_{ps}(\tau, \mathcal{X}_A; \mathtt{i}) = \mathcal{X}_A, \tau \in\, ]0,\infty].$$

The preceding examples show that although necessity of the conditions of Theorem 6.1 through Theorem 6.6 is not proved, these theorems can in a certain case exactly determine the domain $\mathcal{D}_{ps}(\tau, \mathcal{X}_A; \mathcal{I})$.

Notice also that $\mathcal{D}_{ps}(\infty, \mathcal{X}_A; \mathcal{I}) = \mathcal{X}_A$ means that $\mathcal{X}_A$ is a positively invariant set.

## 6.4 Estimate of the domain of practical stability with settling time $\tau_s$

In the analysis of the domain $\mathcal{D}_p(\tau, \tau_s, \mathcal{X}_A, \mathcal{X}_F, \mathcal{I})$ of system practical stability with settling time $\tau_s$, we use both the set $\mathcal{X}_A$ of permitted states at any time $t \in \Re_\tau$ and the set $\mathcal{X}_F$ of allowed states since the settling time $\tau_s$ has passed.

**Theorem 6.7** *Let the system (6.1) have generalised motions* $\underline{X}(t; X_0; \mathrm{i}) \in C_t(\Re_\tau \times \overline{\mathcal{X}}_a \times \mathcal{I})$. *In order for a compact set* $\mathcal{S}$, $\mathcal{S} \subseteq \mathcal{X}_A$, *to be estimate* $\mathcal{E}_p(\tau, \tau_s, \mathcal{X}_A, \mathcal{X}_F, \mathcal{I})$ *of the system practical stability with the settling time* $\tau_s$ *with respect to* $\{\tau, \mathcal{X}_F, \mathcal{I}\}$: $\mathcal{S} = \mathcal{E}_p(\tau, \tau_s, \mathcal{X}_A, \mathcal{X}_F, \mathcal{I})$, *it is sufficient that*

*1)* $\mathcal{S}$ *is connected with non-empty interior:* $\mathring{\mathcal{S}} \neq \emptyset$, *and* $\mathcal{S} \subseteq \mathcal{X}_A$,

*2) there are a connected compact subset* $\mathcal{E}_F$ *of* $\mathcal{X}_F$ *with non-empty connected interior, at most finite subset* $\mathcal{Z}$ *of* $\mathcal{S} \cap \mathcal{E}_F$ *and a function* $v : \Re_\tau \times \Re^n \to (\Re \cup \{-\infty\})$ *such that*

*a)* $v(t, X) \in C[\Re_\tau \times (\overline{\mathcal{X}}_A - \mathcal{Z})]$,

*b)* $v(t, X) \leq v^{\partial}_{m\mathcal{E}}(t) < v(t, Y)$, $\forall (t, X, Y) \in \Re_\tau \times \mathring{\mathcal{E}}_F \times (\overline{\mathcal{X}}_A - \mathcal{E}_F)$,

*c) there is a function* $\psi : \Re_\tau \to \Re$, *which is integrable over* $\Re_\tau$ *and obeys*

*(i)* $\mathcal{D}^+ v(t, X, \mathrm{i}) \leq \psi(t)$ *for all* $(t, X, \mathrm{i}) \in \Re_\tau \times (\overline{\mathcal{X}}_A - \mathcal{Z}) \times \mathcal{I}$,

*(ii)* $\int_0^t \psi(\sigma) d\sigma \leq v^{\partial}_{mA} - v_{\overline{M}S}$ *for all* $t \in \Re_\tau$,

*and*

*(iii)* $\int_0^t \psi(\sigma) d\sigma \leq v^{\partial}_{m\mathcal{E}_F} - v_{\overline{M}S}$ *for all* $t \in \mathcal{T}_{\tau s}$. ∇∇∇

**Proof.** Let all the conditions of the theorem statement hold. Hence, all the conditions of Theorem 6.1 also hold so that $\underline{X}(t; X_0; \mathrm{i}) \in \mathcal{X}_A$, $\forall (t, X_0, \mathrm{i}) \in \Re_\tau \times \mathcal{S} \times \mathcal{I}$; i.e. $\mathcal{S} = \mathcal{E}_{ps}(\tau, \mathcal{X}_A, \mathcal{I})$.

From the conditions under c) after integrating $\mathcal{D}^+ v(t, X)$ we derive

$$v[t, \underline{X}(t; X_0; \mathrm{i})] < v(0, X_0) + v^{\partial}_{m\mathcal{E}_F} - v_{\overline{M}S}, \quad \forall (t, X_0, \mathrm{i}) \in [\tau_s, \tau[ \times \mathcal{S} \times \mathcal{I}.$$

This and $v(0, X_0) \leq v_{\overline{M}S}$, $\forall X_0 \in \mathcal{S}$, imply

$$v[t, \underline{X}(t; X_0; \mathrm{i})] < v^{\partial}_{m\mathcal{E}_F}, \quad \forall (t, X_0, \mathrm{i}) \in [\tau_s, \tau[ \times \mathcal{S} \times \mathcal{I},$$

or

$$\underline{X}(t; X_0; \mathrm{i}) \in \mathcal{E}_F, \quad \forall (t, X_0, \mathrm{i}) \in [\tau_s, \tau[ \times \mathcal{S} \times \mathcal{I},$$

due to the condition b). This result, $\mathcal{E}_F \subseteq \mathcal{X}_F$ and the condition 1) verify all requirements of (b) of Definition 6.1: $\mathcal{S} = \mathcal{E}_{pc}(\tau, \tau_s, \mathcal{X}_F, \mathcal{I})$. This and $\mathcal{S} = \mathcal{E}_{ps}(\tau, \mathcal{X}_A, \mathcal{I})$ prove $\mathcal{S} = \mathcal{E}_{ps}(\tau, \tau_s, \mathcal{X}_A, \mathcal{X}_F, \mathcal{I})$ [(ii-c) of Definition 6.1].

Usually we are interested in a practical system stability with settling time $\tau_s$ with respect to $\{\tau, \mathcal{X}_A, \mathcal{X}_F, \mathcal{I}\}$ so that $\mathcal{D}_p(\tau, \tau_s, \mathcal{X}_A, \mathcal{X}_F, \mathcal{I}) \supset \mathcal{X}_F$ and $\partial\mathcal{D}_p(\tau, \tau_s, \mathcal{X}_A, \mathcal{X}_F, \mathcal{I}) \cap \partial\mathcal{X}_F = \emptyset$. Therefore, we accept

$$\mathcal{E}_F \subseteq \mathcal{X}_F \subset \mathcal{S},\ \partial\mathcal{X}_F \cap \partial\mathcal{S} = \emptyset, \tag{6.17}$$

and

$$\dot{v}_{\overline{M}A\mathcal{E}_F} = sup\left[\mathcal{D}^+ v(t, X, \mathrm{i}) : (t, X, \mathrm{i}) \in \Re_\tau \times (\mathcal{X}_A - \mathcal{E}_F) \times \mathcal{I}\right] \in \Re. \tag{6.18}$$

**Theorem 6.8** *Let the system (6.1) have generalised motions* $\underline{X}(t; X_0; \mathrm{i}) \in C_t(\Re_\tau \times \mathcal{S} \times \mathcal{I})$ *and the sets* $\mathcal{E}_F$, $\mathcal{S}$ *and* $\mathcal{X}_F$ *be compact and connected with non-empty interiors. Let (6.17) and (6.18) hold. In order for the set* $\mathcal{S}$ *to be the estimate* $\mathcal{E}_p(\tau, \tau_s, \mathcal{X}_A, \mathcal{X}_F, \mathcal{I})$ *of the domain of the system practical stability with settling time* $\tau_s$ *with respect to* $\{\tau, \mathcal{X}_A, \mathcal{X}_F, \mathcal{I}\}$: $\mathcal{S} = \mathcal{E}_p(\tau, \tau_s, \mathcal{X}_A, \mathcal{X}_F, \mathcal{I})$, *it is sufficient that*

*1) there exist a function* $v : \Re_\tau \times \Re^n \to \Re$ *and an at most finite subset* $\mathcal{Z}$ *of* $\mathcal{S} \cap \mathcal{E}_F$ *obeying:*

*a)* $v(t, X) \in C[\Re_\tau \times (\mathcal{X}_A - \mathcal{Z})]$,

*and*

*b)* $v(t, X) \leq v^{\partial}_{m\mathcal{E}_F}(t) < v(t, Y),\ \forall (t, X, Y) \in \Re_\tau \times \mathring{\mathcal{E}}_F \times (\mathcal{X}_A - \mathcal{E}_F)$,

*2)* $v_{\overline{M}\mathcal{S}}(0) \leq v^{\partial}_{m\mathcal{A}}(t) - t\dot{v}_{\overline{M}A\mathcal{S}}$ *for all* $t \in \Re_\tau$,

*and*

*3)* $v_{\overline{M}\mathcal{S}}(0) \leq v^{\partial}_{m\mathcal{E}_F}(t) - t\dot{v}_{\overline{M}A\mathcal{E}_F}$ *for all* $t \in [\tau_s, \tau[$. ∇∇∇

**Proof.** Let all the conditions hold. Hence, all conditions of Theorem 6.2 hold, which means that $\mathcal{S} = \mathcal{E}_{ps}(\tau, \mathcal{X}_A, \mathcal{I})$ and $\underline{X}(t; X_0; \mathrm{i}) \in \mathcal{S},\ \forall (t, X_0, \mathrm{i}) \in \Re_\tau \times \mathcal{S} \times \mathcal{I}$. By replacing $\dot{v}_{\overline{M}A\mathcal{S}}$ by $\dot{v}_{\overline{M}A\mathcal{E}}$ in the proof of Theorem 6.2 we show that

$$v[\underline{X}(t; X_0; \mathrm{i})] \leq v(0, X_0) + t\dot{v}_{\overline{M}A\mathcal{E}},\quad \forall (t, X_0, \mathrm{i}) \in \Re_\tau \times \mathcal{S} \times \mathcal{I},$$

which together with the condition 2) yields

$$v[\underline{X}(t; X_0; \mathrm{i})] \leq v(0, X_0) + v^{\partial}_{m\mathcal{E}}(t) - v_{\overline{M}\mathcal{S}}(0) \leq v^{\partial}_{m\mathcal{E}}(t),\quad \forall (t, X_0, \mathrm{i}) \in [\tau_s, \tau[ \times \mathcal{S} \times \mathcal{I}.$$

The last inequality is due to $v_{\overline{M}\mathcal{S}}(0) \geq v(0, X_0),\ \forall X_0 \in \mathcal{S}$. The preceding result and the condition 1b) imply

$$\underline{X}(t; X_0; \mathrm{i}) \in \mathcal{E}_F,\quad \forall (t, X_0, \mathrm{i}) \in [\tau_s, \tau[ \times \mathcal{S} \times \mathcal{I},$$

and

$$\underline{X}(t; X_0; \mathrm{i}) \in \mathcal{X}_F,\quad \forall (t, X_0, \mathrm{i}) \in [\tau_s, \tau[ \times \mathcal{S} \times \mathcal{I},$$

due to $\mathcal{E}_F \subseteq \mathcal{X}_F$. This and the properties of $\mathcal{S}$ to be compact, connected with non-empty interior show that all conditions of b) of Definition 6.1 are satisfied so that $\mathcal{S} = \mathcal{E}_{pc}(\tau, \tau_s, \mathcal{X}_F, \mathcal{I})$. This and $\mathcal{S} = \mathcal{E}_{ps}(\tau, \tau_s, \mathcal{X}_F, \mathcal{I})$ prove $\mathcal{S} = \mathcal{E}_p(\tau, \tau_s, \mathcal{X}_A, \mathcal{X}_F, \mathcal{I})$.

Let

$$Q = (Q_{ij}) \in \Re^{n\times n}, \tag{6.19a}$$

$$Q = \delta_{ij}\, sup\,[h_{ii}(X,\mathtt{i}) : (X,\mathtt{i}) \in (\mathcal{X}_A - \mathring{\mathcal{X}}_F) \times \mathcal{I}] +$$

$$+ (1-\delta_{ij})\, max\,\{0, sup\,[h_{ij}(X,\mathtt{i})\, sign\, x_j : (x,\mathtt{i}) \in (\mathcal{X}_A - \mathring{\mathcal{X}}_F) \times \mathcal{I}]\}, \tag{6.19b}$$

$$\mathcal{X}_F = \{X : \mathtt{b}^T|X| \leq \xi\alpha\},\ \ \mathtt{b} > 0,\ \ \xi \in\, ]0,1[. \tag{6.20}$$

and

$$\mathcal{S} = \{X : \mathtt{b}^T|X| \leq \zeta\alpha\},\ \ \zeta \in\, ]0,1], \tag{6.21}$$

Fig. 6.1.

**Theorem 6.9** *Let the system (6.5) have generalised motions $\underline{X}(t;X_0;\mathtt{i}) \in C_t(\Re_\tau \times \mathcal{S} \times \mathcal{I})$. Let the matrix $Q$ and sets $\mathcal{X}_F$ and $\mathcal{S}$ be defined by (6.19), (6.20) and (6.21), respectively. In order for the set $\mathcal{S}$ to be the estimate $\mathcal{E}_p(\tau,\tau_s,\mathcal{X}_A(6.2),\mathcal{X}_F,\mathcal{I})$ of the domain $\mathcal{D}_p(\tau,\tau_s,\mathcal{X}_A(6.2),\mathcal{X}_F,\mathcal{I}) : \mathcal{S} = \mathcal{E}_p(\tau,\tau_s,\mathcal{X}_A(6.2),\mathcal{X}_F,\mathcal{I})$, it is sufficient that all the conditions of the Theorem 6.3 hold and*

$$\rho t \leq (\xi - \zeta)\alpha,\ \ \forall t \in [\tau_s, \tau[, \tag{6.22}$$

*where*

$$\rho = \alpha \sum_{i=0}^{n} max\left(\frac{q_i}{b_i}, \frac{\xi q_i}{b_i}\right), \tag{6.23a}$$

*and*

$$\mathtt{q} = (q_1\ q_2\ \ldots\ q_n)^T = Q^T\mathtt{b}. \tag{6.23b}$$

∇∇∇

**Proof.** Let all the conditions hold. Therefore, $\mathcal{S} = \mathcal{D}_{ps}(\tau,\mathcal{X}_A,\mathcal{I})$ (Theorem 6.3) and $\underline{X}(t;X_0;\mathtt{i}) \in \mathcal{X}_A$, $\forall(t,X_0,\mathtt{i}) \in \Re_\tau \times \mathcal{S} \times \mathcal{I}$. We accept $\mathcal{E}_F = \mathcal{X}_F$ and use $\mathcal{S}$, (6.21), to find

$$v_{\overline{M}\mathcal{S}}(0) = v_{\overline{M}\mathcal{S}} = \zeta\alpha,\ v^{\partial}_{m\mathcal{E}}(t) = v^{\partial}_{m\mathcal{E}} = v^{\partial}_{mF} = \xi\alpha, \tag{6.24a}$$

and

$$\dot{v}_{\overline{M}AF} = \rho \tag{6.24b}$$

due to (6.19) and (6.23). From (6.22) and (6.24) follows that the condition 2) of Theorem 6.8 is satisfied. All other conditions of the same theorem are also satisfied because the sets $\mathcal{S}$ and $\mathcal{X}_F$ are $\mathcal{O}$-uniquely bounded with the same generating function $u = v$. Altogether, $\mathcal{S} = \mathcal{E}_{pc}(\tau,\tau_s,\mathcal{X}_F,\mathcal{I})$ that together with $\mathcal{S} = \mathcal{E}_{ps}(\tau,\mathcal{X}_A,\mathcal{I})$ proves $\mathcal{S} = \mathcal{E}_p(\tau,\tau_s,\mathcal{X}_A(6.2),\mathcal{X}_F,\mathcal{I})$.

Let (Fig. 6.2)

$$\mathcal{X}_F = \left\{X : X^T PX \leq \xi\alpha\right\}, \ \ \xi \in ]0,1[, \ \ P = P^T = (p_{ij}) > 0, \tag{6.25}$$

$$\mathcal{S} = \left\{X : X^T PX \leq \zeta\alpha\right\}, \ \ \zeta \in ]0,1], \tag{6.26}$$

and

$$M(\cdot) = [m_{ij}(\cdot)] = H^T(\cdot)P + PH(\cdot), \tag{6.27a}$$

$$m_{ij} = \delta_{ij}\, sup\left[w_{ii}(X,\mathtt{i}) : (X,\mathtt{i}) \in (\mathcal{X}_A - \mathring{\mathcal{X}}_F) \times \mathcal{I}\right] +$$

$$+ \ (1-\delta_{ij})\, max\left\{0, sup\left[w_{ij}(X,\mathtt{i})\, sign\,(x_i x_j) : (x,\mathtt{i}) \in (\mathcal{X}_A - \mathring{\mathcal{X}}_F) \times \mathcal{I}\right]\right\}, \tag{6.27b}$$

$$M = (m_{ij}) \in \Re^{n\times n}. \tag{6.27c}$$

**Theorem 6.10** *Let the system (6.5) have generalised motions* $\underline{X}(t;X_0;\mathtt{i}) \in C_t(\Re_\tau \times \mathcal{S} \times \mathcal{I})$ *and Theorem 6.4 hold. Let the sets* $\mathcal{X}_F$*, (6.25), and* $\mathcal{S}$*, (6.26), and the matrix* $M$*, (6.27), obey the following:*

$$\eta t \leq (\xi - \zeta)\alpha, \ \ \forall t \in [\tau_s, \tau[, \tag{6.28a}$$

*with*

$$\eta = \left\{ \begin{array}{ll} \alpha\dfrac{\Lambda(M)}{\lambda(P)} & if\ \Lambda(M) \geq 0, \\ & \\ \xi\alpha\dfrac{\Lambda(M)}{\Lambda(P)} & if\ \Lambda(M) \leq 0. \end{array} \right\} \tag{6.28b}$$

*Then, the set* $\mathcal{S}$ *is estimate* $\mathcal{E}_p(\tau,\tau_s,\mathcal{X}_A(6.7),\mathcal{X}_F,\mathcal{I})$ *of* $\mathcal{D}_p(\tau,\tau_s,\mathcal{X}_F,\mathcal{I})$*:*

$$\mathcal{S} = \mathcal{E}_p(\tau,\tau_s,\mathcal{X}_A(6.7),\mathcal{X}_F,\mathcal{I}).$$

∇∇∇

**Proof.** Under the conditions of the theorem statement we calculate for $v(X) = X^T PX$:

$$v_{\overline{M}\mathcal{S}}(0) = \zeta\alpha, \ \ v^{\partial}_{mF} = \xi\alpha, \ \ \dot{v}_{\overline{M}AF} = \eta,$$

which together with (6.28) verify the condition 2) of Theorem 6.8. Since the sets $\mathcal{S}$ and $\mathcal{X}_F$ are $\mathcal{O}$-uniquely bounded then they satisfy all other conditions of the same theorem, too. Therefore, $\mathcal{S} = \mathcal{E}_p(\tau,\tau_s,\mathcal{X}_A(6.7),\mathcal{X}_F,\mathcal{I})$.

If, (Fig. 6.3),

$$\mathcal{X}_F = \left\{X : max\left(\frac{|x_i|}{b_i} : i = 1,2,\ldots,n\right) \leq \xi\alpha\right\}, \ \ \xi \in ]0,1[, \ \ \mathtt{b} > 0, \tag{6.29}$$

$$\mathcal{S} = \left\{X : max\left(\frac{|x_i|}{b_i} : i = 1,2,\ldots,n\right) \leq \zeta\alpha\right\}, \ \ \zeta \in ]0,1], \tag{6.30}$$

then we use the matrix $Q$ (6.19) and the vector $\mathrm{q}$ (6.23b) in order to specify conditions under which the set $\mathcal{S}$, (6.30), is the estimate of $\mathcal{D}_p(\tau, \tau_s, \mathcal{X}_A, \mathcal{X}_F, \mathcal{I})$.

Let

$$R = (r_{ij}) \in \Re^{n \times n}, \tag{6.31a}$$

$$r_{ij} = \delta_{ij} max\,(\xi Q_{ij}, Q_{ij}) + (1 - \delta_{ij})\, Q_{ij}, \tag{6.31b}$$

$$\mathrm{p} = (p_1\ p_2\ \dots\ p_n)^T = R^T \mathrm{b}. \tag{6.31c}$$

**Theorem 6.11** *Let the system (6.5) have generalised motions $\underline{X}(t; X_0; \mathrm{i}) \in C_t(\Re_\tau \times \mathcal{S} \times \mathcal{I})$, Theorem 6.5 hold, the sets $\mathcal{X}_F$, (6.29), and $\mathcal{S}$, (6.30), and the vector* $\mathrm{p}$, *(6.19), (6.31) obey:*

$$t\,max\left(\frac{p_i}{b_i} : i = 1, 2, \dots, n\right) \leq \xi - \zeta, \quad \forall t \in [\tau_s, \tau[. \tag{6.32}$$

*Then, the set $\mathcal{S}$ is estimate $\mathcal{E}_p(\tau, \tau_s, \mathcal{X}_A(6.9), \mathcal{X}_F, \mathcal{I})$ of $\mathcal{D}_p(\tau, \tau_s, \mathcal{X}_A(6.9), \mathcal{X}_F, \mathcal{I})$:*

$$\mathcal{S} = \mathcal{E}_p(\tau, \tau_s, \mathcal{X}_A(6.9), \mathcal{X}_F, \mathcal{I}).$$

$\nabla\nabla\nabla$

**Proof.** Let all the condition be valid. Theorem 6.5 guarantees $\mathcal{S} = \mathcal{E}_{ps}(\tau, \mathcal{X}_A, \mathcal{I})$ and $\underline{X}(t; X_0; \mathrm{i}) \in \mathcal{X}_A$, $\forall (t, X_0, \mathrm{i}) \in \Re_\tau \times \mathcal{S} \times \mathcal{I}$. In this case:

$$v_{\overline{MS}}(0) = \zeta\alpha, \ v^{\partial}_{MF} = \xi\alpha, \ \dot{v}_{\overline{MAF}} = \alpha\, max\left(\frac{p_i}{b_i} : i = 1, 2, \dots, n\right).$$

These results and (6.32) show that the condition 2) of Theorem 6.8 holds. Since the sets $\mathcal{X}_A$, $\mathcal{X}_F$ and $\mathcal{S}$ are $\mathcal{O}$-uniquely bounded with the same generating function $u = v$, $v(X) = max\left(\frac{|x_i|}{b_i} : i = 1, 2, \dots, n\right)$, and numbers $\alpha$, $\xi$ and $\zeta$ are positive, then the sets satisfy all the conditions of Theorem 6.8 that implies $\mathcal{S} = \mathcal{E}_p(\tau, \tau_s, \mathcal{X}_A(6.9), , \mathcal{X}_F, \mathcal{I})$.

The preceding results are based on properties of $\mathcal{O}$-uniquely bounded sets and are generalised as follows:

**Theorem 6.12** *Let the system (6.5) have generalised motions $\underline{X}(t; X_0; \mathrm{i}) \in C_t(\Re_\tau \times \mathcal{X}_A \times \mathcal{I})$ and all the conditions of Theorem 6.6 hold. Let the sets $\mathcal{X}_A$, $\mathcal{S}$ and $\mathcal{X}_F$ be uniquely bounded with the same generating function $u$, (6.14), and be determined by (6.14), (6.15) and (6.33), respectively,*

$$\mathcal{X}_F = \{X : u(X) \leq \xi\alpha\}, \xi \in ]0, 1[. \tag{6.33}$$

*In order for the set $\mathcal{S}$ to be estimate $\mathcal{E}_p(\tau, \tau_s, \mathcal{X}_A, \mathcal{X}_F, \mathcal{I})$ of the domain $\mathcal{D}_p(\tau, \tau_s, \mathcal{X}_A, \mathcal{X}_F, \mathcal{I})$ of the system practical stability with the settling time $\tau_s$ with respect to $\{\tau, \mathcal{X}_A, \mathcal{X}_F, \mathcal{I}\}$ it is sufficient that*

$$t\,\dot{u}_{MAF} \leq (\xi - \zeta)\alpha, \quad \forall t \in \Re_\tau. \tag{6.34}$$

$\nabla\nabla\nabla$

**Proof.** Let all the conditions hold. Then all the conditions of Theorem 6.8 are satisfied because:

1) unique boundedness of $\mathcal{X}_A$, $\mathcal{X}_F$ and $\mathcal{S}$, and $\zeta, \xi \in \Re^+$, and (6.14), (6.15) and (6.33) guarantee that these sets are compact, connected with non-empty interiors, and obey the condition 1.b) of Theorem 6.8,

2) their generating function $u$ is well defined and continuous on $\mathcal{X}_A$,

3) the condition $t\, \dot{u}_{MAS} \leq (1-\zeta)\alpha$, $\forall t \in \Re_\tau$, of Theorem 6.6 shows that the condition 2) of Theorem 6.8 is fulfilled for $u = v$,

and

4) the condition (6.34) implies validity of condition 3 of Theorem 6.8.

**Example 6.4** Let a second order system of the form (6.1) be defined by

$$\frac{dX}{dt} = -\begin{bmatrix} x_1 + 2x_2 - x_2 sin\left(\frac{\pi}{6}x_1\right) sin\left(\frac{\pi}{6}x_2\right) \\ \\ -2x_1 + x_2 + x_1 sin\left(\frac{\pi}{6}x_2\right) sin\left(\frac{\pi}{6}x_1\right) \end{bmatrix} = f(X).$$

Given:

$$\tau = 20, \;\; \tau_s = 4, \;\; \mathcal{X}_A = \left\{X : ||X||^2 \leq e^{10}\right\}, \;\; \mathcal{X}_F = \left\{X : ||X||^2 \leq e\right\}, \;\; \mathcal{I} = \emptyset.$$

What is the largest number $\zeta \in \Re^+$ such that the set $\mathcal{S} = \{X : ||X||^2 \leq e^\zeta\}$ is estimate $\mathcal{E}_p(20, 4, \{X : ||X||^2 \leq e^{10}\}, \{X : ||X||^2 \leq e\}, \emptyset)$ of the domain $\mathcal{D}_p(20, 4, \{X : ||X||^2 \leq e^{10}\}, \{X : ||X||^2 \leq e\}, \emptyset)$?

Since

$$f(x, \mathtt{i}) \equiv f(X) \in C^{(1)}(\Re^2)$$

then $\underline{X}(t; X_0; 0) \in C_t(\Re_{20} \times \Re^2)$.

The sets

$$\mathcal{X}_A = \left\{X : ||X||^2 \leq e^{10}\right\}, \;\; \mathcal{X}_F = \left\{X : ||X||^2 \leq e\right\},$$

$$\mathcal{S} = \left\{X : ||X||^2 \leq e^\zeta\right\}, \;\; \zeta \in \Re^+, \tag{6.35}$$

are compact, connected and with non-empty interiors. Let

$$v(X) = ln\, ||X||^2, \tag{6.36}$$

so that

$$v(X) \in C(\Re^2 - \{0\}), \tag{6.37}$$

which yields $\mathcal{Z} = \{0\}$. This shows that the function $v$ is not positive definite. But, positive definiteness of the function $v$ is not required in the framework of the practical stability and we may continue with testing it. The results are:

$$\partial\mathcal{X}_A = \left\{X : \|X\|^2 \le e^{10}\right\} \Rightarrow v^{\partial}_{mA} = 10, \tag{6.38a}$$

$$\partial\mathcal{X}_F = \left\{X : \|X\|^2 \le e\right\} \Rightarrow v^{\partial}_{mF} = v^{\partial}_{m\mathcal{E}_F} = 1 \text{ for } \mathcal{E}_F = \mathcal{X}_F, \tag{6.38b}$$

$$\overline{\mathcal{S}} = \mathcal{S} = \left\{X : \|X\|^2 \le e^{\zeta}\right\} \Rightarrow v^{\partial}_{\overline{M}\mathcal{S}}(0) = v^{\partial}_{\overline{M}\mathcal{S}} = \zeta, \tag{6.38c}$$

so that

$$v(X) = ln\,\|X\|^2 < v^{\partial}_{m\mathcal{E}_F} = 1 < v(Y) = ln\,\|Y\|^2, \quad \forall (X,Y) \in \mathring{\mathcal{E}}_F \times (\Re^2 - \mathcal{E}_F). \tag{6.39}$$

Further,

$$\mathcal{D}^+ v(X) = \dot{v}(X) = \frac{1}{\|X\|^2} 2X^T \begin{bmatrix} -x_1 - 2x_2 + x_2 sin\left(\frac{\pi}{6}x_1\right) sin\left(\frac{\pi}{6}x_2\right) \\ +2x_1 - x_2 - x_1 sin\left(\frac{\pi}{6}x_2\right) sin\left(\frac{\pi}{6}x_1\right) \end{bmatrix} = -2,$$

$$\forall (X \neq 0) \in \Re^2,$$

so that

$$\dot{v}_{\overline{M}A\mathcal{E}_F} = -2. \tag{6.40}$$

We use Theorem 6.8 and we will determine the set $\mathcal{S}$ so that all the conditions are satisfied. It was shown above that the sets $\mathcal{X}_A$, $\mathcal{X}_F$ and $\mathcal{S}$ have the required properties. From (6.37) and (6.39) it follows that the function $v$ obeys a) and b) of the condition 1) of the theorem. The set $\mathcal{S} = \{X : \|X\|^2 \le e^{\zeta}\}$ should be determined, i.e. $\zeta$ should be calculated so that the conditions 2) and 3) of the theorem also hold, which take the following form due to (6.38):

*2)* $\left[v_{\overline{M}\mathcal{S}}(0) = \zeta\right] \le \left[v^{\partial}_{mA}(t) \equiv 10\right] - t\left[\dot{v}_{\overline{M}A\mathcal{S}} = -2\right], \quad \forall t \in \Re_{20},$

that is that

$$\zeta \le 10 + 2t, \quad \forall t \in \Re_{20}.$$

Hence, for $(t = 0) \in \Re_{20}$,

$$\zeta \le 10. \tag{6.41}$$

*3)* $\left[v_{\overline{M}\mathcal{S}}(0) = \zeta\right] \le \left[v^{\partial}_{m\mathcal{E}_F}(t) \equiv 1\right] - t\left[\dot{v}_{\overline{M}A\mathcal{E}_F} = -2\right], \quad \forall t \in [4, 20[,$

or

$$\zeta \leq 1 + 2t, \quad \forall t \in [4, 20[.$$

For $(t = 4) \in [4, 20]$ follows

$$\zeta \leq 9. \tag{6.42}$$

Now, (6.41) and (6.42) determine the largest $\zeta = 9$, or equivalently the estimate $\mathcal{S} = \mathcal{E}_p(\tau, \tau_s, \mathcal{X}_A, \mathcal{X}_F, \emptyset)$ of the form $\{X : ||X||^2 \leq e^{\zeta}\}$ as

$$\mathcal{S} = \mathcal{E}_p(20, 4, \{X : ||X||^2 \leq e^{10}\}, \ \{X : ||X||^2 \leq e\}, \emptyset) = \{X : ||X||^2 \leq e^9\}.$$

## 6.5 Conclusion

The practical stability criteria permit an aggregation function $v$ to be non-definite with non-semidefinite Dini derivative. The function $v$ need not be continuous everywhere. The criteria are expressed in a very simple form in terms of extremal values of $v$ and $\mathcal{D}^+ v$ on compact sets. They are suitable for direct calculations.

The criteria show qualitative and quantitative trade-off among the sets $\mathcal{X}_A$, $\mathcal{X}_F$, $\mathcal{E}_{(\cdot)}$, the settling and final time and the form of aggregation function $v$ used. The choice is straightforward in the case of the sets being uniquely bounded, where their generating functions should be taken for the function $v$.

# Chapter 7

# Comparison systems and vector norm-based Lyapunov functions

## 7.1 Introductory comments and definitions

### 7.1.1 Presentation

This last chapter is devoted to the stability study of large scale systems whose models are not completely specified, either because of an undetermined disturbing input, or because the identification of some varying parameters is not possible. This is represented by the equation:

$$\frac{dX}{dt} = g(X, d) \quad g : \Re^n \times \mathcal{S}_d \to \Re^n \tag{7.1}$$

where $X \in \Re^n$, and $d : \Re^k \to \mathcal{S}_d$ is a vector function corresponding to the modelling misreading (parameter or input disturbance).

This book is devoted to nonlinear time-invariant systems: thus the disturbance $d$ is considered as nonlinear time-invariant function, say $d = d(X)$ (however, what follows can be applied to general disturbance $d(X, t)$ without change, provided that a solution $X$ of (7.1) exists). Note that the extension to functional differential equations has also been developed [21], [34], [35], [58].

In the following, $X(t; t_0, X_0; d)$ denotes, as usual, the solution of (7.1) starting from $X_0$ at time $t = t_0$, and for a given function $d \in \mathcal{S}_d$.

For such large scale systems, finding a Lyapunov function that provides conclusions independently of disturbance $d$ is a rather difficult task: the introduction of vector norms (V.N.) aims to construct overvaluations of the system behaviours, so that overvaluations no longer depend on disturbances, and are described by lower-order differential equations.

**Some examples** a) Equation (1.18) of Section 1.2, or (3.4) of Section 3.1:

$$\frac{dX}{dt} = f(X, \mathtt{i}),$$

related to the practical stability concept, belongs to that type of equation (with $k = 1$). In this case, $d = \mathtt{i}(t)$ is an input disturbance, and $\mathcal{S}_d \subset \Re^l$.

b) Equation (1.13), (1.14) of Section 1.1, describing Lurie systems and introducing the absolute stability concept:

$$\frac{dX}{dt} = AX + f(w)$$

$$w = CX + Df(w)$$

is also concerned, with $d = f$, $\mathcal{S}_d = \mathcal{N}_{(\cdot)}(\mathcal{L})$. In the absolute stability problem, the question is to prove asymptotic stability of $x = 0$ independently of the actual variation law of the parameter function $f$ in $\mathcal{S}_d$.

c) In some cases, equation (7.1) can be written as:

$$\frac{dX}{dt} = f(X) + d(X) \tag{7.2}$$

where $f$ is a known function, and $d \in \mathcal{S}_d \subset \Re^n$ is unknown (but $\mathcal{S}_d$ is known).

The general idea of constructing particular Lyapunov functions on the basis of a vector norm $p$ of state $X$ is described by the following line of arguments (1→4):

1) $p(X)$ is a vector of size $k \leq n$, each component of which is a scalar norm of some components of $X$,

2) $p(X)$ obeys the overvaluing relation

$$p(X(t)) \leq Z(t) \quad \text{for any } \; d \; \text{ in } \; \mathcal{S}_d \tag{7.3}$$

with $Z(t)$ governed by

$$\frac{dZ}{dt} = h(Z), \tag{7.4}$$

3) $v = v(Z)$ is a Lyapunov function for system (7.4),

4) then try $v(p(X))$ as a Lyapunov function for system (7.1).

Moreover, studying equilibriums and domains corresponding to (7.4) is a source of information concerning the asymptotic and dynamic behaviours of the actual disturbed large scale system (7.1).

Constructing and working with such overvaluing systems (7.4) is the question we answer in this chapter. As a particularly simple case system (7.4) may be linear time-invariant:

$$\frac{dZ}{dt} = MZ + \mathtt{q} \tag{7.5}$$

and then its explicit solution can easily be used for the evaluation of system (7.1) behaviours.

### 7.1.2 Comparison systems

The notion of comparison system originates in the theory of differential (and integral) inequalities, which inequalities directly follow from the use of Lyapunov's methods. Since the beginning of the century, many works have been devoted to related topics (see for example description and references in [88], Chapter 2).

The "comparison concept" has also been defined as a formal property of qualitative concepts [92] that we shall use in the sequel, under the following reduced expression [162]:

**Definition 7.1** *Let two systems*

$$(S)\ \frac{dX}{dt} = g(X,d) \qquad g: \Re^n \times \mathcal{S}_d \to \Re^n \tag{7.6}$$

$$(C)\ \frac{dZ}{dt} = h(Z,X,d) \qquad h: \Re^k \times \Re^n \times \mathcal{S}_d \to \Re^n \tag{7.7}$$

*and sets $\mathcal{X} \subset \Re^n$ and $\mathcal{Z} \subset \Re^k$.*

*i) Notation $P(C,\mathcal{Z},\mathcal{A}) \Rightarrow Q(S,\mathcal{X},\mathcal{B})$ signifies "If $\mathcal{Z}$ has a property (or concept) $P$ for system $(C)$ with regard to the ordered list of arguments $\mathcal{A}$ (related to concept $P$) then $\mathcal{X}$ has the property (or concept) $Q$ for system $(S)$ with regard to the ordered list of arguments $\mathcal{B}$ (related to concept $Q$)".*

*ii) $(C,\mathcal{Z},\mathcal{A})$ is a comparison system of $(S,\mathcal{X},\mathcal{B})$ with regard to property or concept $P$ if*

$$P(C,\mathcal{Z},\mathcal{A}) \Rightarrow P(S,\mathcal{X},\mathcal{B}).$$

**Example 7.1** The first Lyapunov method can be stated in term of comparison systems, since the asymptotic stability of $Z = 0$ for

$$\frac{dZ}{dt} = AZ \tag{$\alpha$}$$

implies the asymptotic stability of $X = 0$ for

$$\begin{cases} \dfrac{dX}{dt} = AX + b(X), \\ \lim\limits_{X\to 0} \dfrac{b^T(X)b(X)}{X^TX} = 0. \end{cases} \tag{$\beta$}$$

We can say that $((\alpha),\{0\})$ is a comparison system of $((\beta),\{0\})$ with regard to the asymptotic stability property and to the instability property.

**Example 7.2** Let the system $(S)$ be defined by:

$$\frac{dX}{dt} = \begin{bmatrix} 1 - x_1^2 - x_2^2 & -1 \\ 1 & 1 - x_1^2 - x_2^2 \end{bmatrix} X,\ X \in \Re^2, \tag{7.8}$$

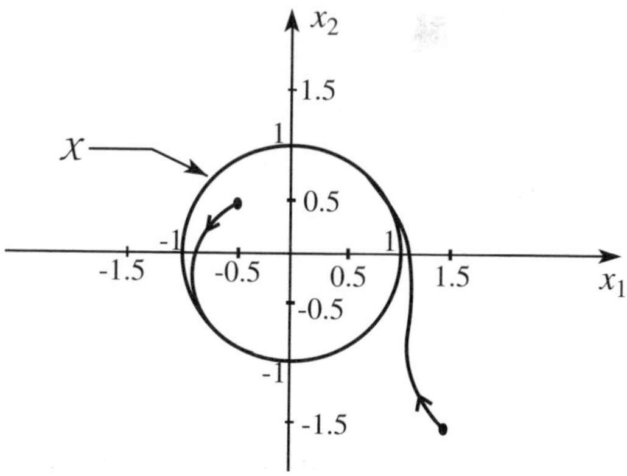


Figure 7.1: Trajectories of system (7.8).

(see Fig. 7.1) and the system $(C)$ be defined by:

$$\frac{dz}{dt} = 2z(1-z), \ z \in \Re. \tag{7.9}$$

Consider the positive definite function

$$z(x) = x_1^2 + x_2^2,$$

then it verifies equation $(C)$.

$z = 0$ is an unstable equilibrium of system $(C)$, and $z_e = 1$ is an asymptotically stable one (see Fig. 7.2), since the solution of $(C)$ is

$$z(t; t_0, z_0) = \frac{z_0}{z_0 + (1 - z_0)\, exp\,(t_0 - t)}. \tag{7.10}$$

$\mathcal{D}(\{z_e\}) = \Re^+$ is the asymptotic stability domain of $z_e = 1$. The convergence of $z$ to $\mathcal{Z} = \{z_e\}$ (for $z_0 > 0$) implies that state $X$ tends to the unit circle $\mathcal{X} = \{X \in \Re^2 : z(X) = 1\}$, for any initial condition $X_0 \in \mathcal{D}(\mathcal{X}) = \Re^2 - \{0\}$.

Then, asymptotic stability of $\mathcal{Z}$ for system $(C)$ with the asymptotic stability domain $\Re^+$ implies asymptotic stability of $\mathcal{X}$ for system $(S)$ with the asymptotic stability domain $\Re^2 - \{0\}$, or asymptotic stability $(C, \{1\}, \Re^+) \Rightarrow$ asymptotic stability $(S, \{X \in \Re^2 : x_1^2 + x_2^2 = 1\}, \Re^2 - \{0\})$ and

1) $(C, \{z_e = 1\})$ is a comparison system of $(S, \{X \in \Re^2 : x_1^2 + x_2^2 = 1\})$ with regard to the asymptotic stability property.

2) $(C, \{z_e = 1\}, \Re^+)$ is a comparison system of $(S, \{X \in \Re^2 : x_1^2 + x_2^2 = 1\}, \Re^2 - \{0\})$ with regard to the asymptotic stability domain concept.

**Example 7.3** We still consider systems (7.8) and (7.9), now related to exponential stability. Equation (7.10) implies that $z_e = 1$ has the domain $\mathcal{D}_e = [0.5, +\infty[$ of exponential stability with regard to $(\beta = 2, \gamma = 1)$ (see Definition 2.8, Chapter 2).

Then $(C, \{z_e = 1\}, \mathcal{D}_e = [0.5, +\infty[, \beta = 2, \gamma = 1)$ is a comparison system of $(S, \{X \in \Re^2 : x_1^2 + x_2^2 = 1\}, \{X \in \Re^2 : x_1^2 + x_2^2 \geq 0.5\}, \beta = 2, \gamma = 1)$ with regard to the exponential stability domain concept.

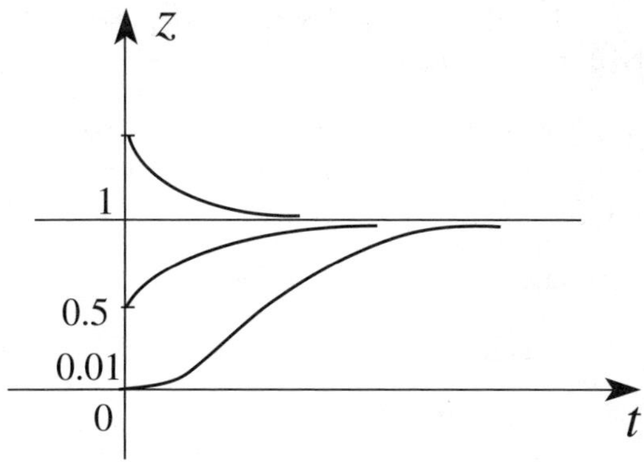


Figure 7.2: Trajectories of system (7.9).

### 7.1.3 Differential inequalities, overvaluing systems

Basic results concerning differential inequalities were provided by Ważewski [206], who gave the necessary and sufficient hypothesis ensuring that the solution of a system described by:

$$\frac{dX}{dt} = F(t, X)$$

with initial condition $X_0$ at $t_0$ with function $F(t, X)$ verifying inequality

$$F(t, X) \leq G(t, X)$$

is overvalued by the solution of the so called "overvaluing system":

$$\frac{dZ}{dt} = G(t, Z)$$

with initial condition $Z_0 \geq X_0$ at $t_0$, or in other words, conditions on function $G(t, Z)$ that guarantee:

$$Z(t; t_0, Z_0) \geq X(t; t_0, X_0).$$

These conditions and other linked results are the topic of this section.

**Definition 7.2** *a) A function $f : \Re^k \to \Re^k$, $f(V) = [f_i(V)]_{i=1\ldots k}$, is locally quasi-increasing on $\mathcal{V} \subset \Re^k$ if and only if it is continuous w.r.t. $V \in \mathcal{V}$, and for any $V = (v_i)$ and $W = (w_i)$ in $\mathcal{V}$, $v_i = w_i$ and $v_j \leq w_j$ $\forall j \in \{1, 2 \ldots, i-1, i+1, \ldots k\}$ implies $f_i(V) \leq f_i(W)$.*

*If $\mathcal{V} = \Re^k$, $f$ is (globally) quasi-increasing.*

*b) A function $f : \Re^n \times \Re^k \to \Re^k$, $f(X, V) = [f_i(X, V)]_{i=1\ldots k}$, is locally quasi-increasing with regard to the variable $V$ on $\mathcal{S} \times \mathcal{V} \subset \Re^n \times \Re^k$ if and only if it is continuous w.r.t. $(X, V) \in \mathcal{S} \times \mathcal{V}$, and for any $X$ in $\mathcal{S}$, $f(X, \cdot) : \Re^k \to \Re^k$ is locally quasi-increasing on $\mathcal{V}$.*

*If $\mathcal{S} = \Re^n$, and $\mathcal{V} = \Re^k$, $f(X, V)$ is (globally) quasi-increasing w.r.t. $V$.*

**Remark:** 1) This definition also exists for time-dependent functions $f(t, X, V)$, that are involved in time-varying systems.

2) "Quasi-increasing" is also called "increasing with respect to the diagonal elements" [92], or "uniformly nonsingular monotone" [88], or "satisfying the Ważewski conditions".

**Example 7.4** a) Any continuous function $f : \Re^n \to \Re^n$

$$f(X) = \begin{pmatrix} h_1(x_2, \ldots, x_n) & |\alpha_{12}| & \ldots & |\alpha_{1n}| \\ |\alpha_{21}| & h_2(x_1, x_3 \ldots, x_n) & \ldots & |\alpha_{2n}| \\ \vdots & \vdots & \ddots & \vdots \\ |\alpha_{n1}| & |\alpha_{n2}| & \ldots & h_n(x_1, \ldots, x_{n-1}) \end{pmatrix} X + q$$

where $\alpha_{ij}$ and $q$ are constant and $h_i$ does not depend on $x_i$, is quasi-increasing (on $\Re^n$).

b) Any continuous function $f : \Re^n \times \Re^k \to \Re^k$

$$f(X, V) = M(X)V + q(X)$$

with $M(X) = (m_{ij}(X))$, $m_{ij}(X) \geq 0$ for $i \neq j$, is quasi-increasing with regard to $V$.

**Definition 7.3** *Let $\alpha$ and $\omega$ be finite or infinite numbers, $\alpha < \omega$, and $\mathcal{T}$ be the time interval $(\alpha, \omega) \subset \Re$.*

*Let $\mathcal{S}$ be a compact connected set, $\mathcal{S} \subset \Re^n$, $\mathcal{S} \neq \emptyset$.*

*i) The set $\mathcal{S}$ is $(\mathcal{T}, \mathcal{S}_d)$-positively invariant for system (7.1),*

$$\frac{dX}{dt} = g(X, d),$$

*iff $\forall X_0 \in \mathcal{S}$, $\forall t_0 \in \mathcal{T}$, $\forall t \in [t_0, \omega)$, $\forall d \in \mathcal{S}_d$,*

$$X(t; t_0, X_0, d) \in \mathcal{S}.$$

*ii) The set $\mathcal{S}$ is $\mathcal{S}_d$-positively invariant for (7.1) if and only if it is $(\Re, \mathcal{S}_d)$-positively invariant.*

**Definition 7.4** *The system (7.13):*

$$\frac{dOS}{dt} = h(OS; d), h : \Re^n \times \mathcal{S}_d \to \Re^n \tag{7.13}$$

*is a $(\mathcal{T}, \mathcal{S}, \mathcal{S}_d)$-local (direct) overvaluing system of system (7.1) iff*

*i) the solutions $OS(t; t_0, OS_0; d)$ of (7.13) and $X(t; t_0, X_0; d)$ of (7.1) exist for $X_0 \subset \mathcal{S}$, $OS_0 \subset \mathcal{S}$, $t \in \mathcal{T}$, $d \in \mathcal{S}_d$, and are continuous with respect to time;*

*ii)* $OS(t; t_0, OS_0; d) \geq X(t; t_0, X_0; d)$ *for any* $OS_0$ *and* $X_0$ *in* $\mathcal{S}$ *such that* $OS_0 = X_0$*, any* $t \in [t_0, \omega)$*, and any* $d \in \mathcal{S}_d$*.*

*iii)* $\mathcal{S}$ *is* $(\mathcal{T}, \mathcal{S}_d)$*-positively invariant for (7.1) and (7.13).*

**Remarks:** 1) The mention "direct", referring to the existence of reverse overvaluing systems that are used in the instability study [162], shall be omitted in the following.

2) In the previous definitions 7.3 – 7.4, arguments $\mathcal{T}, \mathcal{S}$ or $\mathcal{S}_d$ can be omitted:

- if $\mathcal{T} = \Re$ and $\mathcal{S} = \Re^n$, the overvaluing system is $\mathcal{S}_d$-local,
- if $\mathcal{T} = \Re$, the overvaluing system is $(\mathcal{S}, \mathcal{S}_d)$-local,
- if $\mathcal{S} = \Re^n$, the overvaluing system is $(\mathcal{T}, \mathcal{S}_d)$-local,
- if (7.1) and (7.13) do not depend on disturbance $d$ (as in (7.14), (7.15)), then the overvaluing system is $(\mathcal{T}, \mathcal{S})$-local.
- if $\mathcal{T} = \Re$, $\mathcal{S} = \Re^n$ and $\mathcal{S}_d = \emptyset$, the overvaluing system is global.

The two following theorems are direct adaptations of results by [13], Bitsoris himself referring to [206], [141], [142].

**Theorem 7.1** *Let two systems be with time continuous solutions:*

$$\frac{dX}{dt} = g(X)g : \Re^n \to \Re^n \tag{7.14}$$

$$\frac{dOS}{dt} = h(OS)h : \Re^n \to \Re^n \tag{7.15}$$

*and suppose* $h(X) \geq g(X)$, $\forall X \in \Re^n$.

*Then the system (7.15) is a global overvaluing system of system (7.14) iff h is globally quasi-increasing.* $\nabla\nabla\nabla$

**Theorem 7.2** *Consider the disturbed system (7.1), and let* $V(X) : \Re^n \to \Re^k$ *be a continuous function with right-hand gradient, components of which are positive definite functions, such that along the solutions of (7.1) the following holds:*

$$\mathcal{D}_t^+ V(X) = f(X, d) \leq h(X, V(X)) \;\; \forall d \in \mathcal{S}_d,\; \forall X \in \Re^n,\; \forall t \in \Re. \tag{7.16}$$

*Suppose that system (7.17):*

$$\begin{cases} \dfrac{dX}{dt} = g(X, d) & (7.17a) \\[2ex] \dfrac{dV}{dt} = h(X, V) & (7.17b) \end{cases}$$

*has a unique solution, continuous with regard to time, for* $d \in \mathcal{S}_d$*. Then (7.17) is an* $\mathcal{S}_d$*-overvaluing system of system (7.18):*

$$\begin{cases} \dfrac{dX}{dt} = g(X, d) & (7.18a) \\ \mathcal{D}_t^+ V(X) = [\mathcal{D}_X^+ V]^T g(X, d) \leq h(X, V(X)) & (7.18b) \end{cases}$$

*iff $h$ is globally quasi-increasing with regard to the variable $V$.* ∇∇∇

**Proof.** This result follows from [13].

**Definition 7.5** *System (7.17) is a decoupled $\mathcal{S}_d$-overvaluing system of system (7.18) if and only if it is a $\mathcal{S}_d$-overvaluing system of (7.18) with special form (7.19):*

$$\begin{cases} \dfrac{dX}{dt} = g(X, d) & (7.19a) \\ \dfrac{dV}{dt} = h(V(X)) & (7.19b) \end{cases}$$

*which means that (7.16) comes down to*

$$\mathcal{D}_t^+ V(X) = f(X, d) \leq h(V(X)) \;\; \forall d \in \mathcal{S}_d,\; \forall X \in \Re^n,\; \forall t \in \Re.$$

**Corollary 7.1** *If system (7.1) and function $V$ satisfy the hypothesis of Theorem 7.1, then system (7.19) is $\mathcal{S}_d$-decoupled overvaluing system of (7.18) iff $h(V)$ is quasi-increasing.*

*In this case, any motion $Z(t; t_0, Z_0)$ of*

$$\begin{cases} \dfrac{dZ}{dt} = h(Z) \\ Z_0 = V(X_0) \end{cases} \tag{7.20}$$

*verifies*

$$Z(t; t_0, Z_0) \geq V[X(t; t_0, X_0; d)], \;\; \forall d \in \mathcal{S}_d. \tag{7.21}$$

This corollary obviously follows from Theorem 7.2 and Definition 7.4. It gives a way to have a suitable stability study of system (7.1) by comparing it with a simpler system (7.20): simpler, because (7.20) is $k$-order ($k \leq n$) and is not disturbed.

However, this result does not concern general $(\mathcal{S}, \mathcal{T}, \mathcal{S}_d)$-local overvaluing systems, that are to be worked out in the next sections, leading to stability domains estimations. Before this, we have to study some particular quasi-increasing functions.

### 7.1.4 $-M$ matrices

#### Definition and properties

Example 7.4 shows that any function defined by $f(X) = MX + q$ where $M$ is a constant matrix with nonnegative off-diagonal elements, and a constant vector, is quasi-increasing.

Such matrices $M = (m_{ij})_{i,j=1,\ldots,n}$, $m_{i,j} \geq 0$, $\forall i \neq j$, present strong and useful properties that are recalled here, and that shall be useful in the following sections.

**Definition 7.6** [48] *A constant real square matrix $A$ is called:*

*i) a $Z$-matrix if and only if all its off-diagonal elements are non-positive, which is also denoted $A \in \mathbf{Z}$ and a $Z^*$-matrix if and only if off-diagonal elements are negative, this is $A \in \mathbf{Z}^*$.*

*ii) a $-Z$-matrix iff $-A \in \mathbf{Z}$, which is also denoted $A \in -\mathbf{Z}$, and a $-Z^*$-matrix if and only if $-A \in \mathbf{Z}^*$, $A \in -\mathbf{Z}^*$.*

*iii) a $P$-matrix if and only if all its leading principal minors are positive, which is also denoted $A \in \mathbf{P}$,*

$$\mathbf{P} = \left\{ A = (a_{ij})_{i,j=1,\ldots,n} : a_{11} > 0, \ \ldots, \ \begin{vmatrix} a_{11} & \ldots & a_{1n} \\ \vdots & & \vdots \\ a_{n1} & \ldots & a_{nn} \end{vmatrix} > 0 \right\}.$$

*iv) an $M$-matrix, (Metzlerian), or $A \in \mathbf{M}$, if and only if both $A \in \mathbf{Z}$ and $A \in \mathbf{P}$.*

*v) the opposite of an $M$-matrix, or an $-M$-matrix, or $A \in -\mathbf{M}$, if and only if $-A \in \mathbf{M}$.*

A characteristic of $-M$-matrices is to have nonnegative off-diagonal elements, and a negative diagonal.

**Theorem 7.3** [48] *Let $M$ be a constant $n \times n$ matrix with non-negative elements ($M \in -\mathbf{Z}$). Then the proposition "$M$ is a $-M$ matrix" is equivalent to any of the following propositions:*

*1) Any eigenvalue of $M$ has a negative real part.*

*2) Any real eigenvalue of $M$ is negative.*

*3) $M$ verifies the Koteliansky conditions (that is, $-M \in \mathbf{P}$):*

$$m_{11} < 0, \quad \begin{vmatrix} m_{11} & m_{12} \\ m_{21} & m_{22} \end{vmatrix} > 0, \quad \begin{vmatrix} m_{11} & m_{12} & m_{13} \\ m_{21} & m_{22} & m_{23} \\ m_{31} & m_{32} & m_{33} \end{vmatrix} < 0, \ \ldots$$

$$(-1)^k \begin{vmatrix} m_{11} & \dots & m_{1k} \\ \vdots & & \vdots \\ m_{k1} & \dots & m_{kk} \end{vmatrix} > 0 \dots (-1)^n \, det\, M > 0. \qquad (7.24)$$

*4) There exists a constant vector $X \geq 0$ such that $MX < 0$.*

*5) There exists a constant vector $X > 0$ such that $MX < 0$.*

*6) There exists a diagonal matrix $\Delta$ with positive diagonal such that $N = M\Delta$ is a matrix with dominant negative principal diagonal, i.e.:*

$$n_{ii} + \sum_{j \neq i} |n_{ij}| < 0 \ \ \forall i = 1 \dots n \quad (with\ here,\ \ |n_{ij}| = n_{ij}).$$

*7) $M^{-1}$ exists and all its coefficients are non-positive, i.e.: $-M^{-1} \geq 0$*

*8) If $N \in \mathbf{Z}$ and $N \leq M$, then $N^{-1}$ exists.*

*9) For any vector $X \geq 0$, $X \neq 0$ there exists an index $i$ such that if $Y = MX$, $x_i y_i < 0$.*

*10) If $d(A)$ denotes the diagonal of $A$, then for each diagonal matrix $R$ such that $R \geq -d(A)$, the inverse $R^{-1}$ exists and $\sigma(R^{-1}(A - d(A))) < 1$, where $\sigma(\cdot)$ is the spectral radius (i.e. the maximum of the moduli of all eigenvalues).*

*11) There exists a permutation matrix $P$ such that $PMP^T = T_1T_2$, $T_1$ lower-triangular matrix, $T_2$ upper-triangular, $T_1$ and $T_2 \in \mathbf{Z}$ and have strictly negative diagonal.*

$\nabla\nabla\nabla$

**Further properties of $-M$ matrices**

1) If $M \in -\mathbf{M}$, then there exists a negative real eigenvalue $\lambda(M)$ of $M$, called *the importance value* of $M$, such that the real part of any eigenvalue of $M$ is at most $\lambda(M)$.

2) If $M \in -\mathbf{M}$, there exists a nonnegative eigenvector $u(M) \geq 0$ associated to $\lambda(M)$ and called *the importance vector* of $M$.

3) If $M \in -\mathbf{M}$ and $M$ is irreducible, then $u(M) > 0$, and in Theorem 7.3 one can choose $X = u(M)$ (in proposition 5) and $\Delta = diag(u(M))$ (in proposition 6).

**Dynamic systems defined by $-M$ matrices**

We consider here the linear system defined as in (7.22) by

$$\frac{dX}{dt} = MX + \mathsf{q}, \quad M(n \times n) \text{ is a } -M \text{ matrix}, \quad \mathsf{q} \text{ and } X \in \Re^n, \ \mathsf{q} \geq 0. \quad (7.25)$$

Its solution is:

$$\underline{X}(t; t_0, X_0) = exp\left(M(t - t_0)\right)\left(X_0 + M^{-1}\mathsf{q}\right) - M^{-1}\mathsf{q}, \quad (7.26)$$

and the equilibrium point is $X_e = -M^{-1}\mathsf{q}$.

Such a system possesses several invariance properties (see for example [162], [195]), and some of them will be used in the sequel. Let us define, as in the example of Fig. 7.3, for an $n$-vector $\mathsf{a} > 0$ and an $n$-vector $\mathsf{c}$ such that $\mathsf{c} \geq \mathsf{q}$, the sets:

$$\left.\begin{array}{l}
\mathcal{B}_{\mathsf{a}} = \left\{X \in \Re^n : |X| < \mathsf{a}\right\}, \quad \overline{\mathcal{B}}_{\mathsf{a}} = \left\{X \in \Re^n : |X| \leq \mathsf{a}\right\}, \\
\mathcal{B}_{\mathsf{a}}^+ = \left\{X \in \Re_+^n : 0 \leq X < \mathsf{a}\right\}, \quad \overline{\mathcal{B}}_{\mathsf{a}}^+ = \left\{X \in \Re_+^n : 0 \leq X \leq \mathsf{a}\right\}, \\
\mathcal{A} = \overline{\mathcal{B}}_{X_e} = \left\{X \in \Re^n : |X| \leq X_e = -M^{-1}\mathsf{q}\right\}, \\
\mathcal{A}^+ = \overline{\mathcal{B}}_{X_e}^+ = \left\{X \in \Re^n : 0 \leq X \leq X_e = -M^{-1}\mathsf{q}\right\}.
\end{array}\right\} \quad (7.27)$$

Let $u(M) \geq 0$ (respectively $u(M^T) \geq 0$) be an importance vector of $M$ (respectively $M^T$) associated with the importance value $\lambda(M)$. Then, for $\alpha \in \Re^n$, we define two types of closed sets, centred around $X_e$:

$$\left.\begin{array}{l}
\text{rhomboid: } \mathcal{D}_1(\alpha) = \left\{X \in \Re^n : X = X_e + Y, \ u(M^T)^T|Y| \leq \alpha\right\}, \\
\text{parallelepiped: } \mathcal{D}_\infty(\alpha) = \left\{X \in \Re^n : X = X_e + Y, \ |Y| \leq \alpha u(M)\right\}.
\end{array}\right\} \quad (7.28)$$

We note that if $M$ is irreducible, then $u > 0$ and such domains are neighbourhoods of $X_e$.

Finally, let $\mathcal{K}$ be the simplicial cone [195] defined as:

$$\mathcal{K} = \left\{\mathsf{a} \in \Re_+^n : -M\mathsf{a} \geq \mathsf{q}\right\}. \quad (7.29)$$

**Theorem 7.4** *Let the system (7.25) with $M \in -\mathrm{M}$ and $\mathsf{q} \geq 0$. Then:*

*i) The equilibrium point $X_e = -M^{-1}\mathsf{q}$ is globally exponentially stable (with $\gamma = -\lambda(M)$ in Definition 1.8, Chapter 1).*

*ii) $\mathcal{K} \neq \emptyset$, and any $\overline{\mathcal{B}}_{\mathsf{a}}$ or $\overline{\mathcal{B}}_{\mathsf{a}}^+$ for which $\mathsf{a} \in \mathcal{K}$ is positively invariant with respect to motions of (7.25). For example, $\mathcal{A}$ and $\mathcal{A}^+$ are positively invariant.*

*iii) The cone $\mathcal{K}$ is positively invariant with respect to (7.25).*

*iv) Any set $\mathcal{D}_1(\alpha)$ or $\mathcal{D}_\infty(\alpha)$, with $\alpha \geq 0$, is positively invariant with respect to (7.25), and exponentially contractive with coefficient $\lambda(M)$, this is (with $j = 1$ or $\infty$):*

$$\forall X_0 \in \partial\mathcal{D}_j(\alpha), \quad X(t; X_0) \in exp\left(\lambda(M)\,t\right)\partial\mathcal{D}_j(\alpha). \quad (7.30)$$

$\nabla\nabla\nabla$

**Proof.** Let $M = (m_{ij})$, $\mathsf{q} = (q_i)$. First point $(i)$ is obvious.

*Proof of (ii):*

**Existence of $\mathcal{K}$:** Let $\mathsf{u} > 0$ such that $Mu < 0$, $\alpha > 1$ and $\mathsf{a} = -\alpha M^{-1}\mathsf{q} + \mathsf{u}$. Then $\mathsf{a} > \mathsf{q}$ and $-M\mathsf{a} > \mathsf{q}$, which proves $\mathcal{K} \neq \emptyset$.

**Invariance of $\overline{\mathcal{B}}_{\mathsf{a}}$:** Let $\mathsf{a} = (\alpha_1 \ldots \alpha_n)^T > \mathsf{q}$, $\mathsf{b} = -M\mathsf{a} = (\beta_1 \ldots \beta_n)^T \geq \mathsf{q}$, and $X(t) = (x_1 \ldots x_n)^T \in \partial\overline{\mathcal{B}}_{\mathsf{a}}$. Then there is an index $i$ such that at time $t$, $|x_i| = \alpha_i$ and $|x_j| \leq \alpha_j$ for $j \neq i$. Since $M \in -\mathbf{Z}$, we have

$$\mathcal{D}_t^+|x_i| \leq \sum_{i=1}^{n} m_{ij}|x_j| + q_i \leq \sum_{i=1}^{n} m_{ij}\alpha_j + q_i = \beta_i + q_i \leq 0,$$

and $|x_i|$ cannot increase at time $t$. Then, any trajectory of (7.25) crossing $\partial\overline{\mathcal{B}}_{\mathsf{a}}$ enters $\overline{\mathcal{B}}_{\mathsf{a}}$ or lies on $\partial\overline{\mathcal{B}}_{\mathsf{a}}$. This proves the positive invariance of $\overline{\mathcal{B}}_{\mathsf{a}}$.

**Invariance of $\overline{\mathcal{B}}_{\mathsf{a}}^+$:** Suppose that at time $t$, $X(t) \in \partial\overline{\mathcal{B}}_{\mathsf{a}}^+$, and $X(t) \notin \partial\overline{\mathcal{B}}_{\mathsf{a}}$. this means that there is an index $i$ such that at time $t$, $x_i(t) = 0$, and $0 \leq x_j(t) \leq \alpha_j$. Then

$$\frac{dx_i(t)}{dt} = \sum_{j \neq i} m_{ij}x_j + q_i \geq 0,$$

and $X(t)$ cannot go out of $\overline{\mathcal{B}}_{\mathsf{a}}^+$. This, together with positive invariance of $\overline{\mathcal{B}}_{\mathsf{a}}$, proves the positive invariance of $\overline{\mathcal{B}}_{\mathsf{a}}^+$.

*Proof of (iii):* This can be found in [195], in the case $\mathsf{q} = 0$, and is easy to generalise for $\mathsf{q} \geq 0$ by setting $Y = X + M^{-1}\mathsf{q}$, $Y(t, Y_0) = exp\,(Mt)\,Y_0$, and $\mathcal{K}$ comes down to $\{\mathsf{a} \in \Re^n : -M\mathsf{a} > 0\}$.

*Proof of (iv):*

**Domains $\mathcal{D}_1(\alpha)$:** Let the positive (semi)definite function $v_1$ defined by $v_1(X) = u(M^T)^T|Y|$. Then $\mathcal{D}_t^+ v_1(X) \leq u(M^T)^T M|Y| = \lambda(M)v_1(X) \leq 0$, and $v_1(X)$ exponentially tends to zero as $t \to +\infty$.

**Domains $\mathcal{D}_\infty(\alpha)$:** Let $I = \{i \in \{1, 2 \ldots n\}, u_i(M) \neq 0\}$ and $v_\infty(X) = \max_{i \in I} u_i^{-1}|y_i|$ $= u_{i_0}^{-1}|y_{i_0}|$.

$$\begin{aligned}
\mathcal{D}_t^+ v_\infty(X) &\leq u_{i_0}^{-1} \sum_{j \in I} m_{i_0 j}|y_j|\frac{u_j}{u_j} \leq u_{i_0}^{-1} v_\infty(X) \sum_{j \in I} m_{i_0 j}u_j \\
\mathcal{D}_t^+ v_\infty(X) &\leq u_{i_0}^{-1} v_\infty(X)\lambda(M)u_{i_0} = \lambda(M)v_\infty(X).
\end{aligned}$$

Then $v_\infty(X)$ exponentially tends to zero as $t \to +\infty$.

**Example 7.5** Let the linear system be defined by

$$\frac{dX}{dt} = \begin{pmatrix} -2 & 1 \\ 1 & -1 \end{pmatrix} X + \begin{pmatrix} 1 \\ 1 \end{pmatrix}. \tag{7.32}$$

This yields (see Fig. 7.3):

$$X_e = \begin{pmatrix} 2 \\ 3 \end{pmatrix}, \quad \mathcal{K} = \{\mathsf{a} = (a_1, a_2)^T : 1 \leq 1 + a_1 \leq a_2 \leq 2a_1 - 1\},$$

$$\mathcal{A}^+ = \{X = (x_1, x_2)^T : 0 \leq x_1 \leq 2, 0 \leq x_2 \leq 3\},$$

$$\text{For } \mathsf{a} = \begin{pmatrix} 5 \\ 7 \end{pmatrix} \in \mathcal{K}, \overline{\mathcal{B}}_{\mathsf{a}}^{+} = \{X = (x_1, x_2)^T : 0 \leq x_1 \leq 5, 0 \leq x_2 \leq 7\}.$$

An importance vector of $M = M^T$ is $u = u(M) = u(M^T) = (2, 1+\sqrt{5})^T$, associated with $\lambda(M) = \dfrac{-3+\sqrt{5}}{2}$.

$$\mathcal{D}_1(\alpha) = \{X = (2 + y_1, 3 + y_2)^T : 2|y_1| + (1 + \sqrt{5})|y_2| \leq \alpha\},$$

$$\mathcal{D}_\infty(\alpha) = \{X = (2 + y_1, 3 + y_2)^T : |y_1| \leq 2\alpha, |y_2| \leq (1 + \sqrt{5})\alpha\}.$$

Several trajectories of (7.32) are given on Fig. 7.3, which shows invariant domains $\mathcal{K}, \mathcal{A}, \mathcal{A}^+, \overline{\mathcal{B}}_{\mathsf{a}},, \overline{\mathcal{B}}_{\mathsf{a}}^{+}, \mathcal{D}_u(\alpha)$.

### Relation between quasi-increasing functions and $-M$ matrices

The stability theory literatures suggest a tight connection between Wažewski's conditions and $-M$ matrices. However, until [163] (see also [162], [164]), this link always appeared in an implicit manner, and the following Theorem 7.5 aims at clarifying the existing implications.

Let us suppose that $f(X)$ has a bounded Jacobian matrix $f_X(X) = \dfrac{\partial f}{\partial X}(X)$ for $X \in \mathcal{S}$ (connected subset of $\Re^n$), and that the system:

$$\frac{dX}{dt} = f(X)$$

admits an equilibrium point $X_e \in \mathcal{S}$. Then define the following properties (assertions $P1$ to $P4$, referring to Definition 7.6):

$P1$: $f$ is locally quasi-increasing on a neighbourhood $\mathcal{N}$ of $X_e$;

$P2$: $f_X(X_e) \in -\mathbf{Z}$;

$P2'$: $f_X(X_e) \in -\mathbf{Z}^*$ (with (strictly) positive off-diagonal elements);

$P2''$: $f_X(X_e) \in -\mathbf{Z}^*$ for any $X$ in $\mathcal{S}$;

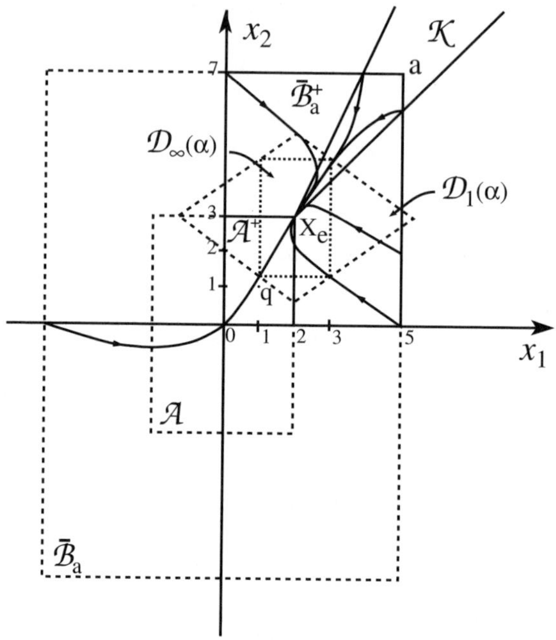


Figure 7.3: Domains for Example 7.5 (eqn. 7.32).

$P3$: $X_e$ is hyperbolic (that is, $f_X(X_e)$ has no eigenvalue with zero real part) and asymptotically stable;

$P4$: $f_X(X_e) \in -\mathrm{M}$.

**Theorem 7.5**

$$P1 \Rightarrow P2$$

$$P2' \Rightarrow P1$$

$$P2'' \Rightarrow P1 \text{ with } \mathcal{N} = \mathcal{S}$$

$$[P1 \text{ and } P3] \Rightarrow P4. \qquad \nabla\nabla\nabla$$

**Corollary 7.2** *If $f(X) = MX + q$ then $f$ is quasi-increasing and $X = -M^{-1}q$ is asymptotically stable, iff $M$ is a $-M$ matrix.*

## 7.2 Vector norm-based comparison systems

### 7.2.1 Definition and aim of vector norms

The concept of vector norm (V.N.) was introduced by Robert (1964) in order to solve numerical analysis problems linked with linear recurrences. It was applied, together with the concept of vector Lyapunov's function (initiated by Bellman (1962) and Matrosov (1962)) to an aggregation of large-scale systems and to their stability analysis, by Borne, Gentina, Laurent, Grujić, Richard (see for example [57], [87], [173]) with further applications to constrained control [169], practical stability [164], and differential-difference equations [34], [35], [21].

**Definition 7.7** *Let $\mathcal{E} = \Re^n$ be a vector space and $\mathcal{E}_1, \mathcal{E}_2, \ldots, \mathcal{E}_k$ vector subspaces of $\mathcal{E}$:*

$$\mathcal{E} = \mathcal{E}_1 + \mathcal{E}_2 + \mathcal{E}_3 + \ldots + \mathcal{E}_k, \quad \mathcal{E}_i = \Re^{n_i}.$$

*Let $X$ be an $n$-vector defined on $\mathcal{E}$ with a projection in the subspace $\mathcal{E}_i$ denoted by $X_i$:*

$$X_i = P_i X, \tag{7.33}$$

*where $P_i$ is a projection operator from $\mathcal{E}$ into $\mathcal{E}_i$. Let $p_i$ be a scalar norm $(i = 1, 2, \ldots, k)$ defined on the subspace $\mathcal{E}_i$, and $p$ be a vector norm (V.N.) of dimension $k$ and with the $i$-th component:*

$$p_i(X) = p_i(X_i) \;\; p(X) : \Re^n \to \Re^k_+.$$

*Then denoting by $Y$ another vector in space $\mathcal{E}$, with $Y_i = P_i Y$, we have:*

$$\begin{aligned} &p_i(X_i) \geq 0, \;\; \forall X_i \in \mathcal{E}_i \;\; \forall i = 1, 2, \ldots, k, \\ &p_i(X_i) = 0 \Leftrightarrow X_i = 0, \;\; \forall i = 1, 2, \ldots, k, \\ &p_i(X_i + Y_i) \leq p_i(X_i) + p_i(Y_i), \;\; \forall X_i, Y_i \in \mathcal{E}_i \;\; \forall i = 1, 2, \ldots, k, \\ &p_i(\lambda X_i) = |\lambda| p_i(X_i), \;\; \forall X_i \in \mathcal{E}_i \;\; \forall i = 1, 2, \ldots, k \;\; \forall \lambda \in \Re. \end{aligned} \tag{7.34}$$

*If $k - 1$ of subspaces $\mathcal{E}_i$ are not sufficient to define the whole space $\mathcal{E}$, the V.N. is surjective. If in addition the subspaces $\mathcal{E}_i$ are in disjoint pairs, the V.N. $p$ is said to be regular:*

$$\mathcal{E} = \mathcal{E}_1 \oplus \mathcal{E}_2 \oplus \mathcal{E}_3 \oplus \ldots \oplus \mathcal{E}_k.$$

In the following, we shall use R.V.N. for "regular vector norm", and

$$\sum_{i=1}^{k} n_i = n, \;\; X_i = (x_j)_{j \in \mathbf{I}_i}, \quad \bigcup_{i=1\ldots k} \mathbf{I}_i = \{1, \ldots, n\}.$$

The aim of this section is to give a general way of computing overvaluing systems that are based on an R.V.N. $p(X)$, in other words, of computing a function $h$ in such a way that system (7.35):

$$\left\{ \begin{aligned} &\frac{dX}{dt} = g(X, d), && \text{(7.35a)} \\ &\frac{dp}{dt} = h(X, p), && \text{(7.35b)} \end{aligned} \right.$$

is a local overvaluing system of

$$\begin{cases} \dfrac{dX}{dt} = g(X, d), & (7.36a) \\ \mathcal{D}_t^+ p(X) = [\mathcal{D}_X^+ p]^T h(X, d). & (7.36b) \end{cases}$$

Then (7.35b) will be used as a comparison system of (7.35a), since its solution verifies $Z(t) \leq p(X(t))$, provided that initial conditions are compatible: $Z(t_0) = p(X(t_0))$. However, the stability study of (7.35b) has to be carried out without solving (7.35a), and this will motivate the use of decoupled overvaluing systems (see Definition 7.5) that are such that $h$ depends only on $Z$:

$$\begin{cases} \dfrac{dX}{dt} = g(X, d), & (7.37a) \\ \dfrac{dZ}{dt} = h(Z). & (7.37b) \end{cases}$$

This could have been obtained by means of the general vector Lyapunov's functions instead of V.N., however the main advantage of V.N. is to give direct and systematic computations.

### 7.2.2 A first statement

**Lemma 7.1** *a) Consider the system (7.1), and let a R.V.N. $p(X)$ and a constant positive vector $\mathsf{c} \in \Re^k$ defining the sets*

$$\mathcal{Z}_c = \{Z \in \Re^k : 0 \leq Z < \mathsf{c}\}, \tag{7.38a}$$

$$\mathcal{X}_c = \{X \in \Re^n : p(X) < \mathsf{c}\} = \{X \in \Re^n : p(X) \in \mathcal{Z}_c\}. \tag{7.38b}$$

*Let $h(X, Z)$ be a continuous function, locally quasi-increasing with regard to the variable $Z$ on $\mathcal{X}_c \times \mathcal{Z}_c$, and such that*

$$\mathcal{D}_t^+(p(X)) \leq h(X, p(X)) \ \forall X \in \mathcal{X}_c. \tag{7.39}$$

*Suppose that the system*

$$\begin{cases} \dfrac{dX}{dt} = g(X, d), & (7.40a) \\ \dfrac{dZ}{dt} = h(X, Z) & (7.40b) \end{cases}$$

*has a unique continuous solution for $d \in \mathcal{S}_d$. Then a sufficient condition for system (7.40) to be a $(\mathcal{X}_c \times \mathcal{Z}_c, \mathcal{S}_d)$-local overvaluing system of*

$$\begin{cases} \dfrac{dX}{dt} = g(X,d), & (7.41a) \\[2ex] \mathcal{D}_t^+ p(X) = [\mathcal{D}_x^+ p]^T g(X,d) & (7.41b) \end{cases}$$

*is that $\mathcal{Z}_c$ is $\mathcal{X}_c$-positively invariant for (7.40b), which means*

$$Z_0 = p(X_0) \in \mathcal{Z}_c \Rightarrow Z(t) \in \mathcal{Z}_c \ \ \forall t \geq t_0, \ \ \forall X \in \mathcal{X}_c. \tag{7.42}$$

*b) If system (7.40) is decoupled, that is, if (7.39) can be written as:*

$$\mathcal{D}_t^+ p(X) \leq h(p(X)), \ \ \forall X \in \mathcal{X}_c, \tag{7.43}$$

*with $h$ a locally quasi-increasing function on $\mathcal{Z}_c$, then the above condition is that $\mathcal{Z}_c$ is positively invariant for (7.44):*

$$\frac{dZ}{dt} = h(Z). \tag{7.44}$$

∇∇∇

**Proof.** a) The proof refers to Definition 7.4, with notations:

$$\left.\begin{aligned} \mathbf{X}(t;t_0,\mathbf{X}_0;d) &= \begin{pmatrix} X(t;t_0,X_0;d) \\[2ex] p(X(t;t_0,X_0;d)) \end{pmatrix} \in \mathcal{S} = \mathcal{X}_c \times \mathcal{Z}_c \subset \Re^{n+k}, \\[4ex] OS(t;t_0,OS_0;d) &= \begin{pmatrix} X(t;t_0,X_0;d) \\[2ex] Z(t;t_0,X_0,Z_0;d) \end{pmatrix} \in \mathcal{S}, \end{aligned}\right\} \tag{7.45}$$

$$Z_0 = p(X_0). \tag{7.46}$$

i) Solution $\mathbf{X}(t;t_0,\mathbf{X}_0;d)$ exists and is continuous by hypothesis.

ii) Since (7.39) holds with $h(X,Z)$ locally quasi increasing w.r.t. $Z$, as long as $X$ remains in $\mathcal{X}_c$ and $Z$ in $\mathcal{Z}_c$, similarly to Theorem 7.2 we have

$$Z_0 = p(X_0) \Rightarrow Z(t;t_0,X_0,Z_0;d) \geq p(X(t;t_0,X_0;d)), \tag{7.47}$$

and for any $t \geq t_0$

$$OS_0 = \mathbf{X}_0 \Rightarrow OS(t; t_0, OS_0; d) \geq \mathbf{X}(t; t_0, \mathbf{X}_0; d) \tag{7.48}$$

supposed that $OS$ and $\mathbf{X}$ remain in $\mathcal{S}$ and $d \in \mathcal{S}_d$.

iii) Let us suppose that (7.42) holds, and consider $Z_0 = p(X_0) < \mathsf{c}$. Then

$$0 \leq Z(t; t_0, X_0, p(X_0); d) < \mathsf{c}, \quad \forall t \geq t_0, \quad \forall d \in \mathcal{S}_d. \tag{7.49}$$

At time $t = t_0$, $X_0$ is in $\mathcal{X}_c$, therefore (7.47) holds until time $t_1$ when $p(X(t_1)) = \mathsf{c}$. However, such a time $t_1$ does not exist because of (7.49). So, (7.47) remains true for any $t \geq t_0$. This proves

$$0 \leq p(X(t; t_0, X_0; d)) \leq Z(t; t_0, X_0, p(X_0); d), \tag{7.50}$$

and $\mathcal{S}_d$-positive invariance of $\mathcal{X}_c$ for system (7.1).

b) The decoupled case appears as a particular case of (a).

### 7.2.3 Computation of overvaluing systems

#### General formula

Let the system (7.1) be decomposed into

$$\left.\begin{array}{l} \dfrac{dX}{dt} = A(X,d)X + b(X,d), \\ A(X,d) \text{ and } v(X,d) \text{ bounded for bounded } X, \quad \text{and any } d \in \mathcal{S}_d. \end{array}\right\} \tag{7.51}$$

This decomposition is obviously non-unique: as we shall see afterwards, it is preferable to work it out in such a way that

$$\frac{dX}{dt} = A(X,d)X \tag{7.52}$$

has its equilibrium $X = 0$ stable, or attractive, or both. Algorithms will also be proposed in order to optimize this decomposition.

Then define for any $i, j \in \{1, 2, \ldots, k\}$ the following notation:

$$\left.\begin{aligned}
&Y \in \Re^n, \quad y_i = P_i Y, \\
&b_i(X,d) = P_i b(X,d), \\
&\kappa_i(X) = \sup_{d \in \mathcal{S}_d} [\mathcal{D}^+_{x_i} p_i(x_i)] b_i(X,d), \\
&\kappa_i = \sup_{x \in \mathcal{S}} \kappa_i(X), \\
&A_{ij}(X,d) = P_i A(X,d) p_j, \\
&m_{ij}(X,Y,d) = \frac{[\mathcal{D}^+_{y_i} p_i(y_i)]^T A_{ij}(X,d) y_j}{p_j(y_j)}, \\
&\mu_{ij}(X) = \sup_{\substack{Y \in \mathcal{S} \\ d \in \mathcal{S}_d}} m_{ij}(X,Y,d), \\
&\mu_{ij} = \sup_{X \in \mathcal{S}} \mu_{ij}(X), \\
&M(X) = [\mu_{ij}(X)]_{i,j \in \{1,\ldots,k\}}, \\
&q(X) = [\kappa_i(X)]_{i \in \{1,\ldots,k\}}, \\
&M = [\mu_{ij}]_{i,j \in \{1,\ldots,k\}}, \\
&q = [\kappa_i]_{i \in \{1,\ldots,k\}}.
\end{aligned}\right\} \qquad (7.53)$$

If $A$ and $b$ are not continuous w.r.t. $X$, then $M(X)$ and $q(X)$ may be discontinuous; then define two functions $\overline{M}$ and $\overline{q}$ continuous on $\mathcal{S}$ such that:

$$M \geq \overline{M}(X) \geq M(X) \quad \text{and} \quad q \geq \overline{q}(X) \geq q(X). \qquad (7.54)$$

**Properties.** By construction, if $p$ is a R.V.N., then for any $X$ in $\mathcal{S}$, the following holds:

1) $q \geq q(X) \geq 0, \quad \overline{q}(X) \geq q(X) \geq 0$ (7.54)

2) $M(X) \in -\mathbf{Z}, \quad \overline{M}(X) \in -\mathbf{Z}, \quad M \in -\mathbf{Z},$ (7.55)

3) $\mathcal{D}^+_t p(X(t)) \leq \overline{M}(X) p(X) + \overline{q}(X)$ (7.56)
is verified for any motion of (7.1) inside $\mathcal{S}$, for $d \in \mathcal{S}_d$.

4) The function defined by

$$h(X,Z) = \overline{M}(X) Z + \overline{q}(X) \qquad (7.57)$$

is locally quasi-increasing with regard to variable $Z$ on $\mathcal{S} \times \Re^k$.

5) The expressions (7.53) provide the smallest expressions of $M(\cdot)$ and $q(\cdot)$ for a given R.V.N. $p(X)$.

**Definition 7.8** *i) Any matrix $\overline{M}(X)$ defined by (7.53) (7.54) is called an $(\mathcal{S} \times \mathcal{S}_d)$-local overvaluing matrix of system (7.1) associated with the R.V.N. $p(X)$.*

*ii) Any vector $\overline{q}(X)$ defined by (7.53) (7.54) is called an $(\mathcal{S} \times \mathcal{S}_d)$-local overvaluing input vector of system (7.1) associated with the R.V.N. $p(X)$.*

*iii) Any pair $(\overline{M}(X), \overline{q}(X))$ defined by (7.53) (7.54) is called an $(\mathcal{S} \times \mathcal{S}_d)$-local overvaluing pair of (7.1) associated with the R.V.N. $p(X)$.*

*iv) If $\mathcal{S} = \Re^n$, the above overvaluing quantities are $\mathcal{S}_d$-local.*

## Simpler formula with the usual Holder's norms and associated matrix measures

Though formulas (7.53), based on the gradient of the vector norm, are general ones, an equivalent calculation of overvaluing pairs $(M, q)$ can be proposed, which is related to the notions of norm and measure of matrices [39].

The norm of a rectangular matrix $A \in \mathcal{C}^{n \times m}$, associated with two norms $||\cdot||$ of vectors in $\mathcal{C}^n$ and $\mathcal{C}^m$, is classically defined as:

$$||A|| = \sup_{X \in \Re^m} \frac{||AX||}{||X||} \in \Re_+. \tag{7.58}$$

The associated measure of a square matrix is:

$$\mu(A) = \lim_{\varepsilon \to 0+} \frac{||I + \varepsilon A|| - 1}{\varepsilon} \in \Re. \tag{7.59}$$

For any matrices $A$ and $B$ of same dimension $n \times n$, we have:

$$\begin{aligned} & Re\, \lambda_i(A) \le \mu(A), \\ & -\mu(A) \le Im\, \lambda_i(A) \le \mu(-jA), \\ & \mu(A + B) \le \mu(A) + \mu(B), \\ & \mu(A) \le ||A||, \\ & \mu(\alpha A) = \alpha \mu(A) \;\; \forall \alpha \ge 0. \end{aligned} \tag{7.60}$$

Considering a matrix $A = (a_{ij}) \in \mathcal{C}^{n \times n}$, and a vector $q = (u_i) \in \mathcal{C}^n$, the following sums up the formulas that concern Holder's norms 1, 2 and $\infty$. The notation $\sum_i$ here obviously means the sum for $i = 1$ to $i = n$, $\max_i$ means the $max$ for $i = 1$ to $n$, and $\lambda(A)$ are the eigenvalues of $A$.

### Holder's norm 1 (sum norm)

$$||u||_1 = \sum_i |u_i|,$$

$$||A||_1 = \max_j \sum_i |a_{ij}|, \tag{7.62}$$

$$\mu_1(A) = \max_j \left( Re\,(a_{jj}) + \sum_{\substack{i \\ i \neq j}} |a_{ij}| \right).$$

**Holder's norm 2 (Euclidean norm)**

$$||u||_2 = \left( \sum_i u_i^2 \right)^{1/2},$$

$$||A||_2 = \left( \max_i \lambda_i(AA^*) \right)^{1/2}, \quad \text{where } \lambda_i(A) \text{ is the } i^{th} \text{ eigenvalue of } A \tag{7.63}$$

$$\mu_2(A) = \frac{1}{2} \max_i \lambda_i(A + A^*).$$

**Holder's norm $\infty$ (max norm)**

$$||u||_\infty = \max_i |u_i|,$$

$$||A||_\infty = \max_i \sum_j |a_{ij}|, \tag{7.64}$$

$$\mu_\infty(A) = \max_i Re\,(a_{ij}) + \sum_{\substack{i \\ i \neq j}} |a_{ij}|.$$

The following inequalities are well known:

$$||X||_\infty \leq ||X||_2 \leq ||X||_1 \leq \sqrt{n}||X||_2 \leq \sqrt{n}||X||_1 \leq \sqrt{n}||X||_\infty.$$

**Assumption 7.1** When all the scalar components $p_i(X)$ of $p(X)$ are of the same Holder's type 1, 2 or $\infty$, the application of formula (7.53) leads to the following formulation of the overvaluing pair $(M(X), q(X))$:

$$\left.\begin{aligned} \mu_{ii}(X) &= \sup_{d \in \mathcal{S}_d} \mu(A_{ii}(X,d)) \\ \mu_{ij}(X) &= \sup_{d \in \mathcal{S}_d} ||A_{ij}(X,d)|| \quad \text{for} \quad i \neq j \\ k_i(X) &= \sup_{d \in \mathcal{S}_d} ||b_i(X,d)|| \end{aligned}\right\} \tag{7.65}$$

where the measure $\mu$ and norms $||\cdot||$ refer to the corresponding scalar Holder's norm.

**Remark.** In the particular case $p(X) = |X|$ (then, $k = n$), the expression is very simple:

$$\left.\begin{aligned} \mu_{ii}(X) &= \sup_{d\in\mathcal{S}_d} a_{ii}(X,d), \\ \mu_{ij}(X) &= \sup_{d\in\mathcal{S}_d} |a_{ij}(X,d)|, \quad \text{for} \quad i \neq j \\ k_i(X) &= \sup_{d\in\mathcal{S}_d} |b_i(X,d)|. \end{aligned}\right\} \tag{7.65bis}$$

**Proof.** We give here only a proof for the Euclidean norm

$$p_i(X_i) = \|X_i\|_2 = \|X_i\| = (X_i^T X_i)^{1/2}, \quad i = 1,\ldots,k.$$

$$\mathcal{D}_t^+ p_i(X_i) = \frac{X_i^T(A_{ii}^T(\cdot) + A_{ii}(\cdot))X_i}{2(X_i^T X_i)^{1/2}} + \frac{X_i^t \sum_{j\neq i} A_{ij}(\cdot)X_j}{(X_i^T X_i)^{1/2}} + \frac{X_i^T b_i(\cdot)}{(X_i^T X_i)^{1/2}}$$

$$\leq \frac{\mu_2(A_{ii}(\cdot))\|X_i\|^2}{\|X_i\|} + \sum_{j\neq i} \frac{\|X_i\|\|A_{ij}(\cdot)\|\|X_j\|}{\|X_i\|} + \frac{\|X_i\|\|b_i(\cdot)\|}{\|X_i\|},$$

that is,

$$\mathcal{D}_t^+\|X_i\| \leq \mu_2(A_{ii}(\cdot))\|X_i\| + \sum_{j\neq i} \|A_{ij}(\cdot)\|\|X_j\| + \|b_i(\cdot)\|.$$

In other cases (norms 1 and $\infty$), analogous demonstrations can be drawn, or deduced from the expressions given in [55] and generalised in [162].

This can again be expressed by developing (7.65):

for $p_i(X_i) = \|X_i\|_1$, $\forall i = 1,\ldots,k$:

$$\left.\begin{aligned} \mu_{ii}(X) &= \sup_{d\in\mathcal{S}_d} \max_{s\in I_i} \left[ a_{rr} + \sum_{\substack{r\in I_i \\ r\neq s}} |a_{rs}| \right], \quad \text{for} \quad i = 1,\ldots,k \\ \mu_{ij}(X) &= \sup_{d\in\mathcal{S}_d} \max_{s\in I_i} \left[ \sum_{r\in I_i} |a_{rs}| \right], \quad \text{for} \quad i,j = 1,\ldots,k,\ i \neq j \\ k_i(X) &= \sup_{d\in\mathcal{S}_d} \sum_{r\in I_i} |b_{rs}|, \end{aligned}\right\} \tag{7.66}$$

for $p_i(X_i) = \|X_i\|_\infty$, $\forall i = 1, \ldots, k$:

$$\left.\begin{array}{l} \mu_{ii}(X) = \sup\limits_{d \in \mathcal{S}_d} \max\limits_{r \in I_i} \left[ a_{rr} + \sum\limits_{\substack{s \in I_i \\ s \neq r}} |a_{rs}| \right], \quad \text{for} \quad i = 1, \ldots, k \\ \mu_{ij}(X) = \sup\limits_{d \in \mathcal{S}_d} \max\limits_{r \in I_i} \left[ \sum\limits_{s \in I_j} |a_{rs}| \right], \quad \text{for} \quad i, j = 1, \ldots, k, \ i \neq j \\ k_i(X) = \sup\limits_{d \in \mathcal{S}_d} \max\limits_{s \in I_j} |b_{rs}|, \end{array} \right\} \tag{7.67}$$

for $p_i(X_i) = \|X_i\|_2$, $\forall i = 1, \ldots, k$.

The expression (7.65) together with (7.63) is not easy to compute if $A$ strongly depends on $X$ or $d$, since eigenvalues are to be calculated. However, one can use the following equations, which are less precise but easier to compute:

$$\left.\begin{array}{l} \mu_{ii}(X) = \sup\limits_{d \in \mathcal{S}_d} \max\limits_{s \in I_i} \left[ a_{ss} + \dfrac{1}{2} \sum\limits_{\substack{r \in I_i \\ r \neq s}} |a_{sr} + a_{rs}| \right] \\ \mu_{ij}(X) = \dfrac{1}{2} \sup\limits_{d \in \mathcal{S}_d} \left( \max\limits_{s \in I_i} [\sum |a_{sr}|] + \max\limits_{r \in I_i} [\sum |a_{rs}|] \right) \\ k_i(X) = \sup\limits_{d \in \mathcal{S}_d} \left( \sum\limits_{s \in I_i} b_i^2 \right)^{1/2}. \end{array} \right\} \tag{7.68}$$

This expression of $\mu_{ij}$ was given in [55] or [87].

### 7.2.4 Overvaluation lemma

The above formulas and properties simplify and make explicit the expression of Lemma 7.1, leading to the following Lemma 7.2.

**Lemma 7.2** *Let the system (7.1) (7.51) have a unique continuous solution* $X(t; t_0, X_0; d)$. *Let a R.V.N.* $p(X)$ *and a constant positive vector* $\mathsf{c} \in \Re^k$, *define the sets* $\mathcal{X}_c$ *and* $\mathcal{Z}_c$ *as in equation (7.38). Let the overvaluing pairs* $(\overline{M}(X), \overline{q}(X))$ *or* $(M, q)$ *be defined by equations (7.53) (7.54) or (7.65), with* $\mathcal{S} = \mathcal{X}_c$. *Then:*

*a) A sufficient condition for the system (7.69):*

$$\left\{ \begin{array}{l} \dfrac{dX}{dt} = g(X, d) = A(X, d)X + b(X, d) \qquad (7.69\text{a}) \\ \dfrac{dZ}{dt} = \overline{M}(X)Z + \overline{q}(X) = h(X, Z) \qquad (7.69\text{b}) \end{array} \right.$$

*to be an* $(\mathcal{X}_c \times \mathcal{Z}_c, \mathcal{S}_d)$*-local overvaluing system of (7.70):*

$$\begin{cases} \dfrac{dX}{dt} = g(X, d) & (7.70a) \\ \\ \mathcal{D}_t^+ p(X) = [\mathcal{D}_X^+ p]^T g(X, d) & (7.70b) \end{cases}$$

*is that* $\mathcal{Z}_c$ *is* $\mathcal{X}_c$*-positively invariant for (7.69b) (then* $X$ *is considered as a disturbance), which means:*

$$Z_0 = p(X_o) \in \mathcal{Z}_c \Rightarrow Z(t) \in \mathcal{Z}_c \ \ \forall t \geq t_0, \ \ \forall X \in \mathcal{X}_c. \qquad (7.71)$$

*b) A sufficient condition for the decoupled system (7.72):*

$$\begin{cases} \dfrac{dX}{dt} = g(X, d) & (7.72a) \\ \\ \dfrac{dZ}{dt} = MZ + q & (7.72b) \end{cases}$$

*to be* $(\mathcal{X}_c \times \mathcal{Z}_c, \mathcal{S}_d)$*-local overvaluing system of(7.70) is that* $M$ *is asymptotically stable and* $\mathsf{c}$ *belongs to the simplicial cone* $\mathcal{K}$ *defined by (7.29) this, i.e.:*

*1)* $M$ *verifies the Koteliansky conditions (Theorem 7.3, proposition 3)*

*2)* $-M\mathsf{c} \geq q$ (7.73)

*c) Defining* $Z_e = -M^{-1}q$, $((7.72b), \mathcal{Z}_c, Z_e)$ *is a comparison system of* $((7.1), \mathcal{X}_c, \mathcal{X}_{Z_e})$ *with regard to the asymptotic stability property with the estimate of asymptotic stability domain, i.e.:*

*"*$Z_e$ *is asymptotically stable for (7.72b), with* $Z_c$ *an estimate of its asymptotic stability domain* $\Re^k$*"*

*implies that:*

*"the set* $\mathcal{X}_c = \{X \in \Re^n : p(X) \leq Z_e\}$ *is asymptotically stable for (7.1), with* $\mathcal{X}_c(-M\mathsf{c} \geq q)$ *an estimate of its asymptotic stability domain".* $\nabla\nabla\nabla$

**Proof.**

a) $\overline{M}(X)$ and $\overline{q}(X)$ are continuous w.r.t. time, whenever $X$ is the continuous solution of (7.1) (7.51). Then (7.69b) admits a unique continuous solution. Properties 3 (7.57) and 7 (7.58) ensure that $h(X, Z)$ fulfils hypothesis of Lemma 7.1 (quasi-increasingness and overvaluation (7.39)) Therefore, Lemma 7.1 (a) proves Lemma 7.2 (a).

b) Solutions of (7.72b) are continuous, and Theorem 7.4 (ii) proves that $\mathcal{Z}_c$ is positively invariant with respect to (7.72b). The function $MZ + q$ is quasi-increasing, and thus Lemma 7.1 (b) proves Lemma 7.2 (b).

c) If $Z_e$ is asymptotically stable, $Z_e$ is invariant, and any motion (7.1) initialised in $\mathcal{X}_c$ remains in $\mathcal{X}_c$ and tends to $\mathcal{X}_{Z_e}$ as $t$ tends to infinity.

**Example 7.6** Consider the system (7.74), with the R.V.N. $p$ of size $n$:

$$\frac{dX}{dt} = A(X)X + b(X) \tag{7.74}$$

$$p(X) = (|x_1|, \dots, |x_n|)^T. \tag{7.75}$$

This yields (see for example (7.65bis)):

$$M(X) = \begin{pmatrix} a_{11}(X) & \dots & |a_{1i}(X)| & \dots & |a_{1n}(X)| \\ \vdots & \ddots & \vdots & & \vdots \\ |a_{i1}(X)| & \dots & a_{ii}(X) & \dots & |a_{in}(X)| \\ \vdots & & \vdots & \ddots & \vdots \\ |a_{n1}(X)| & \dots & |a_{ni}(X)| & \dots & a_{nn}(X) \end{pmatrix} \quad q(X) = \begin{pmatrix} |b_1(X)| \\ \vdots \\ |b_i(X)| \\ \vdots \\ |b_n(X)| \end{pmatrix} \tag{7.76}$$

$M = \sup\limits_{X \in \mathcal{X}_c} M(X), \;\; q = \sup\limits_{X \in \mathcal{X}_c} q(X).$

If, for $\mathcal{X}_c = \Re^n$, $M$ verifies the Koteliansky conditions (7.24), then all motions of (7.74) converge inside of $\{X \in \Re^n : p(X) \leq -M^{-1}q\}$ as $t$ tends to infinite.

**Example 7.7** Let us consider

$$\frac{dX}{dt} = A(X, d)X,$$

$$A(X, d) = \begin{bmatrix} -2 & \cos x_2 & 0.2 & 0.7\delta_1 \\ -0.2 & -1.5 & -0.4 & 0.6 \, sign \, x_3 \\ 0.1\delta_2 & 0.3 & -1.3 & 0.3\delta_2 \\ -0.1 & 0.4 & 1.1 & -2 \end{bmatrix} \tag{7.77}$$

$$d = (\delta_1, \delta_2), \;\; \mathcal{S}_d = \{d, \; ||d||_\infty \leq 1\} \quad (\text{ i.e. } \;\; |\delta_1| \leq 1, \;\; |\delta_2| \leq 1),$$

and define the max-vector norm $p(X)$ as

$$X = \begin{pmatrix} X_1 \\ X_2 \end{pmatrix}, \;\; X_1 = \begin{pmatrix} x_1 \\ x_2 \end{pmatrix}, \;\; X_3 = \begin{pmatrix} x_3 \\ x_4 \end{pmatrix}, \;\; p(X) = \begin{pmatrix} ||X_1||_\infty \\ ||X_2||_\infty \end{pmatrix}. \tag{7.77bis}$$

Then, for $d \in \mathcal{S}_d$

$$\mu_{11}(X) \; = max \, \{-2 + |cos \; x_2|, \; |-0.2| - 1.5\} \leq \mu_{11} = -1,$$

$$\mu_{12}(X,d) = max\,\{|0.2| + |0.7\delta_1|,\ |-0.4| + |0.6\,sign\,x_3|\} \leq \mu_{12} = 1,$$

$$\mu_{21}(X,d) = max\,\{|0.1\delta_2| + |0.3|,\ |-0.1| + |0.4|\} \leq \mu_{21} = 0.5,$$

$$\mu_{22}(X) = max\,\{-1.3 + |0.3\delta_2|,\ |1.1| - 2\} \leq \mu_{22} = -0.9,$$

and the solutions of

$$\frac{dZ}{dt} = \begin{pmatrix} -1 & 1 \\ 0.5 & -0.9 \end{pmatrix} Z = MZ. \tag{7.78}$$

$$Z_0 = p(X_0)$$

verify $p(X(t)) \leq Z(t)$ for any $d$ varying in $\mathcal{S}_d$. In this case, $M$ verifies the Kotelianksy conditions, which leads to the conclusion that $X = 0$ is globally asymptotically stable for (7.77).

## 7.3 Vector norms and Lyapunov stability criteria

The following is a direct elaboration of Lemma 7.2, which constitutes a basic result for vector-norms-based Lyapunov stability criteria.

### 7.3.1 Stability of equilibrium points

In the first part, it is assumed that for any disturbance $d$ in $\mathcal{S}_d$, $X = 0$ is a stationary point of (7.1). Then:

$$g(0,d) = 0 \quad \forall d \in \mathcal{S}_d. \tag{7.79}$$

Then, system (7.1) can be described under the form:

$$\frac{dX}{dt} = A(X,d)X \tag{7.80}$$

with $A(X,d)$ bounded for any $d \in \mathcal{S}_d$ and any $X$ in some neighbourhood $\mathcal{S}$ of $X = 0$. Notice that $X = 0$ is also an equilibrium point of (7.1) and (7.80).

**Theorem 7.6** *Let $\mathcal{S} \subset \Re^n$ be an open connected neighbourhood of the origin $X = 0$. Let $p(X)$ be an R.V.N. of size $k$, and $M$ a constant $(\mathcal{S} \times \mathcal{S}_d)$-local overvaluing matrix of system (7.80) associated with $p$, calculated by means of equations (7.53) or (7.65), or in other words such that:*

$$\mathcal{D}_t^+ p(X) \leq Mp(X) \ \ \forall t \in \Re,\ \forall X \in \mathcal{S},\ \forall d \in \mathcal{S}_d,$$

*and suppose that $M$ is irreducible.*

*Let $\mathsf{c} \in \Re^k$ such that $\mathsf{c} > 0$ and $-M\mathsf{c} > 0$. Let $u(M)$ (resp. $u(M^T)) > 0$ a positive importance vector of $M$ (resp. $M^T$), associated with the importance value $\lambda(M)$, and define the three types of open sets:*

$$\mathcal{X}_c = \{X \in \Re^n : p(X) < \mathsf{c}\},\ -M\mathsf{c} > 0, \tag{7.81}$$

$$\mathcal{D}_1(\alpha) = \{X \in \Re^n : u(M^T)^T p(X) < \alpha\},\ \alpha > 0, \tag{7.82}$$

$$\mathcal{D}_\infty(\alpha) = \{X \in \Re^n : p(X) < \alpha u(M)\},\ \alpha > 0. \tag{7.83}$$

*Then, considering that disturbances d remain in $\mathcal{S}_d$:*

*i) if $M$ is stable, then $X = 0$ is stable for (7.80) and any set $\mathcal{D}_1(\alpha)$ or $\mathcal{D}_\infty(\alpha)$ included in $\mathcal{S}$ is an estimate $\mathcal{E}_s$ of its stability domain;*

*ii) if $M$ is asymptotically stable, then $X = 0$ is asymptotically stable for (7.80) and any set $\mathcal{X}_c$ strictly included in $\mathcal{S}$ is an estimate $\mathcal{E}_{as}$ of its asymptotic stability domain;*

*iii) if $M$ is asymptotically stable, then $X = 0$ is exponentially stable for (7.80) and any set $\mathcal{D}_1(\alpha)$ or $\mathcal{D}_\infty(\alpha)$ strictly included in $\mathcal{S}$ is an estimate $\mathcal{E}_{es}$ of its exponential stability domain with regard to $(\beta, \gamma = -\lambda(M))$;*

*iv) if $\mathcal{S} = \Re^n$, the preceding properties hold with the global sense.* ∇∇∇

**Remark:** Since the definition of exponential stability refers to the Euclidean norm $\|\cdot\|_2$, coefficient $\beta$ in (iii) depends on the relation between the norms $p_i$ and the Euclidean norm. For example, if all $p_i$ ($i = 1 \ldots k$) are of the same type $\|\cdot\|_\omega$ ($\omega = 1, 2$ or $\infty$), we have:

$$\beta = 1 \quad \text{for} \quad p_i = \|\cdot\|_2,$$

$$\beta = \sqrt{n} \quad \text{for} \quad p_i = \|\cdot\|_1,$$

$$\beta = \sqrt{n} \quad \text{for} \quad p_i = \|\cdot\|_\infty.$$

In this case, (iii) means

$$\|X(t; X_0)\|_\omega \leq exp\big(\lambda(M)t\big)\, \|X_0\|_\omega.$$

**Proof.** ii). Consider the set $\mathcal{X}_c$, with $-M\mathsf{c} > 0$. The pair $(M, q = 0)$ is an $(\mathcal{S} \times \mathcal{S}_d)$-overvaluing pair, then it also is a $(\mathcal{X}_c \times \mathcal{S}_d)$-overvaluing pair since $\mathcal{X}_c \subset \mathcal{S}$. Thus, Lemma 7.2 (b and c) ensures that if one defines system

$$\frac{dZ}{dt} = MZ \tag{7.84}$$

then $((7.84), \mathcal{Z}_c, \{0\})$ is a comparison system of $((7.80), \mathcal{X}_c, \{0\})$ with regard to the asymptotic stability property with the estimate of domains. Since $Z = 0$ is asymptotically stable for (7.84) with $\mathcal{Z}_c$ an estimate of its asymptotic stability domain, then point (ii) holds.

iii). Let us consider the set $\mathcal{D}_1(\alpha)\subset_{\neq}\mathcal{S}$ and the positive function

$$v_1(X) = u(M^T)^T p(X), \tag{7.85}$$

that verifies, for any $X$ in $\mathcal{D}_1(\alpha)$:

$$\mathcal{D}^+ v_1(X) \leq u(M^T)^T M p(X) = \lambda(M) v_1(X). \tag{7.86}$$

Now, $\mathcal{D}_1(\alpha) = \{X \in \Re^n : v_1(X) < \alpha\}$. Then, point (iii) holds for $\mathcal{D}_1(\alpha) \subset_{\neq} \mathcal{S}$.

Consider the set $\mathcal{D}_\infty(\alpha)\subset_{\neq}\mathcal{S}$ and the positive definite function

$$v_\infty(X) = \max_{i=1,\dots,k} p_i(X) u_i^{-1} = u_{i_0}^{-1} p_{i_0} \tag{7.87}$$

with notation $i_0 = i_0(X)$, $u(M) = u = (u_i)_{i=1,\dots,k}$. Then,

$$\mathcal{D}^+ v_\infty(X) \leq u_{i_0}^{-1} \sum_{j=1}^{k} m_{i_0 j} p_j(X_j) \frac{u_j}{u_j} \leq u_{i_0}^{-1} v_\infty(X) \sum_{j=1}^{k} m_{i_0 j} u_j$$

and finally

$$\mathcal{D}^+ v_\infty(X) \leq \lambda(M) v_\infty(X). \tag{7.88}$$

Now, $\mathcal{D}_\infty(\alpha) = \{X \in \Re^n : v_\infty(X) < \alpha\}$. Then point (iii) holds for $\mathcal{D}_\infty(\alpha)\subset_{\neq}\mathcal{S}$.

i). We suppose $M$ is stable, then $v_1(X)$ (7.85) and $v_\infty(X)$ (7.87) still are Lyapunov functions, since for any $X$ in $\mathcal{S}$, and from (7.86), (7.88) $\mathcal{D}^+ v_1(X) \leq 0$ and $\mathcal{D}^+ v_\infty(X) \leq 0$; then $X = 0$ is stable, and positive invariance of $\mathcal{D}_1(\alpha) \subset \mathcal{S}$ and $\mathcal{D}_\infty(\alpha) \subset \mathcal{S}$ ensures that both are estimates of its stability domain.

**Theorem 7.7** *Let $\mathcal{S} \subset \Re^n$ be an open connected neighbourhood of the origin $X = 0$. Let $p(X)$ be an R.V.N. of size $k$, and $h : \Re^k \to \Re^k$ a quasi-increasing function that cancels at the origin ($h(0) = 0$), and such that for any $X$ in $\mathcal{S}$ and $d$ in $\mathcal{S}_d$:*

$$\mathcal{D}^+ p(X) \leq M(X) p(X) \leq h(p(X)) \tag{7.89}$$

*where $M(X)$ is calculated with (7.53) applied to (7.80).*

*Let $c > 0$ such that $\mathcal{X}_c \subset \mathcal{S}$ (see eqn. (7.38a)) and the corresponding $\mathcal{Z}_c \subset \Re^k$ (eqn. (7.38b)).*

*Then, considering that the disturbances $d$ remains in $\mathcal{S}_d$:*

*i) if the equilibrium $Z_e = 0$ of*

$$\frac{dZ}{dt} = h(Z) \tag{7.90}$$

*is stable (respectively asymptotically stable), then $X_e = 0$ is a stable equilibrium for (7.80) (respectively asymptotically stable).*

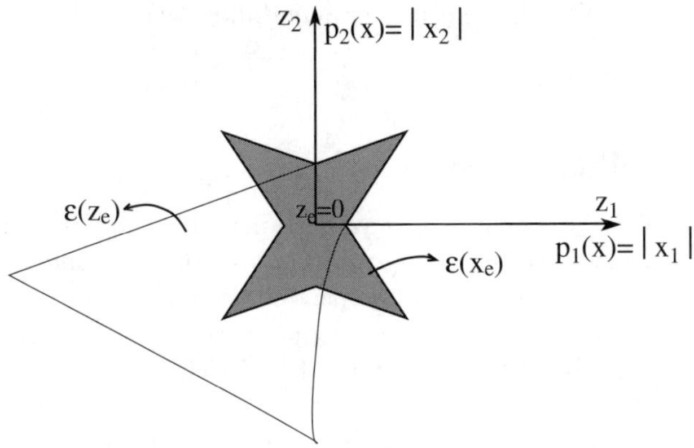


Figure 7.4: Estimates relative to Theorem 7.7.

*ii) if $\mathcal{E}(Z_e)$ is an estimate of the (respectively asymptotic) stability domain of $Z_e = 0$ for (7.90), then the set*

$$\mathcal{E}(X_e) = \{X \in \Re^n : p(X) \in \mathcal{E}(Z_e)\} \tag{7.91}$$

*is an estimate of the (respectively asymptotic) stability domain of $X_e = 0$ for (7.80).* ∇∇∇

**Theorem 7.8 (Borne and Gentina criterion) [87], [173]** *Let $\mathcal{S} \subset \Re^n$ be an open connected neighbourhood of the origin $X = 0$. Let $p(X)$ be an R.V.N. of size $k$, and $M(X)$ an $(\mathcal{S} \times \mathcal{S}_d)$-local overvaluing matrix of system (7.1), or in other words such that:*

$$\mathcal{D}^+ p(X) \leq M(X)p(X),$$

*with the following properties:*

*1) $M(X)$ is irreducible for all $X$ in $\mathcal{S}$;*

*2) there is $\varepsilon > 0$ such that $M(X) + \varepsilon I_k$ is a $-M$-matrix;*

*3) the non constant elements of $M(X)$ are grouped into one unique column. Let $\lambda(M(X_1)) = \sup_{X \in \mathcal{S}} \{\lambda(M(X))\} \leq -\varepsilon < 0$ the upper (with regard to $X \in \mathcal{S}$) importance value of $M(X)$, obtained for $X = X_1$ (possibly not unique).*

*Then, as soon as disturbances d remain in $\mathcal{S}_d$:*

*i) $X = 0$ is exponentially stable for system (7.1) (with rate $\gamma = \varepsilon$) and any set $\mathcal{D}_1(\alpha)$ included in $\mathcal{S}$: $\mathcal{D}_1(\alpha) = \{X \in \Re^n; u^T p(X) < \alpha\} \subset \mathcal{S}$, where $u(M^T(X_1))$ is a positive eigenvector of $M^T(X_1)$ associated with $\lambda(M(X_1))$, is an estimate $\mathcal{E}_{as}$ of its asymptotic stability domain;*

*ii) if $\mathcal{S} = \Re^n$, $X = 0$ is globally exponentially stable.* ∇∇∇

### 7.3.2 Stability of sets

In this part, $X = 0$ may not be an equilibrium point, and if it is, it is not necessarily a stable one. However, the following theorem provides the conditions for the existence and estimation of stable sets and associated domains.

We consider the system described by (7.1), which can be decomposed into (7.51) for computation of overvaluing pairs (7.53), (7.54) or (7.65).

For example, decomposition (7.51) can correspond to the natural decomposition (7.2) in which vector $b(\cdot)$ represents the effect of the disturbance $d \in \mathcal{S}_d$:

$$\frac{dX}{dt} = f(X) + d(X) = A(X)X + d(X). \tag{7.92}$$

In this case, if there is no disturbance, $X = 0$ is an equilibrium, but the influence of disturbance creates a different behaviour (displacement of the equilibrium or limit cycle, or others) that is to be estimated.

**Theorem 7.9 (Borne and Richard criterion)** *Let $\mathcal{S} \subset \Re^n$ be an open connected neighbourhood of the origin $X = 0$. Let $p(X)$ be an R.V.N. of size $k$, and $(M, q)$ a constant $(\mathcal{S} \times \mathcal{S}_d)$-local overvaluing pair of system (7.1), or in other words such that:*

$$\mathcal{D}_t^+ p(X) \leq Mp(X) + q \quad \forall t \in \Re,\ \forall X \in \mathcal{S},\ \forall d \in \mathcal{S}_d, \tag{7.93}$$

*and suppose that $M$ is irreducible and is a $-M$-matrix.*

*Define the set*

$$\mathcal{A} = \left\{X \in \Re^n : p(X) \leq -M^{-1}q\right\} \tag{7.94}$$

*and suppose that $\mathcal{A}$ is included in $\mathcal{S}$:*

$$\mathcal{A} \subset \mathcal{S}. \tag{7.95}$$

*Then, considering that disturbances remain in $\mathcal{S}_d$:*

*i) the set $\mathcal{A}$ is asymptotically stable;*

*ii) any set $\mathcal{X}_c \subset \mathcal{S}$ such that $-M\mathrm{c} > q$ is an estimate of the asymptotic stability domain of $\mathcal{A}$;*

*iii) any set $\mathcal{D}_1(\alpha) = \{X \in \Re^n : u^T(M^T)p(X) < \alpha\} \subset \mathcal{S}$ is an estimate of the asymptotic stability domain of $\mathcal{A}$;*

*iv) if $\mathcal{S} = \Re^n$, (7.95) is verified in a systematic way, and the asymptotic stability of $\mathcal{A}$ is global.* ∇∇∇

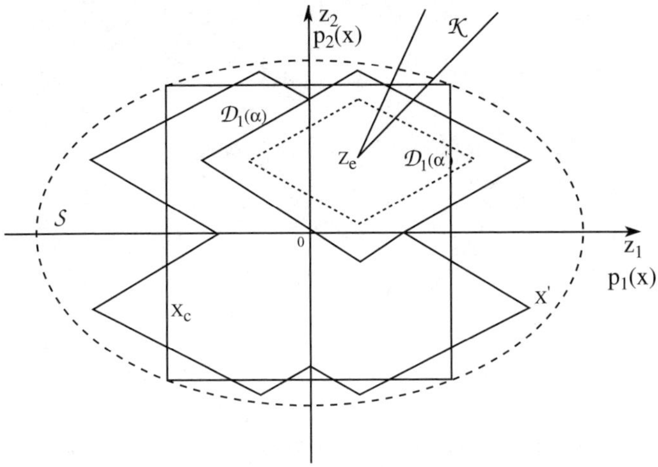


Figure 7.5: Variant of Theorem 7.9.

**Remark:** a variant of this statement can be provided by using the property that $\mathcal{D}_1(\alpha)$ (in eqn. (7.28)) is exponentially contractive for linear systems (variable $Z$).

$$\mathcal{D}_1(\alpha) = \{Z \in \Re^k : Z = -M^{-1}q + Y,\ u^T(M^T)|Y| < \alpha\}. \tag{7.96}$$

This allows the definition of other forms of domains $\mathcal{X}'$, that may be more adapted than $\mathcal{X}_c$ for specific cases of $\mathcal{S}$: these forms (see Fig. 7.5) are constructed on the basis of rhomboids (7.96).

In order for domain $\mathcal{D}_1(\alpha)$ to lead to an invariant domain $\mathcal{X}$ defined by $\mathcal{X} = \{X \in \Re^n : p(X) \in \mathcal{D}_1(\alpha)\} \subset \mathcal{S}$, $\alpha$ has to be great enough to guarantee that it crosses through all axes of $Z$-space (in Fig. 7.4, domain $\mathcal{D}_1(\alpha')$ is not convenient). This gives the condition on $\alpha$ (with $u_i$ the components of $u(M^T)$):

$$Z_e = -M^{-1}q \le \alpha \begin{pmatrix} u_1^{-1} \\ \vdots \\ u_k^{-1} \end{pmatrix}. \tag{7.97}$$

**Theorem 7.10** *Let $\mathcal{S} \subset \Re^n$ be an open connected neighbourhood of the origin $X = 0$. Let $p(X)$ be an R.V.N. of size $k$, and $h : \Re^k \to \Re^k$ a quasi-increasing function such that for any $X$ in $\mathcal{S}$ and d in $\mathcal{S}_d$:*

$$\mathcal{D}^+p(X) \le M(X)p(X) + q(X) \le h(p(X)) \tag{7.98}$$

*where $M(X)$, $q(X)$ are calculated by means of (7.53), (7.54) or (7.65) applied to a decomposition (7.51) of system (7.1).*
*Then, considering that the disturbances d remain in $\mathcal{S}_d$:*

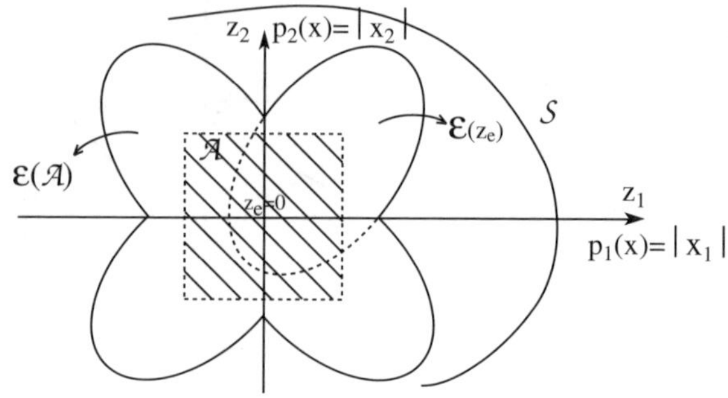


Figure 7.6: Illustration of Theorem 7.10.

*i) if the point $Z_e \geq 0$ is a stable equilibrium (respectively asymptotically stable) of the system:*

$$\frac{dZ}{dt} = h(Z) \tag{7.99}$$

*and if the set $\mathcal{A} \subset \Re^n$ defined by:*

$$\mathcal{A} = \{X \in \Re^n : p(X) \leq Z_e\} \tag{7.100}$$

*is included in $\mathcal{S}$, then $\mathcal{A}$ is stable (respectively asymptotically stable) for system (7.1).*

*ii) if $\mathcal{E}(Z_e)$ is an estimate of the stability (respectively asymptotic stability) domain of $Z_e$ for (7.99), such that:*

$$\mathcal{A} \subset \mathcal{E}(\mathcal{A}) = \{X \in \Re^n : p(X) \in \mathcal{E}(Z_e)\} \subset \mathcal{S} \tag{7.101}$$

*then $\mathcal{E}(\mathcal{A})$ is an estimate of the stability (respectively asymptotic stability) domain of set $\mathcal{A}$ for system (7.1).* $\nabla\nabla\nabla$

**Remark:** in hypothesis (7.101), $\mathcal{A} \subset \mathcal{E}(\mathcal{A})$ is equivalent to $\{Z \in \Re^k_+ : 0 \leq Z < Z_e\} \subset \mathcal{E}(Z_e)$.

**Theorem 7.11 [18]** *Let $\mathcal{S} \subset \Re^n$ be an open connected neighbourhood of the origin $X = 0$. Let $p(X)$ be an R.V.N. of size $k$, and $(M(X), q(X))$ an $(\mathcal{S} \times \mathcal{S}_d)$-local overvaluing pair of system (7.1), or in other words such that:*

$$\mathcal{D}^+ p(X) \leq M(X)p(X) + q(X),$$

*with a matrix $M(X)$ verifying properties 1), 2), 3) of Theorem 7.8, i.e. such that $M(X) + \varepsilon I_k$ ($\varepsilon > 0$) is an irreducible $-M$-matrix, with non constant elements grouped into one unique column.*

*We define $\lambda = \lambda(M(X_1))$ as in Theorem 7.8, $u = u(M^T(X_1))$ a positive associated importance vector of $M^T(X_1)$, and the sets:*

$$\mathcal{A} = \left\{X \in \Re^n : u^T p(X) \leq -\lambda^{-1} \sup_{X \in \mathcal{S}} (u^T q(X))\right\}, \tag{7.102}$$

$$\mathcal{D}_1(\alpha) = \{X \in \Re^n ; u^T p(X) \leq \alpha\} \subset \mathcal{S}. \tag{7.103}$$

*We moreover assume that $\mathcal{A} \subset \mathcal{S}$. Then, considering that the disturbances d remain in $\mathcal{S}_d$:*

*i) $\mathcal{A}$ is exponentially stable for system (7.1) (with rate $\gamma = \varepsilon$);*

*ii) $\mathcal{D}_1(\alpha)$ is an estimate of the asymptotic stability domain of $\mathcal{A}$;*

*iii) if $\mathcal{S} = \Re^n$, then the property is global (and property $\mathcal{A} \subset \mathcal{S}$ is verified in a systematic way).* ∇∇∇

**Proof.** The proof is based on the Lyapunov function:

$$v_1(X) = u^T p(X).$$

We have:

$$\mathcal{D}^+ v_1(X) \le u^T (M(X)p(X) + q(X)),$$

and due to the particular form of $M$, this yields:

$$\mathcal{D}^+ v_1(X) \le \lambda v_1(X) + u^T q(X).$$

The comparison scalar system

$$\dot{\zeta} = \lambda\zeta + \sup_{X \in \mathcal{S}} (u^T q(X))$$

then converges towards $\zeta_e = \lambda^{-1} \sup_{X \in \mathcal{S}} (u^T q(X))$, which defines $\mathcal{A}$.

### 7.3.3 Examples

**Example 7.8 (Theorem 7.9)** Let the disturbed nonlinear system

$$\frac{dX}{dt} = \begin{pmatrix} 1 - x_1^2 - x_2^2 & \delta_1(X) + 0.1 \\ \delta_2(X) + 0.5 & 1 - x_1^2 - x_2^2 \end{pmatrix} = g(X, d) \tag{7.104}$$

$$\text{with } \mathcal{S}_d = \left\{ d(X) = \begin{pmatrix} \delta_1(X) \\ \delta_2(X) \end{pmatrix} : |d(X)| \le \begin{pmatrix} 0.9 \\ 0.5 \end{pmatrix} \right\}. \tag{7.105}$$

(notice that the same study is still available if $\delta_1$ and $\delta_2$ are functions of both state and time $t$ [162]).

Consider $p(X) = |X|$, and define two overvaluing pairs on the basis of a separation of $\Re^2$ in two sets $\mathcal{X}_c$, $\mathsf{c} = (\gamma_1, \gamma_2)^T$:

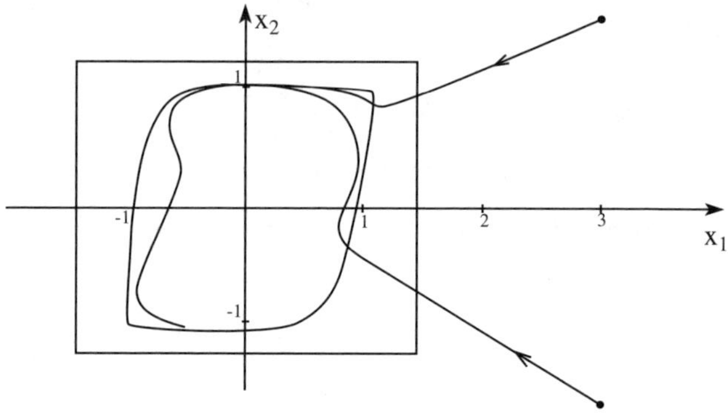


Figure 7.7: Simulations of Example 7.8.

1) if $X \in \overline{\mathcal{X}}_c$, we use the decomposition:

$$g(X,d) = \begin{pmatrix} 1-\gamma_1^2-\gamma_2^2 & \delta_1(X)+0.1 \\ \delta_2(X)+0.5 & 1-\gamma_1^2-\gamma_2^2 \end{pmatrix} X + \begin{pmatrix} x_1(\gamma_1^2 - x_1^2 + \gamma_2^2 - x_2^2) \\ x_2(\gamma_1^2 - x_1^2 + \gamma_2^2 - x_2^2) \end{pmatrix} \tag{7.106}$$

which admits the following $(\overline{\mathcal{X}}_c \times \mathcal{S}_d)$-local overvaluing pair (see formula (7.65bis))

$$M_1 = \begin{pmatrix} 1-\gamma_1^2-\gamma_2^2 & 1 \\ 1 & 1-\gamma_1^2-\gamma_2^2 \end{pmatrix}, \quad q_1 = \frac{2}{3\sqrt{3}}(\gamma_1^2+\gamma_2^2)^{3/2}\begin{pmatrix} 1 \\ 1 \end{pmatrix}. \tag{7.107}$$

2) if $X \notin \overline{\mathcal{X}}_c$, we directly use the writing (7.104) of $g(X,d)$, that leads to the following $(\Re^2 - \mathcal{X}_c \times \mathcal{S}_d)$-local overvaluing pair:

$$M_2 = \begin{pmatrix} 1-\gamma_1^2-\gamma_2^2 & 1 \\ 1 & 1-\gamma_1^2-\gamma_2^2 \end{pmatrix}, \quad q_2 = \begin{pmatrix} 0 \\ 0 \end{pmatrix}. \tag{7.108}$$

Defining $M = max\,\{M_1, M_2\}$ and $q = max\,\{q_1, q_2\}$ leads to the following $\mathcal{S}_d$-local overvaluing pair:

$$M = M_1 = M_2 = \begin{pmatrix} 1-\gamma_1^2-\gamma_2^2 & 1 \\ 1 & 1-\gamma_1^2-\gamma_2^2 \end{pmatrix}, \quad q = q_1. \tag{7.109}$$

$M$ is a $-M$-matrix iff c is such that:

$$\mathrm{c}^T\mathrm{c} = \gamma_1^2 + \gamma_2^2 > 2. \tag{7.110}$$

Under this hypothesis, Theorem 7.9 (iii) ensures that the set $\mathcal{A}$:

$$\mathcal{A} = \left\{ X \in \Re^2 : |X| \leq -M^{-1}q = \frac{2(\gamma_1^2+\gamma_2^2)^{3/2}}{3\sqrt{3}(\gamma_1^2+\gamma_2^2-2)} \begin{pmatrix} 1 \\ 1 \end{pmatrix} \right\} \tag{7.111}$$

is globally asymptotically stable, that is it contains all possible asymptotic behaviours of (7.104) for any $d \in \mathcal{S}_d$ (7.105).

Now, we can carry on the study by minimizing the vector $-M^{-1}q$ with regard to $\mathrm{c} = (\gamma_1, \gamma_2)^T$, under the constraint (7.110). This yields for $\gamma_1^2 + \gamma_2^2 = 6$,

$$\mathcal{A}_{opt} = \left\{ X \in \Re^2 : |X| \leq \begin{pmatrix} \sqrt{2} \\ \sqrt{2} \end{pmatrix} \right\}. \tag{7.112}$$

Simulation of Fig. 7.7 shows particular cases of behaviour corresponding to $d(X) = \left(0.9\, cos\, t,\ 0.5\frac{t^2}{1+t^2}\right)$, and $t_0 = 0$; they are taken from [162].

**Example 7.9 (Theorem 7.9)** Consider the fourth-order nonlinear system defined on Example 7.7 (7.77), with additional input disturbance $d'$:

$$\frac{dX}{dt} = A(X,d)X + d'(X), \quad X \in \Re^4, \tag{7.113}$$

$$(d, d') \in \mathcal{S}_d \times \mathcal{S}_{d'}, \quad \mathcal{S}_{d'} = \{d' \in \Re^4 : |d| < 0.4\}.$$

The R.V.N. $p(X)$ defined as in (7.77bis) as:

$$p(X) = \begin{pmatrix} \|X_1\|_\infty \\ \|X_2\|_\infty \end{pmatrix}$$

leads to the $\mathcal{S}_d$-local overvaluing pair:

$$M = \begin{pmatrix} -1 & 1 \\ 0.5 & -0.9 \end{pmatrix}, \quad q = \begin{pmatrix} 0.4 \\ 0.4 \end{pmatrix}, \tag{7.114}$$

hence:

$$-M^{-1}q = \begin{pmatrix} 1.9 \\ 1.5 \end{pmatrix}, \tag{7.115}$$

and Theorem 7.9 (iii) ensures that the set $\mathcal{A}$:

$$\mathcal{A} = \{X \in \Re^4 : |X|^T \leq (1.9, 1.9, 1.5, 1.5)\}$$

is globally asymptotically stable for $(d, d') \in \mathcal{S}_d \times \mathcal{S}_{d'}$.

**Example 7.10 (Theorem 7.10)** Let us study the 2nd order system:

$$\frac{dX}{dt} = \begin{bmatrix} -2 + \delta_1|x_2| & \delta_2 + 8 \cdot 10^{-7}|x_2|^3 \\ \delta_3 & 1 - x_1^2 \end{bmatrix} X, \tag{7.116}$$

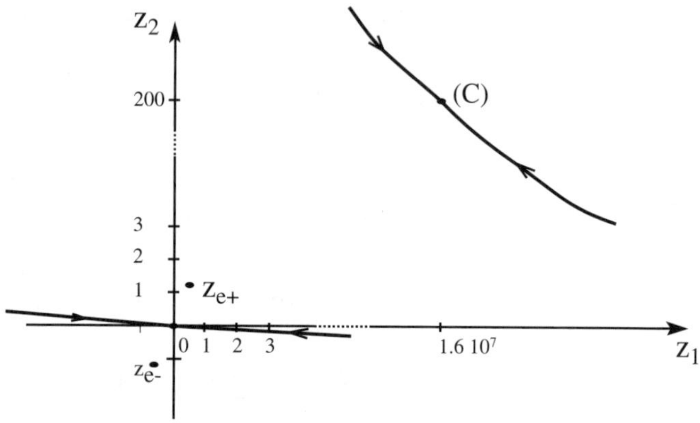


Figure 7.8: Equilibria and phase portrait of system (7.116).

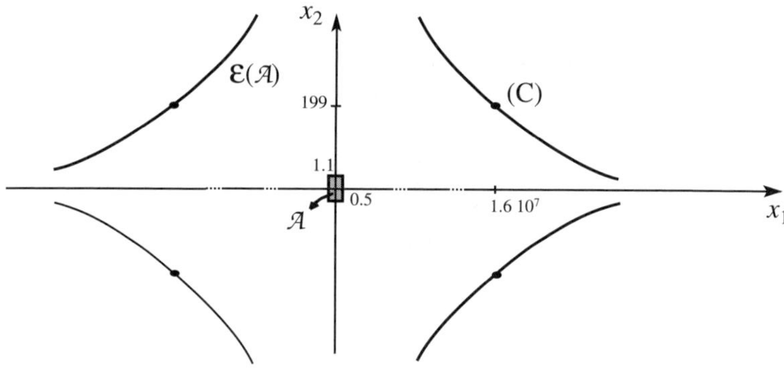


Figure 7.9: Set $\mathcal{A}$ and its asymptotic stability domain estimate, for system (7.116) with any $d \in \mathcal{S}_d$.

$d = (\delta_1(X), \delta_2(X), \delta_3(X))^T$ is the disturbance vector continuous w.r.t. $X$,

$$\mathcal{S}_d = \{d : \Re^2 \to \Re^3 : |\delta_1(X)| < 10^{-2},\ |\delta_2(X)| \le 0.9,\ |\delta_3(X)| \le 0.5\}.$$

Considering $p(X) = |X|$ leads to:

$$h(Z) = \begin{bmatrix} -2 + 10^{-2} z_2 & 0.9 + 8 \cdot 10^{-7} z_2^3 \\ 0.5 & 1 - z_1^2 \end{bmatrix} Z, \tag{7.117}$$

$$\mathcal{D}^+ p(X) \le h(p(X)). \tag{7.118}$$

Moreover, $h$ is quasi-increasing, and cancels at the origin. Then the preliminary hypotheses of Theorems 7.7 and 7.10 are fulfilled. Studying (7.117) shows that $Z = 0$ is an unstable equilibrium, and Theorem 7.7 cannot be applied. However, there exists a positive asymptotically stable equilibrium at

$$Z_{e_+} = \begin{pmatrix} 0.501 \\ 1.107 \end{pmatrix}, \tag{7.119}$$

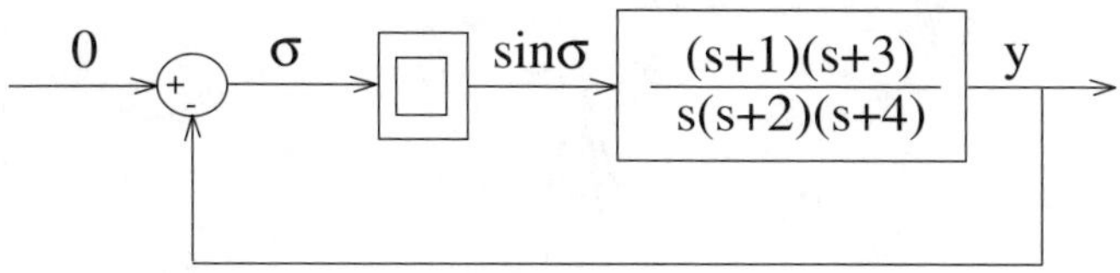


Figure 7.10: Output feedback of Example 7.11.

with asymptotic stability domain described on Fig. 7.8. System (7.117) actually presents two asymptotically stable equilibria $Z_{e_+}$, $Z_e$ (one positive, one negative) and two unstable ones (the origin and $[1.6 \cdot 10^7, 200]^T$), that can be obtained and studied by numerical means since all equilibria are hyperbolic. The manifold $(C)$ defines a boundary of the asymptotic stability domain of $Z_{e_+}$.

Then, Theorem 7.10 leads to the (open) estimate $\mathcal{E}(\mathcal{A})$ (see Fig. 7.9) of the asymptotic stability domain of

$$\mathcal{A} = \{X \in \Re^2 : |x_1| < 0.501,\ |x_2| < 1.107\}. \tag{7.120}$$

**Example 7.11 (Theorem 7.6)** We consider the system described on Fig. 7.10, with the aim that $y$ tends to zero.

It can be represented, using arrow-matrix modelling [12], [172]:

$$\frac{dX}{dt} = A(X)X$$

$$A(X) = \begin{bmatrix} -1 & 0 & 1.5 \\ 0 & -3 & 1.5 \\ 1 & 1 & -2 - \dfrac{\sin\sigma}{\sigma} \end{bmatrix} X = \begin{bmatrix} x_1 \\ x_2 \\ x_3 = \sigma \end{bmatrix}. \tag{7.121}$$

We study the equilibrium $X = 0$.

The R.V.N. $p(X) = |X|$ leads to the following $\mathcal{S}$-local overvaluing matrix (where $\alpha$ is a positive parameter):

$$M = \begin{bmatrix} -1 & 0 & 1.5 \\ 0 & -3 & 1.5 \\ 1 & 1 & -2 - m \end{bmatrix}, m = \inf_{|\sigma|<\alpha} \frac{\sin\sigma}{\sigma} = \frac{\sin\alpha}{\alpha} \ \text{ for } \ \alpha \leq \frac{3\pi}{2} \tag{7.122}$$

$$\mathcal{S} = \{X \in \Re^3 : |x_3| < \alpha\}. \tag{7.123}$$

In order to guarantee that $M$ is asymptotically stable, it is necessary and sufficient that:

$$det\ M = -m = -\frac{\sin\alpha}{\alpha} < 0 \tag{7.124}$$

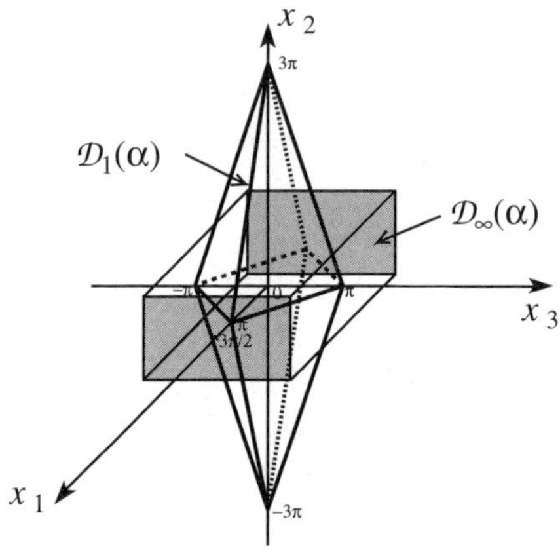


Figure 7.11: Estimates of the asymptotic stability domain for Example 7.11.

thus $\alpha < \pi$, and we define

$$\mathcal{S} = \mathcal{S}(\alpha) = \{X \in \Re^3 : |x_3| < \alpha \leq \pi\}. \tag{7.125}$$

Using Theorem 7.6 (i), we obtain the following domain estimates:

$$\mathcal{D}_1(\alpha) = \{X \in \Re^3 : |x_1| + \frac{1}{3}|x_2| + |x_3| < \pi\} \underset{\neq}{\subset} \mathcal{S}(\pi) \tag{7.126}$$

$$\mathcal{D}_\infty(\alpha) = \left\{ X \in \Re^3 : |X| < \begin{pmatrix} 3\pi/2 \\ \pi/2 \\ \pi \end{pmatrix} \right\} \underset{\neq}{\subset} \mathcal{S}(\pi) \quad \left(\alpha = \frac{\pi}{2}\right) \tag{7.127}$$

that are depicted on Fig. 7.11. Obviously, $\mathcal{D}_1(\alpha) \cup \mathcal{D}_\infty(\alpha)$ is also an estimate.

**Example 7.12 (Theorem 7.8)** Consider the equation (with $f_1$ and $f_2$ piecewise continuous functions)

$$\frac{dX}{dt} = \begin{pmatrix} -1.5 & f_1(X) \\ 2 & f_2(X) \end{pmatrix} X. \tag{7.128}$$

Using $p(X) = |X|$ and Theorem 7.8 shows that $X = 0$ is globally exponentially stable (with $exp\,(-t)$ convergence) if

$$det \begin{pmatrix} -0.5 & |f_1(X)| \\ 2 & 1 + f_2(X) \end{pmatrix} > 0, \ \ \forall X \in \Re^n, \tag{7.129}$$

that is, $f_2(X) < -1 - 4|f_1(X)|$.

It is globally asymptotically stable if, for a sufficiently small $\varepsilon$

$$det \begin{pmatrix} -1.5 & |f_1(X)| \\ 2 & f_2(X) \end{pmatrix} = -1.5\, f_2(X) - 2|f_1(X)| \geq \varepsilon > 0. \tag{7.130}$$

**Example 7.13 (Theorem 7.8)**

$$\frac{dX}{dt} = \begin{pmatrix} -3+\delta_1(X) & \dfrac{x_1^2+x_2^2}{64} \\ \dfrac{x_1^2}{x_1^2+x_2^2} & -1 \end{pmatrix} X + \begin{pmatrix} \delta_2(X) \\ \delta_3(X) \end{pmatrix}, \tag{7.131}$$

$$\mathcal{S}_d = \{(\delta_1, \delta_2, \delta_3) : |\delta_i| < 1 \quad \forall i = 1, 2, 3\}. \tag{7.132}$$

Choosing $p(X) = |X|$, $\mathcal{S} = \{X \in \Re^2,\ X^T X = x_1^2 + x_2^2 \leq 64\}$, leads to

$$\mathcal{D}^+ p(X) \leq \begin{pmatrix} -2 & 1 \\ 1 & -1 \end{pmatrix} p(X) + \begin{pmatrix} 1 \\ 1 \end{pmatrix} \quad \text{for all } X \text{ in } \mathcal{S}, \tag{7.133}$$

which has been studied in Example 7.5, eqn. (7.32). Theorem 7.8 with variant of eqn. (7.96) proves that the set:

$$\mathcal{A} = \{X \in \Re^2 : |x_1| \leq 2,\ |x_2| \leq 3\} \tag{7.134}$$

is asymptotically stable for any disturbance $\delta_i$ in the admissible set, and that the sets $\mathcal{X}_c$, $\mathsf{c} \in \mathcal{K}$, hence the set $\mathcal{E}$ (see Fig. 7.12) union of sets $\mathcal{X}_a$ such that $\mathsf{a}^T \mathsf{a} = 64$, are estimates of the asymptotic stability domain of $\mathcal{A}$. This estimate can be enlarged by using the variant (eqn. (7.96)) of Theorem 7.8, which is represented by domain $\mathcal{E}'$.

## 7.4 Vector norms and practical stability criteria with domains estimation

The definition of a decoupled overvaluing system, related to an R.V.N. $p(X)$ and to a (supposed here) constant overvaluing pair $(M, q)$ allows the comparison of the behaviours of $p(X)$ with the solution:

$$\zeta(t; t_0, p(X_0)) = exp\left(M(t-t_0)\right)\left[p(X_0) + M^{-1}q\right] - M^{-1}q, \tag{7.135}$$

of the linear system:

$$\begin{cases} \dot{\zeta} = M\zeta + q, \\ \zeta(t_0) = p(X_0). \end{cases} \tag{7.136}$$

This yields the following statement:

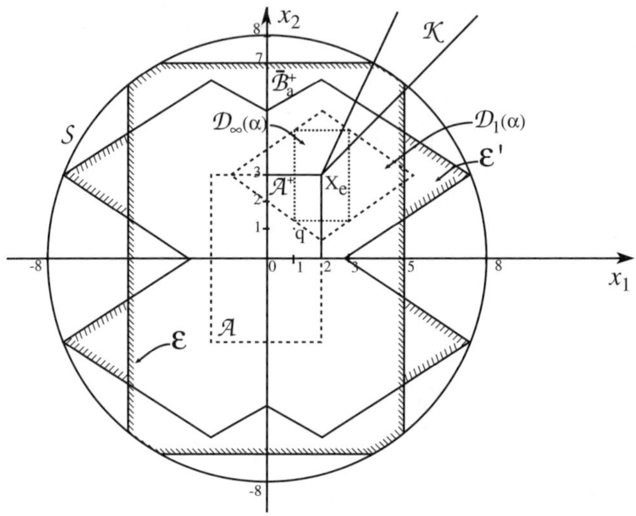


Figure 7.12: Estimates of the asymptotic stability domain for Example 7.13.

**Theorem 7.12 [164]** *Let $\mathcal{S} \subset \Re^n$ be an open connected neighbourhood of the origin $X = 0$. Let $p(X)$ be an R.V.N. of size $k$, and $(M, q)$ a constant $(\mathcal{S} \times \mathcal{S}_d)$-local overvaluing pair of system (7.1), with $M$ an irreducible $-M$-matrix. Define the sets:*

$$\mathcal{X}_0 = \left\{X \in \Re^n : p(X) < p_0,\ p_0 \neq 0,\ p_0 \in \Re_+^k\right\}, \tag{7.137}$$

$$\mathcal{X}_A = \left\{X \in \Re^n : p(X) < p_A,\ p_A \in \Re_+^k,\ -Mp_A > q\right\} \subset \mathcal{S}, \tag{7.138}$$

$$\mathcal{X}_F = \left\{X \in \Re^n : p(X) < p_F,\ p_F \in \Re_+^k,\ -Mp_F > q\right\} \subset \mathcal{S} \tag{7.139}$$

*Then system (7.1) is practically stable with settling time $\tau_s$ with respect to $\{+\infty, \mathcal{X}_0, \mathcal{X}_A, \mathcal{X}_F, \mathcal{S}_d\}$ provided that $p_0$ obeys the following relations:*

$$p_0 \leq p_A, \tag{7.140}$$

$$p_A \geq p_F \geq exp\,(M\tau_s)\left[p_0 + M^{-1}q\right] - M^{-1}q. \tag{7.141}$$

∇∇∇

**Proof.** Lemma 7.2 (b) ensures that $\mathcal{X}_A$ and $\mathcal{X}_F$ are positively invariant, since $M$ verifies the Koteliansky conditions and $\mathsf{c} = p_A$ or $p_F$ verifies: $-M\mathsf{c} > q$. This proves stability aspect. We have for all $d$ in $\mathcal{S}_d$, for all $X_0$ in $\mathcal{S}$, and all $t \geq t_0 = 0$

$$exp\,(Mt)\left[p(X_0) + M^{-1}q\right] - M^{-1}q \geq p(X(t; X_0; d)).$$

Let $X_0 \in \mathcal{X}_0$, then $p(X_0) < p_0$, and for any $t > \tau_s$ we have:

$$p(X(t; X_0; d)) < exp\,(M\tau_s)\left[p_0 + M^{-1}q\right] - M^{-1}q \leq p_F.$$

This proves the contractiveness aspect.

**Remark:** this way of estimating practical stability domains always provides positively invariant domains $\mathcal{X}_0$, which means that setting $\mathcal{X}_A = \mathcal{X}_0$ (i.e. $p_A = p_0$) makes no difference in the calculations, compared to $\mathcal{X}_A \supset \mathcal{X}_0$.

**Example 7.14** Let us consider again the system of Example 7.8:

$$\frac{dX}{dt} = \begin{pmatrix} 1 - x_1^2 - x_2^2 & \delta_1(X) + 0.1 \\ \delta_2(X) + 0.5 & 1 - x_1^2 - x_2^2 \end{pmatrix} X \tag{7.142}$$

with

$$\mathcal{S}_d = \left\{ d = (\delta_1, \delta_2)^T : |\delta_1(X)| \leq 0.9,\ |\delta_2(X)| < 0.5,\ \ \forall X \in \Re^2 \right\}. \tag{7.143}$$

We consider two kinds of studies:

**first study:** $\tau_s$, $\mathcal{X}_A$, $\mathcal{X}_F$ are given, then find $\mathcal{X}_0$.

**second study:** $\tau_s$, $\mathcal{X}_A$, are given, then find the smallest $\mathcal{X}_F$ and then the largest corresponding $\mathcal{X}_0$.

We shall consider the R.V.N. $p(X) = |X|$.

**a) first study:** let $\tau_s = 0.5s$, $\mathcal{X}_A = \Re^2$ or $\mathcal{X}_0$, $p_F = (2\ 2)^T$. The pair $(M, q)$ is defined by (see (7.107)):

$$M(\alpha) = \begin{pmatrix} 1 - \alpha^2 & 1 \\ 1 & 1 - \alpha^2 \end{pmatrix}, \quad q(\alpha) = \frac{2}{3\sqrt{3}} \alpha^3 \begin{pmatrix} 1 \\ 1 \end{pmatrix}, \tag{7.144}$$

with $\alpha > 0$ any scalar parameter such that $M(\alpha)$ is asymptotically stable, i.e.:

$$\alpha^2 > 2 \tag{7.145}$$

Condition $-Mp_F > q$, from $\mathcal{X}_F$ definition (7.139), leads to:

$$\alpha^2 - 2 > \frac{\alpha^3}{3\sqrt{3}} \tag{7.146}$$

which includes condition (7.145).

Condition (7.141) gives the condition relative to $p_0$:

$$\begin{pmatrix} \dfrac{e+1}{2} & \dfrac{e-1}{2} \\ \dfrac{e-1}{2} & \dfrac{e+1}{2} \end{pmatrix} p_0 \leq \left[ 2\,exp\left(\frac{\alpha^2}{2}\right) + \frac{2\alpha^3 \left(e - exp\left(\frac{\alpha^2}{2}\right)\right)}{3\sqrt{3}(\alpha^2 - 2)} \right] \begin{pmatrix} 1 \\ 1 \end{pmatrix}. \tag{7.147}$$

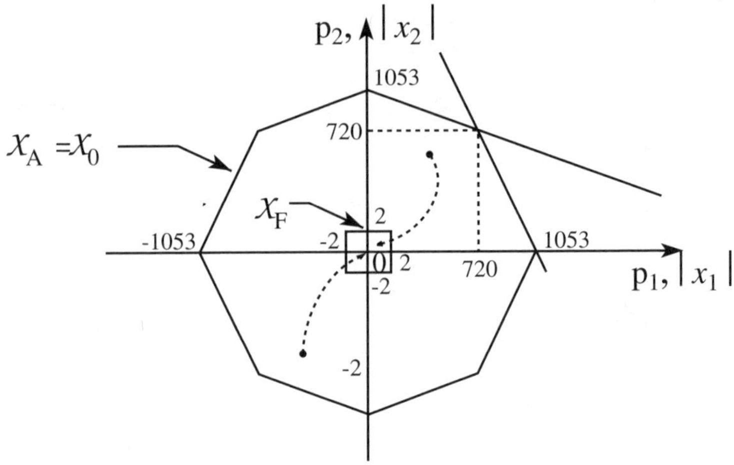


Figure 7.13: Practical stability of (7.142) with the settling time 0.5s, with regard to $\{+\infty, \mathcal{X}_0, \mathcal{X}_A, \mathcal{X}_F, \mathcal{S}_d\}$, first study.

In order to maximize the right-hand side of inequality (7.147) within the constraint (7.146), we have to choose: $\alpha = 4.512$, then $p_0$ has to verify:

$$\begin{pmatrix} 1 & 0.462 \\ 0.462 & 1 \end{pmatrix} p_0 \leq 1052.9 \begin{pmatrix} 1 \\ 1 \end{pmatrix}. \tag{7.148}$$

Any domain $\mathcal{X}_{p_0}$ is convenient as an initial set, and then domain $\mathcal{X}_0$ defined by the union of the $\mathcal{X}_{p_0}$ such than $p_0$ verifies (7.147) is a possible set for practical stability. We also set $\mathcal{X}_A = \mathcal{X}_0$ (see Fig. 7.13).

**b) second study:** let $\tau_s = 0.5s$, $\mathcal{X}_A = \mathcal{X}_0$, and we want to minimize $\mathcal{X}_F$ (then $p_F$). We have now to verify (7.146) and (7.149):

$$\begin{pmatrix} \frac{e+1}{2} & \frac{e-1}{2} \\ \frac{e-1}{2} & \frac{e+1}{2} \end{pmatrix} p_0 \leq exp\left(\frac{\alpha^2}{2}\right) p_F + \frac{2\alpha^3\left(e - exp\left(\frac{\alpha^2}{2}\right)\right)}{3\sqrt{3}(\alpha^2-2)} \begin{pmatrix} 1 \\ 1 \end{pmatrix}. \tag{7.149}$$

Minimization of $p_F$ leads to the same result as in Example 7.8, that is, $\alpha^2 = 6$, $p_F = (\sqrt{2}, \sqrt{2})^T$. The admissible initial set is then characterized by:

$$\begin{pmatrix} 1 & 0.462 \\ 0.462 & 1 \end{pmatrix} p_0 \leq \begin{pmatrix} 2 \\ 2 \end{pmatrix}. \tag{7.150}$$

This leads to the set $\mathcal{X}_0$ (Fig. 7.14), obtained by union of $\mathcal{X}_{p_0}$ such that $p_0$ obeys (7.150).

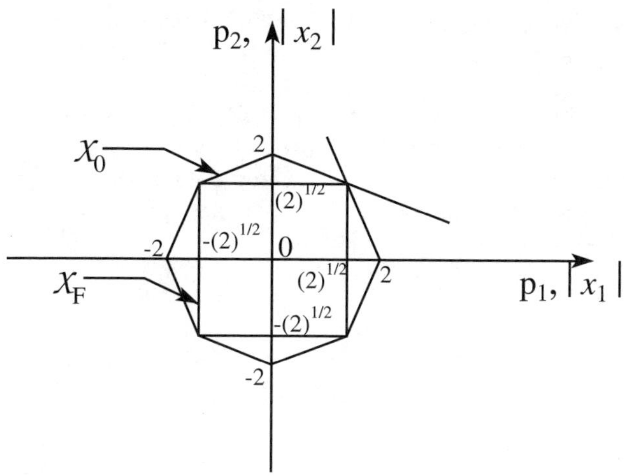


Figure 7.14: Practical stability of (7.142) with the settling time 0.5s, with regard to $\{+\infty, \mathcal{X}_0, \mathcal{X}_A, \mathcal{X}_F, \mathcal{S}_d\}$, second study.

## 7.5 Conclusions

This chapter was devoted to the vector-norms approach. It appears that such a tool addresses the different aspects of stability questions: the stability properties themselves (including practical stability with settling time), but also the related question of the domains estimation and, what was not presented here, the question of constrained control, that appears as a direct application of the preceding invariance properties.

Vector norms constitute a simple case of vector Lyapunov functions, and the main presented results can be enlarged to this more general class, provided the considered vector Lyapunov functions verify the following connectivity property [161]:

- the considered V.L.F. $p$ is regular, radially unbounded, and such that the related following set:

$$\overline{X}_c = \{x \in \Re^n,\ p(x) \leq \mathsf{c}\}$$

  is connected for any positive $\mathsf{c}$,

which property is verified for vector norms because of the triangular inequality.

Therefore, the presented overvaluing comparison systems appear to be a general and practical approach to complex systems:

– On the one hand, the actual remaining drawback of such an approach is the loss of information that may appear if the overvaluing procedure is too strong. The solution of this point is linked to the improvement of the state–space model conditioning [172]. However, the loss of information is encountered in any approach based on sufficient but nonnecessary conditions.

– On the other hand, the main advantage is to allow the unknown disturbing parameters to be taken into account.

Lastly, it is important to remark that this chapter was presented in the framework of nonlinear time-invariant systems: but, the method of vector norms addresses more general classes of models, for example nonlinear *time-varying systems*, or nonlinear *time-delay systems* with possible time-varying delays [21], [35], [58].

# References


1. H.A. Antosiewicz, "A survey of Lyapunov's second method", in *Contributions to the Theory of Nonlinear Oscillations,* Princeton: Princeton Univ. Press, pp. 141–166, 1958.
2. H.A. Antosiewicz, "Recent contributions to Lyapunov's second method", Marseille: *Colloques Internationaux sur les Vibrations Forcées dans les Systèmes Non-linéaires*, pp. 29–37, September, 1964.
3. G. Arienti, C. Sutti and G.P. Szegö, "On the numerical construction of Lyapunov function", *Preprints of the IV-th IFAC Congress*, Warszawa, Paper No. 20.1, pp. 3–20, June 16–21, 1969.
4. E.A. Barbashin, "On construction of Lyapunov functions for nonlinear systems" (in Russian), *Proc. 1st International IFAC Congress: Theory of Continuous Systems – Special Mathematical Problems*, Moscow: Acad. Sci. U.S.S.R., pp. 742–750, 1961.
5. E.A. Barbashin, *Lyapunov Functions* (in Russian), Moscow: Nauka, 1970.
6. E.A. Barbashin and N.N. Krasovskii, "On the Stability of Motion in the Large" (in Russian), Moscow: *Dokl. Akad. Nauk S.S.S.R.*, vol. 86, No. 3, pp. 453–456, 1952.
7. E.A. Barbashin and N.N. Krasovskii, "On the existence of Lyapunov functions in the case of asymptotic stability in the whole" (in Russian), *Prikl. Math. Meh.*, vol. XVIII, pp. 345–350, 1954.
8. R. Bellman, *Stability Theory of Differential Equations*, New York: McGraw Hill, 1953.
9. R. Bellman, "Vector Lyapunov functions", *J. SIAM Control*, Ser. A, vol. 1, No. 1, pp. 32–34, 1962.
10. N.P. Bhatia and G.P. Szegö, *Dynamical Systems: Stability Theory and Applications*, Berlin: Springer Verlag, 1967.
11. A.M. Ben Lashiher and C. Storey, "Final-stability with some applications", *J. Inst. Math. Applics.*, vol. 9, pp. 397–410, 1972.
12. M. Benrejeb, P. Borne and F. Laurent, "Sur une application de la représentation en flèche à l'analyse des processus", *RAIRO Aut. SAC*, vol. 16, No. 2, pp. 133–146, 1982.
13. G. Bitsoris, *Principe de comparaison et stabilité de systèmes complexes*, Thèse d'Etat, University Paul Sabatier, Toulouse, France, No. 818, 1978.
14. P. Borne, *Contribution à l'étude des systèmes discrets nonlinéaires de grande dimension. Application aux systèmes interconnectés*, Thèse de Doctorat es Sciences Physiques, University of Lille 1, France, 1976.

15. P. Borne, M. Benrejeb and J.L. Coquerelle, "On the synthesis of a class of interconnected systems with structural variations – application to an electrical power plant", *MECO 78*, Acta Press, pp. 1–4, 1978.
16. P. Borne and M. Benrejeb, "On the stability of a class of interconnected systems. Application to the forced working conditions", *IVth IFAC Symp. on Multivariable Technolog. Syst.*, Fredericton, pp. 261–265, 1977.
17. P. Borne and M. Benrejeb, "Stability of non-linear composite systems", *1st World Conf. Math. Service of Man*, Barcelona, Techn. Session 4/6, pp. 1–12, 1977.
18. P. Borne, G. Dauphin-Tanguy and J.P. Richard, *Estimation of attractors and invariant domains for perturbed complex and large scale systems*, Stochastic Large Scale Engineering Systems, Chap. 10, pp. 307–316, Ed. S. Tzafestas and K. Watanabe, Marcel Dekker Inc., 1992.
19. P. Borne and J.P. Richard, "On attractivity domains estimation for continuous systems", *IMACS Annals on Computing and Applied Mathematics*, vol. 4, Sect. III, Ed. P. Borne, S. Tzafestas, P. Breedveld and G. Dauphin-Tanguy, J.C. Baltzer AG Sc. Pub. Comp., pp. 179–181, 1989.
20. P. Borne, J.P. Richard and N.E. Radhy, "Stability, stabilisation, regulation using vector norms", *Nonlinear Systems*, vol. II, Chap. 2, 47 pp., Ed. A.J. Fossard and D. Normand-Cyrot, Chapman and Hall, 1996, translated from *Systèmes Non Linéaires*, tome 2 : *Stabilité – Stabilisation*, Chap. 2, pp. 45–90, Masson C^ie^ Editeurs, March 1993.
21. P. Borne, M. Dambrine, W. Perruquetti and J.P. Richard, "Vector Lyapunov functions: nonlinear, time-varying, ordinary and functional differential equations", *Stability Theory at the End of XXth Century*, Chap. 13, pp. 89–112, Ed. A.A. Martynyuk, Taylor & Francis, London, 2002.
22. R.K. Brayton and Č.H. Tong, "Stability of dynamical systems: A constructive approach", *IEEE Trans. Circuits Syst.*, vol. 26, pp. 224–234, April 1979.
23. R.K. Brayton and C.H. Tong, "Constructive stability and asymptotic stability of dynamical systems", *IEEE Trans. Circuits Syst.*, vol. CAS-27, No. 1, pp. 1121–1130, 1980.
24. R.W. Brockett, "The status of stability theory of deterministic systems", *IEEE Trans. Automat. Contr.*, vol. AC-11, No. 3, pp. 596–606, June 1966.
25. G. Burnand and G. Sarlos, "Determination of the domain of stability", *J. of Math. Analysis and Appl.*, vol. 23, pp. 714–722, 1968.
26. T.A. Burton, "On the construction of Lyapunov functions", *SIAM J. Appl. Math.*, vol. 17, No. 6, pp. 1078–1085, November 1969.
27. L. Cesari, *Asymptotic Behavior and Stability Problems in Ordinary Differential Equations,* Berlin: Springer-Verlag, 1959.
28. C.S. Chen and E. Kinnen, "Construction of Lyapunov functions", *J. Franklin Institute*, vol. 289, No. 2, pp. 133–146, February 1970.
29. N.G. Chetaev, *Stability of Motion* (in Russian), GITTL, Moscow, 1955 (English Translation: Oxford: Pergamon Press, 1961).
30. H.D. Chiang and J.S. Thorp, "Stability Regions of Nonlinear Dynamical Systems: A Constructive Methodology", *IEEE Trans. on Automatic Control*, vol. 34, No. 12, pp. 1229–1241, December 1989.

31. P.S.M. Chin, "A general method to derive Lyapunov functions for non-linear systems", *Int. J. Contr.*, vol. 44, No. 22, pp. 381–393, 1986.

32. P.S.M. Chin, "Generalized integral method to derive Lyapunov functions for non-linear systems", *Int. J. Contr.*, vol. 46, No. 3, pp. 933–943, 1987.

33. W.A. Coppel, *Stability and Asymptotic Behavior of Differential Equations*, Boston: D.C. Heath and Co., 1965.

34. M. Dambrine and J.P. Richard, "Stability analysis of time-delay systems", *Dynamic Systems and Applications*, vol. 2, No. 3, pp. 405–414, September 1993.

35. M. Dambrine and J.P. Richard, "Stability and stability domains analysis for nonlinear differential-difference equations", *Dynamic Systems and Applications*, vol. 3, No. 3, pp. 369–378, September 1994.

36. E.J. Davison and E.M. Kurak, "A computational method for determining Lyapunov functions for non-linear systems", *Automatica*, Pergamon Press, vol. 7, pp. 627–636, 1971.

37. B.P. Demidovich, *Lectures on Mathematical Theory of Stability* (in Russian), Moscow: Nauka, 1967.

38. A K. DeSarkar and N.D. Rao, "Zubov's method and transient-stability problems of power systems", *Proc. IEE*, vol. 118, No. 8, pp. 1035–1040, 1971.

39. C. Desoer and M. Vidyasagar, *Feedback Systems: Input–Output Properties*, Academic Press, New York, 1975.

40. M. Djordjević, "Stability analysis of interconnected systems with possibly unstable subsystems", *Systems & Contr. Lett.*, vol. 3, No. 3, pp. 165–169, 1983.

41. M.Z. Djordjević, "Stability analysis of large-scale systems whose subsystems may be unstable", *Large Scale Syst.*, vol. 5, No. 3, pp. 255–262, 1983.

42. M.Z. Djordjević, "Stability analysis of nonlinear systems by the matrix Lyapunov method", *Proc. IMACS-IFAC Symp.*, IDN, Villeneuve d'Ascq, France, pp. 209–212, June 3–6, 1986.

43. P. Dorato, "Short Time Stability in Linear Time-Varying Systems", *IRE Intern. Conv. Rec.*, Pt. 4, pp. 83–87, 1961.

44. P. Dorato, "Short-Time Stability", *IRE Trans. on Automatic Control*, pp. 86, February 1967.

45. N.P. Erugin, "Certain questions of the theory of stability of motion" (in Russian), *Prikl. Mat. Meh.*, vol. XV, pp. 227–236, 1951.

46. F. Fallside and M.R. Patel, "On the application of the Lyapunov method to synchronous machine stability", *Int. J. Contr.*, vol. 4, No. 6, pp. 501–513, 1965.

47. F. Fallside and M.R. Patel, "Control engineering applications of V.I. Zubov's construction procedure for Lyapunov functions", *IEEE Trans. Automat. Contr.*, vol. AC-10, No. 2, pp. 220–222, April 1965.

48. M. Fielder and V. Ptak, "On matrices with nonpositive off-diagonal elements and positive principal minors", *Czech. Nat. Journ.*, vol. 12, No. 87, pp. 382–400, 1962.

49. W.R. Foster and M.L. Davies, "Estimating the domain of attraction for systems with multiple non-linearities", *Int. J. Contr.*, vol. 15, No. 9, pp. 1001–1003, 1972.

50. F.R. Gantmacher, *Theory of Matrices,* Chelsea Publishing Co., New York, 1960.

51. D.P. Garg and M.J. Rabins, "Stability bounds for nonlinear systems designed via frequency domain stability criteria", *Trans. of the ASME – J. of Dynamic Syst., Measurement and Contr.*, vol. 104, No. 3, pp. 262–265, September 1972.
52. W.L. Garrard, "Finite Time Stability in Control Systems Synthesis", *Proc. 1st IFAC World Congress*, Butterworth, London, pp. 21–29, 1961.
53. R. Genesio, M. Cartaglia, and A. Vicino, "On the Estimation of Asymptotic Stability Regions: State of the Art and New Proposals", *IEEE Trans. on Automatic Control*, AC-30, No. 8, pp. 747–755, August 1985.
54. R. Genesio and A. Vicino, "New techniques for constructing asymptotic stability regions for nonlinear systems", *IEEE Trans. Circ. Syst.*, vol. CAS-31, No. 6, pp. 574–581, 1984.
55. J.C. Gentina, *Contribution à l'analyse et à la synthèse des systèmes continus non-linéaires de grande dimension*, Thèse de Doctorat es Sciences Physiques, University of Lille 1, France, 1976.
56. J.C. Gentina, P. Borne, C. Burgat, J. Bernussou and Lj.T. Grujić, "Sur la stabilité des systèmes de grande dimension. Normes vectorielles", *RAIRO Automat. / Syst. Anal. and Contr.*, vol. 13, No. 1, pp. 57–75, 1979.
57. J.C. Gentina, P. Borne and F. Laurent, "Stabilité des systèmes continus non linéaires de grande dimension", *RAIRO*, No. 3, pp. 69–77, August 1972.
58. A. Goubet-Bartholoméüs, M. Dambrine and J.P. Richard, "Stability of perturbed systems with time-varying delays", *System and Control Letters*, vol. 31, pp. 155–163, 1997.
59. M. Gruber, "Path integrals and Lyapunov functionals", *IEEE Trans. Automat. Contr.*, vol. AC-14, No. 5, pp. 465–478, October 1969.
60. Lj.T. Grujić, *Control system synthesis for a rigid body motion through a fluid* (in Serbo-Croatian), M. Sci. Thesis, Department of Electrical Engineering, University of Belgrade, 1970.
61. Lj.T. Grujić, "On Practical Stability", *Intern. J. Control*, vol. 17, No. 4, pp. 881–887, 1973.
62. Lj.T. Grujić, "Non-Lyapunov Stability Analysis of Large-Scale Systems on Time-Varying Sets", *Intern. J. Control*, vol. 21, No. 3, pp. 401–405, 1975.
63. Lj.T. Grujić, "Uniform Practical and Finite-Time Stability of Large-Scale Systems", *Intern. J. Systems Sci.*, vol. 6, No. 2, 1975.
64. Lj.T. Grujić, "Novel development of Lyapunov stability of motion", *Int. J. Contr.*, vol. 22, No. 4, pp. 525–549, 1975.
65. Lj.T. Grujić, "Sets and singularly perturbed systems", *Systems Science* (Wroclaw, Poland), vol. 5, No. 4, pp. 327–338, 1979.
66. Lj.T. Grujić, "Uniquely bounded sets and nonlinear systems", San Diego: *Proc. 1978 IEEE Conf. Decision and Contr.*, vol. 1, pp. 325–333, January 1979.
67. Lj.T. Grujić, "Uniquely bounded sets", Barcelona: 1st Conf. on Math. at the Services of Man, July 11–16, 1977, published: *Proc. of 1st Conf. Math. Serv. Man*, Barcelona, Univ. Politec. Barcelona, vol. 1, pp. 145–161, 1980.
68. Lj.T. Grujić, "On absolute stability and the Aizerman conjecture", *Automatica*, vol. 17, No. 2, pp. 335–349, 1981.
69. Lj.T. Grujić, "Lyapunov-like solutions for stability problems of the most general stationary Lurie–Postnikov systems", *Int. J. Systems Sci.*, vol. 12, No. 7, pp. 813–833, 1981.

70. Lj.T. Grujić, "Concepts of stability domains" (in Serbo-Croatian), Zagreb: *Automatika*, vol. 26, No. 1–2, pp. 5–10, 1985.
71. Lj.T. Grujić, "Stability Domains of General and Large-Scale Systems", *Proc. of the IMACS/IFAC Symposium on Modelling and Simulation for Control of Lumped and Distributed Parameter Systems*, IDN (Lille, France), vol. I, pp. 267–272, 1986.
72. Lj.T. Grujić, "Stability domains of general and large-scale stationary systems", in *Applied Modelling and Simulation of Technological Systems*, Ed. P. Borne and S.G. Tzafestas, Elsevier Science Publishers B.V. (North Holland), IMACS, pp. 317–327, 1987.
73. Lj.T. Grujić, "On large-scale systems stability", *Proc. 12th IMACS World Congress*, Paris, France, vol. 1, pp. 224–229, 1988.
74. Lj.T. Grujić, "On Determination of Lyapunov Functions and the Asymptotic Stability Domain" (in Serbian), *Proc. of the Third Conference SAUM* (Vrnjachka Banja, Serbia, Yugoslavia), pp. 12–20, (October 18–20), 1989.
75. Lj.T. Grujić, "Solutions to Lyapunov Stability Problems: Nonlinear Systems with Globally Differentiable Motions", *The Lyapunov functions method and applications*, P. Borne and V. Matrosov (Editors), J.C. Baltzer AG, Scientific Publishing Co., IMACS (Basel, Switzerland), pp. 19–27, 1990.
76. Lj.T. Grujić, "Solutions to Lyapunov Stability Problems: Nonlinear Systems with Differentiable Motions", *Proc. of the 13th IMACS World Congress on Computation and Applied Mathematics* (Dublin, Ireland), vol. 3, pp. 1228–1231, (July 22–26), 1991; also in *Computational and Applied Mathematics II: Differential Equations*, Ed. W.F. Ames and P.J. van der Houwen.
77. Lj.T. Grujić, "The Necessary and Sufficient Conditions for the Exact Construction of a Lyapunov Function and the Asymptotic Stability Domain", *Proc. of the 30th Conference on Decision and Control* (Brighton, England), vol. 3, pp. 2885–2888, December 11–13, 1991.
78. Lj.T. Grujić, "On Solutions to Lyapunov Stability Problems", *Facta Universitatis*, Series: Mechanics, Automatic Control and Robotics, University of Nish (Nish, Serbia, Yugoslavia), vol. 1, No. 2, pp. 121–138, 1992.
79. Lj.T. Grujić, "Exact determination of a Lyapunov function and the asymptotic stability domain", *Int. J. Systems Sci.*, vol. 23, No. 11, pp. 1871–1888, 1992.
80. Lj.T. Grujić, "Exact both Construction of a Lyapunov Function and Asymptotic Stability Domain Determination", *Proc. 1993 IEEE/SMC International Conference on System, Man and Cybernetics* (Le Touquet, France), vol. I, pp. 331–336, October 17–20, 1993.
81. Lj.T. Grujić, "Solutions to Lyapunov Stability Problems of Sets: Nonlinear Systems with Differentiable Motions", *Int. J. of Mathem. and Mathem. Sci.*, vol. 17, No. 1, pp. 103–112, 1994.
82. Lj.T. Grujić, "Solutions to Lyapunov Stability Problems: Nonlinear Systems with Continuous Motions", *Int. J. Mathem. and Mathematical Sci.*, vol. 17, No.3, pp. 587–596, 1994.
83. Lj.T. Grujić, "Solutions to Lyapunov Stability Problems: Time-Invariant Systems", *Proc. 14th IMACS World Congress* (Atlanta, Georgia, USA), vol. 1, pp. 203–205, July 11–15, 1994.

84. Lj.T. Grujić, "Solutions to Lyapunov Stability Problems via $\mathcal{O}$-uniquely bounded sets", *Control-Theory and Advanced Technology*, vol. 10, No. 4, Part 2, pp. 1069–1091, 1995.
85. Lj.T. Grujić, "New Lyapunov Methodology and Exact Construction of a Lyapunov Function: Exponential Stability", *Nonlinear Analysis Problems in Engineering Systems*, Kazan, Russia, No. 1, pp. 9–16, 1995.
86. Lj.T. Grujić, "Exact Solutions for Asymptotic Stability: Non-Linear Systems", *Int. J. Non-Linear Mechanics*, vol. 30, No. 1, pp. 45–56, 1995.
87. Lj.T. Grujić, J.C. Gentina and P. Borne, "General aggregation of large scale systems by vector Lyapunov functions and vector norms", *Int. J. of Control*, vol. 22, No. 4, pp. 529–550, 1976.
88. Lj.T. Grujić, A.A. Martynyuk and M. Ribbens-Pavella, *Large scale systems stability under structural and singular perturbations* (in Russian), Kiev: Naukova Dumka, 1984 (English edition: New York: Springer Verlag, LNCIS No. 92, 1987).
89. Lj.T. Grujić and A.N. Michel, "Qualitative analysis of neural networks under structural perturbations", *Proc. 1990 IEEE Intern. Symp. on Cyrc. and Syst.*, New Orleans, vol. 1, pp. 391–394, May 1–3, 1990.
90. Lj.T. Grujić and M. Ribbens-Pavella, "Asymptotic stability of large-scale systems: Part 1: domain estimations", *Elec. Power and Energy Systems*, vol. 1, No. 3, pp. 151–157, 1979.
91. R.W. Gunderson, "On Stability over a Finite Interval", *IEEE Transaction on Automatic Control*, vol. AC-12, No. 5, pp. 634–635, 1967.
92. P. Habets and K. Peiffer, "Attractivity concepts and vector Lyapunov functions", *Nonlinear Vibration Problems (Zagadnienia Drgan Nieliniowych)*, pp. 35–52, 1975.
93. T.G. Hallam and V. Komkov, "Application of Liapunov's Functions to Finite Time Stability", *Rev. Roumaine de Math. Pure et Appliqués*, vol. 14, No. 4, pp. 495–501, 1969.
94. W. Hahn, *Stability of Motion*, Berlin: Springer Verlag, 1967.
95. A. Halanay, *Differential Equations*, New York: Academic Press, 1966.
96. J.K. Hale, *Ordinary Differential Equations*, New York: Wiley–Interscience, 1969.
97. C.C. Hang and J.A. Chang, "An algorithm for constructing Lyapunov functions based on variable gradient method", *IEEE Trans. Automat. Contr.*, vol. AC-15, No. 4, pp. 510–512, August 1970.
98. M.A. Hassan and C. Storey, "Numerical determination of domains of attraction for electrical power systems using the method of Zubov", *Int. J. Contr.*, vol. 34, No. 2, pp. 371–381, 1981.
99. J.A. Heinen and S.H. Wu, "Set stability of differential equations", *Int. J. Systems Sci.*, vol. 1, No. 3, pp. 269–277, 1971.
100. J.R. Hewit and C. Storey, "Optimisation of the Zubov and Ingwerson methods for constructing Lyapunov functions", *Electronics Lett.*, vol. 3, No. 5, pp. 211–212, 1967.
101. J.R. Hewit and C. Storey, "Numerical application of Szegö's method for constructing Lyapunov functions", *IEEE Trans. Automat. Contr.*, vol. AC-14, No. 1, pp. 106–108, February 1969.

102. J.R. Hewit and C. Storey, "Comparison of numerical methods in stability analysis", *Int. J. Contr.*, vol. 10, No. 6, pp. 687–701, 1969.

103. J.G. Hocking and G.S. Young, *Topology*, Addison-Wesley Publ. Comp. Inc., Reading, 1961.

104. D.R. Ingwerson, "A modified Lyapunov method for nonlinear stability analysis", *IRE Trans. Automat. Contr.*, vol. AC-6, pp. 199–210, May 1961.

105. H. Jianxun, "On Practical Stability of Discontinuous Systems", *Depart. of Computer Science, Xiamen University*, P.R. China, vol. 8, pp. 1–34, 1982.

106. Lj. Jocić, "Planar regions of attraction", *IEEE Trans. Automat. Contr.*, vol. AC-27, No. 3, pp. 708–710, 1982.

107. P.M. Julich, "On estimating the region of asymptotic stability using Lyapunov's method", *IEEE Trans. Automat. Contr.*, vol. AC-12, No. 6, pp. 787–788, December 1967.

108. R.E. Kalman and J.E. Bertram, "Control system analysis and design via the 'second method' of Lyapunov", Part I, *Trans. of ASME – J. Basic Eng.*, vol. 82, pp. 371–393, 1960.

109. K.A. Karatcharov and A.G. Pilyutik, *Introduction to the technical theory of the motion stability* (in Russian), Fizmatgiz, Moscow, 1962.

110. A.A. Kayande and J.S.W. Wong, "Finite Time Stability and Comparison Principles", *Proc. Camb. Phil. Soc.*, vol. 64, pp. 749–756, 1968.

111. A.A. Kayade, "A Theorem on Contractive Stability", *SIAM J. Applic. Math.*, vol. 21, No. 4, pp. 601–604, 1971.

112. E. Kinnen and C.S. Chen, "Liapunov functions derived from auxiliary exact differential equations", *Automatica*, vol. 4, pp. 195–204, 1968.

113. N.E. Kirin, R.A. Nelepin and V.N. Baidaev, "Construction of the attraction region by Zubov's method" (in Russian), *Diff. Urav.*, vol. 17, No. 8, pp. 1347–1361, 1981. (English translation: pp. 871–880, 1982.)

114. H.W. Knobloch and F. Kappel, *Gewöhnliche Differentialgleichungen*, B. G. Teubner, Stuttgart, 1974.

115. J. Komarnik and C.C. Li, "Decision surface estimate of nonlinear system stability domain by Lie series method", *IEEE Trans. Automat. Contr.*, vol. AC-18, No. 5, pp. 666–669, October 1972.

116. N.N. Krasovskii, *Stability of Motion*, Stanford: Stanford University Press, 1963.

117. Y.H. Ku and N.N. Puri, "On Liapunov functions of high order non-linear systems", *J. Franklin Institute*, vol. 276, No. 5, pp. 349–364, 1963.

118. V. Lakshmikantham and S. Leela, *Differential and Integral Inequalities*, New York: Academic Press, 1969.

119. J.P. LaSalle, "Some extensions of Liapunov's second method", *IRE Trans. on Circuit Theory*, vol. CT-7, pp. 520–527, December 1960.

120. J.P. LaSalle, *The Stability of Dynamical Systems*, Philadelphia. SIAM, 1976.

121. J.P. La Salle and S. Lefschetz, *Stability by Lyapunov's Direct Method*, Academic Press, New York, 1961.

122. S. Lefschetz, *Stability of Nonlinear Control Systems*, New York: Academic Press, 1965.

123. W. Leighton, "On the construction of Liapunov functions for certain autonomous nonlinear differential equations", *Contributions to Differential Equations*, vol. II, New York: John Wiley & Sons (ed. J.P. LaSalle), pp. 367–383, 1963.

124. K.A. Loparo and G.L. Blankenship, "Estimating the domain of attraction of nonlinear feedback systems", *IEEE Trans. Automat. Contr.*, vol. AC-23, No. 4, pp. 602–608, August 1978.

125. A.I. Lurie and V.N. Postnykov, "On the stability theory of control systems" (in Russian), *Appl. Math. and Mech.*, vol. 8, pp. 246–248, 1944.

126. A.I. Lurie and E.W. Rozenvasser, "On methods for generating Lyapunov functions in the theory of nonlinear regulating systems" (in Russian), Moscow: *Proc. 1st Internat. IFAC Congress: Theory of Continuous Systems – Special Mathematical Problems*, Acad. Sci. U.S.S.R., pp. 709–717, 1961.

127. A.M. Lyapunov, *The General Problem of Stability of Motion* (in Russian), Kharkov: Kharkov Mathematical Society, 1892; *Academician A.M. Lyapunov: "Collected Papers"*, Moscow: U.S.S.R. Academy of Science, vol. II, pp. 5–263, 1956. (French translation: "Problème général de la stabilité du mouvement", *Ann. Fac. Toulouse*, vol. 9, pp. 203–474, 1907; also in: *Annals of Mathematics Study*, No. 17, Princeton: Princeton University Press, 1949; English translation: *Intern J. of Control*, vol. 55, pp. 531–773, 1992; also the book, Taylor and Francis, London, 1992).

128. S.M. Madani-Esfahani, S. Hui and S.H. Żak, "On the estimation of sliding domains and stability regions of variable structure control systems with bounded controllers", Tampa: *Proc. 28th Conference on Decision and Control*, pp. 793–794, December 1989.

129. I.G. Malkin, "To the question on reversibility of the theorem by Lyapunov on the asymptotic stability" (in Russian), *Prikl. Mat. Meh.*, vol. XVIII, pp. 129–138, 1954.

130. I.G. Malkin, *Theory of Stability of Motion* (in Russian), Moscow: Gostehizdat (the second edition: Moscow: Nauka, 1968).

131. S.G. Margolis and W.G. Vogt, "Control engineering applications of V.I. Zubov's construction procedure for Lyapunov functions", *IEEE Trans. Automat. Contr.*, vol. AC-8, No. 2, pp. 104–113, April 1963.

132. R. Marino and S. Nicosia, "Hamiltonian-type Lyapunov functions", *IEEE Trans. Automat. Contr.*, vol. AC-28, No. 11, pp. 1055–1057, 1983.

133. A.A. Martynyuk, "Practical Stability and Stabilization of Control Processes" (in Russian), *Applied Mechanics*, vol. XVII, No. 10, pp. 3–19, 1981.

134. A.A. Martynyuk, "The Lyapunov matrix function", *Nonlinear Analysis, Theory and Applications*, vol. 8, No. 10, pp. 1223–1226, 1984.

135. A.A. Martynyuk, "On matrix Lyapunov functions and stability of motions" (in Russian), *Proc. Acad. Sci. U.S.S.R.*, vol. 280, No. 5, pp. 1062–1066, 1985.

136. A.A. Martynyuk, "Lyapunov matrix-function and stability theory", *Proc. IMACS-IFAC Symp.*, IDN, Villeneuve d'Ascq, France, pp. 261–265, June 3–6, 1986.

137. A.A. Martynyuk and R. Gutovski, *Integral Inequalities and Stability of Motion* (in Russian), Kiev: Naukova Dumka, 1979.

138. J.L. Massera, "On Liapunov's conditions of stability", *Annals of Mathematics*, vol. 50, No. 3, pp. 705–721, July 1949.
139. J.L. Massera, "Contributions to stability theory", *Annals of Mathematics*, vol. 64, No. 1, pp. 182–206, July 1956.
140. V.M. Matrosov, "To the theory of stability of motion" (in Russian), *Prikl. Mat. Meh.*, vol. 25, No. 5, pp. 885–895, 1962.
141. V.M. Matrosov, "Comparison principle and vector Lyapunov function I-II" (in Russian), *Diff. Urav.*, vol. 4, No. 8, pp. 1374–1386, Vol. 10, pp. 1739–1752, 1968.
142. V.M. Matrosov, "Vector Lyapunov functions in the analysis of nonlinear interconnected systems", *Symp. Math.*, Academic Press, New York, vol. 6, pp. 209–242, 1971.
143. E.J. McShane, *Integration*, Princeton University Press, Princeton 1944.
144. A.N. Michel, "On the bounds of the trajectories of differential systems", *Int. J. Control*, vol. 10, No. 5, pp. 593–600, 1969.
145. A.N. Michel, "Stability, transient behaviour and trajectory bounds of interconnected systems", *Int. J. Control*, vol. 11, No. 4, pp. 703–715, 1970.
146. A.N. Michel, "Quantitative Analysis of Simple and Interconnected Systems: Stability, Boundedness, and Trajectory Behavior", *IEEE Trans. on Circuit Theory*, vol. CT-17, No. 3, pp. 292–301, 1970.
147. A.N. Michel and R.K. Miller, *Qualitative Analysis of Large-Scale Dynamical Systems*, New York: Academic Press, 1977.
148. A.N. Michel, R. Miller and B.H. Nam, "Stability analysis of interconnected systems using computer generated Lyapunov functions", *IEEE Trans. on Circuits and Systems* vol. CAS-29, No. 7, pp. 431–440, 1982.
149. A.N. Michel, B.H. Nam and V. Vittal, "Computer generated Lyapunov functions for interconnected systems: Improved results with application to power systems", *IEEE Trans. Circuits and Systems*, vol. CAS-31, No. 2, pp. 189–198, February 1984.
150. A.N. Michel and D.W. Porter, "Practical Stability and Finite-Time Stability of Discontinuous Systems", *IEEE Trans. on Circuit Theory*, vol. CT-19, No. 2, pp. 123–129, 1972.
151. A.N. Michel, N.R. Sarabudla and R.K. Miller, "Stability analysis of complex dynamical systems", *Circuits, Systems, Signal Processing*, vol. 1, No. 2, pp. 171–202, 1982.
152. R.K. Miller and A.N. Michel, *Ordinary Differential Equations*, Academic Press, New York, 1982.
153. D. Mitra and H.C. So, "Existence conditions for $L_1$ Lyapunov functions for a class of nonautonomous systems", *IEEE Trans. Circ. Theory*, vol. CT-19, No. 6, pp. 594–598, July 1972.
154. H. Miyagi and T. Taniguchi, "Construction of Lyapunov function for power systems", *Proc. IEE*, vol. 124, No. 12, pp. 1197–1202, 1977.
155. T. Nagaraja and V.V. Chalam, "Generation of Lyapunov functions. A new approach", *Int. J. Contr.*, vol. 19, No. 4, pp. 781–787, 1974.
156. R.M. Nasyirov, "Stability over a Finite Time Interval in Case of Two Zero Roots" (in Russian), *Izv. Vyssh. Uchebn. Zav. (Matematika)*, vol. 57, No. 2, pp. 48–56, 1967.

157. V.V. Nemytskii and V.V. Stepanov, *Qualitative Theory of Differential Equations*, Princeton University Press, Princeton, 1960.
158. E. Noldus, A. Galle and L. Josson, "The computation of stability regions for systems with many singular points", *Int. J. Contr.*, vol. 17, No. 3, pp. 644–652, 1973.
159. M.A. Pai and C.L. Narayana, "Finite regions of attraction for multinonlinear systems and its application to the power system stability problem", *IEEE Trans. Automat. Contr.*, vol. 21, No. 5, pp. 716–721, October 1976.
160. P.C. Parks and A.J. Pritchard, "On the construction and use of Liapunov functionals", Warszawa: *Proc. 4th IFAC Congress*, Tech. Session, No. 20, Paper No. 20.5, pp. 59–76, June 16–21, 1969.
161. W. Perruquetti, J.P. Richard and P. Borne, "Vector Lyapunov functions: recent developments for stability, robustness, practical stability, and constrained control", *Nonlinear Times and Digest*, vol. 2, pp. 227–258, 1995.
162. W. Perruquetti, *Sur la stabilité et l'estimation des comportements non linéaires non stationnaires perturbés*, Doctoral Thesis, University of Lille 1, France, No. 1286, 25 February 1994.
163. W. Perruquetti and J.P. Richard, "Connecting Wazewski's conditions with M-matrices: application to constrained stabilization", *Dynamic Systems and Applications*, vol. 5, No. 1, pp. 81–96, March 1996.
164. W. Perruquetti, J.P. Richard, Lj.T. Grujić and P. Borne, "On practical stability with the settling time via vector norms", *Int. J. of Control*, vol. 62, No. 1, pp. 173–189, 1995.
165. H. Poincaré, "Sur les courbes définies par une équation différentielle", *Journal de Mathématiques*, serie 7, pp. 375–422, 1881.
166. H. Poincaré, "Sur les courbes définies par une équation différentielle", *Journal de Mathématiques*, serie 8, pp. 251–296, 1882.
167. J.P. Ponzo, "On the stability of certain nonlinear differential equations", *IEEE Trans. Automat. Contr.*, vol. AC-10, No. 5, pp. 470–472, October 1965.
168. V.M. Popov, *Hyperstability of Control Systems*, Springer-Verlag, Berlin, 1973.
169. N.E. Radhy, P. Borne and J.P. Richard, "Regulation of nonlinear time-varying continuous systems with constrained state", *IMACS Annals on Computing and Applied Mathematics*, The Lyapunov Functions Method and Applications, J.C. Baltzer AG, Sc. Pub. Co., vol. 8, Sect. IV, pp. 81–88, 1990.
170. R. Reiss and G. Geiss, "The construction of Liapunov functions", *IEEE Trans. Automat. Contr.*, vol. AC-8, No. 4, pp. 382–383, 1963.
171. M. Ribbens-Pavella, "Critical survey of transient stability studies of multimachine power systems by Liapunov's direct method", Urbana-Champaign: Univ. of Illinois, *Proc. 9th Allerton Conf. Circ. and Syst. Theory*, pp. 151–167, 1971.
172. J.P. Richard and P. Borne, "State–space modelling and transformations", *Systems & Control Encyclopedia*, Ed. M.G. Singh, Pergamon Press, pp. 4540–4547, 1987.
173. J.P. Richard, P. Borne and J.C. Gentina, "Estimation of stability domains by use of vector norms", *Information and Decision Technologies*, North-Holland, vol. 14, pp. 241–251, 1988.

174. J.J. Rodden, "Numerical applications of Lyapunov stability theory", *Proc. JACC, Session IX*, Paper 2, pp. 261–268, 1972.
175. H.H. Rosenbrock, "A Lyapunov function with applications to some non-linear physical systems", *Automatica*, vol. 1, pp. 31–53, 1962.
176. H.H. Rosenbrock, "A Lyapunov function for some naturally-occurring linear homogeneous time-dependent equations", *Automatica*, vol. 2, pp. 97–109, 1963.
177. N. Rouche, P. Habets and M. Laloy, *Stability Theory by Liapunov's Direct Method*, New York: Springer-Verlag, 1977.
178. N. Rouche and J. Mawhin, *Equations Différentielles Ordinaires*, Masson C[ie] Editeurs, Paris, 1973.
179. V.P. Rudakov, "Solution Estimate and Stability of Pseudolinear Systems over a Finite Time Interval" (in Russian), *Differential Equations*, NAU, vol. 5, No. 8, pp. 1385–1389, 1969.
180. E. Sarti, "Approximate determination of the stability domain for nonlinear systems", Warszawa: *Proc. 4th IFAC World Congress*, Technical Session 20, pp. 43–58, June 16–21, 1969.
181. V.R. Sastry, "Finite regions of attraction for the problem of Lur'e", *Int. J. Contr.*, vol. 14, No. 4, pp. 789–790, 1971.
182. D.G. Schultz, "The generation of Liapunov functions", in *Advances in Control Systems*, ed. C. Leondes, vol. 2, New York: Academic Press, pp. 1–64, 1965.
183. D.G. Schultz and J.E. Gibson, "The variable gradient method for generating Liapunov functions", *IEEE Trans. Appl. Ind.*, vol. 81, pp. 203–210, 1962.
184. D.N. Shields and C. Storey, "The behaviour of optimal Lyapunov functions", *Int. J. Contr.*, vol. 21, No. 4, pp. 561–573, 1975.
185. D.D. Šiljak, *Large-Scale Dynamic Systems*, New York: North Holland, 1978.
186. D.D. Šiljak and S. Weissenberger, "Regions of exponential stability for the problem of Lur'e", *Regelungstechnik*, vol. 17, No. 1, pp. 27–29, 1969.
187. D.D. Šiljak and S. Weissenberger, "Regions of exponential boundedness for the problem of Lur'e", *Regelungstechnik and Prozess Datenverarbeitung*, vol. 18, No. 2, pp. 69–71, 1970.
188. A.S. Skidmore, "On the stability of solutions of a differential equation of fourth order", *J. London Math. Soc.*, vol. 41, Part 4, No. 164, pp. 649–661, 1966.
189. M. Spivak, *Calculus on Manifolds*, W.A. Benjamin, Inc., New York, 1965.
190. D.F. Stewart and E.T. Wall, "A matrix transformation for the direct generation of Liapunov state equations", *Acta Technica ČSAV*, No. 6, pp. 667–682, 1969.
191. G.P. Szegö, "On a new partial differential equation for the stability analysis of time invariant control systems", *J. SIAM Control*, Ser. A, vol. 1, No. 1, pp. 63–75, 1962.
192. G.P. Szegö, "New methods for constructing Liapunov functions for time-invariant control systems", Basel: *Proc. 2nd IFAC World Congress*, pp. 584–589, 1963 (published in London: Butterworths, 1964).
193. G.P. Szegö and G.R. Geiss, "A remark on 'A new partial differential equation for the stability analysis of time invariant control systems'", *J. SIAM Control*, Ser. A, vol. 1, No. 3, pp. 369–376, 1963.

194. C. Sy-In, "On Stability over a Finite Time Interval" (in Russian), *Applied Mathematics and Mechanics*, vol. XXIII, pp. 230–238, 1959.

195. S. Tarbouriech and C. Burgat, "Positively invariant sets for constrained continuous-time systems with cone properties", *IEEE Trans. Autom. Control*, vol. 39, No. 2, pp. 401–405, 1994.

196. J. Texter, "Numerical algorithm for implementing Zubov's construction in two-dimensional systems", *IEEE Trans. Automat. Contr.*, vol. AC-20, No. 1, pp. 62–63, February 1974.

197. I. Troch, "The evaluation of the domain of attraction of nonlinear control systems with hybrid computing systems", Paris: *Proc. 5th IFAC World Congress*, Paper No. 32.1, pp. 1–5, 1972.

198. A. Vanelli and M. Vidyasagar, "Maximal Lyapunov Functions and Domains of Attraction for Autonomous Nonlinear Systems", *Automatica (IFAC)*, vol. 21, No.1, pp. 69–80, 1985.

199. E.T. Wall, "A topological approach to the generation of Liapunov functions", *Acta Technica ČSAV*, No. 2, pp. 159–177, 1968.

200. E.T. Wall, "The generation of Liapunov functions in control theory by an energy metric algorithm", *Proc. 1968 J.A.C.C.*, pp. 172–179, 1968.

201. E.T. Wall, "A synthesis of Liapunov functions for non-linear time-varying control systems", *Int. J. Syst. Science*, vol. 4, No. 4, pp. 565–575, 1973.

202. E.T. Wall and M.L. Moe, "An energy metric algorithm for the generation of Liapunov functions", *IEEE Trans. Automat. Contr.*, vol. AC-13, pp. 121–122, 1968.

203. E.T. Wall and M.L. Moe, "Generation of Liapunov functions for time-varying nonlinear systems", *IEEE Trans. Automat. Contr.*, vol. AC-14, pp. 211, 1969.

204. J.A. Walker and H.H. McClamroch, "Finite regions of attraction for the problem of Lur'e", *Int. J. Contr.*, vol. 6, No. 4, pp. 331–336, 1967.

205. J.W. Watson and A.R. Sturberud, "Stability of Systems Operating in a Finite Time Interval", *IEEE Transact. on Automatic Control,* vol. AC-12, No. 1, p. 116, 1967 (see also: "Correction to Stability of Systems Operating in a Finite Time Interval", *IEEE Trans. on Automatic Control,* vol. AC-12, No. 5, p. 635, 1967).

206. T. Ważewski, "Systèmes des équations et des inégalités différentielles ordinaires aux seconds membres monotones et leurs applications", *Ann. Soc. Polon. Math.*, vol. 23, pp. 112–166, 1950.

207. L. Weiss, "Converse theorems for finite time stability", *Proc. 1st Asilomar Conf. on Circuits and Systems,* pp. 1006–1014, 1967.

208. L. Weiss and E.F. Infante, "On the stability of systems defined over a finite time interval", *Proc. Nat. Ac. Sci. USA: Mathematics,* vol. 54, No.1, pp. 44–48, 1965.

209. L. Weiss and E.F. Infante, "Finite time stability under perturbing forces and on product spaces", *IEEE Trans. on Automatic Control,* vol. AC-12, No. 1, pp. 54–59, 1967.

210. S. Weissenberger, "Stability-boundary approximations for relay-control systems via a steepest-ascent construction of Lyapunov functions", *Trans. of ASME – Journal of Basic Eng.*, vol. 88, No. 2, pp. 419–428, June 1966.

211. S. Weissenberger, "Application of results from the absolute stability problem to the computation of finite stability domains", *IEEE Trans. Automat. Contr.*, vol. AC-13, No. 1, pp. 124–125, February 1968.
212. S. Weissenberger, "Comments on 'Application of results from the absolute stability problem to the computation of finite stability domains'", *IEEE Trans. Automat. Contr.*, vol. AC-14, No. 1, pp. 109, February 1969.
213. S. Weissenberger, "Design for stability region maximization", *IEEE Trans. Automat. Contr.*, vol. AC-14, No. 3, pp. 314–315, June 1969.
214. S. Weissenberger, "Piecewise-quadratic and piecewise-linear Lyapunov functions for discontinuous systems", *Int. J. Contr.*, vol. 10, No. 2, pp. 171–180, 1969.
215. S. Weissenberger, "Control synthesis for stability region improvement", *Int. J. Contr.*, vol. 12, No. 3, pp. 489–496, 1970.
216. J.L. Willems, "The computation of finite stability regions by means of open Liapunov surfaces", *Int. J. Contr.*, vol. 10, No. 5, pp. 537–544, 1969.
217. T. Yoshizawa, *Stability Theory by Lyapunov's Second Method*, Tokyo, The Mathematical Society of Japan, 1966.
218. V.I. Zubov, *Methods of A.M. Liapunov and Their Applications* (in Russian), Leningrad: Leningrad Gos. University, (English translation: Groningen: P. Noordhoff Ltd., 1964).

# Author index

# Subject index